Edited by
Vladimir M. Mirsky and
Anatoly K. Yatsimirsky

**Artificial Receptors
for Chemical Sensors**

Related Titles

Fritzsche, W.
Molecular Plasmonics
2011
ISBN: 978-3-527-32765-2

Li, S., Singh, J., Li, H., Banerjee, I. A. (eds.)
Biosensor Nanomaterials
2011
ISBN: 978-3-527-32841-3

Kumar, C. S. S. R. (ed.)
Series: Nanomaterials for the Life Sciences (Volume 7)
Biomimetic and Bioinspired Nanomaterials
2010
ISBN: 978-3-527-32167-4

Haus, J.
Optical Sensors
Basics and Applications
2010
ISBN: 978-3-527-40860-3

Schalley, C. A., Springer, A.
Mass Spectrometry of Non-Covalent Complexes
Supramolecular Chemistry in the Gas Phase
2009
ISBN: 978-0-470-13115-2

Brand, O., Fedder, G. K., Hierold, C., Korvink, J. G., Tabata, O. (eds.)
Carbon Nanotube Devices
Properties, Modeling, Integration and Applications
2008
ISBN: 978-3-527-31720-2

Meijer, G. (ed.)
Smart Sensor Systems
2008
ISBN: 978-0-470-86691-7

Ottow, E., Weinmann, H. (eds.)
Nuclear Receptors as Drug Targets
2008
ISBN: 978-3-527-31872-8

Diederich, F., Stang, P. J., Tykwinski, R. R. (eds.)
Modern Supramolecular Chemistry
Strategies for Macrocycle Synthesis
2008
ISBN: 978-3-527-31826-1

Marks, R. S., Lowe, C. R., Cullen, D. C., Weetall, H. H., Karube, I. (eds.)
Handbook of Biosensors and Biochips
2 Volume Set
2007
ISBN: 978-0-470-01905-4

Mirkin, C. A., Niemeyer, C. M. (ed.)
Nanobiotechnology II
More Concepts and Applications
2007
ISBN: 978-3-527-31673-1

Edited by
Vladimir M. Mirsky and Anatoly K. Yatsimirsky

Artificial Receptors for Chemical Sensors

WILEY-VCH Verlag GmbH & Co. KGaA

The Editors

Prof. Dr. Vladimir M. Mirsky
Lausitz University of Applied Sciences
Nanobiotechnology - BCV
Grossenhainer str. 57
01968 Senftenberg
Germany

Prof. Dr. Anatoly K. Yatsimirsky
Universidad Nacional Autónoma de México
Facultad de Química
04510 México D.F.
Mexico

Cover
Dr. N.V. Roznyatovskaya is acknowledged for her assistance in the design of the book cover.

All books published by **Wiley-VCH** are carefully produced. Nevertheless, authors, editors, and publisher do not warrant the information contained in these books, including this book, to be free of errors. Readers are advised to keep in mind that statements, data, illustrations, procedural details or other items may inadvertently be inaccurate.

Library of Congress Card No.: applied for

British Library Cataloguing-in-Publication Data
A catalogue record for this book is available from the British Library.

Bibliographic information published by the Deutsche Nationalbibliothek
The Deutsche Nationalbibliothek lists this publication in the Deutsche Nationalbibliografie; detailed bibliographic data are available on the Internet at http://dnb.d-nb.de.

© 2011 WILEY-VCH Verlag & Co. KGaA, Boschstr. 12, 69469 Weinheim, Germany

All rights reserved (including those of translation into other languages). No part of this book may be reproduced in any form – by photoprinting, microfilm, or any other means – nor transmitted or translated into a machine language without written permission from the publishers. Registered names, trademarks, etc. used in this book, even when not specifically marked as such, are not to be considered unprotected by law.

Composition Thomson Digital, Noida, India
Printing and Binding Fabulous Printers Pte. Ltd., Singapore
Cover Design Schulz Grafik-Design, Fußgönheim

Printed in Singapore
Printed on acid-free paper

ISBN: 978-3-527-32357-9
ePDF ISBN: 978-3-527-63249-7
oBook ISBN: 978-3-527-63248-0
ePub ISBN: 978-3-527-63250-3

Contents

Preface *XI*
List of Contributors *XIII*

1 **Quantitative Characterization of Affinity Properties of Immobilized Receptors** *1*
Vladimir M. Mirsky
1.1 Introduction *1*
1.2 Measurements Under Equilibrium Conditions *3*
1.3 Kinetic Measurements *7*
1.4 Analysis of Temperature Dependencies *10*
1.5 Experimental Techniques *12*
References *14*

2 **Selectivity of Chemical Receptors** *17*
Hans-Jörg Schneider and Anatoly K. Yatsimirsky
2.1 Introduction *17*
2.2 Some General Considerations on Selectivity *18*
2.3 Correlation Between Selectivity and Affinity *21*
2.4 Crown Ether and Cryptand Complexes: Hole Size Fitting and Other Effects *24*
2.5 Recognition of Transition and Heavy Metal Ions *29*
2.6 Recognition via Ion Pairing *31*
2.7 Hydrogen Bonded Complexes and Solvent Effects *37*
2.8 Lewis Acid Receptors *41*
2.8.1 Associations with Transition Metal Complexes *41*
2.8.2 Associations with Other Lewis Acids *45*
2.9 Complexes with Stacking and van der Waals Interactions *46*
2.10 Multifunctional Receptors for Recognition of Complex Target Molecules *48*

2.10.1	Complexation of Amino Acids and Peptides 48
2.10.2	Complexation of Nucleotides and Nucleosides 54
2.11	Conclusions 60
	References 61

3	**Combinatorial Development of Sensing Materials** 67
	Radislav A. Potyrailo
3.1	Introduction 67
3.2	General Principles of Combinatorial Materials Screening 68
3.3	Opportunities for Sensing Materials 69
3.4	Designs of Combinatorial Libraries of Sensing Materials 71
3.5	Discovery and Optimization of Sensing Materials Using Discrete Arrays 73
3.5.1	Radiant Energy Transduction Sensors 73
3.5.2	Mechanical Energy Transduction Sensors 79
3.5.3	Electrical Energy Transduction Sensors 84
3.6	Optimization of Sensing Materials Using Gradient Arrays 92
3.6.1	Variable Concentration of Reagents 92
3.6.2	Variable Thickness of Sensing Films 93
3.6.3	Variable 2D Composition 94
3.6.4	Variable Operation Temperature and Diffusion Layer Thickness 95
3.7	Emerging Wireless Technologies for Combinatorial Screening of Sensing Materials 97
3.8	Summary and Outlook 102
	References 104

4	**Fluorescent Cyclodextrins as Chemosensors for Molecule Detection in Water** 113
	Hiroshi Ikeda
4.1	Introduction 113
4.2	Pyrene-Appended Cyclodextrins 114
4.2.1	Pyrene-Appended γ-Cyclodextrins as Molecule Sensors 114
4.2.2	Pyrene-Appended γ-Cyclodextrin as a Bicarbonate Sensor 115
4.2.3	Bis-Pyrene-Appended γ-Cyclodextrins 116
4.3	Fluorophore–Amino Acid–CD Triad Systems 118
4.3.1	Dansyl-Leucine-Appended CDs 118
4.3.2	Dansyl-Valine-Appended and Dansyl-Phenylalanine-Appended CDs 122
4.4	Molecular Recognition by Regioisomers of Dansyl-Appended CDs 123
4.4.1	6-O, 2-O, and 3-O-Dansyl-γ-CDs 123
4.4.2	6-O, 2-O, and 3-O-Dansyl-β-CDs 125
4.5	Turn-On Fluorescent Chemosensors 125
4.6	Effect of Protein Environment on Molecule Sensing 127
4.7	CD–Peptide Conjugates as Chemosensors 128

4.8	Immobilized Fluorescent CD on a Cellulose Membrane	*130*
4.9	Conclusion	*132*
	References	*132*

5	**Cyclopeptide Derived Synthetic Receptors**	*135*
	Stefan Kubik	
5.1	Introduction	*135*
5.2	Receptors for Cations	*138*
5.3	Receptors for Ion Pairs	*149*
5.4	Receptors for Anions	*150*
5.5	Receptors for Neutral Substrates	*157*
5.6	Conclusion	*160*
	References	*161*

6	**Boronic Acid-Based Receptors and Chemosensors**	*169*
	Xiaochuan Yang, Yunfeng Cheng, Shan Jin, and Binghe Wang	
6.1	Introduction	*169*
6.2	*De Novo* Design	*172*
6.3	Combinatorial Approaches	*177*
6.4	Template Directed Synthesis	*181*
	References	*185*

7	**Artificial Receptor Compounds for Chiral Recognition**	*191*
	Thomas J. Wenzel and Ngoc H. Pham	
7.1	Introduction	*191*
7.2	Cyclodextrins	*191*
7.2.1	Alkylated Cyclodextrins	*192*
7.2.2	Acylated and Mixed Acylated/Alkylated Cyclodextrins	*194*
7.2.3	Carbamoylated Cyclodextrins	*194*
7.2.4	2-Hydroxypropylether Cyclodextrins	*195*
7.2.5	*Tert*-butyldimethylsilyl Chloride-Substituted Cyclodextrins	*195*
7.2.6	Anionic Cyclodextrins	*196*
7.2.7	Cationic Cyclodextrins	*197*
7.2.8	Miscellaneous Cyclodextrins	*198*
7.3	Crown Ethers	*200*
7.3.1	1,1′-Binaphthalene-Based Crown Ethers	*201*
7.3.2	Carbohydrate-Based Crown Ethers	*202*
7.3.3	Tartaric Acid-Based Crown Ethers	*204*
7.3.4	Crowns Ethers with Phenol Moieties	*206*
7.3.5	Crown Ethers with Pyridine Moieties	*208*
7.4	Calixarenes	*209*
7.5	Calix[4]resorcinarenes	*218*
7.6	Miscellaneous Receptor Compounds	*225*
7.7	Metal-Containing Receptor Compounds	*235*
	References	*237*

8	**Fullerene Receptors Based on Calixarene Derivatives** *249*	
	Pavel Lhoták and Ondřej Kundrát	
8.1	Introduction *249*	
8.2	Calixarenes *251*	
8.3	Solid State Complexation by Calixarenes *252*	
8.4	Complexation in Solution *257*	
8.5	Calixarenes as Molecular Scaffolds *263*	
8.6	Outlook *268*	
	References *269*	
9	**Guanidinium Based Anion Receptors** *273*	
	Carsten Schmuck and Hannes Yacu Kuchelmeister	
9.1	Introduction *273*	
9.2	Instructive Historical Examples *275*	
9.3	Recent Advances in Inorganic Anion Recognition *283*	
9.4	Organic and Biological Phosphates *287*	
9.5	Polycarboxylate Binding *292*	
9.6	Amino Acid Recognition *297*	
9.7	Dipeptides as Substrate *301*	
9.8	Polypeptide Recognition *303*	
9.9	Conclusion *312*	
	References *313*	
10	**Artificial Receptors Based on Spreader-Bar Systems** *319*	
	Thomas Hirsch	
	References *331*	
11	**Potential of Aptamers as Artificial Receptors in Chemical Sensors** *333*	
	Bettina Appel, Sabine Müller, and Sabine Stingel	
11.1	Introduction *333*	
11.2	Generation and Synthesis of Aptamers *334*	
11.2.1	Selection of Aptamers from Combinatorial Libraries (SELEX) *334*	
11.2.2	SELEX Variations *336*	
11.2.2.1	SELEX Using Modified Oligonucleotide Libraries *336*	
11.2.2.2	PhotoSELEX *336*	
11.2.2.3	Automated SELEX *337*	
11.2.3	Alternative Approaches for Selection of Aptamers from Combinatorial Libraries *338*	
11.2.3.1	Capillary Electrophoresis Techniques *338*	
11.2.3.2	AFM Techniques *339*	
11.2.4	Synthesis of Aptamers and Stabilization *339*	
11.2.4.1	Chemical Synthesis of Aptamers *339*	
11.2.4.2	Stabilization of Aptamers *340*	
11.3	Aptamer Arrays *341*	
11.4	Techniques for Readout of Ligand Binding to the Aptamer *343*	

11.4.1	Conformational Effects 343
11.4.2	Aptazymes 344
11.4.3	Methods of Sensing 345
11.4.3.1	Optical Sensing 345
11.4.3.2	Electrochemical Sensing 350
11.4.3.3	Acoustic Sensing 353
11.4.3.4	Quartz Crystal Microbalance Based Sensing 354
11.4.3.5	Cantilever Based Sensing 354
11.5	Outlook/Summary 354
	References 355

12	**Conducting Polymers as Artificial Receptors in Chemical Sensors** 361
	Ulrich Lange, Nataliya V. Roznyatovskaya, Qingli Hao, and Vladimir M. Mirsky
12.1	Introduction 361
12.2	Transducers for Artificial Receptors Based on Conducting Polymers 362
12.3	Intrinsic Sensitivity of Conducting Polymers 366
12.3.1	Sensitivity to pH Changes 366
12.3.2	Affinity to Inorganic Ions 368
12.3.3	Affinity to Gases and Vapors 368
12.4	Conducting Polymers Modified with Receptor Groups 369
12.4.1	Conducting Polymers with Receptor Groups Attached to the Monomer 369
12.4.1.1	Receptors for Ions 369
12.4.1.2	Receptors for Organic/Bioorganic Molecules 381
12.4.2	Conducting Polymers Doped with Receptor 382
12.4.3	Molecular Imprinting of Conducting Polymers 383
12.5	Conclusion 383
	References 384

13	**Molecularly Imprinted Polymers as Artificial Receptors** 391
	Florian Meier and Boris Mizaikoff
13.1	Introduction 391
13.2	Fundamentals of Molecular Imprinting 393
13.2.1	What are MIPs? 393
13.2.2	Approaches toward Molecular Imprinted Polymers 394
13.2.2.1	Noncovalent Imprinting (Self-Assembly Approach) 394
13.2.2.2	Covalent Imprinting (Preorganized Approach) 395
13.2.2.3	Semi-Covalent Imprinting 395
13.2.2.4	Advantages and Disadvantages of Different Imprinting Approaches 396
13.2.3	Reagents and Solvents in Molecular Imprinting Technology 396
13.2.3.1	Functional Monomers 396
13.2.3.2	Crosslinkers 397

13.2.3.3	Radical Initiators	*401*
13.2.3.4	Solvents	*402*
13.2.4	How are MIPs Prepared?	*404*
13.3	Polymer Formats and Polymerization Methods for MIPs	*405*
13.3.1	Bulk Polymers	*405*
13.3.2	Micro- and Nanobeads	*407*
13.3.3	MIP Films and Membranes	*408*
13.3.4	Comparison of MIP Formats Prepared by Different Polymerization Methods	*412*
13.4	Evaluation of MIP Performance – Imprinting Efficiency	*414*
13.4.1	Binding Capacity and Binding Affinity	*414*
13.4.2	Binding Selectivity	*418*
13.5	MIPs Mimicking Natural Receptors	*420*
13.5.1	Comparison of MIPs and Antibodies	*421*
13.5.2	MIPs as Artificial Antibodies in Pseudo-Immunoassays	*421*
13.5.3	MIPs as Catalysts with Enzymatic Activity	*425*
13.5.4	Quantitative Data on the Binding Properties of MIPs	*427*
13.6	Conclusions and Outlook	*427*
	References	*430*
14	**Quantitative Affinity Data on Selected Artificial Receptors**	*439*
	Anatoly K. Yatsimirsky and Vladimir M. Mirsky	
14.1	Structures of Receptors	*449*
	References	*458*

Index *461*

Preface

One of the great achievements in modern chemistry is the development of artificial synthetic receptors, which are typically low molecular weight compounds that perform selective binding (recognition) of a compound of interest. Such compounds are used for the design and chemically addressed assembly of supramolecular structures. Another important application of these compounds is the development of chemical sensors. Over the past few few decades affinity assays have been based mainly on immunological techniques: antibodies are used routinely in clinical applications or in food analysis. However, they are expensive, unstable, and cannot be prepared for some types of analytes. For example, one cannot immunize animals with highly toxic compounds or with compounds that are present in all animals (such as glucose, sodium ion). The solution is to use artificial receptors. Recent years have seen intensive development in this field. Traditional approaches based on the chemical synthesis of small molecules with high affinity have been completed by molecularly imprinted polymerization and by biotechnological preparation and selection of natural macromolecules with such properties.

Syntheses, structures, and recognition properties of artificial receptors are touched on in many monographs in the field of supramolecular chemistry. The book *Functional Synthetic Receptors* edited by Thomas Schräder and Andrew D. Hamilton (Wiley-VCH Verlag GmbH, 2005) provides deep insight into fundamental aspects of the subject. There is little literature, however, discussing artificial receptors with the emphasis on their practical applications as components of chemical sensors and assays. This book intends to fill this gap.

The book starts with two chapters discussing the most relevant quantitative characteristics of receptors – binding affinity and selectivity. Chapter 3, on the combinatorial development of sensing materials, deals with an advanced technological approach to the discovery of new receptors. Chapters 4–9 discuss particular types of receptors (cyclodextrins, cyclopeptides, boronic acids, chiral receptors, calixarenes, and guanidinium derivatives) of significant current importance. Finally, Chapters 10–13 deal with receptors based on organized (spreader-bar approach) or polymeric (aptamers, conducting polymers, and molecular imprinting) structures.

Chapter 14 provides an extensive compilation of the affinity properties of different receptors taken from current literature.

We hope that this book will be useful as a handbook for scientists (from universities and industry) and graduate and post-graduate students working in analytical and supramolecular chemistry, chemical sensors and biosensors, and in material science. It will also be of interest to experts and students working/studying surface chemistry, physical chemistry, and in some fields of organic chemistry, pharmacology, medical diagnostics, biotechnology, chemical technology, food, and environmental monitoring.

Finally, we express our gratitude to the authors of this book and hope that readers also find these contributions enjoyable, interesting, and useful.

February 2010

Vladimir M. Mirsky, Senftenberg
and
Anatoly K. Yatsimirsky, Mexico City

List of Contributors

Bettina Appel
Ernst-Moritz-Arndt Universität
Greifswald
Institut für Biochemie
Felix-Hausdorff-Str. 4
17487 Greifswald
Germany

Yunfeng Cheng
Georgia State University
Department of Chemistry and Center
for Biotechnology and Drug Design
Atlanta, GA 30302-4098
USA

Qingli Hao
Nanjing University of Science
and Technology
Key Laboratory of Soft Chemistry
and Functional Materials
Ministry of Education
210094 Nanjing
China

Thomas Hirsch
University of Regensburg
Institute of Analytical Chemistry
Chemo- and Biosensors
Universitätsstraße 31
93040 Regensburg
Germany

Hiroshi Ikeda
Tokyo Institute of Technology
Graduate School of Bioscience
and Biotechnology
Department of Bioengineering
4259-B-44 Nagatsuta-cho, Midori-ku
Yokohama 226-8501
Japan

Shan Jin
Georgia State University
Department of Chemistry and Center
for Biotechnology and Drug Design
Atlanta, GA 30302-4098
USA

Stefan Kubik
Technische Universität Kaiserslautern
Fachbereich Chemie - Organische
Chemie
Erwin-Schrödinger-Strasse
67663 Kaiserslautern
Germany

Hannes Yacu Kuchelmeister
University of Duisburg-Essen
Institute for Organic Chemistry
Universitätsstraße 7
45141 Essen
Germany

Ondřej Kundrát
Prague Institute of Chemical
Technology
Department of Organic Chemistry
Technická 5
166 28 Prague 6
Czech Republic

Ulrich Lange
University of Regensburg
Institute of Analytical Chemistry,
Chemo- and Biosensors
Universitätsstraße 31
93040 Regensburg
Germany

Pavel Lhoták
Prague Institute of Chemical
Technology
Department of Organic Chemistry
Technická 5
166 28 Prague 6
Czech Republic

Florian Meier
University of Ulm
Institute of Analytical and Bioanalytical
Chemistry
Albert-Einstein-Allee 11
89061 Ulm
Germany

Vladimir M. Mirsky
Lausitz University of Applied Sciences
Nanobiotechnology-BCV
Grossenhainer str. 57
01968 Senftenberg
Germany

Boris Mizaikoff
University of Ulm
Institute of Analytical and Bioanalytical
Chemistry
Albert-Einstein-Allee 11
89069 Ulm
Germany

Sabine Müller
Ernst-Moritz-Arndt Universität
Greifswald
Institut für Biochemie
Felix-Hausdorff-Str. 4
17487 Greifswald
Germany

Ngoc H. Pham
Bates College
Department of Chemistry
S. Andrews Road
Lewiston, ME 04240
USA

Radislav A. Potyrailo
General Electric Global Research
Chemical and Biological Sensing
Laboratory
Chemistry Technologies and Material
Characterization
1 Research Circle
Niskayuna, NY 12309
USA

Nataliya V. Roznyatovskaya
Fraunhofer Institute of Chemical
Technology
Joseph-von-Fraunhofer-Straße 7
76327 Pfinztal-Berghausen
Germany

Carsten Schmuck
University of Duisburg-Essen
Institute for Organic Chemistry
Universitätsstraße 5
45141 Essen
Germany

Hans-Jörg Schneider
Universität des Saarlandes
FR Organische Chemie
66041 Saarbrücken
Germany

Sabine Stingel
Ernst-Moritz-Arndt Universität
Greifswald
Institut für Biochemie
Felix-Hausdorff-Str. 4
17487 Greifswald
Germany

Binghe Wang
Georgia State University
Department of Chemistry and Center
for Biotechnology and Drug Design
Atlanta, GA 30302-4098
USA

Thomas J. Wenzel
Bates College
Department of Chemistry
S. Andrews Road
Lewiston, ME 04240
USA

Xiaochuan Yang
Georgia State University
Department of Chemistry and Center
for Biotechnology and Drug Design
Atlanta, GA 30302-4098
USA

Anatoly K. Yatsimirsky
Universidad Nacional Autónoma
de México
Facultad de Química
04510 México D.F.
México

1
Quantitative Characterization of Affinity Properties of Immobilized Receptors

Vladimir M. Mirsky

1.1
Introduction

Affinity as a tendency of molecules (ligands) to associate with another type of molecules or polymers (receptor) can be described by a set of kinetic and thermodynamic parameters. These parameters include the adsorption (or binding, or association) constant, which can be recalculated as the free energy of binding, binding enthalpy and entropy, kinetic adsorption and desorption constants, and activation energies for binding and for dissociation.

There are several reasons for a quantitative characterization of affinity. The first is due to possible applications of these receptors in affinity sensors. The sensors are intended to measure *volume* concentrations of analytes. However, transducers of affinity sensors (refractometric, interferometric, mechano-acoustical, capacitive, and others) provide information on the *surface* concentration of analytes (ligands) on a layer of immobilized receptors (Figure 1.1). Therefore, it is important to obtain a calibration curve – the dependence between volume concentration of an analyte and its surface concentration. Such relations are named in physical chemistry as adsorption isotherms (the binding is usually performed at constant temperature). Several adsorption isotherms can be obtained from simple physical models. A mathematical description of these isotherms allows one not only a better understanding of binding process but also provides a mathematical basis for interpolation and extrapolation of the calibration curve, which is of importance for analytical applications.

The second reason includes material science aspects. Quantitative information can be used to make an appropriate choice of synthetic receptors for different applications and for prediction of the detection limit and selectivity of analytical devices based on these receptors. Additionally, these data can be used as descriptors for combinatorial optimization and for discovery of new sensing materials (the combinatorial approach is discussed in Chapter 3) [1–3].

The association can be investigated in the bulk phase (in the solution or suspension of the ligand and receptor molecules) or on a surface. Ligand–receptor interactions

Artificial Receptors for Chemical Sensors. Edited by V.M. Mirsky and A.K. Yatsimirsky
Copyright © 2011 WILEY-VCH Verlag GmbH & Co. KGaA, Weinheim
ISBN: 978-3-527-32357-9

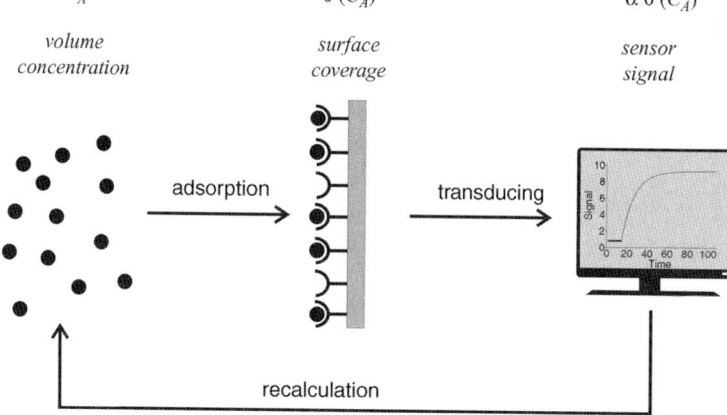

Figure 1.1 Transducers of affinity sensors provide information on the surface concentration of bound analytes. For practical applications, information on the volume concentration is required.

can be investigated in a bulk phase by titration and application of any analytical technique that is sensitive to the concentration of free or bound ligands or receptors. For example, the binding of dye molecules can be determined by colorimetry or fluorescence, while ions or redox-active species can be measured by potentiometric techniques. One of the most commonly used techniques that can be applied for various ligand–receptor systems is a NMR titration. Another general approach is based on isothermal calorimetry; in this case the heat produced during the ligand–receptor interaction is measured. This technique has the advantage of providing the enthalpy of binding in a single titration experiment. Analysis of ligand–receptor binding in bulk phases is well described in literature [4–7] and, therefore, is not covered in this chapter.

Applications of artificial receptors for chemical sensors require an immobilization of these receptors on a surface. In the 1980s to the beginning of 1990s there were many attempts to induce physical immobilization of chemoreceptors, for example, by using the Langmuir–Blodgett technique or electrostatically driven adsorption. Nowadays, receptor immobilization is performed mainly by formation of chemical bonds with some surface groups (e.g., peptide bond) or by introduction of surface anchoring groups into the receptor molecules. Chemical aspects of covalent receptor immobilization have been described [8–15]. The most widely used technology of chemical immobilization is based on the formation of an amide bond through activation by either EDC [1-ethyl-3-(3-dimethylaminopropyl)carbodiimide)] [14, 15] or an EDC–NHS (N-hydroxysuccinimide) mixture. The most widely used surface anchor group is the thiol group, which forms, spontaneously, an extremely strong bond with gold, silver, palladium, copper, and some other materials [13, 16–19].

Chemical immobilization of a receptor may influence its affinity properties. This chapter is focused on the characterization of affinity properties of immobilized receptors.

1.2
Measurements Under Equilibrium Conditions

Binding of a ligand (A) with a receptor (B) leading to the formation of a complex can be considered using a formal kinetic approach:

$$A + B \rightleftharpoons AB$$

The rate of association (adsorption onto the receptor layer) is proportional to the ligand concentration (c_L) and to the fraction of uncoated binding sites ($1 - \theta$), where θ is the fraction of occupied binding sites. Therefore, the association rate is $k_{ads}c_L(1 - \theta)$, where k_{ads} is the kinetic adsorption constant (also indicated as the kinetic association constant or kinetic binding constant). The rate of dissociation is $k_{des}\theta$, where k_{des} is the kinetic constant of desorption (also indicated as the kinetic constant of dissociation). At equilibrium conditions:

$$k_{ads} \cdot c_L(1-\theta_{eq}) = k_{des} \cdot c_L \theta_{eq} \tag{1.1}$$

therefore:

$$\theta_{eq} = \frac{c_L K}{1 + c_L K} \tag{1.2}$$

where:

$$K = k_{ads}/k_{des} \tag{1.3}$$

This model, well known as the Langmuir adsorption isotherm, is valid for most cases of ligand–receptor binding. The model is based on the following assumptions: (i) all binding sites are equivalent; (ii) the ability of a ligand molecule to bind a binding site is independent of the occupation of neighboring sites; (iii) the number of binding sites is limited; and (iv) there is a dynamic equilibrium between bound molecules and free molecules in aqueous environment.

At low ligand concentration ($c_L K \ll 1$) this equation is linearized: $\theta = c_L K$ (Henry adsorption isotherm). Physically, it corresponds to binding at low surface coverage.

Analysis of the literature on chemical sensors, biosensors, and supramolecular chemistry demonstrates that the Langmuir adsorption isotherm is valid for most cases. The main reasons for possible deviations from this simple model are non-homogeneity of binding sites [deviation from postulate (i)] and an influence of occupation of binding sites on affinity properties of neighboring sites [deviation from postulate (ii)].

Non-homogeneity of binding sites is a typical case for artificial receptors formed by molecularly imprinted polymerization (artificial receptors based on molecularly imprinted polymers are discussed in Chapter 13) [20, 21]. For a receptor layer consisting of several different types of receptors, the Langmuir adsorption isotherm has the form:

$$\theta_{eq} = \sum_i \gamma_i \frac{c_L K_i}{1 + c_L K_i} \tag{1.4}$$

where γ_i is the molar fraction of the receptor with binding constant K_i. For a continuous distribution of binding constants the sum in Eq. (1.4) can be replaced by an integral. In practice, Eq. (1.4) can be applied only for $i = 2$ or 3 and only if the difference between K_i is at least several-fold. In some cases the heterogeneity of the receptor properties and distribution through the sensing surface can be described by fractal geometry [22]. An exponential distribution of energy of the binding sites is described by Freundlich adsorption isotherm:

$$\theta = A \sqrt[B]{c} \tag{1.5}$$

As a criterion for this isotherm, a linearization versus logarithmic concentration scale is used.

In general, although non-homogeneity of the receptor properties can be explained and described mathematically, this leads to serious practical problems in applications of such receptors for analytical purposes. In this case the results depend strongly on the initial occupation of binding states of the receptor layer, and non-complete desorption of ligands leads to a quite different adsorption isotherm.

The distance between receptors immobilized on a surface is usually large enough to exclude an influence of the occupation of one binding site on the affinity of neighboring binding sites. However, such effects have been reported in several studies. Numerous models have been suggested to describe adsorption under such conditions. The most widely used is the Frumkin isotherm.

$$Kc = \frac{\theta}{1-\theta} \exp(-2a\theta) \tag{1.6}$$

This is an extension of the Langmuir isotherm, taking into account an interaction of adsorbed molecules. The interaction is described by parameter a, which is positive for attraction and negative for repulsion between adsorbed molecules (Figure 1.2). For $a = 0$ the Frumkin isotherm coincides with the Langmuir isotherm.

A signal of affinity sensors (S) is proportional to their surface coverage (θ) (this fact is often considered as a definition of affinity sensors); therefore, for binding according to the Langmuir isotherm the signal has the following dependence on ligand concentration:

$$S = \alpha \frac{c_L K}{1 + c_L K} \tag{1.7}$$

where α is a proportionality coefficient depending on the type of transducer and on an amplification factor during the signal processing. For small concentrations ($c_L K \ll 1$) this leads to a linear equation:

$$S = \alpha c_L K \tag{1.8}$$

The main question in the analysis of experimental data obtained under quasi-equilibrium conditions is the extraction of the binding constant K. In the case of a linear dependence of the signal of an affinity sensor on concentration [Eq. (1.8)],

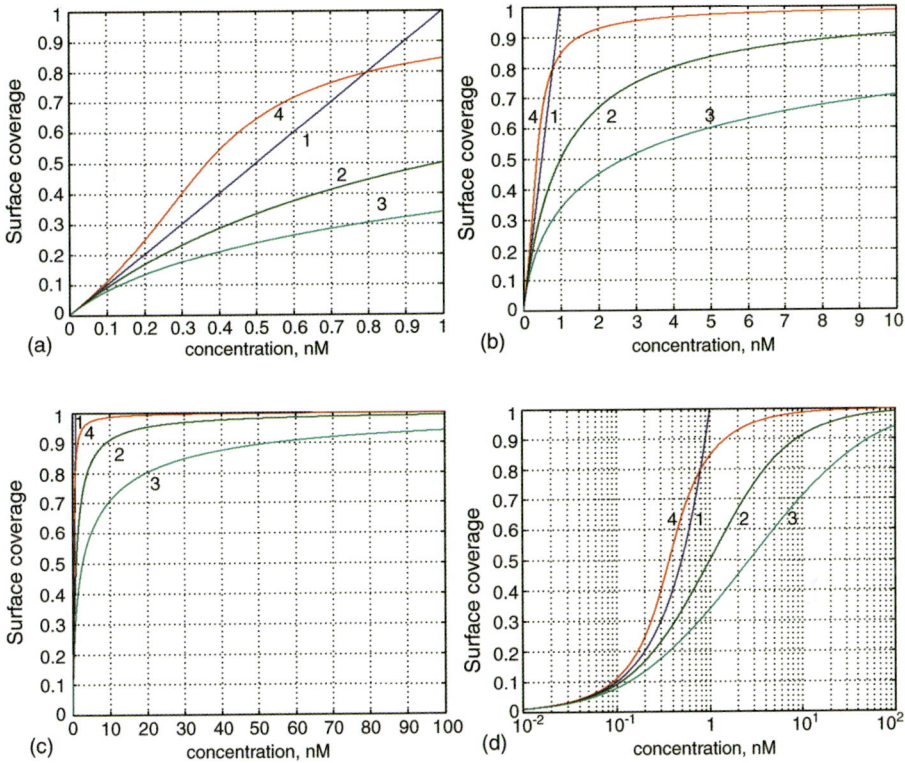

Figure 1.2 Simulated adsorption isotherms according to the models of Henry (curve 1), Langmuir (curve 2) and Frumkin (curves 3 and 4) for attraction ($a = +1$, curve 4) and repulsion ($a = -1$, curve 3) of adsorbed molecules at different concentration scales (a)–(c) and as a semilogarithmic plot (d). Binding constant in all the curves is $1/(1\,\text{nM})$.

such analysis would require information on the coefficient α, which is not always available and in general can be different for different receptors. However, even for unknown but constant values of α, a comparison of slopes of the curves measured at different conditions or for modified receptors provides valuable information on comparison of binding constants.

Binding curves reaching their saturation range are more informative. An analysis of such dependence can provide information not only on the binding constant but also on the homogeneity of the binding sites (or on deviations from the Langmuir model) and on the value of the proportionality factor α corresponding to the saturated value of the signal. The quality of data analysis improves with increasing deviation from linear dependence. However, the maximal concentration that can be used for experimental tests is usually limited (e.g., by solubility or by available amount of the ligand).

Principally, the parameters K and α can be extracted from experimental data by direct nonlinear fitting using Eq. (1.7). Another approach, providing a better

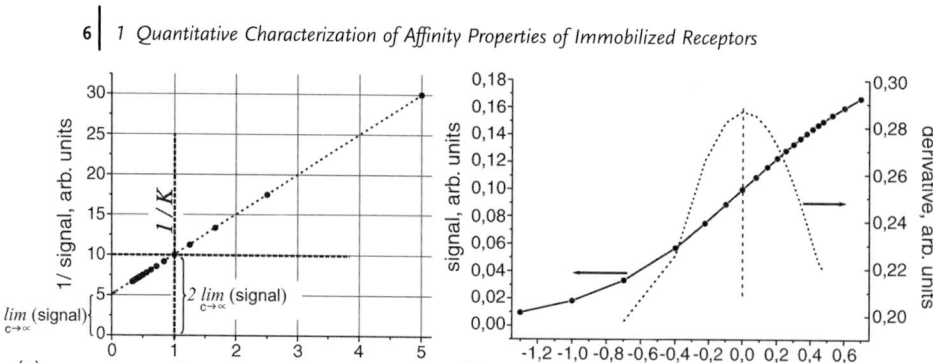

Figure 1.3 Extraction of binding constant by linearization of Langmuir isotherm in double reciprocal plots (a) and by symmetrization in semilogarithmic plots (b). The symmetry point can be found more exactly from the derivative.

visualization of the type of possible deviation from the Langmuir model, is based on linearization of this dependence in a double reciprocal plot (Lineweaver–Burk plot):

$$\frac{1}{S} = \frac{1}{\alpha K}\frac{1}{c_L} + \frac{1}{\alpha} \tag{1.9}$$

Extrapolation of the line gives information on both parameters of Eq. (1.7) (Figure 1.3a). Although this approach is sometimes considered as a relic of the before-computer era, it has several advantages. For example, it may indicate a systematic deviation from the Langmuir model or heterogeneity of binding sites. In this case, n types of binding sites with different affinity give $n-1$ breaks of the straight line. Practically it is applicable only for $n=2$ or 3 and only when the binding constants are very different (about one order or more), and the measurements are performed over the wide concentration range.

Another approach to obtain the binding constant from experimental data measured before saturation is based on data presentation versus a logarithmic concentration scale (Figure 1.2d, curve 2 and Figure 1.3b). In this case the reciprocal binding constant corresponds to the concentration at the symmetry point of the curve. This point also corresponds to the maximal slope or to the point of change of the sign of the curvature. Therefore, for data measured in a small concentration range, where the symmetry point is not clearly visible, it can be also found from analysis of the first or second derivatives (Figure 1.3b).

The obtained binding constant can be then recalculated into the molar standard Gibbs free energy (Gibbs energy) of the binding reaction:

$$\Delta G^0 = -RT \ln K \tag{1.10}$$

The binding constant in Eq. (1.10) characterizes the ratio of activities of corresponding species and is therefore a dimensionless value. By application of usual approximations of activities by concentrations, the concentration values should be converted into dimensionless ones by normalization of the concentrations to $1\,\mathrm{mol\,l^{-1}}$.

In practice, the binding constants of immobilized receptors based on equilibrium between volume and surface concentrations are measured under quasi-stationary conditions, as a response of an affinity sensor with this receptor to stepwise increasing ligand concentration. Typically, the measurement is performed 10–40 min after the concentration increase, when the signal "looks" constant. There is no strong criterion on reaching of equilibrium, and this uncertainty may be a reason for a systematic error.

An analysis of binding and selection of an appropriate binding isotherm carry a risk of misinterpretation and require very critical consideration. For example, logarithmic conversion, used for validation of the Freundlich isotherm, leads to strong compression of the variation range and therefore to some data linearization. A break in the double-reciprocal plot of the Langmuir isotherm for two binding sites is observed only for a sufficiently high difference of binding constants of these sites. In addition, data analysis based on a Scatchard plot, which is widely used for affinity experiments in bulk phase, is often misinterpreted [23, 24]. An uncertainty in the definition of quasi-equilibrium was discussed above. In many cases, statements on the selection of adsorption models can be checked by numerical simulations.

1.3
Kinetic Measurements

Binding constants can be also obtained from kinetic measurements. Numerous investigations have demonstrated that the values obtained from kinetic and equilibrium measurements are almost identical. Kinetic measurements of binding constants can be performed much faster than the measurements under quasi-equilibrium conditions. Moreover, for some receptor–ligand pairs the kinetics are so slow that the measurements in quasi-equilibrium conditions are impossible. Another advantage of kinetic measurements is that they provide information not only on the equilibrium binding constant but also on the kinetic constants of binding (k_{ads}) and dissociation (k_{des}).

As is usual for each surface reaction, a process of ligand to receptor binding includes two kinetic steps: diffusion of ligand to the receptor and ligand–receptor binding. A model used for analysis of kinetic measurements is based on the following assumptions: (i) the kinetic limiting step of the whole process is the binding and (ii) the binding can be described by the Langmuir isotherm. It is not a rare case when assumption (i) is not valid. Its validation will be described below.

According to the formal kinetics, the detected adsorption rate $d\theta/dt$ can be described as an algebraic sum of the rates of adsorption and desorption:

$$\frac{d\theta}{dt} = k_{ads}(1-\theta)c_L - k_{des}\theta \qquad (1.11)$$

or:

$$\frac{d\theta}{dt} = -\theta k_S + k_{ads} c_L \quad (1.12)$$

and:

$$\frac{dS}{dt} = -S k_S + k_{ads} c_L \quad (1.13)$$

where:

$$k_S = k_{ads} c_L + k_{des} \quad (1.14)$$

The solution of the differential equation (1.12) is:

$$\theta(t) = [1 - \exp(-k_S t)]\, \theta_{eq} \quad (1.15)$$

where θ_{eq} is the surface coverage in equilibrium and can, therefore, be found from Eq. (1.2).

The desorption kinetics observed after replacement of the ligand solution by the solution without ligand can be obtained from the equation:

$$\frac{d\theta}{dt} = -k_{des} \theta \quad (1.16)$$

The solution with the corresponding starting condition is:

$$\theta(t) = \theta_{eq} \exp(-k_{des} t) \quad (1.17)$$

It follows from Eq. (1.12) that k_S is an apparent kinetic constant of adsorption. This value can be obtained from experimental data. It is paradoxical that the desorption constant is one of the components contributing to the adsorption constant in Eq. (1.14).

The dependence of k_S on ligand concentration allows one to obtain values of k_{ads} and k_{des}. The value k_S can be obtained by exponential fitting of adsorption kinetics with Eq. (1.15). Instead of nonlinear fitting, the dependence of dS/dt versus S can be analyzed: a linearity of this dependence verifies the exponential character of the signal kinetics while its slope gives k_S (Figure 1.4). The desorption rate constant k_{des} can also be extracted from fitting of desorption kinetics with Eq. (1.17). This provides a principal way of obtaining kinetic constants of adsorption and desorption and of estimating the binding constant from single experiment. However, desorption kinetics for many receptors are very slow, and k_{ads} cannot be measured directly. In addition, more reliable data can be obtained by repetition of the measurements at different concentrations and by analysis of the concentration dependence of k_S according to Eq. (1.14). Values of k_{ads} and k_{des} obtained from kinetic measurements can be used to calculate the binding constant *(K)* [Eq. (1.3)]. A deviation of the dependence of k_S versus ligand concentration from a linear dependence may indicate that the process is not limited by the binding step. The quality of this analysis depends strongly on the signal/noise ratio.

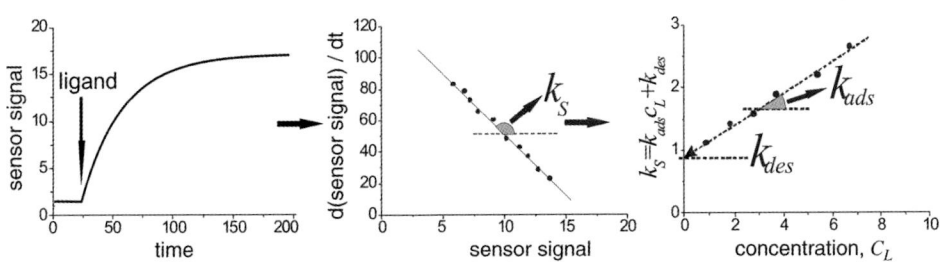

Figure 1.4 Extraction of kinetic constants of adsorption and desorption.

As mentioned above, it is not a rare case that the binding process is limited by diffusion. It is more typical for receptor layers with a three-dimensional matrix. Ignoring this fact and a mechanistic application of the analysis based on the Langmuir model to the diffusion limited processes leads to large errors in the determination of k_{ads} and k_{des} and may lead to considerable errors in binding constants. Therefore it is important to formulate simple criteria to distinguish diffusion and reaction control of the binding process. A deviation of the binding kinetics from mono-exponential dependence is a sign of diffusion limitation; however, it is difficult to provide a reasonable quantitative criterion for this deviation. Moreover, diffusion limited processes can also display a mono-exponential kinetics.

For diffusion controlled processes an adsorption rate depends on the solution viscosity. The viscosity can be modified by addition of sugar, glycerol, or other compounds. The influence of viscosity on binding kinetics should be independent of the chemical nature of the viscosity modifier – a criterion that excludes a chemical influence of the viscosity modifier on the binding process.

Another approach to distinguish diffusion and reaction kinetics is based on variation of thickness of the diffusion boundary layer. It can be performed in devices that are similar to the rotating electrodes widely used in electrochemical experiments. In the case of a reaction limited process, no influence of the rotation rate is expected. For a diffusion limited process, the observed adsorption rate should be increased with increasing rotation rate. The thickness of the hydrodynamic equivalent of the diffusion boundary layer (δ) depends linearly on the square root of reciprocal rotation rate [25]:

$$\delta = 1.61 \cdot D^{\frac{1}{3}} \eta^{\frac{1}{6}} \omega^{-\frac{1}{2}} \tag{1.18}$$

where

D is the diffusion coefficient,
η is the solvent viscosity,
ω is the circular frequency of the electrode rotation.

The diffusion flux is proportional to the reciprocal value of thickness of the diffusion boundary layer. Therefore, for a diffusion controlled process one can expect a linear dependence between the adsorption rate and the square root of the rotation rate. In flow-through cells this approach can be realized by variation of the flow rate.

In cells without through-flow some variation of the thickness of diffusion layer can be obtained by modification of stirring intensity. An application of a rotating receptor surface in immunoassay has been described [26]. Identification of the limiting kinetic step based on modification of viscosity and on investigation of the dependence of binding kinetics on rotation rate of disk electrode has been employed [27]; however, such analysis is time consuming and too difficult for routine applications. A simpler test is an investigation of the influence of flow rate on binding kinetics; a quantitative analysis and experimental validation of this approach has been described [28–31].

1.4
Analysis of Temperature Dependencies

Affinity properties can be further characterized by investigation of the temperature dependencies.

Spontaneous adsorption means that the free energy of the process is negative. Adsorption leads to a decrease in freedom. Taking into account that $\Delta H = \Delta G + T\Delta S$, this means that the enthalpy of adsorption is also negative. Therefore, a binding of a ligand with a receptor leads to heat production. The entropy production can be determined quantitatively by investigation of the temperature dependence of the binding constant. Substitution of $\Delta G = \Delta H - T\Delta S$ in Eq. (1.10) leads to the van't Hoff relationship, which allows one to obtain the molar enthalpy of the binding reaction from the temperature dependence of the binding constant:

$$\ln K = \frac{1}{R}\left(\Delta S^0 - \frac{\Delta H^0}{T}\right) \tag{1.19}$$

Therefore, a slope of the dependence of the logarithm of the equilibrium constant as a function of reciprocal temperature gives the value of standard reaction enthalpy while extrapolation to zero value of the reciprocal temperature gives the value of reaction entropy. A deviation of this dependence was explained by a contribution of a temperature-dependent heat capacity [32]. Assuming a constant difference in heat capacity between free and associated ligand and receptor (ΔC_p) in the temperature range between the temperature of the standard state T_0 (usually 25 °C) and the current temperature T, one can rewrite Eq. (1.19) more exactly:

$$\ln K = \frac{1}{R}\left[\Delta S^0 - \frac{\Delta H^0}{T} + \Delta C_p\left(\frac{T-T_0}{T} - \ln\frac{T}{T_0}\right)\right] \tag{1.20}$$

This provides, in principle, the possibility of determining not only the binding enthalpy and binding entropy but also ΔC_p. However, a numerical simulation demonstrated that the fitting of experimental data is not very sensitive to this parameter, and typical values of experimental errors makes quantitative determination of ΔC_p very uncertain [33].

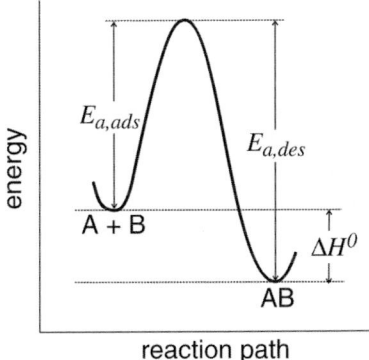

Figure 1.5 Activation energies for binding and dissociation ($E_{a,ads}$ and $E_{a,des}$, respectively) allows us to calculate the molar enthalpy of the binding reaction.

An investigation of temperature dependencies of the kinetic constants of binding (k_{ads}) and dissociation (k_{des}) allows us to obtain activation energies for binding and dissociation, $E_{a,ads}$ and $E_{a,des}$, respectively:

$$k_{ads} \sim \exp(-E_{a,ads}/RT); \quad k_{des} \sim \exp(-E_{a,des}/RT) \tag{1.21}$$

The difference between $E_{a,ads}$ and $E_{a,des}$ can be interpreted as the molar enthalpy of the binding reaction (Figure 1.5):

$$E_{a,ads} - E_{a,des} \approx \Delta H^0 \tag{1.22}$$

The transition state theory developed by Eyring and coauthors [34] provides the principal means of making a more detailed analysis of the reaction profile and to separate the contributions of enthalpy and entropy in the activation process. This theory gives the following equation for the kinetic constant of binding:

$$k_{ads} = \frac{\varkappa k_B T}{h} \exp(-\Delta H^{\#0}_{ads}/RT) \exp(-\Delta S^{\#0}_{ads}/R) \tag{1.23}$$

where

the superscript # indicates that the corresponding thermodynamic potential refers to an activation process,
k_B is the Boltzmann constant,
h is the Plank constant,
$\varkappa$ is a transmission factor, which is considered to be between 0.5 and 1.

Linearization in coordinates $\ln(k/T)$ versus $1/T$ (Eyring plots) can be used to extract values of standard activation enthalpies and entropies; however, a determination of entropy requires an assumption on the value of $\varkappa$ (usually, as $\varkappa = 1$).

The risk of misinterpretation of temperature dependencies of equilibrium and kinetic data is very high. Critical analysis of many aspects of this analysis has been performed [33, 35].

1.5
Experimental Techniques

While affinity analysis in the bulk phase can be performed by any analytical technique providing measurements of concentration of free or occupied ligand or receptor molecules, affinity analysis of immobilized receptors requires special methods that provide measurements of surface concentrations. Figure 1.6 gives a short review of the main label-free techniques.

An application of an impedometric approach to study adsorption phenomena came into analytical chemistry from classical electrochemistry. This approach is also known as capacitive detection because, usually, an imaginary component of impedance is used. Obviously, this approach can be applied only for conducting surfaces. In ideal case, the interface should display pure capacitive properties, that is, should block any charge transfer.

The specific electrical capacitance (C) of an interface coated by some dielectric layer with relative dielectric constant ε and thickness d is:

$$C = \frac{\varepsilon \varepsilon_0}{d} \tag{1.24}$$

where ε_0 is the dielectric constant of a vacuum. There are two ways to apply capacitive binding detection: to detect changes of either the layer thickness or dielectric constant. The first approach has been used, for example, in spreader-bar systems (the spreader-bar technique is described in Chapter 10) [36, 37]: binding of ligands to receptors leads to an increase of the thickness of the dielectric layer. The second approach has been applied for molecularly imprinted polymers (Chapter 13) [38, 39]:

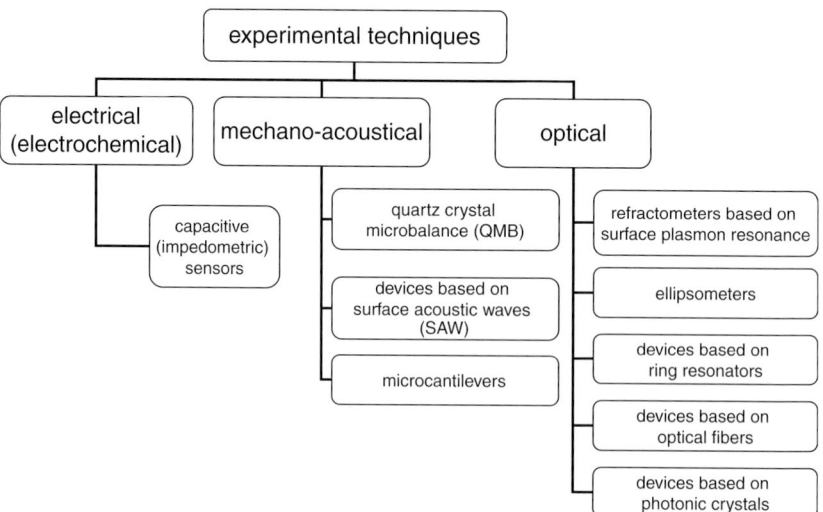

Figure 1.6 Main experimental techniques used for investigation of affinity properties of immobilized receptors.

binding of ligands into the pores of the polymer leads to the replacement of highly polarizable water molecules by lower polarizable ligand molecules and to a corresponding decrease of dielectric constant of the polymer. Linearity between capacitive effect and surface concentration of ligand has been proved for binding of detergents [40]; however, one can expect a strong deviation for large molecules with complex shape [41].

The main types of mechano-acoustic transducers for investigation of ligand–receptor interaction include the quartz crystal microbalance (QCM) [42], devices based on surface acoustic waves (SAWs), and systems based on microcantilevers. The QCM is based on changes of its resonance frequency due to an increase of the mass of the receptor layer caused by binding of ligand. In the gas phase this frequency shift Δf can be described by the Sauerbrey equation [43]:

$$\Delta f = -a f_0^2 \Delta m \tag{1.25}$$

where

f_0 is the resonance frequency of the quartz,
Δm is the change in mass,
a is a constant that depends on the quartz material.

The behavior of QCM in viscous phases is more complicated. Commercial QCM devices have been reviewed [44]. According to Eq. (1.25), the sensitivity of a QCM increases as the second power of the resonance frequency; therefore, an increase of this frequency is favorable. However, its increase over 40 MHz requires the formation of crystals that are too thin to provide the required mechanical stability. This limitation can be overcome in SAW-devices, which can operate at hundreds of MHz [45]. Conversely, microcantilevers have a lower Q-factor, lower resonance frequency, and therefore lower sensitivity in dynamic mode. This can probably be compensated for by their high sensitivity in deflection mode. Analytical features of microcantilevers have been reviewed [46].

Sensitive optical methods for label-free detection of ligand interaction with immobilized receptors include surface plasmon resonance (SPR) [47, 48] and several interferometric techniques [49]. The present sensitivity of SPR devices is about 10^{-7}–10^{-6} refractive index units and is limited by the high temperature dependence of the refractive index of water. Numerous efforts have been made to increase the sensitivity of SPR and to minimize a contribution of bulk water phase.

The methods for analysis of binding properties of immobilized receptors can be applied to obtain complete characterization of the affinity of artificial receptors, including binding constants, adsorption and desorption rate, activation energies for adsorption and desorption, to perform analysis of binding enthalpy and entropy, or even to obtain enthalpy and entropy contributions to activation process during the binding. The main drawback influencing applications of these techniques is the relatively low sensitivity of analytical methods for detection of binding with immobilized receptors. For this reason it may be difficult to obtain reliable results for small ligands.

References

1 Potyrailo, R.A. and Mirsky, V.M. (eds) (2009) *Combinatorial Methods for Chemical and Biological Sensors*, Springer.
2 Potyrailo, R.A. and Mirsky, V.M. (2008) *Chem. Rev.*, **108**, 770–813.
3 Mirsky, V.M., Kulikov, V., Hao, Q., and Wolfbeis, O.S. (2004) *Macromol. Rapid Commun.*, **25**, 253–258.
4 Hirose, K. (2007) Determination of binding constants, in *Analytical Methods in Supramolecular Chemistry* (ed. C.A. Schalley), Wiley-VCH Verlag GmbH, Weinheim, pp. 17–54.
5 Schmidtchen, F.P. (2007) Isothermal titration calorimetry in supramolecular chemistry, in *Analytical Methods in Supramolecular Chemistry* (ed. C.A. Schalley), Wiley-VCH Verlag GmbH, Weinheim, pp. 55–78.
6 Hirose, K. (2001) *J. Inclusion Phenom. Macrocyclic Chem.*, **39**, 193–209.
7 Connors, K.A. (1986) *Binding Constants: The Measurement of Molecular Complex Stability*, Wiley-Interscience.
8 Hermanson, G.T. (1996) *Bioconjugate Techniques*, Academic Press.
9 Albers, W.M., Vikholm, I., Viitala, T., and Peltonen, J. (2001) Interfacial and materials aspects of the immobilization of biomolecules onto solid surfaces in *Handbook of Surfaces and Interfaces of Materials* (ed. H.S. Nalwa), vol. 5, Academic Press, pp. 1–31.
10 Wong, L.S., Khan, F., and Micklefield, J. (2009) *Chem. Rev.*, **109**, 4025–4053.
11 Sethi, D., Gandhi, R.P., Kuma, P., and Gupta, K.C. (2009) *Biotechnol. J.*, **4**, 1513–1529.
12 Novick, S.J. and Rozzell, J.D. (2005) *Methods Biotechnol.*, **17**, 247–271.
13 Mirsky, V.M., Riepl, M., and Wolfbeis, O.S. (1997) *Biosens. Bioelectron.*, **12**, 977–989.
14 Wrobel, N., Schinkinger, M., and Mirsky, V.M. (2002) *Anal. Biochem.*, **305**, 135–138.
15 Nakajima, N. and Ikada, Y. (1995) *Bioconjugate Chem.*, **6**, 123–130.
16 Love, J.C., Estroff, L.A., Kriebel, J.K., Nuzzo, R.G., and Whitesides, G.M. (2005) *Chem. Rev.*, **105**, 1103–1170.
17 Ulman, A. (1996) *Chem. Rev.*, **96**, 1533–1554.
18 Mirsky, V.M. (2002) *Trends Anal. Chem.*, **21**, 439–450; Schreiber, F. (2004) *J. Phys.: Condens. Matter*, **16**, R881–R900.
19 Finklea, H.O. (1996) Electrochemistry of organized monolayers of thiols and related molecules on electrodes, in *Electroanalytical Chemistry: A Series of Advances*, vol. 19 (eds A.J. Bard and I. Rubinstein), Marcel Dekker, New York, pp. 110–317.
20 Umpleby, R.J. II, Baxter, S.C., Bodea, M., Berch, J.K. Jr., Shaha, R.N., and Shimizua, K.D. (2001) *Anal. Chim. Acta*, **435**, 35–42.
21 Kim, H., Kaczmarski, K., and Guiochon, G. (2006) *J. Chromatogr. A*, **1101**, 136–152.
22 Sadana, A. (2006) *Binding and Dissociation Kinetics for Different Biosensor Applications Using Fractals*, Elsevier.
23 Nørbyb, J.G., Ottolenghib, P., and Jensen, J. (1980) *Anal. Biochem.*, **102**, 318–320.
24 Zierler, K. (1977) *Biophys. Struct. Mech.*, **3**, 275–289.
25 Levich, V.G. (1962) *Physico-Chemical Hydrodynamics*, Prentice-Hall International., London.
26 Wang, G., Driskell, J.D., Porter, M.D., and Lipert, R.J. (2009) *Anal. Chem.*, **81**, 6175–6185.
27 Beck Erlach, M., and Mirsky, V.M. (2010) submitted.
28 Glaser, R.W. (1993) *Anal. Biochem.*, **213**, 152–161.
29 Lok, B.K., Cheng, Y.-L., and Robertson, C.R. (1983) *J. Colloid Interface Sci.*, **91**, 104–116.
30 Filippov, L.K. and Filippova, N.L. (1997) *J. Colloid Interface Sci.*, **189**, 1–16.
31 Svitel, J., Boukari, H., Van Ryk, D., Willson, R.C., and Schuck, P. (2007) *Biophys. J.*, **92**, 1742–1758.
32 Naghibi, H., Tamura, A., and Sturtevant, J.M. (1995) *Proc. Natl. Acad. Sci. USA*, **92**, 5597–5599.
33 Zhukov, A. and Karlsson, R. (2007) *J. Mol. Recognit.*, **20**, 379–385.
34 Wynn-Jones, W.F.K. and Eyring, H. (1935) *J. Phys. Chem.*, **3**, 492–502.
35 Winzor, D.J. and Jackson, C.M. (2006) *J. Mol. Recognit.*, **19**, 389–407.

36 Mirsky, V.M., Hirsch, T., Piletsky, S.A., and Wolfbeis, O.S. (1999) *Angew. Chem. Int. Ed.*, **111**, 1179–1181.
37 Hirsch, T., Kettenberger, H., Wolfbeis, O.S., and Mirsky, V.M. (2003) *Chem. Commun.*, 432–433.
38 Prodramidis, M., Hirsch, T., Wolfbeis, O.S., and Mirsky, V.M. (2003) *Electroanalysis*, **15**, 1–4.
39 Delaney, T.L., Zimin, D., Rahm, M., Weiss, D., Wolfbeis, O.S., and Mirsky, V.M. (2007) *Anal. Chem.*, **79**, 3220–3225.
40 Krause, C., Mirsky, V.M., and Heckmann, K.D. (1996) *Langmuir*, **12**, 6059–6064.
41 Terrettaz, S., Stora, T., Duschl, C., and Vogel, H. (1993) *Langmuir*, **9**, 1361–1369.
42 Steinem, C. and Janshoff, A. (ed.) (2007) *Piezoelectric Sensors*, Springer, Heidelberg.
43 Sauerbrey, G. (1959) *Z. Phys.*, **155**, 206–222.
44 Handley, J. (2001) *Anal. Chem.*, **73**, 225A–229A.
45 Länge, K., Rapp, B.E., and Rapp, M. (2008) *Anal. Bioanal. Chem.*, **391**, 1509–1520.
46 Goeders, K.M., Colton, J.S., and Bottomley, L.A. (2008) *Chem. Rev.*, **108**, 522–542.
47 Schasfoort, R.B.M. and Tudos, A.J. (ed.) (2008) *Handbook of Surface Plasmon Resonance*, RSC Publishing, Cambridge.
48 Homola, J. (2008) *Chem. Rev.*, **108**, 462–493.
49 Fan, X., White, I.M., Shopova, S.I., Zhu, H., Suter, J.D., and Sun, Y. (2008) *Anal. Chim. Acta*, **620**, 8–26.

2
Selectivity of Chemical Receptors
Hans-Jörg Schneider and Anatoly K. Yatsimirsky

2.1
Introduction

Selectivity is obviously a primary issue in the design of sensors, in addition to sensitivity. The theoretical correlation between the two aims is discussed first. Noncovalent interactions are the basis of most sensing techniques; we therefore highlight the achievement of selective synthetic receptors according to the underling intermolecular binding mechanisms, and illustrate them with some typical examples. Noncovalent interactions have the advantage of fast response times and reversibility, which is a significant issue for the re-usability and lifetime of sensors. Nevertheless, covalent-bond making and breaking also plays an important role, provided such reactions are fast enough. Such approaches are particularly promising if noncovalent interactions are too weak and, thereby, also barely selective in a given medium. Thus, hydrogen bonding in an aqueous surrounding competes with the bulk water, for which reason one uses for carbohydrate sensing, preferentially, rapid ester formation with boronic acids (Chapter 6).

High selectivity is a hallmark of biological receptors, and has always been a most important aim of synthetic supramolecular chemistry [1]. At the same time, high affinity, which ensures high sensitivity for a chosen analyte or reagent, is an equally significant goal. Possible correlations between affinity and selectivity have been emphasized only recently [2, 3], and will be a focus in the present chapter, along with the consequences of multivalency in supramolecular complexes. The underlying principles are relatively well understood for complexation of metal ions [4], and aspects of selectivity in metal complex formation related to strain and stress as well as to fit and misfit have been reviewed [5]. In this chapter we discuss typical examples of host–guest complexes, arranged according to their interaction mechanism.

The mechanistic basis for most selective molecular recognition poses many problems, which will be addressed in the present chapter. In traditional supramolecular chemistry selectivity is characterized by the hole-size fitting concept [1], which, however, requires significant modifications. As has been shown in particular

Artificial Receptors for Chemical Sensors. Edited by V.M. Mirsky and A.K. Yatsimirsky
Copyright © 2011 WILEY-VCH Verlag GmbH & Co. KGaA, Weinheim
ISBN: 978-3-527-32357-9

with coordination complexes, an optimal affinity is reached if bonds between transition metal ions and the donor functions are formed with a minimum of strain energy [4, 5]. Strain may originate from the formation of bonds between receptor and analyte, as well as from deformation of the analyte or receptor molecule by an induced fit. Another limitation is that, due to solvation and likely entropic factors, a higher binding constant is often observed if the guest molecule enjoys more freedom, and occupies only part of the space inside a host cavity [6]. The design of suitable host geometries can be significantly supported by calculational approaches [7]; free energy perturbation (FEP) methods allow explicit consideration also of solvation [8]. Although the selectivity aspect is always considered in reviews and monographs on supramolecular chemistry, it has rarely been a subject of special discussion [3, 9]. The obvious importance of this aspect for the design of sensors, of discriminators for separation processes, of many supramolecular devices, and last not least of drug compounds justifies an attempt to highlight the basic principles and possible new approaches in this area.

2.2
Some General Considerations on Selectivity

When considering the selectivity of chemical receptors designed for using as sensors one naturally turns to the field of analytical chemistry, where the following definition has been recommended: "Selectivity refers to the extent to which the method can be used to determine particular analytes in mixtures or matrices without interferences from other components of similar behavior" [10]. Different approaches to improve and to quantify the selectivity have been developed [11], in particular for multicomponent analysis, when a simultaneous determination of several analytes is performed on the basis of, for example, spectrophotometric measurements of mixtures of absorbing compounds at different wavelengths [12]. Extensively developed new methods use sensor arrays ("electronic tongues, or noses," etc.) [13]; here selectivity is achieved by a certain pattern of signals for a given analyte from a number of less specific receptors [14]. Of course, when the selectivity of a method is concerned, many tools, such as the use of masking reagents, preliminary separation by, for example, extraction or chromatography, and so on, can be employed to improve it. The selectivity of a single receptor can be due to the type of signal it produces on interaction with the analyte, for example, the absorption wavelength of the complex, or due to the different degree of binding of analytes. In the latter case optimal binding and the correlation between affinity and selectivity is the critical issue, which is the subject of this section.

Some general aspects of binding selectivity of a given receptor R toward analytes X and Y can be illustrated by assuming a simple 1:1 binding isotherm (or the Langmuir adsorption isotherm for heterogeneous systems) for the interaction of R with a mixture of X and Y with respective binding constants K_{RX} and K_{RY}. Let us assume that the signal is proportional to the concentration of the receptor–analyte complex and so the binding selectivity is given by the ratio of concentrations of the complexes with both analytes, that is, [RX]/[RY]. In some cases the receptor is applied at very low concentrations, so that $[R] \ll [X]$ and $[R] \ll [Y]$. This is typically the case when R is a strongly absorbing or highly fluorescent signaling compound. In such a situation the fraction of bound analyte is always small; its equilibrium concentration is approximately equal to its total concentration and the selectivity is given by Eq. (2.1):

$$[RX]/[RY] = K_{RX}[X]/K_{RY}[Y] = (K_{RX}/K_{RY}) \times ([X]/[Y]) \qquad (2.1)$$

It follows from Eq. (2.1) that the selectivity is independent of the concentration of R, and if one has, for example, an equimolar mixture of analytes the selectivity is given solely by the ratio of the binding constants at any total concentration of analytes. However, for practical applications it is preferable that the signal is proportional to the analyte concentration. This means that the total concentration of analytes must be low enough to avoid saturation of the receptor as then the signal does not depend on the concentration of the analyte. The situation is illustrated graphically in Figure 2.1 for a hypothetical case when increasing concentrations of guests X and Y with binding constants $K_{RX} = 1 \times 10^4 \, M^{-1}$ and $K_{RY} = 1 \times 10^3 \, M^{-1}$ taken at constant molar ratio 1: 1 are added to receptor R at a constant concentration 2×10^{-5} M. Note that "saturation" occurs in this case simultaneously for both guests, although in the titration curves for individual guests "saturation" with stronger bound guest occurs earlier than with the weaker bound guest. The selectivity given by the ratio $[RX]/[RY]$ (dashed line) is slightly lower than the ratio of the binding constants (10) because the calculation takes into account the mass balance for each guest, which is ignored in the simplified Eq. (2.1), and is approximately constant, as predicted with Eq. (2.1).

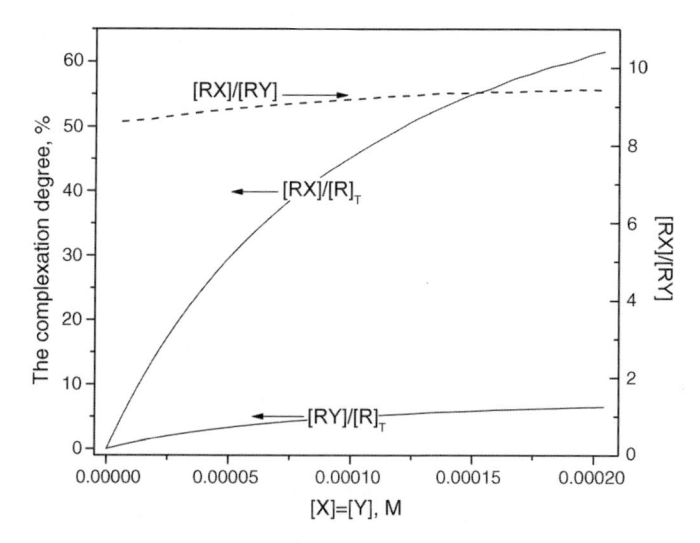

Figure 2.1 Simulated species distribution plot for the titration of 2×10^{-5} M receptor R by a 1: 1 mixture of guests X and Y with the ratio of binding constants $K_{RX}/K_{RY} = 10$. Solid lines show percentage of conversion of the receptor into complexes [RX] and [RY] (left-hand axis), while the dashed line shows the selectivity [RX]/ RY] (right-hand axis).

We now consider the opposite situation, when one applies a high excess of the receptor over analytes. Such conditions are convenient when the receptor has low affinity and the signal is produced by the analyte; they are often employed with, for example, cyclodextrin receptors. In such a situation the fraction of bound receptor is always small, its equilibrium concentration is approximately equal to its total concentration, and the selectivity is given by Eq. (2.2), where [X] and [Y] are total concentrations of analytes:

$$[RX]/[RY] = (K_{RX}/K_{RY}) \times ([X]/[Y]) \times \{(1 + K_{RY}[R])/(1 + K_{RX}[R])\} \quad (2.2)$$

Under these conditions the signal is always proportional to the analyte concentration and the signal is actually largest at saturation, but one again needs to avoid saturation conditions, now because at saturation the binding selectivity is completely lost. Indeed, as follows from Eq. (2.2), when $K_{RY}[R] \gg 1$ and $K_{RX}[R] \gg 1$ the binding constants cancel out and the ratio [RX]/[RY] equals the ratio of analyte concentrations [X]/[Y]. This situation is illustrated graphically in Figure 2.2 for a hypothetical case when a 1:1 mixture of guests X and Y with binding constants $K_{RX} = 1 \times 10^4 \, M^{-1}$ and $K_{RY} = 1 \times 10^3 \, M^{-1}$ taken at a constant concentration $2 \times 10^{-5} \, M$ is titrated with receptor R. Note that in this case "saturation" occurs for each guest separately – earlier for the stronger bound guest than with the weaker bound guest. The selectivity given by the ratio [RX]/[RY] (dashed line) decreases rapidly with increasing receptor concentration.

Besides binding or thermodynamic selectivity one may consider also the selectivity created by differences in reaction rates, that is, kinetic selectivity. This is the basis of so-called kinetic methods of analysis [15]. It has been shown that selectivity of affinity sensors can be significantly improved by using them in non-equilibrium conditions

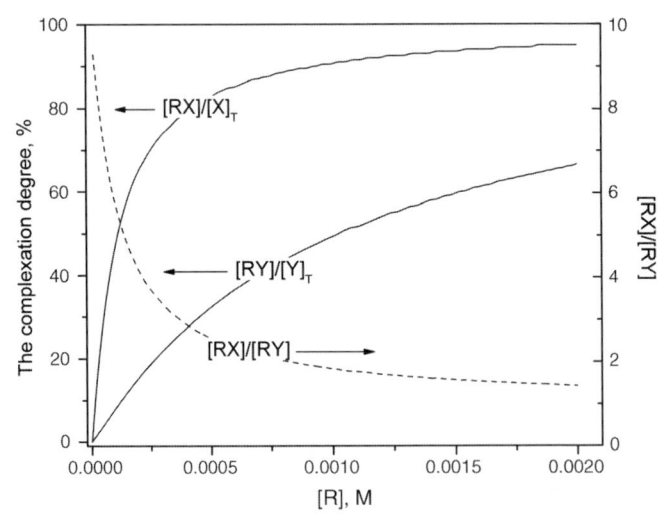

Figure 2.2 Simulated species distribution plot for the titration of a 1:1 mixture of 2×10^{-5} M guests X and Y with receptor R with the ratio of binding constants $K_{RX}/K_{RY} = 10$. Solid lines show percentage of conversion of analytes into complexes [RX] and [RY] (left-hand axis), while the dashed line shows the selectivity [RX]/[RY] (right-hand axis).

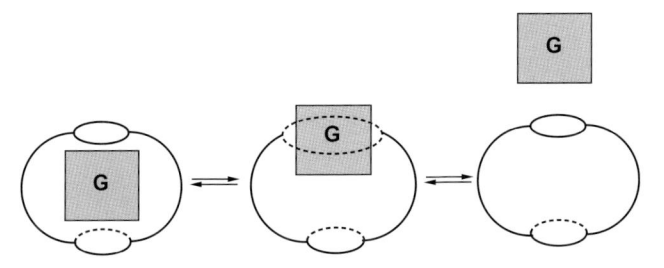

Scheme 2.1 Dissociation process for host–guest complex formation requiring a conformational change to allow the guest to pass through a small aperture of the host capsule.

due to differences in adsorption and/or desorption rates of analytes [16]. Kinetic selectivity is of great importance for the proper functioning of biological systems. For example, the amazing selectivity of protein synthesis when the probability of incorporation of a wrong amino acid into the protein molecule is less than 1:10 000 even for amino acids that differ just by one methylene group, for example, valine versus isoleucine, is achieved by a unique proofreading mechanism of aminoacyl-tRNA synthetases, providing selectivity beyond the classical lock-and-key principle [17].

For most synthetic receptors the kinetics of host–guest interactions are fast and equilibrium is reached immediately after mixing of reagents. Slow rates are observed in cases when encapsulation of a guest occurs inside a host that does not have a sufficiently wide portal to allow rapid entrance of the guest (Scheme 2.1).

In this situation the host and/or guest must undergo a conformational change to allow the entrance/exit of the guest and this creates a potential barrier leading to slow kinetics. Mechanisms of such encapsulation reactions for hemicarcerands, cucurbiturils, hydrogen-bonded assemblies, and metal–ligand assemblies have been reviewed recently [18]. Interestingly, the factors that affect the rate of encapsulation are not the same as those that affect the complexation free energy, for example, the bulkiness of the guest is much more important for kinetics than for equilibrium; solvent effects are also different. Therefore, kinetic guest discrimination might have a completely different selectivity than the thermodynamic discrimination, but we are not aware of practical applications of this principle for discrimination of different analytes.

2.3
Correlation Between Selectivity and Affinity

For reactions of a receptor R with either ligand X or ligand Y, with equilibrium constants K_{RX} and K_{RY}, one can write the familiar Eqs (2.3) and (2.4):

$$K_{RX} = \exp(-\Delta G_{RX}/RT) \quad (2.3)$$
$$K_{RY} = \exp(-\Delta G_{RY}/RT) \quad (2.4)$$

The binding selectivity now can be expressed by Eq. (2.5):

$$K_{RX}/K_{RY} = \exp[(\Delta G_{RY} - \Delta G_{RX})/RT] \quad (2.5)$$

In most general terms, with larger absolute values of interaction free energies one also may expect to observe larger differences between binding selectivities and therefore an increased selectivity with increased affinity. This is particularly true for several types of intermolecular interactions for which the binding free energy (or log K) can be expressed as a product of certain physicochemical properties (P) of receptor R and ligand L, Eq. (2.6):

$$\Delta G_{RL} = aP_R P_L + b \tag{2.6}$$

Such behavior is observed for ionic association, where P_R and P_L are charges of receptor and ligand, for hydrogen bonding, where P_R and P_L are acidity and basicity parameters, respectively, of H-donor and H-acceptor groups of receptor and ligand molecules, and in a more complicated form for Lewis acid–base (metal–ligand) interactions (Section 2.5). Combining Eqs. (2.5) and (2.6) one obtains the following expressions for selectivity:

$$\Delta G_{RY} - \Delta G_{RX} = aP_R(P_Y - P_X) \tag{2.7}$$

$$K_{RX}/K_{RY} = \exp[aP_R(P_Y - P_X)/RT] \tag{2.8}$$

Obviously, for a given pair of ligands X and Y the ratio of binding constants will increase for receptors possessing larger P_R values and larger affinity to both ligands.

Most synthetic and biological receptors are polydentate; the total binding free energy then is usually the sum of single interaction energies. This additivity principle has been shown to be valid for a large number of supramolecular complexes, which use ion pairing, hydrogen bonds, and *van der Waals* forces as intermolecular binding contributions, provided of course that the interacting functions are in a geometrically matching position [19]. An important particular case of this situation is involvement of a number n of the same kind of interactions provided by a multidentate receptor for binding of a given ligand. In this case one obtains the following expression for ΔG_{RL}:

$$\Delta G_{RL} = n\Delta\Delta G_i \tag{2.9}$$

where $\Delta\Delta G_i$ is the pair-wise binding free energy increment. Obviously, these increments will be different for ligands X and Y, so that one obtains for the free energy difference and for the ratio of binding constants Eqs (2.10) and (2.11), respectively:

$$\Delta G_{RY} - \Delta G_{RX} = n(\Delta\Delta G_{iY} - \Delta\Delta G_{iX}) \tag{2.10}$$

$$K_{RX}/K_{RY} = \exp[n(\Delta\Delta G_{iY} - \Delta\Delta G_{iX})/RT] \tag{2.11}$$

Since n is a property of the complex with receptor R, Eqs. (2.10) and (2.11) apparently are similar to Eqs (2.7) and (2.8), and predict an increase in selectivity by multivalency on the basis of a total binding free energy increase.

Almost perfect correlations between selectivity and affinity in terms of Eqs. (2.10) and (2.11) can be seen, for example, in simple coordination complexes with linear polyamine ligands (Figure 2.3).

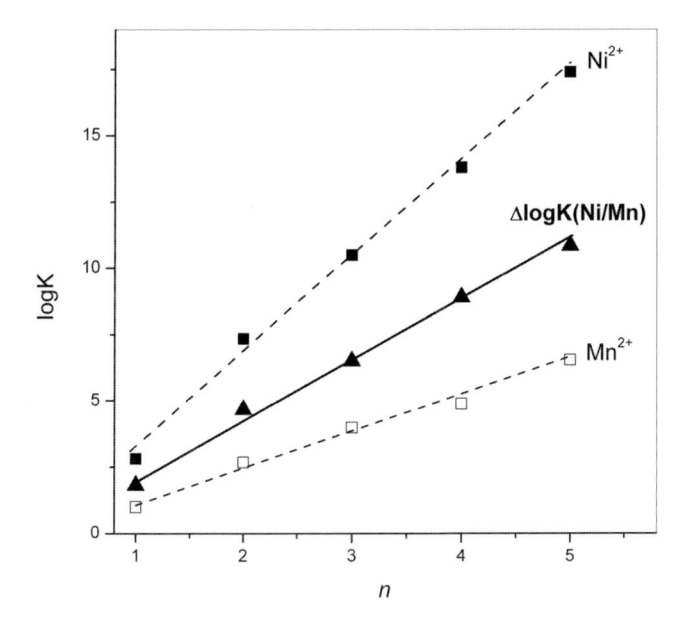

Figure 2.3 Logarithms of stability constants of Mn^{2+} and Ni^{2+} complexes with linear polyamine ligands $H_2N(CH_2CH_2NH)_{n-1}H$ as a function of total number n of nitrogen donor atoms. As n increases from 1 to 5 the selectivity of binding of Ni^{2+} over Mn^{2+} increases in terms of K_{Ni}/K_{Mn} by nine orders of magnitude. Reproduced from Reference [3] by permission of The Royal Society of Chemistry.

The additivity of multisite interactions does not necessarily lead to a selectivity–affinity correlation. Often a host possesses one primary binding site, which may secure a high affinity to the guest through binding to a complementary guest interaction site, and a second, often geometrically distant, site that secures selectivity by weaker or even repulsive interactions. Figure 2.4 illustrates that if two guest molecules X and Y feel with primary sites A or B of hosts A and B different interaction free energies $\Delta G_{AA'}$ and $\Delta G_{BA'}$, but the same interaction free energies ΔG_{SX} and ΔG_{SY} at the discrimination site S, that the selectivity ratio K_X/K_Y can be the same, even if the total binding constants may differ enormously. For example, if $\Delta G_{AA'} = -60\,kJ\,mol^{-1}$ and $\Delta G_{BA'} = -30\,kJ\,mol^{-1}$ with $\Delta G_{SX} = -10\,kJ\,mol^{-1}$ and $\Delta G_{SY} = -4\,kJ\,mol^{-1}$, the selectivity $K_X/K_Y = 10$ for both hosts A and B, but the relative affinity $K_A/K_B = 1.6 \times 10^5$.

Many host systems employed for the discrimination of anisotropic guest molecules make use of at least two different interaction sites (examples can be found in following sections), for which reason we can expect only for simple isotropic systems such as metal cations or halide anions a linear correlation between selectivity and affinity of complexation. Statistical deviations from linear correlations are also expected with complexes of very high affinities: here the binding strength can reach, for instance with inhibitor drugs, several hundreds of $kJ\,mol^{-1}$; the $\Delta\Delta G$ necessary for an acceptable selectivity of, for instance, $K_A/K_B \approx 100$ is only in the order of

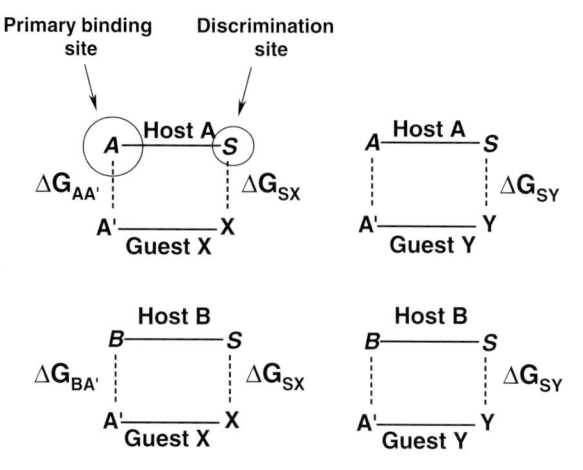

Figure 2.4 Host–guest complexes with separate binding sites for high affinity (A-A′ or B-A′) and for selectivity (S-X or S-Y). The same selectivity will be observed with either large total binding constants (host A) or smaller binding constants (host B).

$\Delta\Delta G \approx 10\,\text{kJ mol}^{-1}$, so that small fluctuations of the total binding energy will mask the selectivity contributions [20].

2.4
Crown Ether and Cryptand Complexes: Hole Size Fitting and Other Effects

The choice of suitable binding units for metal ions follows the known principles of coordination chemistry. Alkali and alkali earth cations are preferentially complexed with oxygen-containing ligands; with macrocyclic systems such as crown ethers one can achieve selectivity by shape complementarity. Figure 2.5 illustrates that with a closed crown ether cavity one can attain selectivity between K^+ and Na^+ ions surpassing that of natural ionophores such as valinomycin [21]; at the same time one observes here the theoretically expected correlation between affinity and selectivity [3].

Selectivity of complexation by crown ethers and cryptands has been discussed extensively from the very beginning of supramolecular chemistry. Although initially it was interpreted as a simple hole-size fitting phenomenon, later studies emerged with a more complex picture. Preorganization and the presence of additional binding sites (lariat ethers) strongly affect both the affinity and the selectivity. In addition, solvation plays a very significant role, as discussed below.

With ion complexes and electroneutral ligands one can expect the major variation to arise from solvation or desolvation of the most polar partner. Indeed, it has long been realized that the hole-size peak selectivity of metal ion complexation by crown ethers and related ligands is to a great extent due to the cation solvation effect. Complexation between potassium salts and 18-crown-6 in 14 different solvents has shown a linear correlation (with $R \approx 0.95$) of ΔG

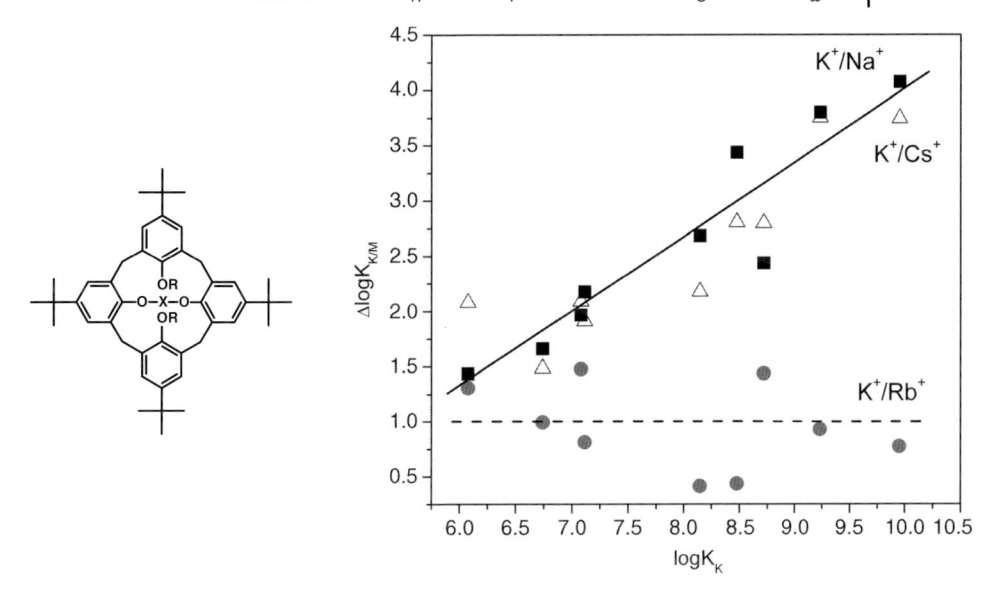

Figure 2.5 Selectivity Δlog K with calixarene-crown ether complexes (X = CH$_2$CH$_2$(OCH$_2$-CH$_2$)$_n$, n = 3 or 4, R = Me, Et, n-C$_3$H$_7$, i-C$_3$H$_7$ or CH$_2$C$_6$H$_5$) for K$^+$ versus Na$^+$ (filled squares), K$^+$ versus Cs$^+$ (open triangles), and K$^+$ versus Rb$^+$ (filled circles); experimental data from Reference [20]. Reproduced from Reference [3] by permission of The Royal Society of Chemistry.

with standard Gibbs transfer energies ΔG_t^o of the metal ion from water to the given solvent [19a]. Comparison with some other cations and ligands, including [222] cryptand, has also revealed that the complexation constant changes are essentially a linear function of the cation desolvation free energies. Less meaningful correlations ($R \approx 0.9$) are obtained with values characterizing the electron-donor capacity of the solvent. Parameters characterizing the solvent polarity, such as E_T, are extremely poor descriptors ($R = 0.3$). Reaction enthalpies ΔH vary much more than ΔG, for instance from 12 kJ mol^{-1} (in MeCN) to 68 kJ mol^{-1} (in Me$_2$CHOH), without meaningful correlations to any known solvent properties [19a].

Both calculational [22] and experimental gas-phase measurements [23] show that in the absence of solvent the affinity of cations to crown ethers of different sizes always parallels the cation charge density, that is, the highest affinity is always observed for Li$^+$ among alkali metal ions and for Mg^{2+} among alkaline-earth cations. Measurements of stability constants for alkali metal ions in the poorly solvating nitromethane indeed show the expected order Na$^+$ > K$^+$ > Rb$^+$ > Cs$^+$ regardless of the macrocycle size [24].

As discussed above, measurements in different solvents usually show a decreased affinity in more polar solvents due to stronger solvation of metal ions. This may lead to observation of an affinity–selectivity correlation, which indeed can be found in

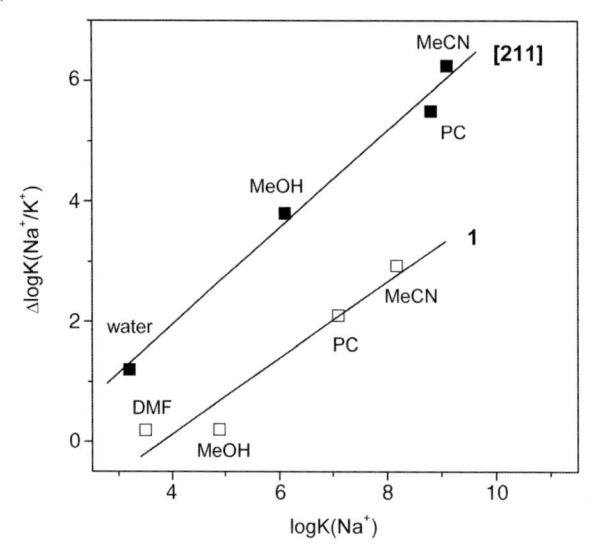

Figure 2.6 Correlations between differences in log K values for complexation of Na^+ and K^+ with [211] cryptand (filled squares) or with lariat ether **1** (open squares) and log K for Na^+ in different solvents. Reproduced from Reference [3] by permission of The Royal Society of Chemistry.

several cases, for example, for complexation of alkali metal ions with [221] cryptand and with a bibracchial lariat ether (**1**) (Figure 2.6) [25].

Results on solvent effects on cation and anion recognition with large polyfunctional receptors point to the importance of co-complexation of solvent molecules with guests for affinity and selectivity. Studies on the cation binding by calixarene **2** [26] in four solvents [MeCN, MeOH, DMF, and propylene carbonate (PC)] show that in MeOH and DMF one observes the binding of only two cations, Ag^+ and Hg^{2+}, surprisingly with inversed selectivity ($K_{Ag} \gg K_{Hg}$ in MeOH, but $K_{Hg} > K_{Ag}$ in DMF). However, in MeCN the receptor binds more cations (Li^+, Na^+, Ag^+, Ca^{2+}, Pb^{2+}, Hg^{2+}, and Cu^{2+}) with strongly increased affinity compared to Hg^{2+}, but with decreased affinity to Ag^+ as compared to MeOH and DMF. It has been shown that inclusion of a MeCN molecule in the cavity formed by aromatic rings of **2** causes an allosteric effect inducing a widening of the cavity formed by polar chains, where metal ion binding occurs; this allows complexation of a large number of cations in this solvent.

Binding constants of 18C6 with protonated amines in water, 2-propanol, *tert*-butyl alcohol, *n*-octanol, DMF, DMSO, pyridine, and HMPT vary by factors of up to 1000; the solvent effects can be described with values characterizing the electron-donor capacity of the solvent. Although no linear dependence of the difference $\Delta\log K$ between benzylammonium and anilinium chloride on the total affinity was observed, somewhat larger $\Delta\log K$ values were found with less competing solvents [27].

One important application of crown ethers is selective extraction and transmembrane transport of alkali metal ions. A priori one expects the selectivity of extraction to follow the selectivity of complexation in homogeneous solution. In fact the extraction selectivity may be different and in addition it depends on the counterion. Figure 2.7 shows the selectivity factors observed for extraction of alkali metal ions by benzocrown ethers with chloroform from aqueous picrate and iodide solutions [28]. With the monobenzo derivative the selectivity is approximately the same, but with di- and especially tribenzo ethers the selectivity to K over Rb and Cs disappears in iodide solutions. The major difference in these anions is the ability of picrate to form adducts with aromatic groups of the ligand and this affects, although in a non-evident way, the selectivity of extraction.

The extraction equilibria are closely related to the functioning of carrier-based ion-selective electrodes and bulk optodes, two types of sensors experiencing rapid growth in practical applications [29]. In both types of devices a lipophilic carrier (receptor) is incorporated in a solid, most often a poly(vinyl chloride) membrane, and creates either a phase boundary potential (ion-selective electrodes) or an optical response in the case of an ionic guest, by a respective chromoionophore or a fluoroionophore in the bulk membrane (bulk optodes). The selectivity depends on several factors,

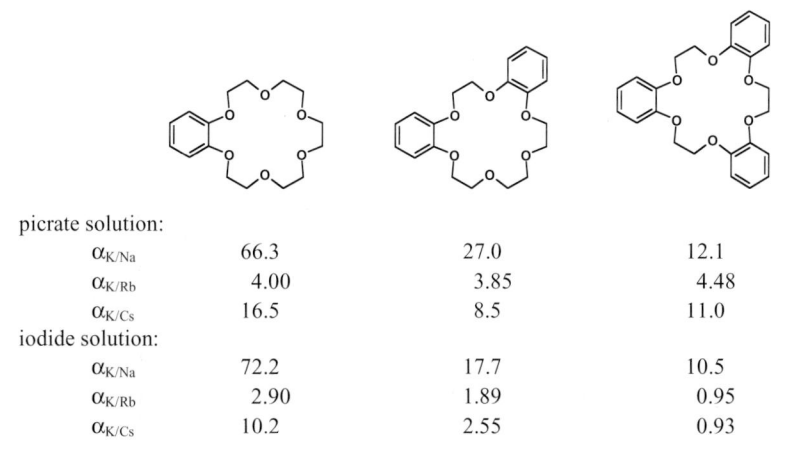

picrate solution:			
$\alpha_{K/Na}$	66.3	27.0	12.1
$\alpha_{K/Rb}$	4.00	3.85	4.48
$\alpha_{K/Cs}$	16.5	8.5	11.0
iodide solution:			
$\alpha_{K/Na}$	72.2	17.7	10.5
$\alpha_{K/Rb}$	2.90	1.89	0.95
$\alpha_{K/Cs}$	10.2	2.55	0.93

Figure 2.7 Separation factors (α) for competitive alkali metal ion extractions from aqueous picrate and iodide solutions of 18-crown-6 derivatives with $CHCl_3$.

principally on ion exchange and complex formation properties, which also are responsible for the extraction selectivity. Several extensive compilations of selectivity coefficients for ion-selective electrodes are available [30].

Crown ethers are often employed as components of more sophisticated devices like switches and logical gates. Some of these systems demonstrate significantly altered selectivity towards alkali metal ions. The fluorophore shown in Figure 2.8 binds potassium ions in basic solutions within ring A, and in acidic solution sodium ions in ring B, due to N-protonation, and represents a logical gate function [31]. As shown,

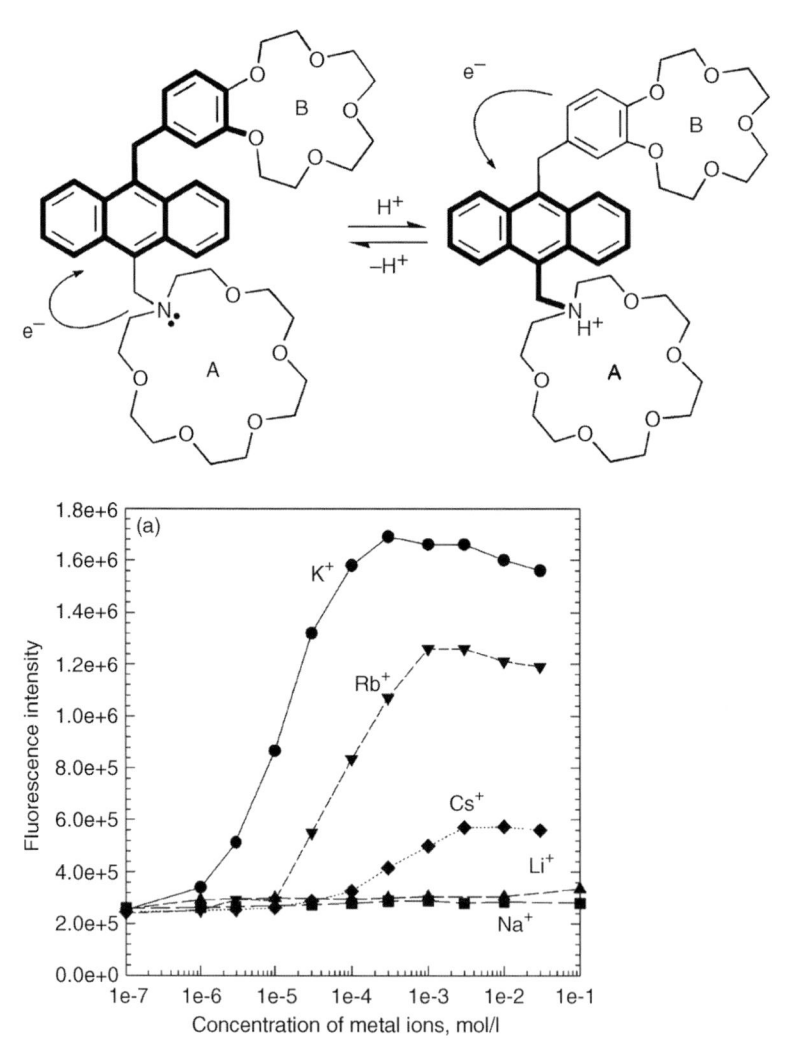

Figure 2.8 A crown ether derivative for selective alkali ion detection; pH-dependent fluorescence emission, under basic condition (a); under acidic condition (b). Reproduced from Reference [31] by permission of The Royal Society of Chemistry.

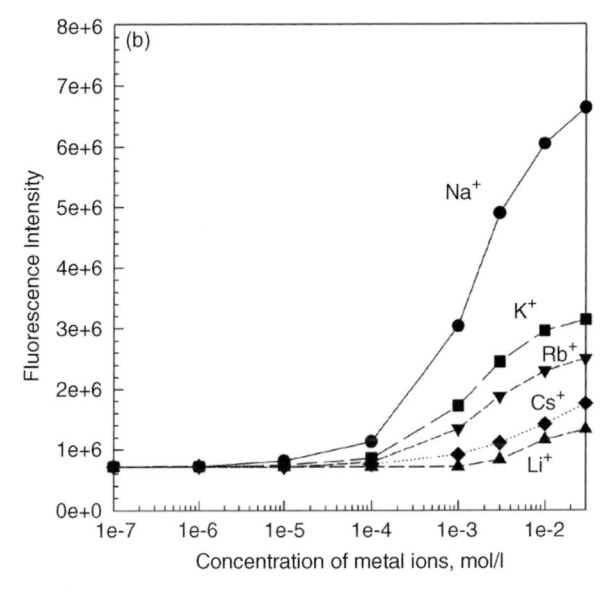

Figure 2.8 (Continued)

two PET (photo-induced electron transfer) processes are switched by protonation, leading to distinct fluorescence emission, which amounts to a 7.2-fold enhancement with K^+ ions in basic solutions, and under acidic conditions to a 9.4-fold enhancement with Na^+ ions. Depending on pH, other cations such as Rb^+ also induce a sizeable emission. As is evident from Figure 2.8, the K^+/Na^+ selectivity in this system thus inverts on going from acid to basic conditions.

2.5
Recognition of Transition and Heavy Metal Ions

Selectivity of complexation of transition and heavy metal ions is dictated principally by the chemical nature of donor groups. For simple 1:1 complexation of metal ions (Lewis acids, A) with monodentate ligands (Lewis bases, B) the binding constants K_1 in water can be successfully predicted with Eq. (2.12), where parameters E and C reflect the tendency of A and B to form ionic and covalent bonds, respectively, and D reflects steric effects [4]:

$$\log K_1 = E_A E_B + C_A C_B - D_A D_B \tag{2.12}$$

Equation (2.12) is a quantitative expression of the HSAB (hard-soft acid-base) principle. The ratio E/C, called "ionicity," is close to Pearson's hardness parameter, where C stands for the soft-soft combination. Table 2.1 provides numerical values of parameters for some of the most common cations and ligands.

Ligands employed in practice for recognition of metal ions are polydentate. Additional factors that influence the selectivity of complexation with polydentate

Table 2.1 Parameters for use in Eq. (2.12) for selected cations and ligands in water [4b].

Cation	E_A	C_A	D_A	Ligand	E_B	C_B	D_B
Ag^+	−1.53	0.190	0.0	F^-	1.00	0	0
Hg^{2+}	1.35	0.826	0.0	AcO^-	0.0	4.76	0
Cu^{2+}	1.25	0.466	6.0	OH^-	0.0	14.00	0
H^+	3.07	1.009	20.0	NH_3	−1.08	12.34	0.0
Cd^{2+}	0.99	0.300	0.6	$HOCH_2CH_2S^-$	−3.78	39.1	0.9
Zn^{2+}	1.33	0.312	4.0	Py	−0.74	7.0	0.1
Pb^{2+}	2.76	0.413	0.0	Cl^-	−1.04	10.4	0.6
La^{3+}	3.90	0.379	0.0	Br^-	−1.54	14.2	1.0
Mg^{2+}	1.86	0.178	1.5	I^-	−2.43	20.0	1.7
Ca^{2+}	0.98	0.081	0.0	$(H_2N)_2CS$	−2.46	18.2	0.6
Li^+	0.58	0.026	1.0	CN^-	−4.43	30.0	0.38

ligands are the number and size of chelate cycles and the strain induced by complexation. A general rule regarding the effect of chelate ring size is that five-membered rings are more favorable for larger cations and six-membered rings fit better to smaller cations. Modern calculation methods allow one to estimate the complexation-induced strain and so predict the way ligands are preorganized [5].

An important area that requires high selectivity of complexation of metal ions is removal of long-lived, highly toxic isotopes of minor actinides from radioactive waste. The actinides are removed in several steps, mostly performed by solvent extraction, one of which is separation of small amounts of the actinides (An) from a much larger mass of fission product lanthanides (Ln) [32, 33]. The problem is challenging because both An and Ln typically are in the trivalent state, have similar ionic radii and charge density, similar electronic configurations, and cations of both groups are classified as hard, which means that they should prefer similar ligands with hard donor atoms, typically oxygen. In fact, An(III) has a somewhat greater tendency to covalent bonding than Ln(III), that is, to behave as a softer cation and indeed show some preference for softer donor atoms such as S or aromatic N. At present, the most successful extractants are ligands derived from triazine and pyridine fragments, such as TPTZ or DPTP, which possess relatively soft nitrogen donor atoms and are also capable of some back-bonding with An(III), which theoretically is considered one of the factors important for selective binding of actinides [32]. Separation factors for Ac(III)/Eu(III) for these ligands are in the range 10–100, with extraction coefficients, depending on conditions, of from less than 1 to about 300 [32].

TPTZ

DPTP

In contrast, numerous preorganized ligands based on macrocyclic platforms (calix-arenes, cavitands) as well as podands have been prepared with the aim of selective extraction of An(III) [33]. These ligands usually have an Ac(III)/Eu(III) selectivity factors below 10, which is inferior to that of triazine-pyridine ligands, although they often have higher extraction coefficients of up to 300. Thus, presently, chemical selectivity provides a better answer to the problem than the "supramolecular" approach.

An interesting approach to regulate the metal ion selectivity by changing the conformation of the calixarene scaffold has been demonstrated by the development of highly sensitive and highly selective fluorogenic sensors for Pb(II) [34]. The dansyl-containing calixarene 3 in cone conformation was reported previously as a selective receptor for Hg(II). It also binds Pb(II), but with lower affinity. A detailed NMR study of the conformations of Pb(II) and Hg(II) complexes of 3 led the authors to conclude that changing its conformation from cone to partial cone should be unfavorable for complexation of Hg(II), but favorable for complexation with Pb(II). Indeed, the respective receptor 4 demonstrated an excellent selectivity for Pb(II) in 50% aqueous MeCN. The signal is produced by formation of complex 5 with deprotonated ligand, which possesses strongly enhanced fluorescence.

2.6
Recognition via Ion Pairing

The equilibrium constants for a large number of ionic association reactions in water can be satisfactorily fitted to Eq. (2.13), where z_R and z_L are total charges of R and L [35]:

$$\log K_{RL} = a z_R z_L + b \tag{2.13}$$

Parameters a and b vary for different types of charged species, for example, they are slightly different for inorganic anions and carboxylates, but within a series of given chemical species Eq. (2.13) provides a reasonably good correlation. Alternatively, the free energy of association can be estimated by multiplication of a constant binding increment of -5.5 ± 1 kJ mol^{-1} (at ionic strength 0.1 M) by the number of all pairwise ionic contacts between R and L (surprisingly independent of the nature of the ions) [19a]. The expression for selectivity that follows from Eqs. (2.13) and (2.8) takes the form of Eq. (2.14):

$$\Delta \log K_{RL} = a z_R (z_Y - z_X) \tag{2.14}$$

This provides perhaps the simplest case of selectivity when ions of different charges can be differentiated by interaction with oppositely charged receptor, where the degree of differentiation increases with the increase in receptor charge. Obviously, ions of similar charge ($z_Y = z_X$) cannot be discriminated in this way.

Figure 2.9 shows the experimental results for ion association of five inorganic anions with differently protonated forms of spermine, which follow Eq. (2.13) with $a = 0.612$ and $b = 0$. The respective theoretical profiles are shown as dashed lines. As expected, the separation between binding constants for, for example, the tripolyphosphate pentaanion and the hydrophosphate dianion increases from $\Delta \log K = 1.9$ to 6.6 on going from monoprotonated to tetraprotonated spermine [the $\Delta \log K$ values predicted from Eq. (2.14) are 1.8 and 7.3, respectively].

Classic examples of selective binding due to shape differences between spherical and non-spherical anions, or between dicarboxylic acids of varying length, mainly due to Lehn *et al.*, can be found in books [1] and reviews, particularly on anion complexation [36].

Shape selectivity can be observed in the recognition even of ionic compounds in sufficiently rigid systems. This has been demonstrated by stronger complexation of some rigid tricarboxylates as compared to flexible citrate with protonated forms of the [21]aneN$_7$ macrocycle [37]. More recent examples of such selectivity are discussed below. Selective recognition of isomeric tricarboxylates **6a–c** was achieved by complexation with polyammonium macrocycles **7a** and **7b** [38].

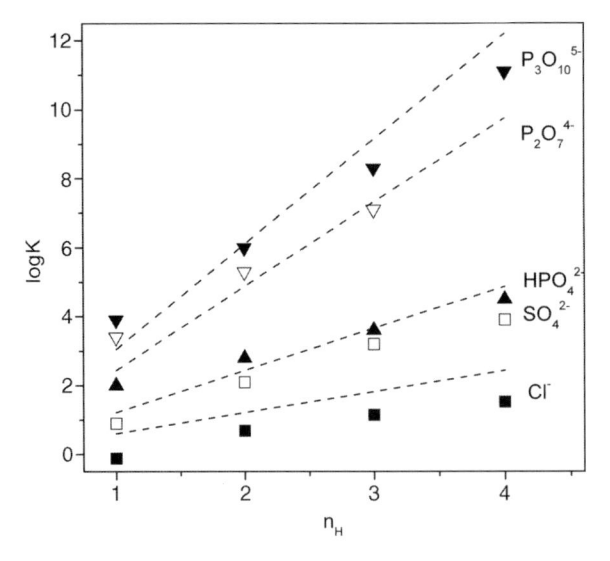

Figure 2.9 Logarithms of binding constants of different anions to protonated forms of spermine in water at zero ionic strength [34]; n_H is the number of protons bound to spermine; dashed lines are theoretical profiles calculated in accordance with Eq. (2.13) for anion charges 1, 2, 4, and 5. Reproduced from Reference [3] by permission of The Royal Society of Chemistry.

Figure 2.10 shows logarithms of stepwise binding constants of trianions **6a–c** with protonated forms of both macrocycles as a function of degree of protonation. Plots are roughly linear with positive slopes in agreement with primarily electrostatic nature of complex formation, but for a given degree of protonation of the macrocycle the binding constants for different isomers are significantly different despite the similar charge of all guest anions. Binding of symmetrical trianion **6c** is stronger than that of other guests with both macrocycles. The receptor **7b** (Figure 2.10b) can also discriminate isomers **6a** and **6b**. The selectivity in this system is essentially independent of affinity. Similar slopes of correlations for different anions indicate similar binding increments for each additional salt bridge and, therefore, similar electrostatic contribution to total binding. This means that the selectivity can be attributed entirely to the better geometric fit of **6c**, creating larger non-electrostatic binding (hydrophobic, van der Waals, hydrogen bonding) contributions. In favor of this is the fact that binding of completely protonated neutral guests to partially protonated receptors, attributed to hydrogen bonding of COOH groups to remaining unprotonated amino groups, is significant and has the same selectivity as for anions, for example, log K for neutral forms of **6a–c** and octaprotonated **7a** equal 5.17, 3.61, and 6.65 respectively. Thus, in general terms, this system operates in accordance with the mechanism outlined in Figure 2.4: discriminating non-electrostatic interactions are independent of principal ion-pairing interactions, which provide growing affinity with conserved selectivity.

Recognition of isomeric dicarboxylates **8a–c** by receptor **9** is another example of shape selectivity [39].

Figure 2.11 shows a plot for binding constants of dianions **8a–c** to protonated forms of **9**. Here the situation resembles to some extent that in Figure 2.3: the selectivity grows together with affinity and each guest has its own binding

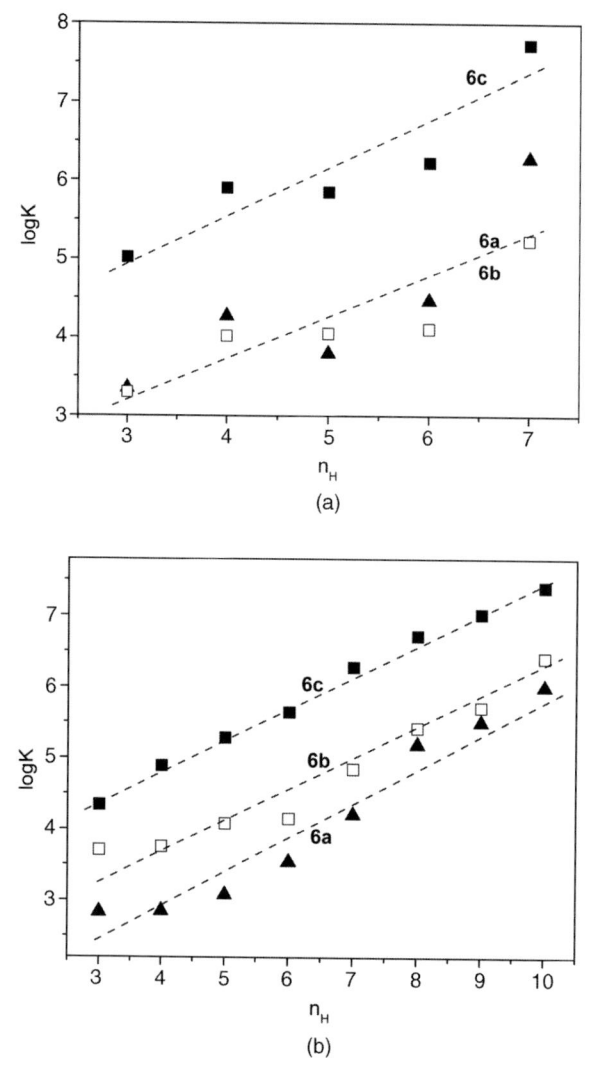

Figure 2.10 Logarithms of stability constants for complexes of trianions **6a** (filled triangles), **6b** (open squares) and **6c** (filled squares) with protonated forms of **7a** (a) and **7b** (b) versus number of protons n_H bound to the macrocycle [37].

increment per salt bridge. Therefore, the selectivity in this case seems to be related principally to the better fit of the guest carboxylate groups to the macrocycle ammonium groups.

The actual selectivity of recognition is better analyzed by using so-called observed or conditional stability constants K_{cond}, which are defined in terms of total bound and free concentrations of components (Eq. (2.15) where A is the anion and R is the polyammonium receptor) [37]:

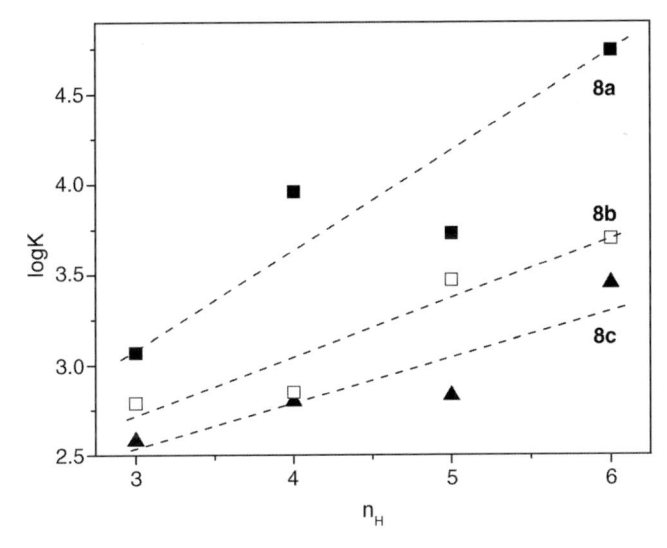

[Figure: plot of logK vs n_H with data points for 8a, 8b, 8c]

Figure 2.11 Logarithms of stability constants for complexes of dianions **8a–c** with protonated forms of **9** versus number of protons n_H bound to the macrocycle [38].

$$K_{cond} = \frac{\Sigma[H_{i+j}RA]}{(\Sigma[H_iA])(\Sigma[H_jR])} \qquad (2.15)$$

The advantage of this parameter is that it represents the binding constant that would in reality be observed experimentally in, for example, a spectroscopic titration of the receptor at a given pH. The K_{cond} values are based on the actual distribution of differently protonated species of receptors and guests at a given pH while the stepwise association constants correspond to interactions between individual species. Under conditions when the complexation degree is directly proportional to the binding constant, that is, far from saturation, the selectivity is simply the ratio of K_{cond} for different guests.

The calculated values of K_{cond} for anions **6a–6c** are plotted versus pH in Figure 2.12. The first observation is that although stepwise binding constants to **7a** and **7b** at the same degree of protonation are about one order of magnitude larger for **7a** (Figure 2.10), the K_{cond} values are very close to each other for both receptors. The reason for this is that smaller macrocycles of **7** provide a higher charge density and bind anions more strongly, but the host **7b** is more basic and at a given pH it is protonated to a larger degree. One also observes that the largest discrimination is not reached at the pH that is optimal for affinity: in more acid solutions the electrostatic attraction is smaller and predominant interactions are of van der Waals/hydrogen bonding type, which are responsible for the selectivity.

The selective and sensitive detection of the biologically important analyte pyrophosphate has been reviewed recently [40, 41a]. An old yet impressive example is an ion pair complex with exactly matching cation–anion distances based on a anthracene derived host **10**, exhibiting in water high binding constants of $K \approx 10^6\,M^{-1}$, with a 2200-fold increase in pyrophosphate/phosphate discrimination [42].

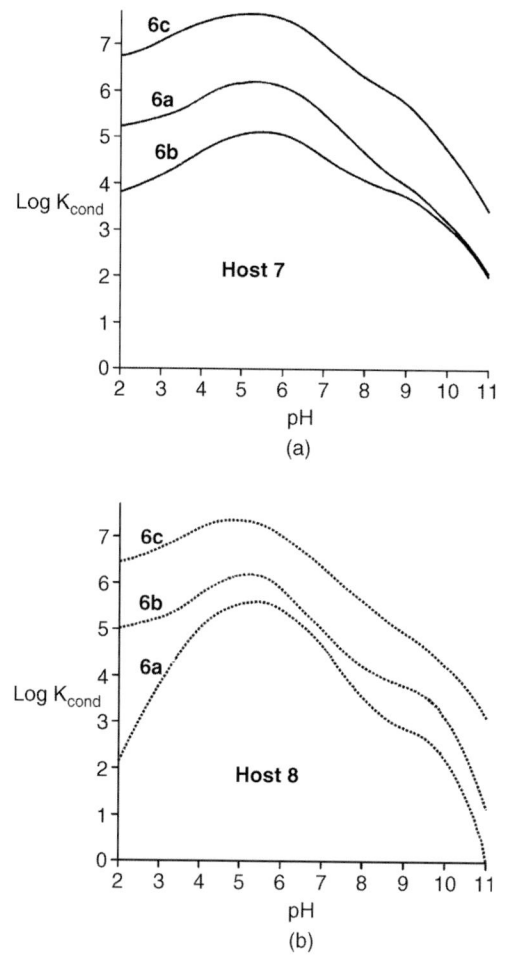

Figure 2.12 Calculated log K_{cond} [see Eq. (2.15)] for anions **6a–6c** and hosts **7a** (a) and **7b** (b) versus pH. Adapted with permission from Reference [38]. Copyright 2005 The American Chemical Society.

Low Fluorescence ⇌ High Fluorescence

10

11

Zwitterionic macrocyclic hosts such as **11** [$X=(CH_2)_n$, $n=6$ or 8] form stable complexes with halides and with cyanide K_{assoc} are between $6 \times 10^3 \, M^{-1}$ and $3 \times 10^5 \, M^{-1}$ in water with the selectivity defined by the confined internal space, where bromide and iodide binding is enthalpically driven [36d]. In summary, selectivity in complexations where ion pairing dominates is essentially dictated by geometric matching; otherwise significant selectivities can be reached if additional non-ionic mechanisms contribute.

2.7
Hydrogen Bonded Complexes and Solvent Effects

Hydrogen bonds are important elements in selective host–guest complex formation, in particular as in contrast to ion pairs these interactions are directional [43]. However, in the first place hydrogen bond strength also depends on polar interactions. These can be described for species possessing H-donor (D) and H-acceptor (A) sites in terms of Eqs. (2.16) or (2.17) [44]:

$$\Delta G_{RL} = 2.43 C_A C_D + 5.70 \quad (2.16)$$

$$\log K_{RL} = 7.354 \alpha_2^H \beta_2^H - 1.094 \quad (2.17)$$

The acidity and basicity factors α and β, or C_A and C_D, respectively, are obtained from equilibrium measurements of many complexes in, for instance, CCl_4 solution. If the receptor is, for example, a proton donor the selectivity for ligands with proton acceptor sites is given by Eqs. (2.18) or (2.19):

$$\Delta\Delta G_{RL} = 2.43 C_D(R)\{C_A(X) - C_B(Y)\} \quad (2.18)$$

$$\Delta\log K_{RL} = 7.354 \alpha_2^H(R)\{\beta_2^H(X) - \beta_2^H(Y)\} \quad (2.19)$$

For a receptor bearing an aliphatic OH group ($\alpha_2^H = 0.4$) or alternatively a carboxyl group ($\alpha_2^H = 0.6$) the discrimination between basic guest compounds will, according to Eq. (2.19), increase by a factor of 1.5 in terms of $\Delta\log K_{RL}$.

Many examples of hydrogen bonding receptors in low polar media, where bulk water will not interfere with the host–guest interaction, have been discussed already with respect to comparison of selectivity and affinity [3]. Here we discuss in more detail solvent effects.

The selectivity of hydrogen bonding receptors is generally greater in solvents with smaller donor or acceptor capacities, although a simple correlation between affinity and selectivity is rarely seen. Jeong *et al.* have recently demonstrated how proper choice of the solvent can lead to high selectivity, reaching with the simple bis-indole host **12a**, for example, $K_{Cl}/K_{Br} = 6$ in MeCN, but 16 in THF [45a]. With the tetraindole **12b** not only are the affinities significantly increased, but the selectivity reaches peak values of $K_{Cl}/K_{Br} = 140$ in MeCN, and 120 even in the polar medium acetone. The binding constants show the expected increased selectivities on comparing the bis-indole with

Table 2.2 Complexation affinity[a] and selectivity[b] of diphenylurea with different halides in different solvents. Data: K. Kato and H.-J. Schneider, unpublished results; data in DMSO from Reference [45b].

	$(CD_3)_2CO$			CD_3CN			DMSO-d_6		
	Cl	Br	I	Cl	Br	I	Cl	Br	I
K_{av}[a]	14 800	3820	260	2140	580	65	40	4	<1
$-\Delta G$[a]	23.7	20.3	13.7	18.9	15.7	10.2	9.2	3.4	>0
K_{Cl}/K_{Br}	—	4	15	—	4	9	—	10	<4

a) Average values K_{av} and $-\Delta G$ (kJ mol^{-1}) from single K values for each signal with data weight according to the CIS values.
b) Ratios between the K values of chloride and bromide or values of bromide and iodide.

tetraindole as host, but the correlation between affinity and selectivity is much less clear in the comparison of different reaction media. Halide anion complexation with diphenylurea as host (Table 2.2) also shows the highest selectivity (with $K_{Cl}/K_{Br} = 15$) in acetone, illustrating again the importance of specific solvation effects.

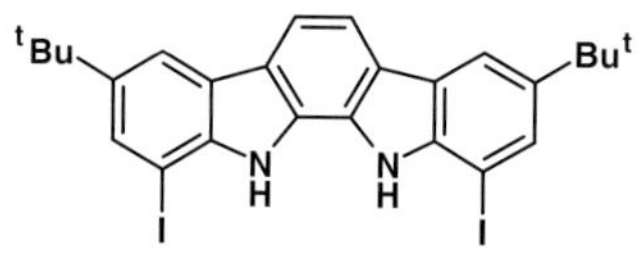

12a
12b 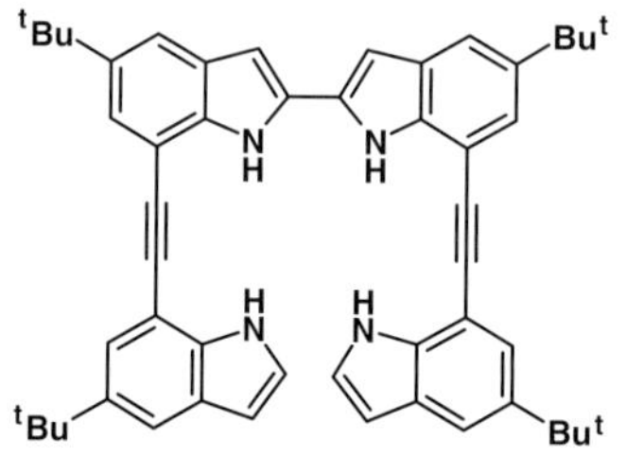

Simultaneous action of salt bridges and discriminating hydrogen bonds (complex **13**) can amount in methanol to stability constants above 10^5 M^{-1} [46]. Several 2,5-bisamidopyrroles like **14** show selective oxo-anion complexation while bis-amides containing dipyrrolylmethane groups, for example, **15**, form strong complexes with dihydrogen phosphate anions in DMSO–water mixtures [47]. Chemically related diindolylureas such as **16** show high selectivity to dihydrogen-phosphate anion in DMSO containing 10% water [48].

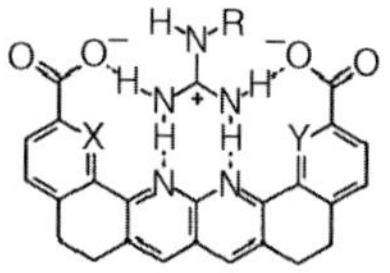

13
14 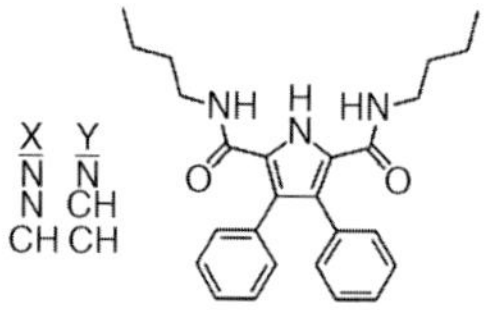
15 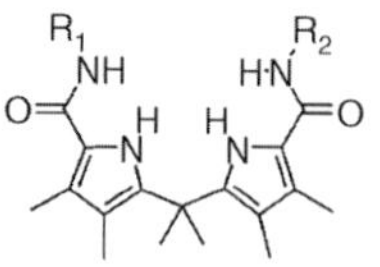
16 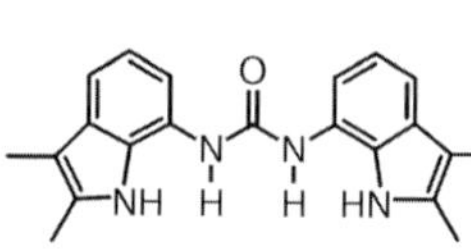

In line with Eqs. (2.16) or(2.17) the selectivity of anion recognition by hydrogen bonding receptors is usually determined by stronger binding of more basic anions, but in some cases significant and rather unexpected deviations can occur. Table 2.3 shows the binding constants for different anions to monocationic receptor **17a** and

Table 2.3 Binding constants for 1:1 complexes for receptors **17a** and **17b** in MeCN [49].

Anion	Log K	
	17a	17b
F$^-$	6.19	5.04
Cl$^-$	3.20	>7
Br$^-$	2.48	6.65
I$^-$	<2	4.55
NO$_3^-$	2.58	5.59
AcO$^-$	4.68	4.37
OH$^-$	6.42	5.58

the corresponding tricationic trifurcate receptor **17b** in MeCN [49]. A remarkable feature of these receptors is their ability to form hydrogen bonded complexes with an hydroxo anion. The selectivity of **17a** is principally dictated by the anion basicity, but receptor **17b** is highly specific to Cl$^-$ and binds Br$^-$, I$^-$, and NO$_3^-$ more strongly than much more basic AcO$^-$. The reason for the altered selectivity is a specific pattern of up to six hydrogen bonds provided by the receptor (three NH and three CH proton donors) to the anion binding.

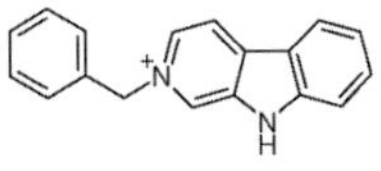

17a 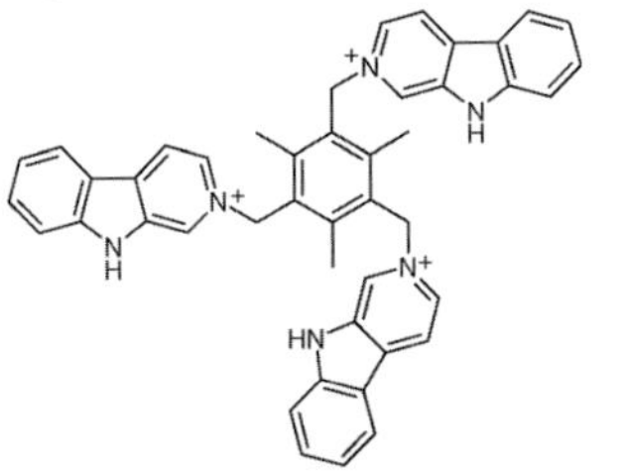 **17b**

Another recently developed chloride-selective receptor is the rotaxane **18**, which binds chloride preferably over more basic hydrophosphate and acetate in a medium containing 10% water in mixture with ethanol–chloroform (1:1) [50].

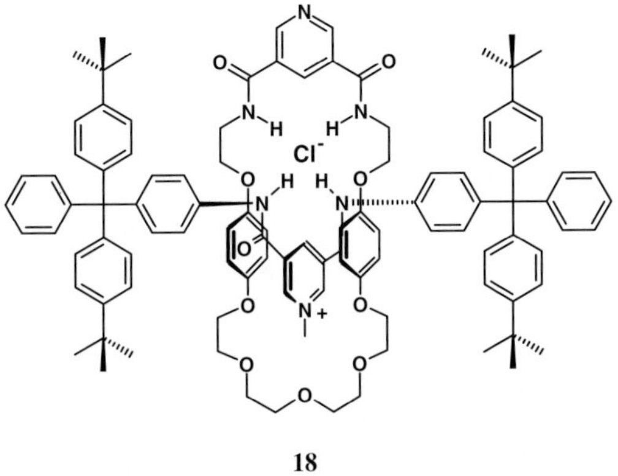

18

Notably, hydrogen bonds are crucial for the selective recognition of nucleic bases by artificial ligands with the help of geometrically matching donor–acceptor functions [51], such as shown with structure **19** (X = N or X = CH); such associations are also useful for selective formation of new materials [51c].

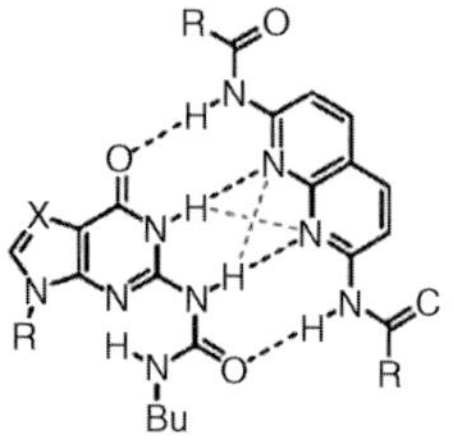

19

Selective recognition of adenine over *all* other nucleobases has been achieved recently with receptors **20a,b**, which exploit for nucleobase recognition both Watson–Crick and Hoogsteen-type interactions, as shown in structure **21** [52].

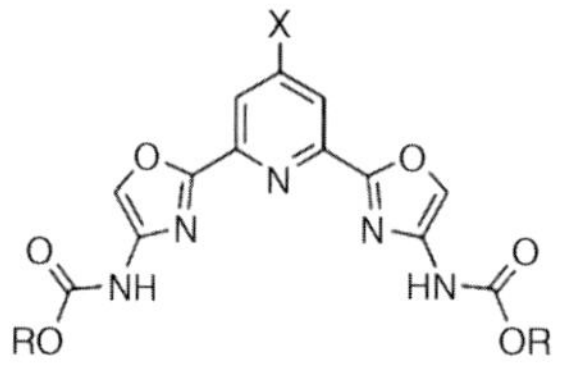

20

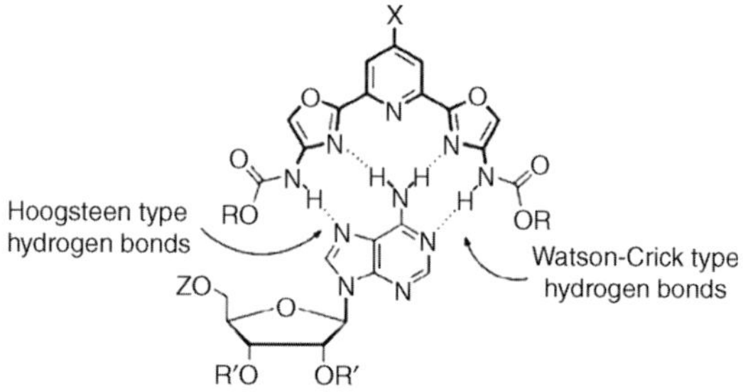

(a) X=H, R=CH$_3$(CH$_2$)$_5$
(b) X=OCH3, R=CH$_3$(CH$_2$)$_5$

21

Table 2.4 shows the formation constants for complexes of receptors **20a** and **20b** with five O-protected nucleosides **22** in chloroform, demonstrating the very large selectivity of both receptors towards adenine.

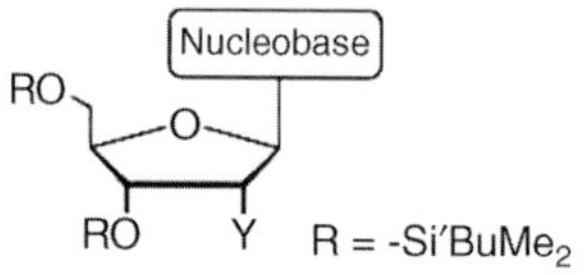

R = -Si^{t}BuMe$_2$

22

Table 2.4 Formation constants (M^{-1}) for complexes of receptors **20a** and **20b** with protected nucleosides **22** in CHCl$_3$ [52].

Nucleobase in 22	Host 20a	Host 20b
Adenine	12 400	10 500
Guanine	37	62
Cytosine	104	163
Uracil	<10	<10
Thymine	27	27

2.8
Lewis Acid Receptors

Many important analytes (all anions, amines, thiols, nucleobases, esters and amides, ureas, etc.) are Lewis bases. Traditionally hydrogen bond donor molecules are used as receptors for such compounds, but they also can be recognized by using Lewis acids as receptors. An advantage of using Lewis acids is that they retain a significant affinity even in protic solvents, including water, where hydrogen bonding practically disappears. Mostly, transition metal ions are employed as Lewis acid centers in receptors, but there is also some interest in other acids like Sn(IV) and B(III).

2.8.1
Associations with Transition Metal Complexes

Coordinatively unsaturated metal complexes can provide strong binding sites to Lewis base units of organic ligands. Introduction of additional organic residues in the metal complex allows the design of selective host systems [53]. The metal ion center alone will of course already discriminate, for example, basic amino acids or peptides from others. For example, Cu(II) complexes of cyclic bis-diene with the fluorescence dye eosin ($K = 10^7 \, M^{-1}$) show, as result of the Cu(II) ion, total quenching of fluorescence, which is regenerated only by His, in contrast to other amino acids (Figure 2.13) [54].

The relatively weak binding of crown ethers to NH_3-groups of peptides in water can be significantly enlarged by a pendant copper imidodiacetic acid complex, which coordinates with high affinity to histidine, **23**. This allows selective detection of only His-containing peptides [55].

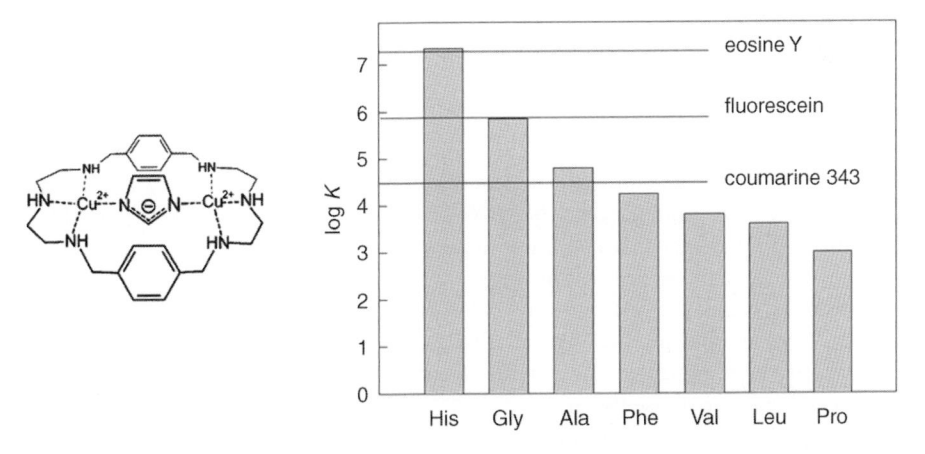

Figure 2.13 Equilibrium constants for the interaction of the Cu(II)-diene host with amino acids (bars) and for comparison with fluorescent indicators alone (horizontal solid lines). The position of the horizontal line with respect to the bars determines the selectivity of the chemosensing ensemble diene/Cu(II)/indicator towards the chosen amino acid. Reprinted with permission from Reference [54]. Copyright 2003 The American Chemical Society.

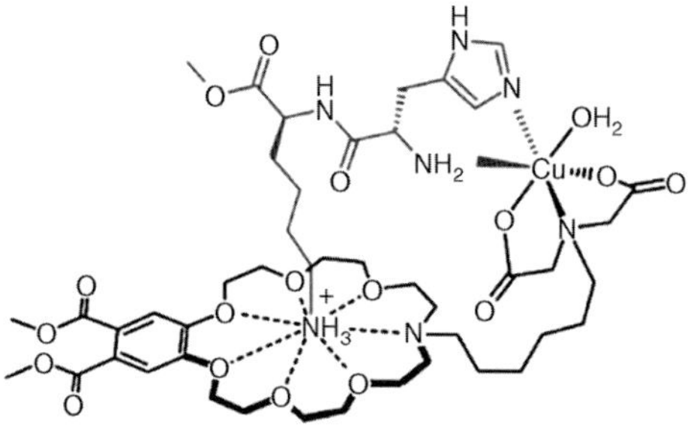

23

Placement of several metal centers at distances designed, for example, with the help of computer-aided molecular modeling allows by geometric matching distinction of, for example, peptides and even proteins containing, for example, lysine or histidine. Obviously, an optimal match will lead to optimal strength and selectivity, in particular if the interactions are not "soft," as in the case of ion pairing or coulombic forces, but fall off significantly with non-ideal distances between interacting sites. Since such a situation is more typical for metal complexes one may expect affinity–selectivity correlations to hold more often than with other noncovalent interactions. For example, two Zn(II) bispicolylamine units can, for example, be attached to spacers such as biphenyl (**24**) as host for helical oligopeptides containing His in different positions; the binding constants in water are in the range $K = 10^4$ to $10^5 \, M^{-1}$, and depend on the geometrical matching between the Zn^{2+} and His location (see, for example, Reference [56]). Selective phosphate complexation with Zn(II)-dipicolylamine host compounds has been reviewed recently [41a].

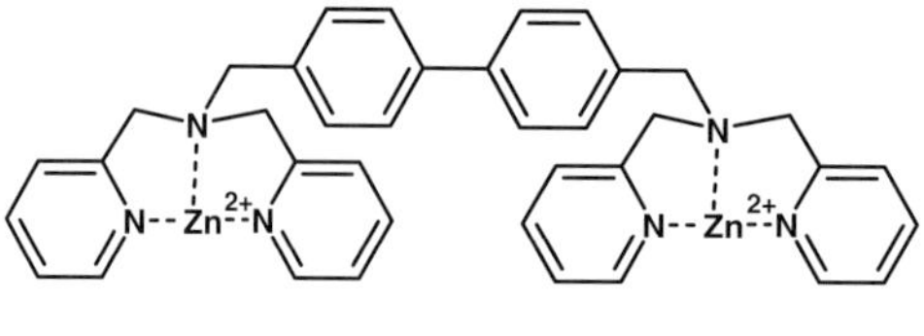

24

For eight His-containing synthetic and quite flexible peptides (Figure 2.14b) Cu(II) complexes with iminodiacetate [(HOOC-CH$_2$-)$_2$NH] derivatives R1-R5 (Figure 2.14a) show affinities reaching binding constants above $K = 10^6 \, M^{-1}$ for the most exact geometric match between metal centers of the receptors and imidazole groups of the peptides. Figure 2.15 illustrates that indeed the highest selectivity is obtained with the Cu complex showing the highest affinity [57]. Based on the same principle it is possible to recognize unique patterns of surface-exposed histidines in proteins such as carbonic anhydrase [58].

The dicopper complex of cryptand **25**, $[Cu_2^{II}(\mathbf{25})]^{4+}$ (Figure 2.16) [59], binds carboxylates, phosphates, sulfate, and so on also by ion pairing with the metal centers; the affinities of the anions do not follow stability constants of their Cu(II) complexes. Basicity, shape, and size of anions are not important factors. The best

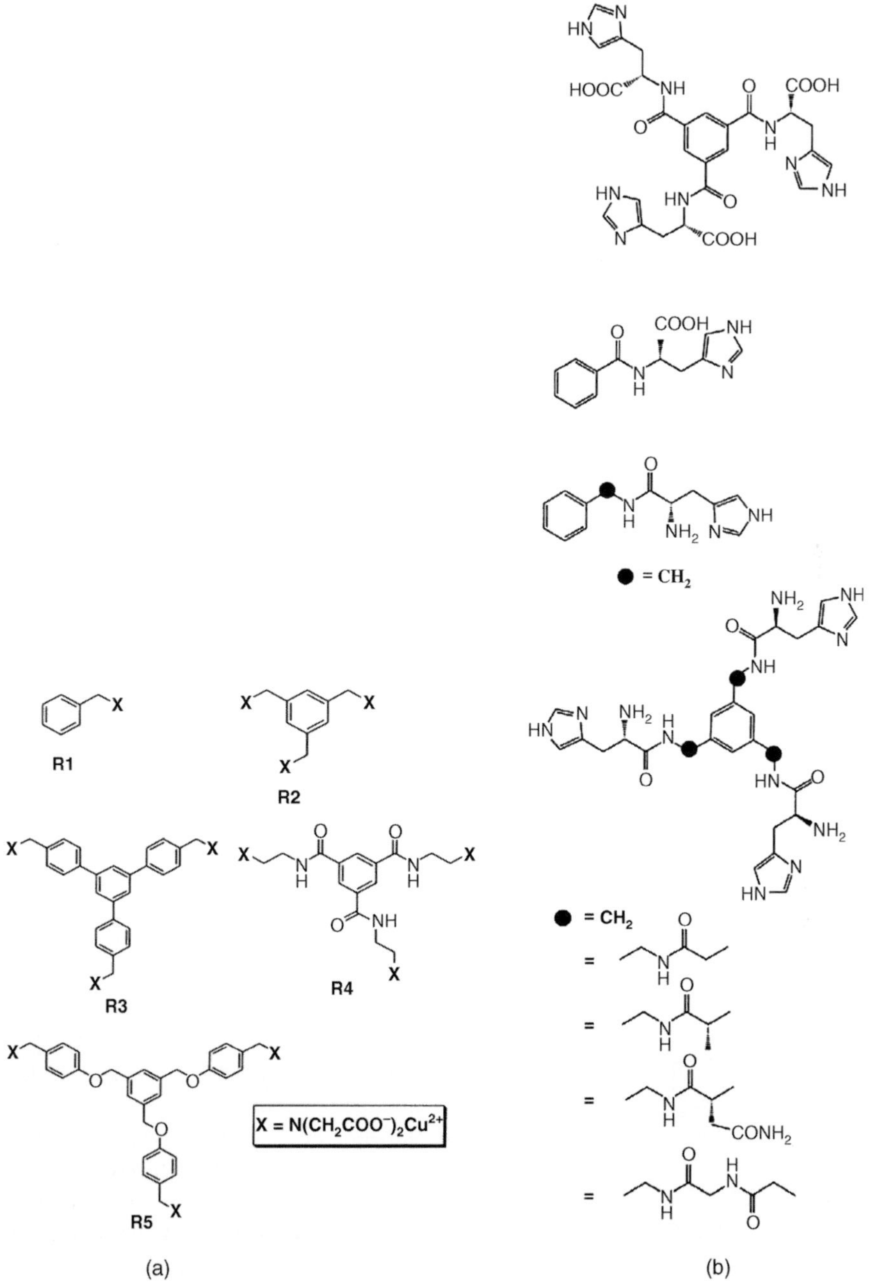

Figure 2.14 (a) Five Cu(II) receptors used for recognition of peptides; (b) a library of peptides recognized with derivatives of iminodiacetate-Cu(II) complexes [57]. Reprinted with permission from Reference [57]. Copyright 2002 The American Chemical Society.

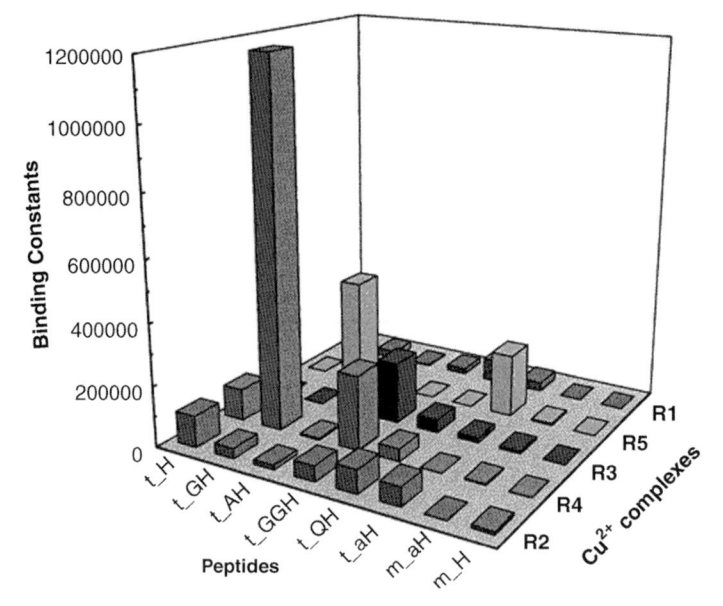

Figure 2.15 Binding constants of eight peptides (Figure 2.14b) with Cu(II) complexes of iminodiacetate derivatives (Figure 2.14a). Reprinted with permission from Reference [57]. Copyright 2002 The American Chemical Society.

correlation is observed with the "bite length" of anions, defined as a distance between two terminal donor atoms of the anion. The highest affinity is observed for anions with a bite length fitting the distance between two Cu(II) ions in axial positions of the cryptand.

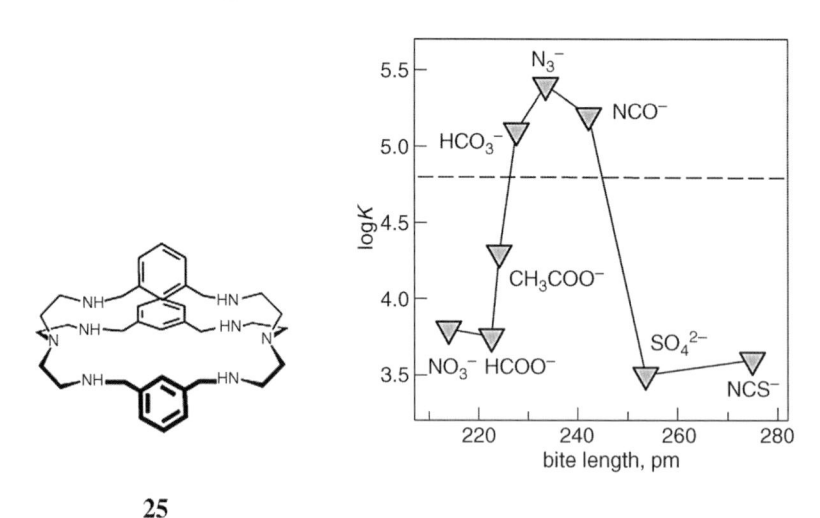

25

Figure 2.16 Selectivity of anion binding by cryptate $[Cu_2^{II}(25)]^{4+}$ in water. Reprinted from Reference [59]. Copyright 2006, with permission from Elsevier.

Of particular interest are receptors in which a Lewis acid center cooperates with hydrogen bonding in a polar, even aqueous, solution. One of the first demonstrations of the utility of this approach was the design of an "artificial guanine," a Cd(II) complex (**26**) for recognition of cytidine in a manner imitating the Watson–Crick classical G-C base pair (**27**) [60]. The cadmium complex **26** selectively binds cytidine over adenosine, guanosine, and thymine in DMSO with a stability constant of 117 M^{-1}.

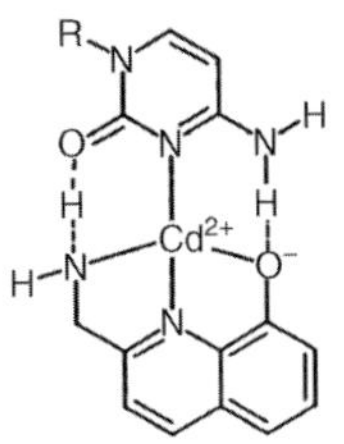

26 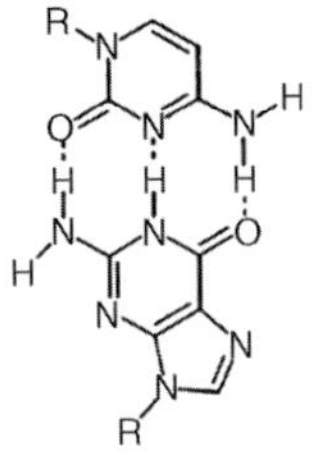   G-C (**27**)

Four acetamido groups in the binuclear Zn(II) complex **28** provide additional binding to pyrophosphate, which is bound to receptor **28** in water with $K = 5.39 \times 10^{10}$ M^{-1} and with high selectivity over other anions [61]. Notably, a selective sensing of pyrophosphate in the presence of ATP is important because pyrophosphate is the product of the DNA polymerase reaction, and its detection allows one a real-time monitoring of DNA sequencing [40]. Further examples of metal coordination–hydrogen bonding cooperation can be found in a recent review [62].

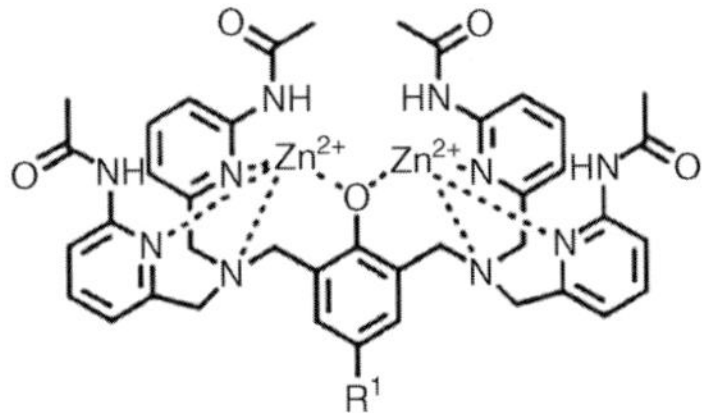

28

2.8.2
Associations with Other Lewis Acids

Although transition metal ions remain the most popular Lewis acid centers, other Lewis acids, in particular organic derivatives of Sn(IV), have attracted growing interest as receptors specific to hard Lewis bases. Initial attempts to introduce organotin compounds as macrocyclic anion hosts of structure **29** demonstrated a very low affinity to tested anions (chloride, bromide) even in solvents like chloroform and a surprising absence of any macrocyclic effect [63]. However, more recently, several more successful tin-based systems were reported. The macrocycle **30** tightly binds chloride in dichloromethane to afford complex **31**; a similar reaction occurs also with fluoride [64]. Tin crown ether derivative **32** affords a ditopic recognition of NaSCN [65].

Tin(IV) complexes of N-confused porphyrins of the type **33** ("w" is a coordinated water molecule, which is substituted with chloride in the recognition process) are highly sensitive fluorescence chloride receptors [66]. Bidentate Sn(IV) Lewis acids **34** show strong affinity to halide ions [67].

Other Lewis acid centers employed for anion recognition involve B(III) and Hg(II) as metal ions. The bidentate receptor **35** binds fluoride anion in aqueous tetrahydrofuran (1: 9 v/v) with $K = 6.2 \times 10^4$ M^{-1} and with high selectivity over such anions as Cl$^-$, Br$^-$, NO$_3^-$, H$_2$PO$_4^-$, and HSO$_4^-$ and with only weak interference with AcO$^-$ [68].

2.9
Complexes with Stacking and van der Waals Interactions

Many associations of biologically important analytes contain aromatic units [69] with stacking or edge-to-face binding involving coulombic and cation–π as well as dispersive interactions. Such complexations are usually confined to aqueous media, where hydrophobic forces can also contribute, and where van der Waals interactions are not decreased by the larger polarizability of organic solvents in comparison to

2.9 Complexes with Stacking and van der Waals Interactions

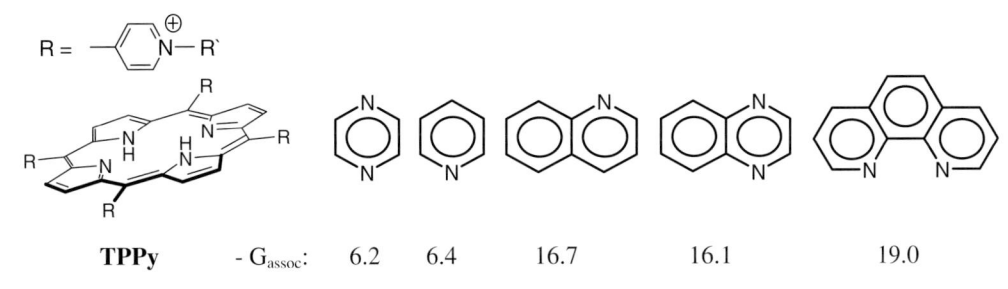

Figure 2.17 Free energies of associations ($-\Delta G$, kJ mol^{-1}, in water) of a cationic water-soluble porphyrin (TPPy) with aromatic guest molecules of increasing size [70].

| TPPy | $-G_{assoc}$: | 6.2 | 6.4 | 16.7 | 16.1 | 19.0 |

water. The selectivity of complexes dominated by either van der Waals or by hydrophobic contributions is discussed together in one section, as these mechanisms are often difficult to separate.

All of the above-mentioned mechanisms can lead to an affinity increase with the contact surface between host and guest. Selectivity with respect to the size of, for example, an aromatic guest molecule can therefore be achieved not only by inclusion in geometrically matching host cavities but simply also by interaction with a flat host surface, as illustrated by affinity differences in associations with water-soluble porphyrins (Figure 2.17) [70]. Heteroatoms as part of a π-system make little difference, since the polarizability of the aryl system does not change much; they do, however, lead to sizeable changes if they occur in substituents, as illustrated in Table 2.5, again with simple porphyrin complexes. Thus, binding constants between, for example, p-halogen and p-nitro benzoates differ by a factor of up to about 10, due to their polarizability differences. Regioisomers with heterosubstituents either in the meta- or para- position are not distinguished, whereas those in the ortho-position lead to steric distortions and thus to affinity changes. Contributions of aliphatic groups here are obviously negligible, but can be sizeable in complexes dominated by hydrophobic interactions such as cyclodextrins (see below). The absence of binding contributions of saturated systems in the porphyrin complexes, and a roughly linear correlation of binding energies with the guest polarizability, indicates that indeed the decisive mechanism here involves dispersive interactions [19c].

In line with dominating stacking interactions, complexation of an electron-rich cyclophane shows only moderate selectivity of, for example, phenyl derivatives with

Table 2.5 Free energies of associations $-\Delta G$ (kJ mol^{-1}, in water) between the large π-surface of porphyrin TPPy [70] (see Figure 2.17) or α-cyclodextrin (α-CD) and anions of benzoic acids, para-X-C$_6$H$_4$COO$^-$ with different substituents X; values with α-CD from Reference [71].

X	H	Me	CH=CH$_2$	F	Cl	Br	I	NO$_2$
TPPy	13.8	13.8	16.0	13.9	15.1	15.8	16.55	19.3
α-CD	6.5	4.8		6.5	12.7			Approx. 15[a]

a) Literature values of K_{assoc} vary between 70 and 360 M^{-1}.

Figure 2.18 Complexation free energies ΔG (in water, kJ mol^{-1}) with an electron-rich cyclophane [69].

X	Y	$-\Delta G$, kJ/mol
COOMe	COOMe	28.5
Me	NO$_2$	25.1
Me	Me	22.3

different substituents (Figure 2.18). Remarkably, the selectivity in terms of enthalpy differences $\Delta\Delta H$ is much more pronounced than that in terms of $\Delta\Delta G$, which are to large degree compensated by adverse entropic contributions. Therefore, one can expect here a selectivity increase with higher temperature.

Aromatic clefts or tweezers such as shown in Figure 2.19 bear a significant negative charge at the concave inside. Therefore, electrostatic forces rule the stabilities of these complexes with electron-deficient neutral and cationic substrates [71]. Phosphonate appended dianionic tweezers can be used for the selective binding of lysine and arginine in water, with enhanced cation–π interactions [72].

2.10
Multifunctional Receptors for Recognition of Complex Target Molecules

2.10.1
Complexation of Amino Acids and Peptides

With natural analytes such as amino acids and peptides significant affinity and selectivity of complexation can be achieved only if several interaction mechanisms

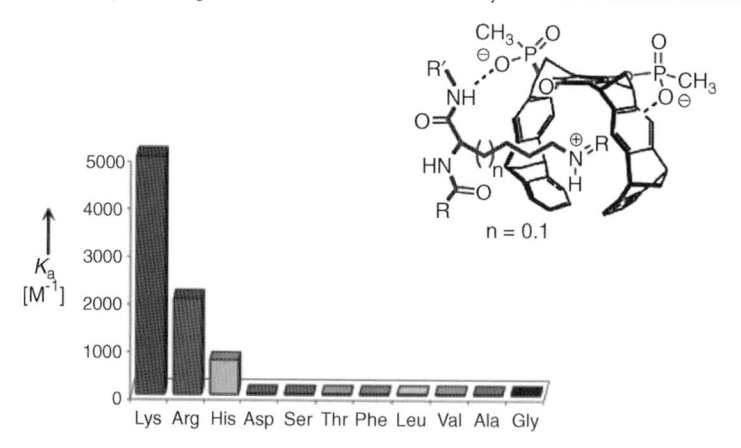

Figure 2.19 Selectivity of the tweezer host for N/C-protected amino acids; K_a values in buffered aqueous solution. Reprinted with permission from Reference [72]. Copyright 2005 The American Chemical Society.

operate simultaneously. Basic and acidic amino acid residues can of course be distinguished by ion pairing with cationic receptors [73]. Calixarenes bearing phosphoryl- or sulfonato-groups rely on ion pairing with the charged termini between the $-^+NH_3$-terminus and the anionic functions [74]. Calixarene **36** binds in methanol many amino acids with log K values above 3, with a stability that increases with the hydrophobicity of the amino acids [75]; in particular, aromatic residues are bound within the calixarene cavity [76]. The complex **37** illustrates how, besides ion pairing, hydrogen bonding and stacking contribute to binding of, for example, phenylalanine within a protected Ac-Phe-Ala-OMe dipeptide by a nitroarene-modified aminopyrazole host, although non-aqueous solvents must be used with such systems [77]. Selective binding of aromatic peptide residues can be improved by implementation of a nitro group into the host **37**, which is in line with the dispersive effects discussed above, enhancing the binding constant by an order of magnitude.

Stacking interactions lead with the peptidocalix[4]arene **38** to significant selectivity for aryl- over alkyl carboxylates (Table 2.6). Solubility reasons prevent the use of water as solvent, where larger selectivity can be expected. The presence of chiral amino acids in the host leads to moderate enantioselectivity with N-protected amino acids as guest molecules [78].

Table 2.6 Binding constants of carboxylates to **38** and N-protected amino acids in acetone [79].

Guest	$10^{-4}K$ (M^{-1})
CH$_3$COO$^-$	1.05
C$_6$H$_5$COO$^-$	4.01
N-Ac-Gly-COO$^-$:	0.62
N-Ac-L-Ala-COO$^-$	0.49
N-Ac-D-Ala-COO$^-$	0.57
N-Ac-L-Phe-COO$^-$	0.79
N-Ac-D-Phe-COO$^-$	1.05

In cucurbiturils like **39**, binding with $K = 3 \times 10^7\,\mathrm{M}^{-1}$ is observed, for example, with Phe-Gly, due to strong hydrogen bonding of the $^+\mathrm{NH}_3$ with the polarized carbonyl groups of the host, even in water [79]. This interaction is absent with the inverted sequence Gly-Phe, which exhibits $K = 1300\,\mathrm{M}^{-1}$. With a larger cucurbituril (**40**) methyl viologen (**41**) is bound simultaneously with amino acids or peptides, which also allows detection by a UV signal [80]. Figure 2.20 shows the binding constants obtained, and Table 2.7 illustrates the selectivity that can be reached with respect to the amino acid residues.

Another frequently used cavity for sensing is that of cyclodextrins [81], which, particularly after substitution with charged groups (e.g., **42** R = H, X = $^+\mathrm{NH}_2\mathrm{Me}$ or R = X = $^+\mathrm{NH}_2\mathrm{Me}$), provide stronger binding to anionic guests. Relatively strong

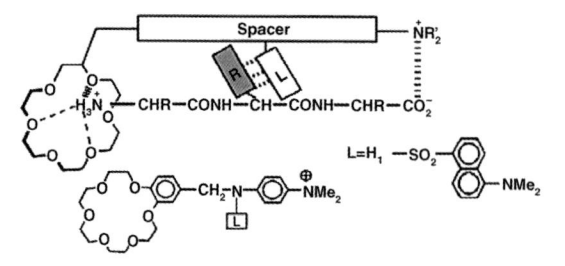

Figure 2.20 Receptor for length- and sequence-selective detection of tripeptides; Reprinted with permission from Reference [86]. Copyright 1998 The American Chemical Society.

Table 2.7 Binding constants of peptides and amino acids with the cucurbituril **40** together with methyl viologen (**41**) [81].

Guest	K (M^{-1})
Trp-Gly-Gly	1.3×10^5
Gly-Trp-Gly	2.1×10^4
Gly-Gly-Trp	3.1×10^3
Trp	4.3×10^4
Phe	5.3×10^3
Tyr	2.2×10^3
His	No binding

complexation of di- und tripeptides with monoamino-β-CDs is only observed with lipophilic amino acids, the residues of which are bound within the cavity, with preference for this at the C-terminus [82]. Unprotected di- and tripeptides associate with cyclodextrins bearing one charge at the rim (e.g., **42** R = H, X = $^+$NH$_2$Me) with K up to 200 M^{-1}, and after protection at the N terminus with up to 680 M^{-1} [82]. A moderate sequence selectivity is observed with monosubstituted cyclodextrins like **42** (R = H, X = $^+$NH$_2$Me), but less with the heptaamino-β- cyclodextrin such as **42** (R = X = $^+$NH$_2$Me), where ion pairing dominates. Substituted cyclodextrins also often lead to improved enantio- and diastereoselective detection of amino acids and peptides [83]. Protected amino acids are also bound weakly to unsubstituted cyclodextrins; for example, N-acetyl amino acids with γ-cyclodextrin bind in water with only $K \approx 50$ M^{-1}; the N-carbobenzyloxy derivative reacts with $K \approx 100$ M^{-1}, due to insertion of the benzyl residue in the cavity [84].

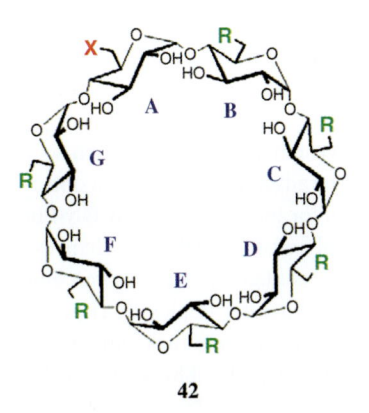

42

Cyclodextrins with pendant aromatic side chains have been used for fluorometric assays of many analytes (Chapter 4). Detailed NMR analyses [85] with such host compounds have already provided clear evidence for self-inclusion of the pendant

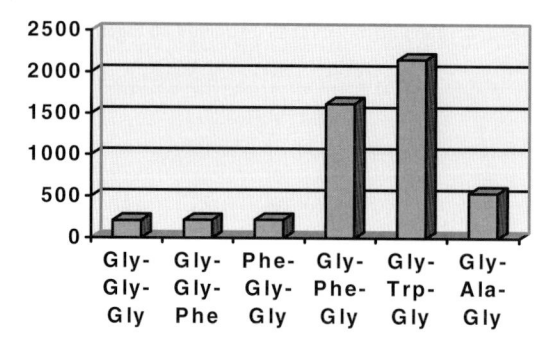

2500					
2000					
1500					
1000					
500					
0					
Gly-Gly-Gly	Gly-Gly-Phe	Phe-Gly-Gly	Gly-Phe-Gly	Gly-Trp-Gly	Gly-Ala-Gly

Figure 2.21 Binding constants of the receptor shown in figure 2.20 to tripeptides; with different amino acids and sequences, di- or tetrapeptides the receptor exhibits K values below 50.

aromatic side chain within the cyclodextrin cavity, which can be expelled by addition of stronger binding analytes; alternatively, binding of a smaller analyte together with the fluorophore is possible inside the cavity. In both cases large changes of the fluorescence occur.

A general strategy for length- and sequence-selective detection of native peptides in water is based on linear receptors, which bear a permanent charged group for binding to the $-COO^-$ terminus, a crown ether for association to the $-NH_3^+$ terminus, and a selector group L for the different amino acid residues. A crown ether instead of a negatively charged end group is necessary to avoid self-association of the receptor (Figure 2.21). In addition, the lone pairs of the crown ether oxygens lead to quenching of fluorescence emission of a fluorophor like a dansyl group, which at the same time can serve as selector L by, for example, stacking with an amino acid residue; occupation of the crown unit by the NH_3 terminus of a peptide then leads to strong fluorescence emission. Placement of the stacking unit RH in the host, which is also length-selective, allows a favorable binding contribution only if the complementary amino acid with a side chain RG is in the center position of the peptide; other positions are discriminated by up to ten-times smaller binding constants (Figure 2.21) [86].

A combination of stacking and binding to crown ether can also be used with porphyrin-based receptors such as **43**, which have the advantage of a built-in optical signal in the form of the Soret bands, and associate, for example, with Gly-Gly-Phe with log $K = 4.4$ in water [87]. Amino acids with special functions in the side chain can be recognized by specific reactions. Imidazole in His-containing peptides can bind preferentially with, for example, Cu(II) ions, as shown above in complex **23**. The large surface of porphyrins and their ability to implement metal ions make such compounds promising candidates also for complexation of peptides [88]. With suitable substituents as well as in the form of dimeric hosts they can be used for chiral discrimination, for example, of lysine [89]. Gable-type bisporphyrins promote binding of amines and oligopeptides, with $K = 9.4 \times 10^5 \, M^{-1}$ for the protected peptide H-His-Leu-His-NHBz [90].

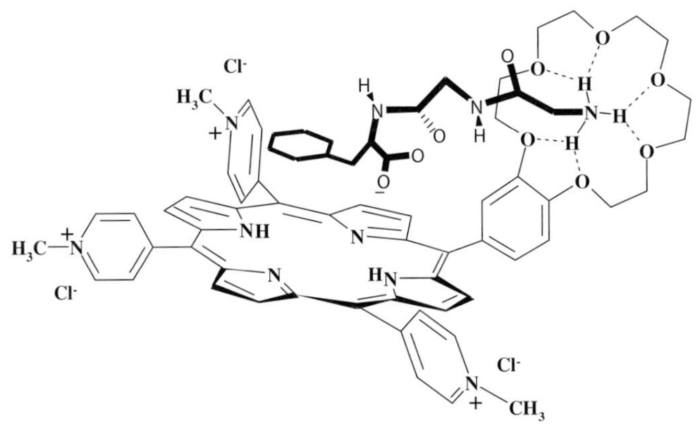

2.10 Multifunctional Receptors for Recognition of Complex Target Molecules

43

Ditopic receptors **44** and **45**– which combine a Zn(II) binuclear recognition site for phosphorylated serine with either a guanidinium group (**44**) or with another binuclear Zn(II) (**45**) site capable of recognizing a complementary anionic carboxylate or neutral coordinating imidazole groups respectively – have been designed for specific recognition of peptide sequences of human STAT (signal transducer and activator of transcription) proteins [91].

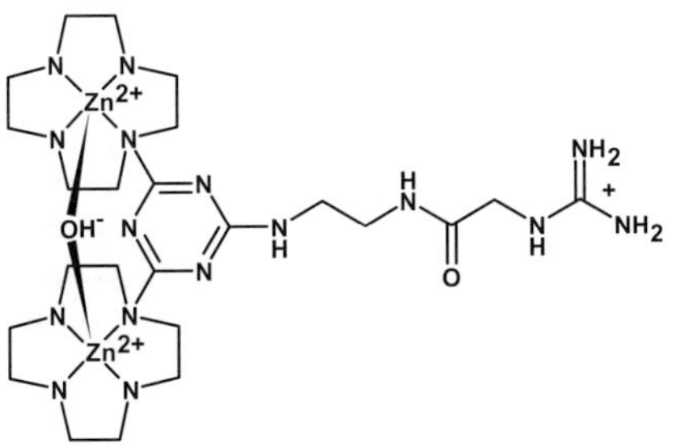

44 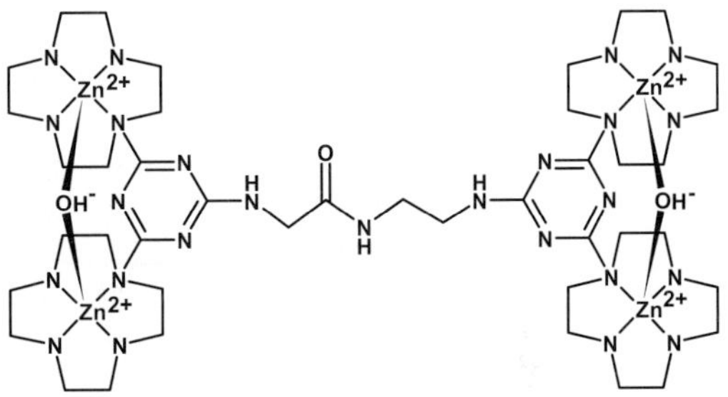 **45**

Model fluorescein labeled peptides **46** and **47** have been employed in titration experiments, which revealed a very high affinity of receptors together with expected selectivity. Thus with receptor **44** log K values are 8.0 and 4.5 for peptides **46** and **47**, respectively, but with receptor **45** the respective values are 5.0 and 7.5.

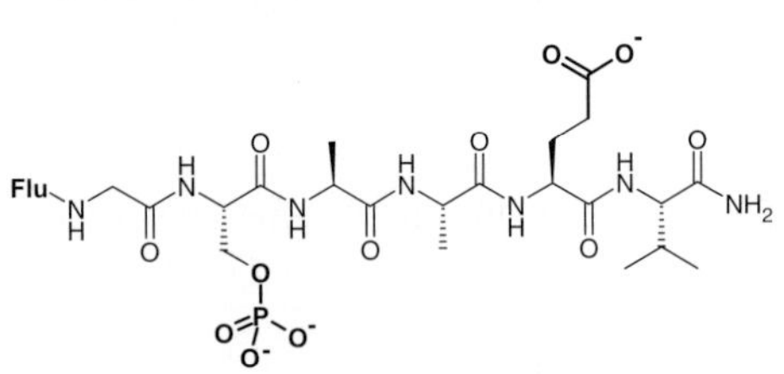

46 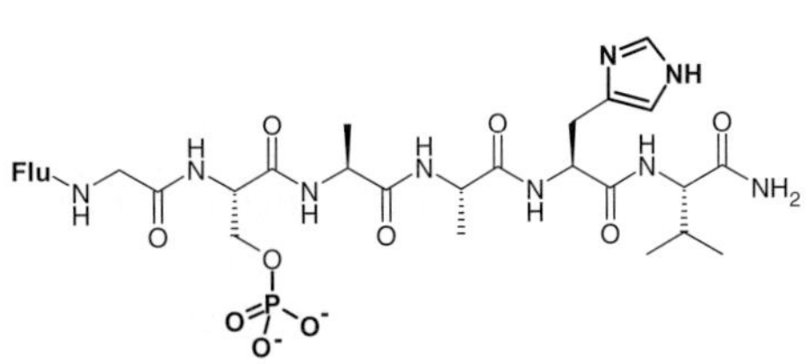 **47**

A highly selective multiparameter sensor array for detection of amino acids has been developed by combining enzymatic degradation of amino acids by the corresponding amino acid decarboxylases to amines, which substitute a dye molecule from the cavity of cucurbituril **39**, thus generating an optically detectable signal [92]. The signal appears only when both the amino acid and the respective specific for it decarboxylase enzyme are present together in the solution.

2.10.2
Complexation of Nucleotides and Nucleosides

This section focuses on the application of noncovalent bonding via fully synthetic organic receptors towards the optical detection of nucleotides and nucleosides in aqueous media. Again, several mechanisms can be used, primarily ion pairing, which obviously distinguishes mono- from di- and triphosphates.

The fluorescent cavitand **48**, bearing imidazolium and pyrene groups, shows a moderate selectivity towards GTP over ATP and CTP with binding constants of 7.38×10^4, 1.4×10^4, and 7.7×10^3 M^{-1} for GGP, ATP, and CTP, respectively, all in DMSO–water (6: 4, 0.02 M HEPES buffer pH 7.4) [93]. The stoichiometry of the complexes is 1: 1. For related examples see Reference [94]. In receptor **49**, two porphyrin units are held by an o-dioxymethylphenyl group in the shape of a cleft. Six pyridinium groups attached to the receptor provide solubility in water. This dimeric porphyrin is able to complex nucleotides and nucleosides in water or in aqueous buffer. Moreover, even electroneutral nucleosides like cytidine bind, reaching K values up to 270 000 M^{-1} compared with 430 000 M^{-1} for GTP. The proximity of the values indicates that the predominant interaction is stacking [95].

48   **49**

Protonated macrocyclic polyamines are water-soluble and, due to their positive charge, they can interact with anions such as phosphate (Section 2.6). The introduction of acridine moieties in macrocyclic polyamine **50** promotes more π–π interactions with adenine and nicotinamide groups. While the UV-vis spectra of **51** and **52** in the presence of one equivalent of ATP (0.5 M, 5 mM Tris acetate, pH 7.6) shows no change in absorbance, the fluorescence intensity increases by 150% for **51** and by 250% for **52**, forming a stable 1: 1 complex [96].

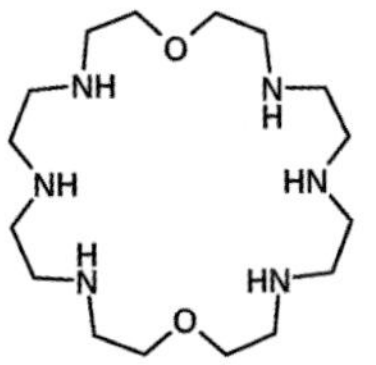

2.10 Multifunctional Receptors for Recognition of Complex Target Molecules

50

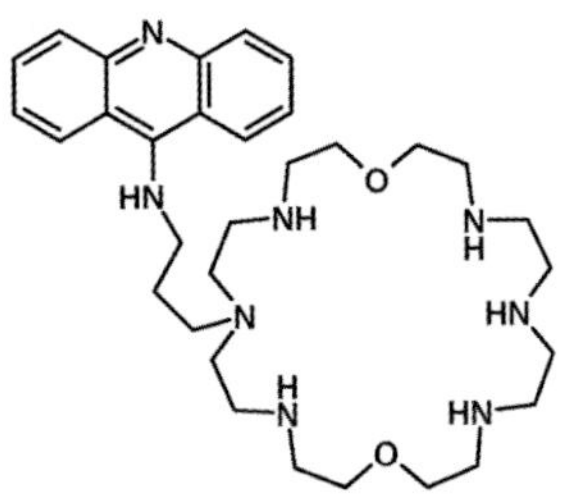

51

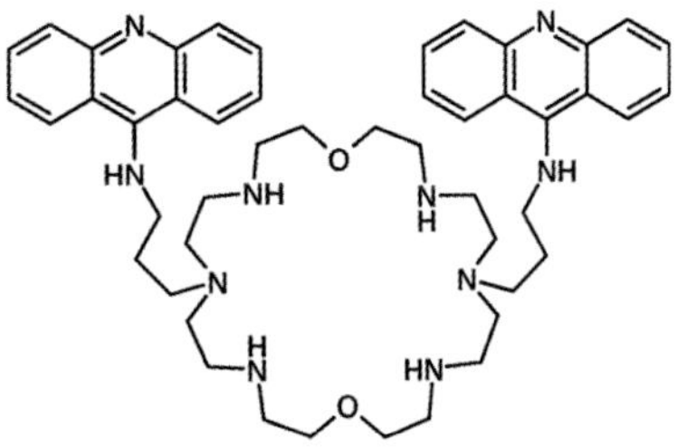

52

Cyclophane **53**, consisting of pyrene and azoniacrown groups, exhibits good interactions with nucleotides, but small selectivity with respect to the nucleobase. The binding constants indicate, as expected, a stronger preference for triphosphates (ATP, GTP, CTP, and UTP) over mono- and di- phosphates (Table 2.8). The UV-vis titration of **53** (1.0×10^{-5} M) with ATP (0–1.4 equiv) in water shows a decrease in absorbance. Fluorescence titration shows quenching with an excitation wavelength of 370 nm [97].

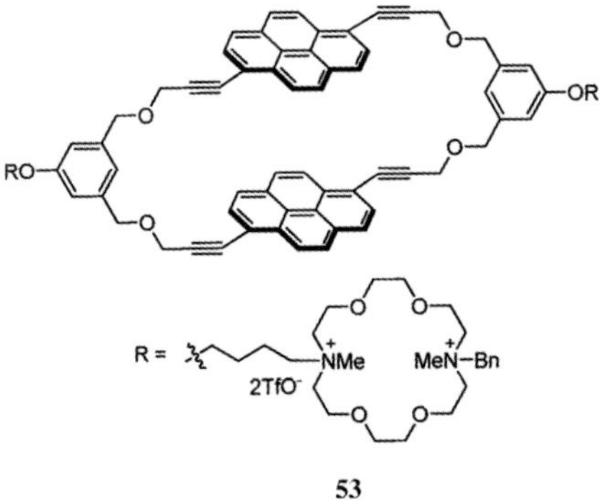

53

Table 2.8 Association constants (K) and free-energy changes for the complexation of **53** with nucleotides [98].

Nucleotide	K (M^{-1})	$-\Delta G_{293}$ (kJ mol^{-1})
ATP	1.0×10^6	34
ADP	5.3×10^3	21
AMP	1.9×10^3	18
GTP	1.3×10^6	34
GDP	4.0×10^3	20
GMP	2.7×10^3	19
CTP	2.6×10^5	30
CDP	7.7×10^3	22
CMP	4.0×10^3	20
UTP	7.7×10^5	33
UDP	2.2×10^4	24
UMP	3.0×10^3	20

Another related example is cyclophane receptor **54**, which consists of two quinacridine units bridged with alkylamino groups [98]. The binding constants for nucleosides (guanosine, adenosine, cytidine, and uridine) are $\leq 1 \times 10^2\,M^{-1}$; the maximum value for nucleotides corresponds to GTP^{2-} $6.31 \times 10^6\,M^{-1}$. All the monophosphates produce 1:2 complexes, while diphosphates and triphosphates produce 1:1 complexes. For more examples see References [99].

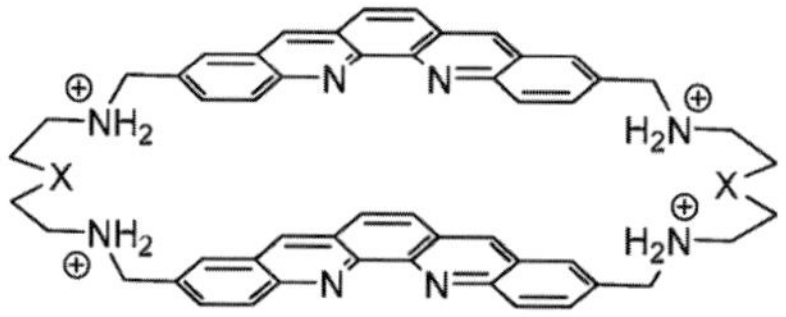

X = NH, O, CH$_2$-N$^+$H(CH$_3$)-CH$_2$

54

Peptide-based receptors correspond to interactions between nucleotides and proteins (i.e., enzymes) found in nature. For example, peptide **55**, consisting of twelve residues, has been designed for ATP recognition. The peptide sequence is acetyl-Arg-Trp-Val-Lys-Val-Asn-Gly-Orn-Trp-Ile-Lys-Gln-NH$_2$ [100].

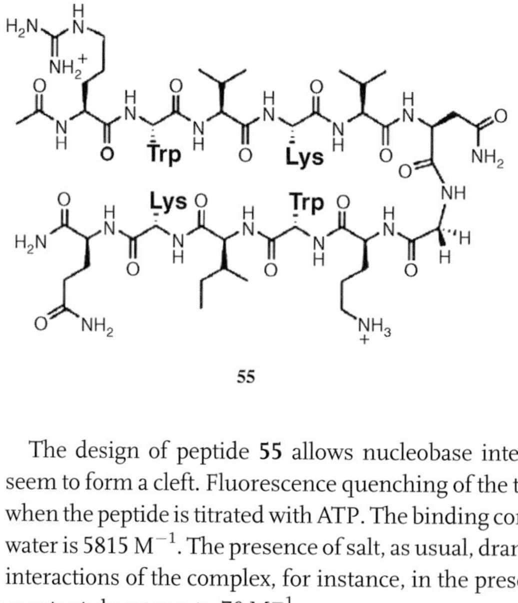

55

The design of peptide **55** allows nucleobase intercalation and the Trp residues seem to form a cleft. Fluorescence quenching of the tryptophan moieties is observed when the peptide is titrated with ATP. The binding constant of the peptide and ATP in water is $5815\,M^{-1}$. The presence of salt, as usual, dramatically lowers the electrostatic interactions of the complex, for instance, in the presence of NaCl 0.2 M the binding constant decreases to $70\,M^{-1}$.

Receptor **56** has been obtained by combinatorial methods and developed a strong fluorescence response towards ATP (0.2 M HEPES buffer pH 7.4) with a binding constant for the complex with ATP of $3.4 \times 10^3\,M^{-1}$ [101]. Fluorescence titrations of **56** with AMP and GTP showed practically no fluorescence change, indicating that, comparing ATP and GTP, it is the peptide sequence Ser-Tyr-Ser that imparts nucleobase specificity. This peptide combination would promote π-stacking between the phenolic moiety in tyrosine and the adenine group and hydrogen bonding interactions between the ribose or adenine moieties and the OH of serine.

56

Generally, one observes with positively charged host compounds the expected binding strength increase from mono- to di- and tri-phosphates. Some receptors, in which stacking contributions dominate, such as **49**, show a remarkable independence of the guest charge. Similarly, a bis(phenanthridinium) receptor binds nucleotides and nucleosides with similar strength (log $K = 5$ to 6) [102]. If the nucleobase is less exposed to the aqueous environment, as in the complexes with the amino-cyclodextrins **42** ($R = X = {}^+NH_2Me$), which binds ATP with $K = 3 \times 10^6\,M^{-1}$, one finds a slightly increased selectivity; for instance, for AMP $K = 12.5$, in comparison to GMP (4.0), UMP (2.0), and CMP (8.7); all in $10^4\,(M^{-1})$ units [103]. Improved discrimination must rely on a combination of stacking and hydrogen bonding, which, however, in water is usually too weak.

Recently an X-ray crystal structure of a nucleotide–polyammonium receptor complex has been reported, which allows better assignment of the interactions involved in recognition [104]. The receptor **57** in a tetraprotonated selectively binds a triphosphate nucleotide TTP (log $K = 4.57$ versus 3.58 for ATP). The crystal structure of the $57 \cdot H_4$-TTP complex (**58**) indicates formation of three salt bridges between ammonium groups of the receptor and guest phosphate groups, as well as hydrogen bonding and stacking interactions with the nucleobase, which are responsible for the binding selectivity (Figure 2.22).

58 | *2 Selectivity of Chemical Receptors*

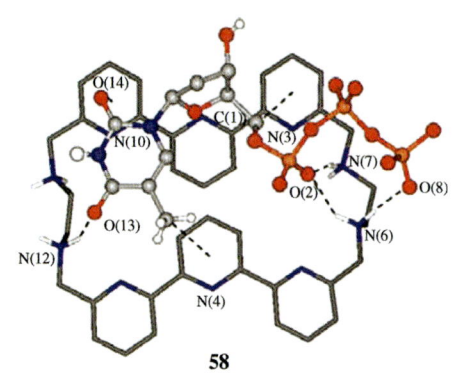

58

Figure 2.22 Crystal structure of the **57** · H$_4$-TTP complex (**58**). Reprinted with permission from Reference [104]. Copyright 2008 The American Chemical Society.

57   TTP

In this context, notably, a binuclear Zn(II) complex (**59**) has been proposed as a sensitive fluorescent sensor for nucleotides, while the lack of a secondary binding site for nucleobase recognition leads to essentially zero selectivity: log K values with this receptor are 6.23, 6.11, 5.83, and 6.23 for binding ADP, ATP, CTP, and GTP respectively [105].

59

Thus, we see that receptors designed for selective recognition of nucleotides use the principle outlined in Figure 2.4: a receptor capable of discriminating nucleotides

2.10 Multifunctional Receptors for Recognition of Complex Target Molecules

with different nucleobases posses a primary binding site to attract nucleotide anions by ion pairing (mostly by ammonium centers), which provides an affinity amounting to $K = 10^4$–10^5 M^{-1}, and a secondary weak binding site for recognition of the nucleobase via stacking interactions and, if possible, hydrogen bonding.

In the particular case of thymidine nucleotides the secondary interaction may be achieved by strong binding of the nucleobase to a Zn(II) polyamine complex, since such complexes are known to bind thymidine via deprotonation of the imide nitrogen. This approach has been realized by using a two-component receptor assembly (Figure 2.23), where the recognition of TTP via the triphosphate moiety is accomplished by a non-selective receptor similar to **59** and the nucleobase recognition is accomplished by a Zn(II) cyclen unit, which interacts with the primary receptor by an energy transfer mechanism [106].

The problem of nucleobase recognition of course can be solved more easily in non-aqueous media, where complementarity based on hydrogen bonding can be employed. This approach has been used to design carrier-based ion-selective electrodes for nucleotides. One of the first successful systems for this was a lipophilic triamine bearing a cytosine moiety (**60**) as carrier, which demonstrated good selectivity to 5′-GMP over 5′-AMP and simple inorganic anions [107]. The recognition mechanism involves ion-pairing of the phosphate moiety of the nucleotide with protonated amino groups of the receptor, and simultaneous Watson–Crick recognition of the guanine, as shown schematically in the structure **61**.

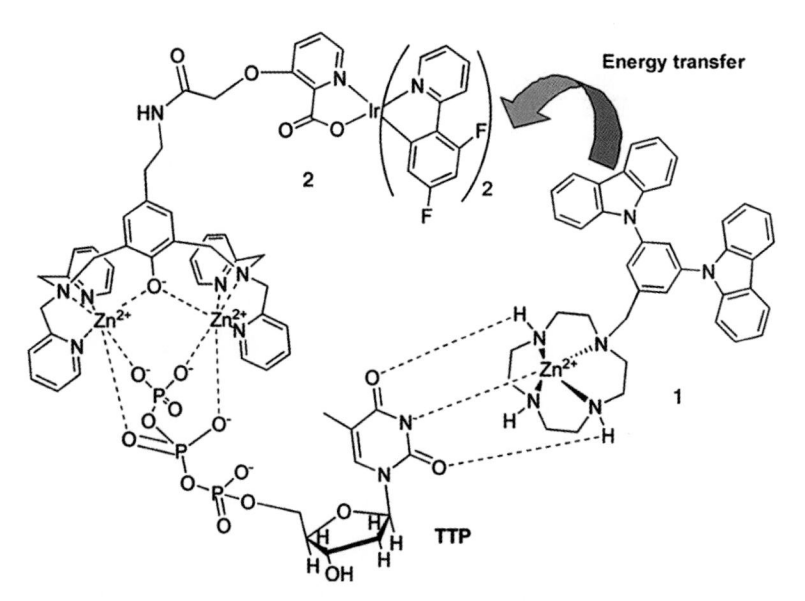

Figure 2.23 Selective recognition of TTP in a two-component receptor assembly. Reproduced with permission from Reference [106]. Copyright 2008 Wiley-VCH Verlag GmbH & Co. KGaA.

60 (R=(CH$_2$))$_5$CH$_3$ **61**

Further development of this approach was based on the use of nucleobase-substituted sapphyrins and calix[4]pyrroles like **62** or **63** in which the macrocycle serves as the phosphate recognition element [108].

62   **63**

The selectivity pattern in this system is rather complicated. Thus, in model through-membrane transport experiments the receptor **63** shows the expected selectivity toward 5′-GMP over 5′-AMP and 5′-ACP, but receptor **62** functionalized with the same cytosine nucleobase shows the highest rate of transport for 5′-ACP.

Remarkably, even neutral carriers can afford a nucleobase-selective response when applied in membrane electrodes. For example, receptor **64** recognizes 5′-GMP via formation of a complex resembling **61**, but without an ion-pairing contribution [109]. An efficient 5′-AMP selective electrode has also been prepared with receptor **20** [52].

64

2.11
Conclusions

In this chapter we have provided examples of selectivity patterns of chemical receptors, according to the type of noncovalent interactions employed for the guest recognition. When only one kind of interaction participates, one can expect to see the theoretically expected linear free energy correlation between selectivity and total affinity. Both can gain from multivalency by augmentation of the number of binding sites, even in poorly preorganized flexible receptors. The contribution of secondary interaction sites for discrimination between different analytes is the most important limitation of linear affinity–selectivity correlations. At the same time, the introduction of such secondary interaction sites holds much promise for the development of highly selective supramolecular hosts, where a primary interaction site can simultaneously and independently provide for high affinities. Intelligent choice of different solvents (or membrane materials) can lead to increased affinity, and thus also selectivity; different media can modulate the participating binding mechanisms, possibly with the consequence of gaining selectivity at the expense of affinity.

Since the observed selectivity is a balance of numerous aspects affecting the binding, the influence of such factors as solvent effects or co-existence of multiple species of, for example, differently protonated forms of a given receptor cannot be underestimated. The situation is even more complicated if one considers selectivity in experimental reality in terms of different complexation degrees under the chosen conditions, rather than the simple ratio of binding constants. Nevertheless, this "simplistic" expression of selectivity is of a primary importance since in practice it is preferable to use receptors as analytical sensors far from saturation, that is, under conditions when the ratio of binding constants is directly proportional to the ratio of concentrations of receptor–analyte complexes.

References

1 (a) Lehn, J.-M. (1995) *Supramolecular Chemistry. Concepts and Perspectives*, Wiley VCH Verlag GmbH, Weinheim; (b) Beer, P.D., Gale, P.A., and Smith, D.K. (1999) *Supramolecular Chemistry (Oxford Chemistry Primers, 74)*, Oxford University Press;(c) Schneider, H.-J. and Yatsimirsky, A. (2000) *Principles and Methods in Supramolecular Chemistry*, John Wiley & Sons, Ltd., Chichester; (d) Steed, J.W. and Atwood, J.L. (2000) *Supramolecular Chemistry*, John Wiley & Sons, Inc., New York.

2 Clare, J.P., Ayling, A.J., Joos, J.B., Sisson, A.L., Magro, G., Perez-Payan, M.N., Lambert, T.N., Shukla, R., Smith, B.D., and Davis, A.P. (2005) *J. Am. Chem. Soc.*, **127**, 10739.

3 Schneider, H.-J. and Yatsimirsky, A.K. (2008) *Chem. Soc. Rev.*, **37**, 263–277.

4 (a) Martell, A.E. and Hancock, R.D. (1996) *Metal Complexes in Aqueous Solution*, Plenum Press, New York, London; (b) Hancock, R.D. and Martell, A.E. (1995) *Adv. Inorg. Chem.*, **42**, 89.

5 Comba, P. and Schiek, W. (2003) *Coord. Chem. Rev.*, **238/239**, 21.

6 Optimal binding in capsules (review: Hof, F., Craig, S.L., Nuckolls, C., and Rebek, J. (2002) *Angew. Chem., Int. Ed.*, **41**, 1488) occurs if only about 50% of the available volume is used.

7 (a) Hay, B.P., Firman, T.K., and Moyer, B.A. (2005) *J. Am. Chem. Soc.*, **127**, 181; (b) Comba, P. (2000) *Coord. Chem. Rev.*, **123**, 1; (c) Comba, P. and Hambley, T.W. (2001) *Molecular Modeling of Inorganic Compounds*, 2nd edn, Wiley-VCH Verlag GmbH, Weinheim; (d) Deeth, R.J. (2004) *Molecular Mechanics and Comparison of DFT, AOM, and Ligand Field Approaches*, in *Comprehensive Coordination Chemistry II: Volume 2: Fundamentals: Physical Methods*,

Theoretical Analysis, and Case Studies (vol. ed. A.B.P. Lever) (series eds J.A. McCleverty and T.J. Meyer), Elsevier, Amsterdam, pp. 457–465, 643–650.

8 (a) Jorgensen, W.L. (1989) *Acc. Chem. Res.*, **22**, 124; (b) Kollman, P. (1993) *Chem. Rev.*, **93**, 2395; (c) Aqvist, J., Luzhkov, V.B., and Brandsdal, B.O. (2002) *Acc. Chem. Res.*, **35**, 358; (d) Foloppe, N. and Hubbard, R. (2006) *Curr. Med. Chem.*, **13**, 3583.

9 Schmidtchen, F.P. (2006) *Coord. Chem. Rev.*, **250**, 2918, and references therein.

10 Vessman, J., Stefan, R.I., van Staden, J.F., Danzer, K., Lindner, W., Burns, D.T., Fajgelj, A., and Müller, H. (2001) *Pure Appl. Chem.*, **73**, 1381.

11 (a) Holzbecher, Z., Diviš, L., Král, M., Sucha, L., and Vlačil, F. (1975) *Organic Reagents in Inorganic Analysis*, SNTL, Praha; (b) Massart, D.L., Dijkstra, A., and Kaufman, L. (1978) *Evaluation and Optimization of Laboratory Methods and Analytical Procedures*, Elsevier, Amsterdam; (c) Inczédy, J. (1982) *Talanta*, **29**, 595.

12 Faber, N.M., Ferré, J., Boqué, R., and Kalivas, J.H. (2003) *Trends Anal. Chem.*, **22**, 352.

13 Vlasov, Yu., Legin, A., Rudnitskaya, A., Di Natale, C., and D'Amico, A. (2005) *Pure Appl. Chem.*, **77**, 1965.

14 (a) Rakow, N. and Suslick, K. (2000) *Nature*, **406**, 710–712; (b) Drew, S., Janzen, D., and Mann, K. (2002) *Anal. Chem.*, **74**, 2547–2555; (c) Lavigne, J. and Anslyn, E. (2001) *Angew. Chem., Int. Ed.*, **40**, 3118–3130.

15 Müller, H. (1995) *Pure Appl. Chem.*, **67**, 601–613.

16 Mirsky, V.M. (2001) *Sensors*, **1**, 13–17.

17 Cramer, F. and Freist, W. (1987) *Acc. Chem. Res.*, **20**, 79.

18 (a) Pluth, M.D. and Raymond, K.N. (2007) *Chem. Soc. Rev.*, **36**, 161; (b) Biros, S.M. and Rebek, J. (2007) *Chem. Soc. Rev.*, **36**, 93; (c) Jasat, A. and Sherman, J.C. (1999) *Chem. Rev.*, **99**, 931; (d) Warmuth, R. and Yoon, Y. (2001) *Acc. Chem. Res.*, **34**, 95.

19 (a) Schneider, H.-J. (1994) *Chem. Soc. Rev.*, **22**, 227; (b) Schneider, H.J. (1991) *Angew. Chem., Int. Ed. Engl.*, **30**, 1417; (c) Schneider, H.J. (2009) *Angew. Chem., Int. Ed.* **48**, 2082.

20 Schneider, H.-J. (2008) *Eur. J. Med. Chem.*, **43**, 2307–2315.

21 Ghidini, E., Ugozzoli, F., Ungaro, R., Harkema, S., El-Fald, A.A., and Reinhoudt, D.N. (1990) *J. Am. Chem. Soc.*, **112**, 6979.

22 Glendening, E.D. and Feller, D. (1996) *J. Am. Chem. Soc.*, **118**, 6052.

23 (a) Armentrout, P.B. (1999) *Int. J. Mass Spectrom.*, **193**, 227; (b) Anderson, J.D., Paulsen, E.S., and Dearden, D.V. (2003) *Int. J. Mass Spectrom.*, **227**, 63.

24 Katsuta, S., Ito, Y., and Takeda, Y. (2004) *Inorg. Chim. Acta*, **357**, 541.

25 Lucas, J.B. and Lincoln, S.F. (1994) *J. Chem. Soc., Dalton Trans.*, 423.

26 Danil de Namor, A.F., Chahine, S., Castellano, E.E., and Piro, O.E. (2004) *J. Phys. Chem. B*, **108**, 11384.

27 Solov'ev, V.P., Kazachenko, V.P., Raevsky, O.A., Schneider, H.-J., and Rüdiger, V. (1999) *Eur. J. Org. Chem.*, 1847.

28 Talanova, G.G., Elkarim, N.S.A., Talanov, V.S., Hanes, R.E., Hwang, H., Bartsch, R.A., and Rogers, R.D. (1999) *J. Am. Chem. Soc.*, **121**, 11281.

29 (a) Bakker, E., Bühlmann, P., and Pretsch, E. (1997) *Chem. Rev.*, **97**, 3083–3132; (b) Bühlmann, P., Pretsch, E., and Bakker, E. (1998) *Chem. Rev.*, **98**, 1593–1687.

30 (a) Umezawa, Y. (ed.) (1990) *Handbook of Ion-Selective Electrodes: Selectivity Coefficients*, CRC Press, Boca Raton, FL; (b) Umezawa, Y., Bühlmann, P., Umezawa, K., Tohda, K., and Amemiya, S. (2000) *Pure Appl. Chem.*, **72**, 1851–2082; (c) Umezawa, Y., Umezawa, K., Bühlmann, P., Hamada, N., Aoki, H., Nakanishi, J., Sato, M., Xiao, K.P., and Nishimura, Y. (2002) *Pure Appl. Chem.*, **74**, 923–994; (d) Umezawa, Y., Bühlmann, P., Umezawa, K., and Hamada, N. (2002) *Pure Appl. Chem.*, **74**, 995–1099.

31 Xu, H., Xu, X.H., Dabestani, R., Brown, G.M., Fan, L., Patton, S., and Ji, H.F. (2002) *J. Chem. Soc. Perkin Trans. 2*, 636–643.

32 Kolarik, Z. (2008) *Chem. Rev.*, **108**, 4208.

33 Dam, H.H., Reinhoudt, D.N., and Verboom, W. (2007) *Chem. Soc. Rev.*, **36**, 367.

34 Buie, N.M., Talanov, V.S., Butcher, R.J., and Talanova, G.G. (2008) *Inorg. Chem.*, **47**, 3549.

35 De Robertis, C., De Stefano, C., Foti, O., Giuffre, O., and Sammartano, S. (2001) *Talanta*, **54**, 1135, and references therein.

36 (a) Bianchi, A., Bowman-James, K., and García-España, E. (eds) (1997) *The Supramolecular Chemistry of Anions*, Wiley-VCH Verlag GmbH, Weinheim; (b) Amendola, V. and Fabbrizzi, L. (2006) *Acc. Chem. Res.*, **39**, 343; (c) Bowman-James, K. (2005) *Acc. Chem. Res.*, **38**, 671; (d) Schmidtchen, F.P. and Berger, M. (1997) *Chem. Rev.*, **97**, 1609; (e) Schmidtchen, F.P. (2005) *Top. Curr. Chem.*, **255**, 1.

37 (a) García-España, E., Díaz, P., Llinares, J.M., and Bianchi, A. (2006) *Coord. Chem. Rev.*, **250**, 2952; (b) Bencini, A., Bianchi, A., Burguete, M.I., Dapporto, P., Doménech, A., García-España, E., Luis, S.V., Paoli, P., and Ramírez, J.A. (1994) *J. Chem. Soc., Perkin Trans. 2*, 569.

38 Bazzicalupi, C., Bencini, A., Bianchi, A., Borsari, L., Giorgi, C., Valtancoli, B., Anda, C., and Llobet, A. (2005) *J. Org. Chem.*, **70**, 4257–4266.

39 Arbuse, A., Anda, C., Martínez, M.A., Pérez-Mirón, J., Jaime, C., Parella, T., and Llobet, A. (2007) *Inorg. Chem.*, **46**, 10632–10638.

40 Kim, S.K., Lee, D.H., Hong, J.I., and Yoon, J. (2009) *Acc. Chem. Res.*, **42**, 23.

41 For other recent papers on pyrophosphate recognition see (a) Sakamoto, T., Ojida, A., and Hamachi, I. (2009) *Chem. Commun.*, 141; (b) Du, J., Wang, X.Y., Jia, M., Li, T.J., Mao, J.F., and Guo, Z.J. (2008) *Inorg. Chem. Commun.*, **11**, 999.

42 Vance, D.H. and Czarnik, A.W. (1994) *J. Am. Chem. Soc.*, **116**, 9397.

43 For recent references for use of hydrogen bonds in supramolecular complexes see (a) Cooke, G. and Rotello, V.M. (2002) *Chem. Soc. Rev.*, **31**, 275; (b) Choi, K.H. and Hamilton, A.D. (2003) *Coord. Chem. Rev.*, **240**, 101.

44 (a) Abraham, M.H. (1993) *Chem. Rev.*, **22**, 73; (b) Raevsky, O.A. (1990) *Russ. Chem. Rev.*, **59**, 219, and references cited therein.

45 (a) Chang, K.-J., Jeong, K.-S., Moon, D., and Lah, M.S. (2005) *Angew. Chem. Int. Ed.*, **44**, 7926; (b) Werner, F., Schneider, H.-J. (2000) *Helv. Chim. Acta*, **83**, 465.

46 Bell, T.W., Khasanov, A.B., and Drew, M.G.B. (2002) *J. Am. Chem. Soc.*, **124**, 14092.

47 Gale, P.A. (2005) *Chem. Commun.*, 3761.

48 Caltagirone, C., Gale, P.A., Hiscock, J.R., Brooks, S.J., Hursthouse, M.B., and Light, M.E. (2008) *Chem. Commun.*, 3007–3009.

49 Amendola, V., Boiocchi, M., Fabbrizzi, L., and Palchetti, A. (2005) *Chem. Eur. J.*, **11**, 5648–5660.

50 Hancock, L.M. and Beer, P.D. (2009) *Chem. Eur. J.*, **15**, 42–44.

51 (a) Schmuck, C. and Wienand, W. (2001) *Angew. Chem., Int. Ed.*, **40**, 4363; (b) Sivakova, S. and Rowan, S.J. (2005) *Chem. Soc. Rev.*, **34**, 9; (c) Park, T. and Zimmerman, S.C. (2006) *J. Am. Chem. Soc.*, **128**, 11582.

52 Hisamatsu, Y., Hasada, K., Amano, F., Tsubota, Y., Wasada-Tsutsui, Y., Shirai, N., Ikeda, S., and Odashima, K. (2006) *Chem. Eur. J.*, **12**, 7733–7741.

53 (a) Reichenbach-Klinke, R. and König, B. (2002) *J. Chem. Soc. Dalton Trans.*, 121; (b) Kruppa, M. and König, B. (2006) *Chem. Rev.*, **106**, 352.

54 Hortala, M.A., Fabbrizzi, L., Marcotte, N., Stomeo, F., and Taglietti, A. (2003) *J. Am. Chem. Soc.*, **125**, 20.

55 Kruppa, M., Mandl, C., Miltschitzky, S., and König, B. (2005) *J. Am. Chem. Soc.*, **127**, 3362.

56 (a) Ojida, A., Mitooka, Y., Sada, K., and Hamachi, I. (2004) *J. Am. Chem. Soc.*, **126**, 2454; (b) Geduhn, J., Walenzyk, T., König, B. (2007) *Curr. Org. Synth.* **4**, 390.

57 Sun, S., Fazal, M.A., Roy, B.C., Chandra, B., and Mallik, S. (2002) *Inorg. Chem.*, **41**, 1584.

58 (a) Fazal, M., Roy, B.C., Sun, S., Mallik, S., and Rodgers, K.R. (2001) *J. Am. Chem. Soc.*, **123**, 6; (b) see also Ojida, A., Miyahara, Y., Kohira, T., and Hamachi, I. (2004) *Biopolymers*, **76**, 177.

59 Amendola, V., Bonizzoni, M., Esteban-Gómez, D., Fabbrizzi, L., Licchelli, M., Sancenón, F., and Taglietti, A. (2006) *Coord. Chem. Rev.*, **250**, 1451.

60 Mancin, F. and Chin, J. (2002) *J. Am. Chem. Soc.*, **124**, 10946.

61 Lee, J.H., Park, J., Lah, M.S., Chin, J., and Hong, J.-I. (2007) *Org. Lett*, **9**, 3729.

62 Natale, D. and Mareque-Rivas, J.C. (2008) *Chem. Commun.*, 425.

63 Blanda, M.T., Horner, J.H., and Newcomb, M. (1989) *J. Org. Chem.*, **54**, 4626.

64 Schulte, M., Schürmann, M., and Jurkschat, K. (2001) *Chem. Eur. J.*, **7**, 347.

65 Reeske, G., Schürmann, M., Costisella, B., and Jurkschat, K. (2007) *Organometallics*, **26**, 4170.

66 Xie, Y., Morimoto, T., and Furuta, H. (2006) *Angew. Chem. Int. Ed.*, **45**, 6907.

67 Altmann, R., Jurkschat, K., and Schürmann, M. (1998) *Organometallics*, **17**, 5858.

68 Lee, M.H. and Gabbaï, F.P. (2007) *Inorg. Chem.*, **46**, 8132.

69 Meyer, E.A., Castellano, R.K., and Diederich, F. (2003) *Angew. Chem., Int. Ed.*, **42**, 1210.

70 Schneider, H.J., Liu, T., Sirish, M., and Malinovski, V. (2002) *Tetrahedron*, **58**, 779.

71 Klärner, F.G. and Kahlert, B. (2003) *Acc. Chem. Res.*, **36**, 919.

72 Fokkens, M., Schrader, T., and Klärner, F.-G. (2005) *J. Am. Chem. Soc.*, **127**, 14415.

73 See, for example, (a) Schmuck, C., Heil, M., Scheiber, J., and Baumann, K. (2005) *Angew. Chem., Int. Ed.*, **44**, 7208; (b) Wehner, M. and Schrader, T. (2002) *Angew. Chem., Int. Ed.*, **41**, 1751; (c) Peczuh, M.W. and Hamilton, A.D. (2000) *Chem. Rev.*, **100**, 2479–2493.

74 Arena, G., Contino, A., Gulino, F.G., Magri, A., Sansone, F., Sciotto, D., and Ungaro, R. (1999) *Tetrahedron Lett.*, **40**, 1597–1600, and references cited therein.

75 Zielenkiewicz, W., Marcinowicz, A., Cherenok, S., Kalchenko, V.I., and Poznanski, J. (2006) *Supramol. Chem.*, **18**, 167–176.

76 Douteau-Guevel, N., Perret, F., Coleman, A.W., Morel, J.P., and Morel-Desrosiers, N. (2002) *J. Chem. Soc. Perkin Trans. 2*, 524–532, and references cited therein.

77 Wehner, M., Janssen, D., Schäfer, G., and Schrader, T. (2006) *Eur. J. Org. Chem.*, 138–153.

78 Sansone, F., Baldini, L., Casnati, A., Lazzarotto, M., Ugozzoli, F., and Ungaro, R. (2002) *Proc. Nat. Acad. Sci. USA*, **99**, 4842.

79 Rekharsky, M.V., Yamamura, H., Ko, Y.H., Selvapalam, N., Kim, K., and Inoue, Y. (2008) *Chem. Commun.*, 2236–2238.

80 Bush, M.E., Bouley, N.D., and Urbach, A.R. (2005) *J. Am. Chem. Soc.*, **127**, 14511–14517.

81 Shahgaldian, P. and Pieles, U. (2006) *Sensors*, **6**, 593–615.

82 Hacket, F., Simova, S., and Schneider, H.J. (2001) *J. Phys. Org. Chem.*, **14**, 159–170.

83 Kano, K. (1997) *J. Phys. Org. Chem.*, **10**, 286–291.

84 Rekharsky, M.V., Yamamura, H., Kawai, M., and Inoue, Y. (2003) *J. Org. Chem.*, **68**, 5228–5235.

85 (a) Impellizzeri, G., Pappalardo, G., D'Alessandro, F., Rizzarelli, E., Savioano, M., Iacovino, R., Benedetti, E., and Pedone, C. (2000) *Eur. J. Org. Chem.*, **6**, 1065–1076; (b) Salvatierra, D., Sanchez-Ruiz, X., Garduno, R., Cervello, E., Jaime, C., Vergili, A., and Sanchez-Ferrando, F. (2000) *Tetrahedron*, **56**, 3035–3041; (c) Uccello-Barretta, G., Balzano, F., Cuzzola, A., Menicagli, R., and Salvadori, P. (2000) *Eur. J. Org. Chem.*, **3**, 449–453.

86 Hossain, M.A. and Schneider, H.J. (1998) *J. Am. Chem. Soc.*, **120**, 11208.

87 (a) Sirish, M. and Schneider, H.-J. (1999) *Chem. Commun.*, 907–908; (b) Sirish, M., Chertkov, V.A., and Schneider, H.-J. (2002) *Chem. Eur. J.*, **8**, 1181–1188.

88 Ogoshi, H. and Mizutani, T. (1998) *Acc. Chem. Res.*, **31**, 81–89.

89 Hayashi, T., Aya, T., Nonoguchi, M., Mizutani, T., Hisaeda, Y., Kitagawi, S., and Ogoshi, H. (2002) *Tetrahedron*, **58**, 2803–2811.

90 Mizutani, T., Wada, K., and Kitagawa, S. (2002) *Chem. Commun.*, 1626–1627.

91 Grauer, A., Riechers, A., Ritter, S., and König, B. (2008) *Chem. Eur. J.*, **14**, 8922–8927.

92 Bailey, D.M., Hennig, A., Uzunova, V.D., and Nau, W.M. (2008) *Chem. Eur. J.*, **14**, 6069–6077.

93 Kim, S.K., Moon, B.-S., Park, J.H., Seo, Y.I., Koh, H.S., Yoon, Y.J., Lee, K.D., and Yoon, J. (2005) *Tetrahedron Lett.*, **46**, 6617–6620.

94 (a) Shi, Y. and Schneider, H.-J. (1999) *J. Chem. Soc., Perkin Trans.* 2, 1797–1803; (b) Zadmard, R. and Schrader, T. (2006) *Angew. Chem. Int. Ed.*, **45**, 2703–2706.

95 Schneider, H.-J. and Sirish, M. (2000) *J. Am. Chem. Soc.*, **122**, 5881–5882.

96 Fenniri, H., Hosseini, M.W., and Lehn, J.-M. (1997) *Helv. Chim. Acta*, **80**, 786–803.

97 Abe, H., Mawatari, Y., Teraoka, H., Fujimoto, K., and Inouye, M. (2004) *J. Org. Chem.*, **69**, 495–504.

98 Baudoin, O., Gonnet, F., Teulade-Fichou, M.P., Vigneron, J.P., Tabet, J.C., and Lehn, J.-M. (1999) *Chem. Eur. J.*, **5**, 2762–2771.

99 (a) Moreno-Corral, R. and Lara, K.O. (2008) *Supramol. Chem.*, **20**, 427–435; (b) Oshovsky, G.V., Reinhoudt, D.N., and Verboom, W. (2007) *Angew. Chem. Int. Ed.*, **46**, 2366–2393.

100 (a) Butterfield, S.A., Sweeney, M.M., and Waters, M.L. (2005) *J. Org. Chem.*, **70**, 1105–1114; (b) Butterfield, S.M., Sweeney, M.M., and Waters, M.L. (2003) *J. Am. Chem. Soc.*, **125**, 9580–9581.

101 Schneider, S.E., O'Neil, S.N., and Anslyn, E.V. (2000) *J. Am. Chem. Soc.*, **122**, 542–543.

102 Žinić, M., Vigneron, J.-P., and Lehn, J.-M. (1995) *Chem. Commun.*, 1073–1075.

103 Eliseev, A.V. and Schneider, H.-J. (1994) *J. Am. Chem. Soc.*, **116**, 6081–6088.

104 Bazzicalupi, C., Bencini, A., Bianchi, A., Faggi, E., Giorgi, C., Santarelli, S., and Valtancoli, B. (2008) *J. Am. Chem. Soc.*, **130**, 2440.

105 Ojida, A., Takashima, I., Kohira, T., Nonaka, H., and Hamachi, I. (2008) *J. Am. Chem. Soc.*, **130**, 12095.

106 Kwon, T.-H., Kim, H.J., and Hong, J.-I. (2008) *Chem. Eur. J.*, **14**, 9613.

107 Tohda, K., Tange, M., Odashima, K., Umezawa, Y., Furuta, H., and Sessler, J.L. (1992) *Anal. Chem.*, **64**, 960.

108 Sessler, J.L., Král, V., Shishkanova, T.V., and Gale, P.A. (2002) *Proc. Natl. Acad. Sci. USA*, **99**, 4848–4853.

109 Amemiya, S., Bühlmann, P., Tohda, K., and Umezawa, Y. (1997) *Anal. Chim. Acta*, **341**, 129.

3
Combinatorial Development of Sensing Materials
Radislav A. Potyrailo

3.1
Introduction

Rational design of sensing materials based on prior knowledge is a very attractive approach because it could avoid time-consuming synthesis and testing of numerous materials candidates [1–3]. However, to be quantitatively successful, rational design [4–9] requires detailed knowledge regarding relation of intrinsic properties of sensing materials to a set of their performance properties. This knowledge is typically obtained from extensive experimental and simulation data. However, with the increase of structural and functional complexity of materials, the ability to rationally define the precise requirements that result in a desired set of performance properties becomes increasingly limited [10]. Thus, in addition to rational design, various sensing materials ranging from dyes and ionophores, to biopolymers, organic, and hybrid polymers, and to nanomaterials have been discovered using detailed experimental observations or simply by chance [11–19]. Such an approach in development of sensing materials reflects a more general situation in materials design that is "still too dependent on serendipity" with only limited capability for rational materials design [20].

Conventionally, detailed experimentation with sensing materials candidates for their screening and optimization consumes a tremendous amount of time and project cost. Thus, developing sensing materials is a recognized challenge because it requires extensive experimentation not only to achieve the best short-term performance, but also long-term stability, manufacturability, and other practical requirements. Numerous practical challenges in rational sensing material design provide tremendous prospects for the application of combinatorial methodologies for the development of sensing materials.

This chapter demonstrates the broad applicability of combinatorial technologies in discovery and optimization of new sensing materials. We discuss general principles of combinatorial materials screening, followed by discussion of the opportunities facilitated by combinatorial technologies for discovery and optimization of new sensing materials. We further critically analyze results of materials development using discrete and gradient materials arrays and provide examples from a wide

---

Artificial Receptors for Chemical Sensors. Edited by V.M. Mirsky and A.K. Yatsimirsky
Copyright © 2011 WILEY-VCH Verlag GmbH & Co. KGaA, Weinheim
ISBN: 978-3-527-32357-9

variety of sensors based on various energy-transduction principles that involve radiant, mechanical, and electrical types of energy.

3.2
General Principles of Combinatorial Materials Screening

Combinatorial materials screening is a process that couples the capability for parallel production of large arrays of diverse materials together with different high-throughput measurement techniques for various intrinsic and performance properties followed by navigation in the collected data for identifying "lead" materials [21–29]. The terms "combinatorial materials screening" and "high-throughput experimentation" are typically interchangeably applied for all types of automated parallel and rapid sequential evaluation processes of materials and process parameters that include truly combinatorial permutations or their selected subsets.

Individual aspects of accelerated materials development have been known for decades. These include combinatorial and factorial experimental designs [30], parallel synthesis of materials on a single substrate [31, 32], screening of materials for performance properties [33], and computer data processing [34, 35]. However, in 1970, Hanak suggested an integrated materials-development workflow [36]. Its key aspects included (i) complete compositional mapping of a multicomponent system in one experiment, (ii) simple rapid non-destructive all-inclusive chemical analysis, (iii) testing of properties by a scanning device, and (iv) computer data processing. In 1995, Xiang, Schultz, and coworkers initiated applications of combinatorial methodologies in materials science [37]. Since then, combinatorial tools have been employed to discover and optimize a wide variety of materials (Table 3.1 [37–59]).

Figure 3.1 outlines a typical combinatorial materials development cycle [63, 64]. Compared to the initial idea of Hanak [36], the modern workflow has several new important aspects such as design/planning of experiments, data mining, and scale up. In combinatorial screening of materials, concepts originally thought as highly automated, have been recently refined to have more human input, with only an appropriate level of automation. For the throughput of 50–100 materials formulations per day, it is acceptable to perform certain aspects of the process manually [65, 66]. To address numerous materials-specific properties, various high-throughput characterization tools are required. Characterization tools are used for rapid and automated assessment of single or multiple properties of the large number of samples fabricated together as a combinatorial array or "library" [25, 67, 68].

In addition to the parallel synthesis and high-throughput characterization instrumentation that significantly differ from conventional equipment, the data management approaches also differ from conventional data evaluation [29]. In an ideal combinatorial workflow, one should "analyze in a day what is made in a day" [69] and that requires significant computational assistance. In a well-developed combinatorial workflow, design and syntheses protocols for materials libraries are computer assisted, materials synthesis and library preparation are carried out with computer-controlled manipulators, and property screening and materials characterization

Table 3.1 Examples of materials developed using combinatorial screening techniques.

Materials examples	Reference
Superconductor materials	[37]
Ferroelectric materials	[38]
Magnetoresistive materials	[39]
Luminescent materials	[40]
Agricultural materials	[41]
Structural materials	[42]
Hydrogen storage materials	[43]
Organic light-emitting materials	[44]
Ferromagnetic shape-memory alloys	[45]
Thermoelastic shape-memory alloys	[46]
Heterogeneous catalysts	[47]
Homogeneous catalysts	[48]
Polymerization catalysts	[49]
Electrochemical catalysts	[50]
Electrocatalysts for hydrogen evolution	[51]
Polymers	[52]
Zeolites	[53]
Metal alloys	[54]
Materials for methanol fuel cells	[55]
Materials for solid oxide fuel cells	[56]
Materials for solar cells	[57]
Automotive coatings	[58]
Waterborne coatings	[59]
Vapor-barrier coatings	[60]
Marine coatings	[61]
Fouling-release coatings	[62]

are also software controlled. Further, materials synthesis data as well as property and characterization data are collected into a materials database. This database contains information on starting components, their descriptors, process conditions, materials testing algorithms, and performance properties of libraries of sensing materials. Data in such a database is not just stored but also processed with the proper statistical analysis, visualization, modeling, and data-mining tools. Combinatorial synthesis of materials provides a good possibility for formation of banks of combinatorial materials [64]. Such banks can be used, for example, for further re-investigation of the materials of interest for some new applications or as reference materials.

3.3
Opportunities for Sensing Materials

Figure 3.2 shows the development process of a sensor system with a new sensing material can be described using technology readiness levels (TRLs). The concept of TRLs is an accepted way to assess technology maturity [70]. These TRLs provide

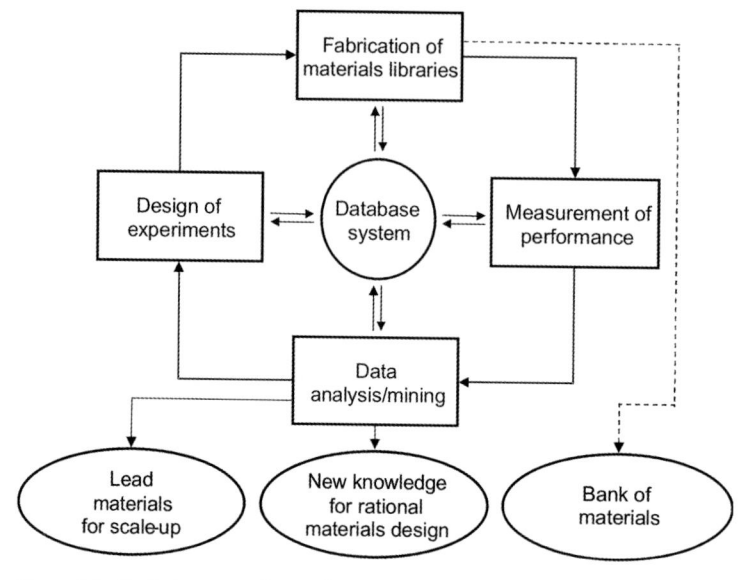

Figure 3.1 Typical process for combinatorial materials development.

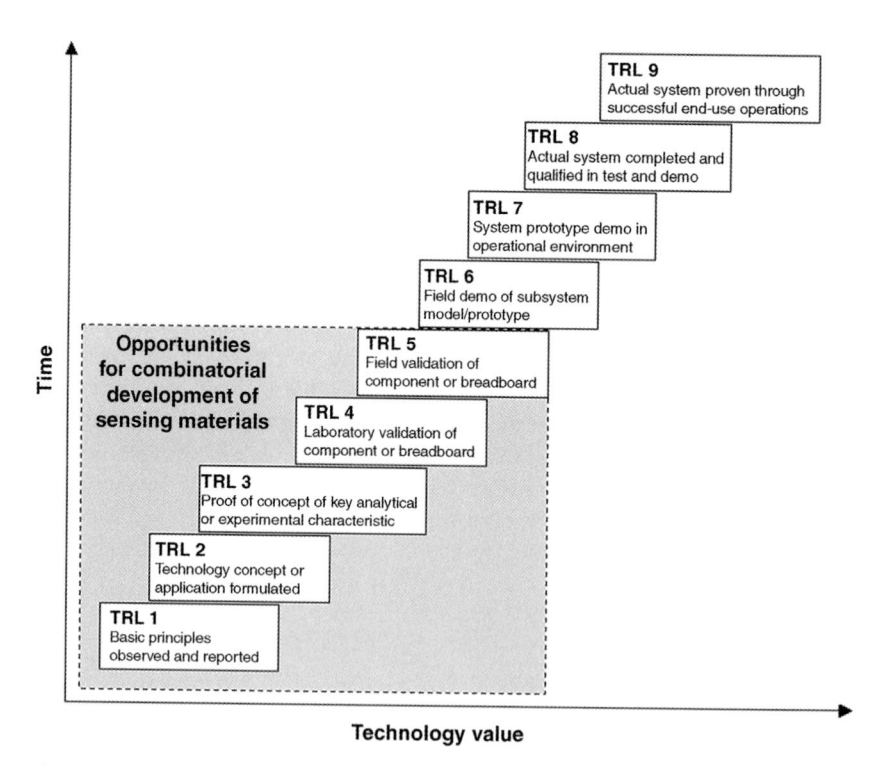

Figure 3.2 Opportunities for combinatorial development of sensing materials across the technology readiness levels.

a scale from TRL 1 (least mature) to TRL 9 (most mature) that describes the maturity of a technology with respect to a particular use. Sensor development includes several phases such as discovery with initial observations, feasibility experimentation, and laboratory-scale detailed evaluation (TRLs 1–4), followed by validation of components and the whole system prototype in the field (TRLs 5 and 6), and followed by the testing of the system prototype in the operational environment (TRL 7) and tests and end-use operation of the actual system (TRLs 8 and 9).

At the initial concept stage, performance of the sensing material is matched with the appropriate transducer for the signal generation. The stage of the laboratory-scale evaluation is very labor-intensive because it involves a detailed testing of sensor performance. Several key aspects of this evaluation include optimization of the sensing material composition and morphology, its deposition method, detailed evaluation of response accuracy, stability, precision, selectivity, shelf-life, long-term stability of the response, and key noise parameters (e.g., material instability because of temperature, potential poisons). Thus, as illustrated in Figure 3.2, combinatorial methodologies for the development of sensing materials have broad opportunities in TRLs 1–5.

3.4
Designs of Combinatorial Libraries of Sensing Materials

The broad goals of combinatorial development of sensing materials are to discover and optimize performance parameters and to optimize fabrication parameters of sensing materials. Figure 3.3 outlines the key performance and fabrication parameters of sensing materials. Factors affecting performance of sensing material films are also summarized in Figure 3.3 and can be categorized as those originating from the sample, sample–film interface, the bulk of the film, and the film–substrate interface. Depending on the real-world application, the qualities of the sensing materials can be often weighted differently. For example, response speed with millisecond time resolution is critical in gas sensors for intensive care while a much slower response speed is sufficient in home blood glucose biosensors [71, 72]. Specific requirements for medical *in vivo* sensors and bioprocess sensors include sample compatibility [73–75]. Resistance to γ-radiation during sterilization, the drift-free performance, and cost are the most critical specific requirements for sensors in disposable bioprocess components [75].

Combinatorial experimentation is performed by arranging materials candidates as discrete and gradient sensing materials arrays. A wide variety of array fabrication methods have been reported (Table 3.2) [76–104]. A specific type of library layout depends on the required density of space to be explored, available library-fabrication capabilities, and capabilities of high-throughput characterization techniques. Upon array fabrication, the array is exposed to an environment of interest and steady-state or dynamic measurements are acquired to assess materials performance. Serial scanning mode of analysis (e.g., optical or impedance spectroscopies) is often performed to provide more detailed information about materials property over parallel analysis (e.g., imaging). When monitoring a dynamic process (e.g., response/recovery time,

72 | *3 Combinatorial Development of Sensing Materials*

Goals of combinatorial screening of sensing materials

Discovery and optimization of performance parameters:
- Sensitivity
- Selectivity
- Dynamic range
- Accuracy
- Response speed
- Recovery speed
- Shelf life
- Long term stability
- Sample compatibility
- Mechanical robustness
- Resistance to poisoning
- Temperature range of operation
- Sterilizability

Optimization of fabrication parameters:
- Hazard-free material
- Initial cost
- Manufacturability

Examples of factors affecting performance of sensing materials

Sample:
- Temperature stability
- Contaminating particulates
- [Interferences]/[analyte] ratio

Sample/film interface:
- Initial morphology
- Long-term surface contamination
- Long-term surface aging

Film bulk:
- Initial film composition, microstructure
- Long-term material stratification
- Long-term aging of components

Film/substrate interface:
- Initial contact
- Long term delamination

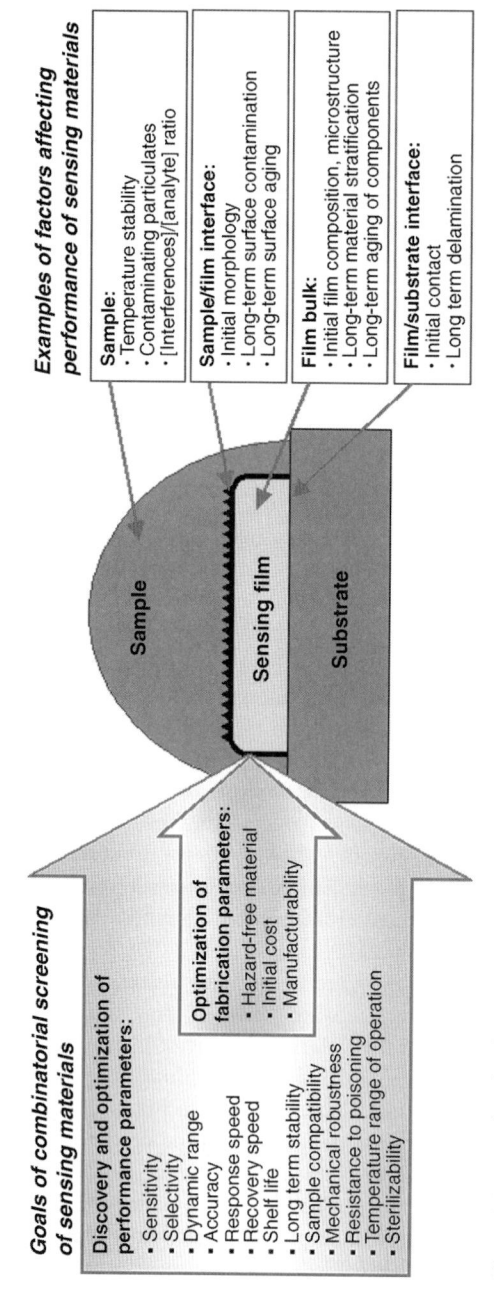

Figure 3.3 Broad goals of combinatorial development of sensing materials and examples of factors affecting materials performance.

Table 3.2 Examples of fabrication methods of discrete and gradient materials arrays.

Types of arrays of sensing materials	Fabrication methods	Reference
Discrete arrays	Ink jet printing	[76–78]
	Robotic liquid dispensing	[79, 80]
	Robotic slurry dispensing	[81]
	Microarraying	[82]
	Automated dip-coating	[83]
	Electropolymerization	[84, 85]
	Chemical vapor deposition	[86]
	Pulsed-laser deposition	[87]
	Spin coating	[88, 89]
	Screen printing	[90]
	Electrospinning	[91]
Gradient arrays	*In situ* photopolymerization	[92]
	Micro-extrusion	[93–95]
	Solvent casting	[96–98]
	Colloidal self-assembly	[99]
	Surface-grafted orthogonal polymerization	[100]
	Ink jet printing	[101]
	Temperature-gradient chemical vapor deposition	[86]
	Thickness-gradient chemical vapor deposition	[102]
	2D thickness gradient evaporation of two metals	[103]
	Gradient surface coverage and gradient particle size	[104]

aging) of sensing materials arranged in an array with a scanning system, the maximum number of elements in sensor library that can be measured with the required temporal resolution can be limited by the data-acquisition ability of the scanning system [98]. In addition to measurements of materials performance parameters, it is important to characterize intrinsic materials properties [105].

To demonstrate the broad applicability of combinatorial technologies in discovery and optimization of sensing materials, in the following sections we critically analyze results of materials development using discrete and gradient materials arrays and provide examples from a wide variety of sensors based on various energy-transduction principles that involve radiant, mechanical, and electrical types of energy.

3.5
Discovery and Optimization of Sensing Materials Using Discrete Arrays

3.5.1
Radiant Energy Transduction Sensors

Sensors based on radiant energy transduction can be categorized on the basis of the five parameters that completely describe a light-wave: its amplitude, wavelength, phase, polarization state, and time-dependent waveform. Most development of sensing materials for these types of sensors relies on the colorimetric and fluorescent materials properties. While at present organic fluorophores dominate sensing applications because of the diversity of their functionality and well-understood methods of their synthesis, new semiconducting nanocrystal labels have several advantages (photostability, relatively narrow emission spectra, and broad excitation spectra [106, 107]) over organic fluorophores. Thus, finding a solution to complement the existing organic fluorescent reagents with more photostable yet chemically or biologically responsive nanocrystals is very attractive. Various photoluminescent materials are sensitive to the local environment [108]. In particular, polished or etched bulk CdSe semiconductor crystals [109, 110] and nanocrystals [111, 112] have been shown to be sensitive to environmental changes. To better understand the environmental sensitivity of semiconductor nanocrystals upon their incorporation into polymer films, mixtures of multi-size CdSe nanocrystals have been incorporated into numerous rationally selected polymeric matrices (Table 3.3) to produce thin films. These films were further screened for their photoluminescence (PL) response to vapors of different polarity upon an excitation with a 407-nm laser [113–115].

It was discovered that CdSe nanocrystals of different size (2.8 and 5.6 nm diameter) and passivated with tri-*n*-octylphosphine oxide had dramatically different PL response patterns upon exposure to methanol and toluene after incorporation into polymeric matrices (Figure 3.4a). As an example, Figure 3.4b shows response patterns of gas-dependent PL of the two-size CdSe nanocrystals in poly(methyl methacrylate) (PMMA) sensor film. The difference in response patterns of the

Table 3.3 Polymer matrices for incorporation of different size CdSe nanocrystals [115].

Polymer #	Polymer type	Rationale for selection as sensor matrix
1	Poly(trimethylsilyl-propyne)	Polymer with largest known solubility of oxygen, candidate for efficient oxidation of CdSe nanocrystals
2	Poly(methyl methacrylate)	Polymer for solvatochromic dyes
3	Silicone block polyimide	Polymer with very high partition coefficient for sorbing organic vapors
4	Polycaprolactone	Polymer for solvatochromic dyes
5	Polycarbonate	Polymer with high T_g for sorbing of organic vapors
6	Polyisobutylene	Polymer with low T_g for sorbing of organic vapors
7	Poly(dimethylaminoethyl methacrylate)	Polymer for surface passivation of semiconductor nanocrystals
8	Polyvinylpyrrolidone	Polymer for sorption of polar vapors
9	Styrene–butadiene ABA block copolymer	Polymer for sorbing nonpolar vapors

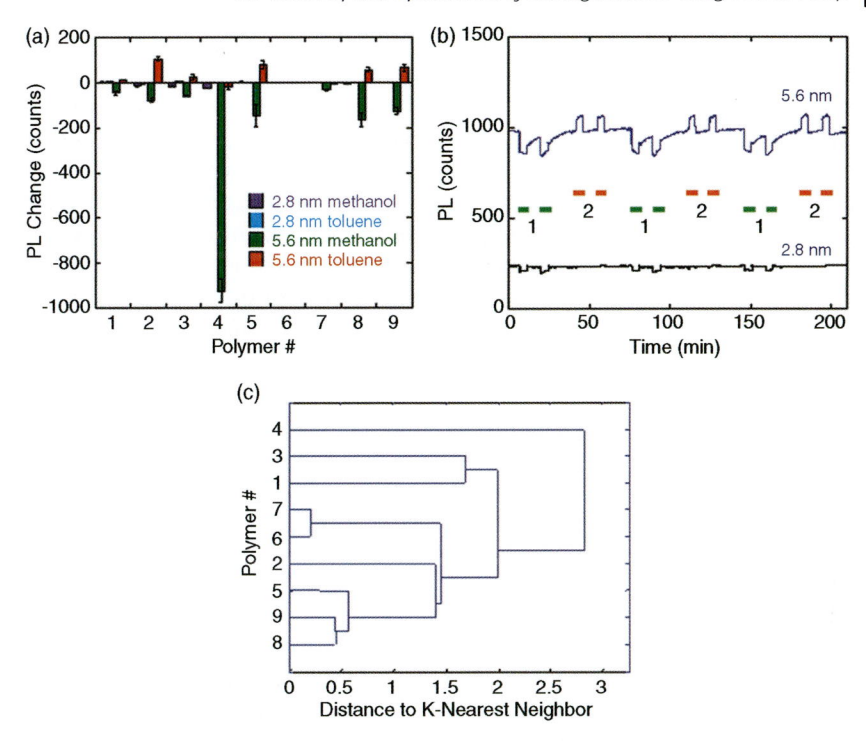

Figure 3.4 Diversity of steady-state PL response of two-size (2.8 and 5.6 nm) mixtures of CdSe nanocrystals to polar (methanol) and nonpolar (toluene) vapors. (a) Magnitude of PL change in nine polymer matrices listed in Table 3.3. (b) Gas-dependent PL of the two-size CdSe nanocrystals sensor film (polymer #2) with emission of 2.8-nm nanocrystals at 511 nm and emission of 5.6-nm nanocrystals at 617 nm. (c) Results of KNN cluster analysis of PL response patterns upon exposure to methanol and toluene after incorporation into nine polymer matrices. Numbers 1 and 2 in (b) are replicate exposures of sensor film to methanol (6% vol.) and toluene (1.5% vol.), respectively. (a) Reprinted with permission from Reference [115]. Copyright 2006 Materials Research Society. (b) Reprinted with permission from Reference [114]. Copyright 2006 American Institute of Physics.

nanocrystals was attributed to the combined effects of the dielectric medium surrounding the nanocrystals, their size, and surface oxidation state. The sensing films were tested for 16 h under a continuous laser excitation and exhibited a high stability of PL intensity [116].

To quantitatively evaluate polymer matrices, the K-nearest neighbor (KNN) cluster analysis was employed as a data-mining tool. Cluster analysis as one of the data-mining tools is often used in assessing the diversity of materials by compositional or performance properties and in developing structure–property relationship models [117, 118]. In the KNN analysis, links are made between nearest neighbors of adjoining clusters. A measure that accounts for the different scales of variables and their correlations is the Mahalanobis distance [117]. Results of cluster analysis of PL response patterns upon exposure to methanol and toluene after incorporation into polymeric matrices are demonstrated in the dendrogram in Figure 3.4c. The dendrogram was constructed by performing principal component analysis on the data from Figure 3.4a and further using the Mahalanobis distance on three principal components (PCs). From this dendrogram, it is clear that polymers 6 and 7 were the most similar in their vapor response with studied CdSe nanocrystals, as demonstrated by a very small distance to K-nearest neighbor between them. Polymer 4 was the most different from the rest of polymers, as indicated by the largest diversity distance to K-nearest neighbor. Such data-mining tools provide a means to quantitatively evaluate polymer matrices. When coupled with quantitative structure–property relationships simulation tools that will incorporate molecular descriptors, new knowledge generated from high-throughput experiments may provide additional insights for the rational design of gas sensors based on incorporated semiconductor nanocrystals. In future, such work promises to complement existing solvatochromic organic dye sensors with more photostable and reliable sensor materials.

Optimizing formulated sensor materials is a cumbersome process because theoretical predictions are often limited by practical issues, such as poor solubility and compatibility of formulation components [79, 119–121]. These practical issues represent significant knowledge gaps that prevent a more efficient rational design of formulated sensor materials. Thus, combinatorial methodologies have been demonstrated for the development of multicomponent formulated sensor materials for gaseous [79, 89, 122, 123] and ionic [80, 97, 98, 124, 125] species. Because polymer matrices [19, 92, 126–131] and plasticizers [131–135] affect the response of sensors for gases and liquids, an automated screening was applied to determine which polymers and plasticizers were best to construct oxygen-sensing materials based on Ru(4,7-diphenylphenanthroline) fluorophore. Following the initial study of screening of polymer matrices **1–9** (Figure 3.5a), focused libraries were constructed with plasticizers **10–13** to tune sensor sensitivity (Figure 3.5b). While in general the sensitivity of the sensor coatings increased with the plasticizer concentration due to an increase in the permeability of oxygen in the polymer matrix, it was unexpectedly found that plasticizers **12** and **13** showed an initial decrease of sensitivity at their low concentrations. By combining manual and automated steps in the preparation of discrete sensor film arrays, it was possible to reduce the time needed to screen sensor materials by at least 1000-fold [79].

Applying polymers with an intrinsic conductivity also permits development of chemical and biological sensors [136–140]. Various conjugated organic monomers readily undergo polymerization and form linear polymers. For example, acetylene, *p*-phenylenevinylene, *p*-phenylene, pyrrole, thiophene, furan, and aniline form conducting polymers that are widely employed in sensors [137, 141–143]. However, as prepared, conducting polymers lack selectivity and often are unstable. Thus, such polymers are chemically modified to reduce these undesirable effects. Modification methods include side-group substitution of heterocycles, doping of polymers, charge compensation upon polymer oxidation by incorporation of functionalized counterions, formation of organic–inorganic hybrids, incorporation of various biomaterials (e.g., enzymes, antibodies, nucleic acids, cells), and others

3.5 Discovery and Optimization of Sensing Materials Using Discrete Arrays

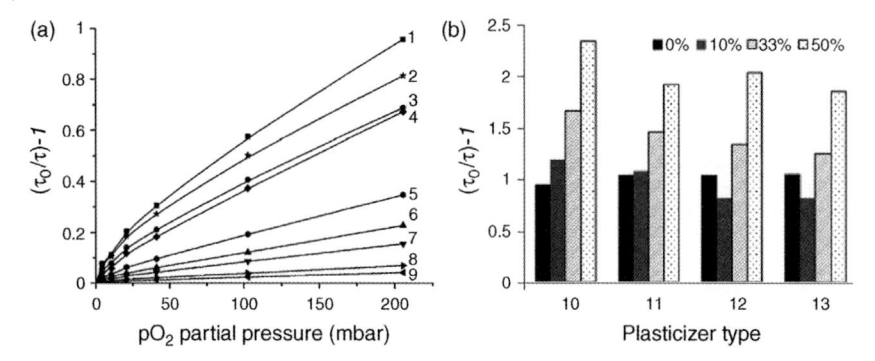

Figure 3.5 Results of combinatorial screening of steady-state responses of formulated optical gas sensor materials. (a) Stern–Volmer plots of oxygen quenching of Ru(4,7-diphenylphenanthroline)-based fluorophore in polymers **1–9** as changes in the fluorescence decay time; (b) effect of type and concentration of plasticizers **10–13** on the sensitivity of fluorescent oxygen-sensing materials in polymer **1**. Reprinted with permission from Reference [79]. Copyright 2004 American Chemical Society.

[141, 144, 145]. Variations in polymerization conditions (e.g., oxidation potential, oxidant, temperature, solvent, electrolyte concentration, monomer concentration, etc.) can be also employed to produce diverse polymeric materials from the same monomer because polymerization conditions affect sensor-related polymers properties (e.g., morphology, molecular weight, connectivity of monomers, conductivity, band gap, etc.) [137, 146].

Recently, a combinatorial approach for the colorimetric differentiation of organic solvents has been developed [91]. A polydiacetylene (PDA)-embedded electrospun fiber mat, prepared with aminobutyric acid-derived diacetylene monomer (PCDA-ABA), displayed colorimetric stability when exposed to common organic solvents. In contrast, a fiber mat prepared with the aniline-derived diacetylene (PCDA-AN) exhibited a solvent-sensitive color transition. Arrays of PDA-embedded microfibers were constructed by electrospinning poly(ethylene oxide) solutions containing various ratios of two diacetylene monomers. Unique color patterns were developed when the conjugated polymer-embedded electrospun fiber arrays were exposed to common organic solvents in a manner that enabled direct colorimetric differentiation of the tested solvents. Figure 3.6 presents results of these experiments. The scanning electron microscopy (SEM) images of electrospun fiber mats encapsulated with DA monomers prepared from pure PCDA-ABA, pure PCDA-AN, and 1 : 1 molar mixture of PCDA-ABA and PCDA-AN are presented in Figure 3.6a. No significant morphological differences were observed among these electrospun fiber mats and polymer fibers with an average diameter of $\sim 1\,\mu m$. The color patterns of the combinatorial arrays of fiber mats derived from different combinations of DA monomers (Figure 3.6b) demonstrated the significance of the combinatorial approach for sensor development. This methodology enables the generation of a compositionally diverse array of sensors, starting with only two DA monomers, for the visual differentiation of organic solvents.

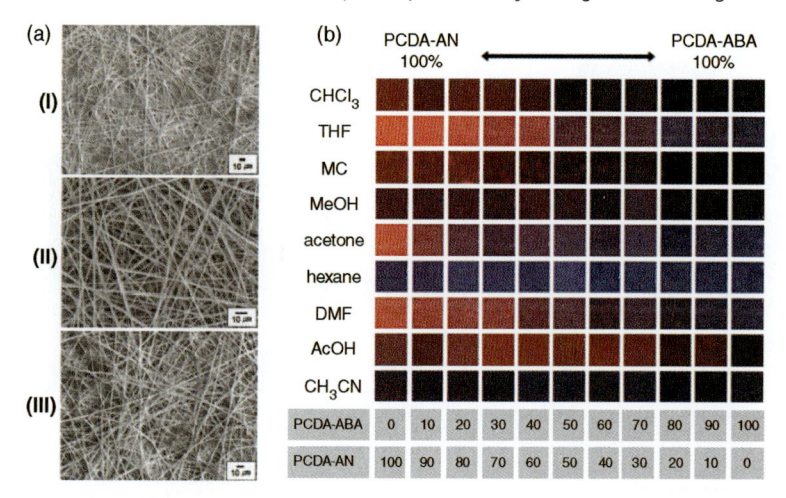

Figure 3.6 Combinatorial approach for colorimetric differentiation of organic solvents based on conjugated polymer-embedded electrospun fibers. (a) SEM images of electrospun fiber mats embedded with (I) PCDA-ABA, (II) PCDA-AN, and (III) 1 : 1 molar ratio of PCDA-ABA and PCDA-AN after UV irradiation. (b) Photographs of the polymerized PDA-embedded electrospun fiber mats after exposure to organic solvents at 25 °C for 30 s. Reprinted with permission from Reference [91]. Copyright 2009 Wiley-VCH Verlag GmbH.

3.5.2
Mechanical Energy Transduction Sensors

Sensors based on mechanical energy transduction can be categorized on the basis of the transducer functionality, which includes cantilevers and acoustic-wave devices. The mass loading and/or changes in the viscoelastic properties of the sensing materials lead to the transducer response.

A 2D multiplexed cantilever array platform has been developed for an elegant combinatorial screening of vapor responses of alkane thiols with different functional end groups [147, 148]. The cantilever sensor array chip (size 2.5 × 2.5 cm) had ∼720 cantilevers and was fabricated using surface and bulk micromachining techniques. The optical readout has been developed for parallel analysis of deflections from individual cantilevers. Figure 3.7a illustrates the general view of the developed 2D cantilever array system. To evaluate the performance of this 2D sensor array for screening of sensing materials, nonpolar and polar vapors such as toluene and water vapor were selected as analytes. The screening system was tested with three candidate alkane thiol materials as sensing films with different functional end groups: mercaptoundecanoic acid (MUA) $SH-(CH_2)_{10}-COOH$, mercaptoundecanol (MUO) $SH-(CH_2)_{11}-OH$, and dodecanethiol (DOT) $SH-(CH_2)_{11}-CH_3$. Each type of sensing films had a different chemical and physical property because -COOH group is acidic in nature and can dissociate to give a -COO- group, a -OH group does not dissociate easily but can form hydrogen bonds with polar molecules, and a -CH₃ group would be inert to polar molecules and the

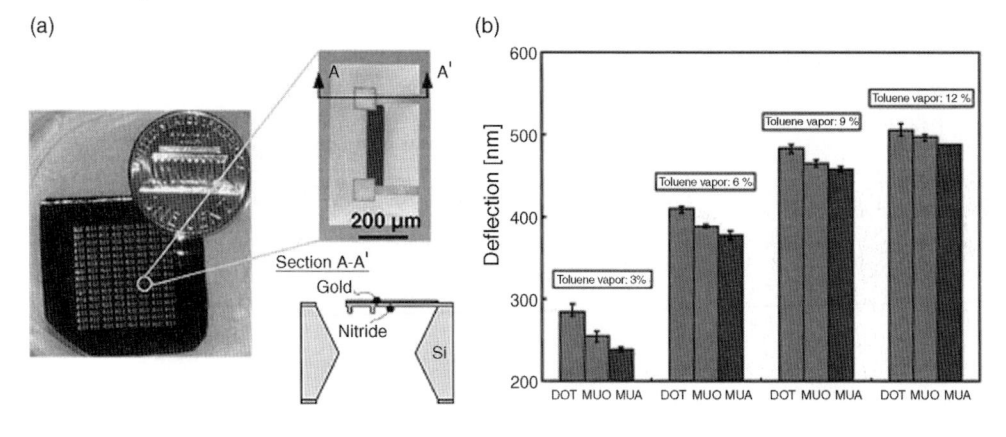

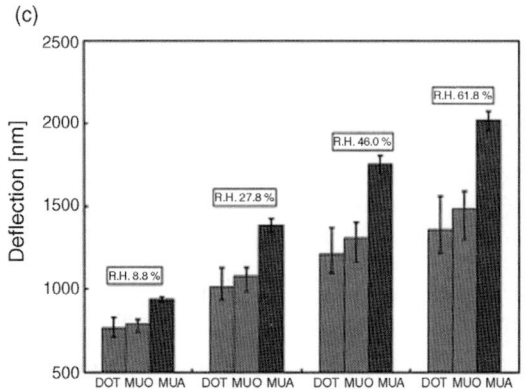

Figure 3.7 Combinatorial vapor-response screening of alkane thiols with different functional end groups using a 2D multiplexed cantilever array system. (a) General view of the fabricated cantilever array chip; (b) steady state deflection values of cantilevers upon exposure to toluene vapor at four concentration levels (3, 6, 9, and 12% by mass); (c) steady state deflection values of cantilevers upon exposure to water vapor at four concentration levels (8.8, 27.8, 46.0, and 61.8% RH). Reprinted with permission from Reference [147]. Copyright 2006 Elsevier.

only interactions that it can have are from van der Waals and hydrophobic effects. Figure 3.7b, c presents results of these experiments.

Since toluene is an organic solvent, it is likely to interact via van der Waals interaction with the thiol film on the gold side. Thus, van der Waals intermolecular interaction is generally an attractive one; it would bring thiol molecules close to toluene molecules. This in turn will bring thiol molecules closer to each other inducing shrinkage in the gold layer. This would result in upward deflection as shown in Figure 3.7b. In the case of DOT, the $-CH_3$ group being nonpolar, it would have maximum contact area with toluene to exhibit van der Waals interaction. This tendency would reduce as the end groups become more polar in nature. Hence, the $-OH$ group of MUO will have a higher van der Waals interaction than $-COOH$ group of MUA. As a result, the induced stress in the gold layer would be maximum for DOT, medium for MUO, and least for MUA. Figure 3.7c shows water vapor experiment results for various relative humidity (RH) levels. An upward deflection was recorded for all thiols, indicating that the gold film was under compression. The response ranking of three thiols to water vapor was opposite to that of the response to toluene. The largest response was of cantilevers coated with MUA, followed by those coated with MUO, and the smallest response was of cantilevers coated with DOT. This multiplexed cantilever sensor platform was further proposed as a search tool for sensing materials with improved selectivity [147].

Polymeric materials are widely used for sensing because they provide the ability for room temperature sensor operation, rapid response and recovery times, and long-term stability over several years [18, 149, 150]. In gas sensing with polymeric materials, polymer–analyte interaction mechanisms include dispersion, dipole induction, dipole orientation, and hydrogen bonding [151, 152]. These mechanisms facilitate a partial selectivity of response of different polymers to diverse vapors. An additional molecular selectivity in response is added by applying molecular imprinting of target vapor molecules into polymers and formulating polymers with molecular receptors. While there have been several models developed to calculate polymer responses [153–157], the most widely employed model is based on the linear solvation energy relationships (LSERs) [153, 154]. This LSER method has been applied as a guide to select a combination of available polymers to construct an acoustic wave sensor array based on thickness shear mode (TSM) resonators for determination of organic solvent vapors in the headspace above groundwater [158]. Field testing of the sensor system [159] demonstrated that its detection limit with available polymers was too high (several ppm) to meet the requirements for detection of groundwater contaminants. However, a new polymer has been found for sensing (silicone block polyimide **14**) that had a partition coefficient of >200 000 to part per billion (ppb) concentrations of trichloroethylene (TCE) and provided at least 100-times more sensitive response for detection of chlorinated organic solvent vapors than other known polymers [18, 160].

14

For development of materials for more selective ppb detection of chlorinated solvent vapors in presence of interferences, six families of polymeric materials have been fabricated based on polymer **14**. Performance of these polymeric materials was evaluated with respect to the differences in partition coefficients to analytes per-chloroethylene (PCE), trichloroethylene (TCE), and *cis*-dichloroethylene (cis-DCE), and to interferences (carbon tetrachloride, toluene, and chloroform). For quantitative screening of sensing materials candidates, a 24-channel TSM sensor system was built that matched a 6×4 microtiter well-plate format (Figure 3.8a, b). The sensor array was further positioned in a gas flow-through cell and kept in an environmental chamber. A comprehensive materials screening was performed with three levels

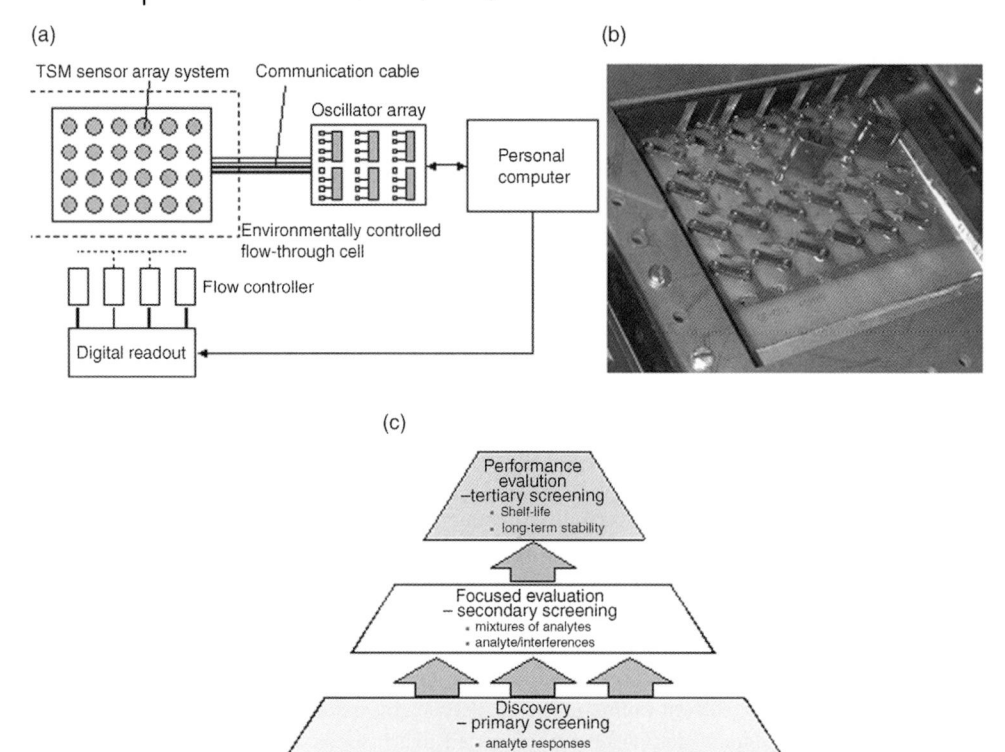

Figure 3.8 Approach for high-throughput evaluation of sensing materials for field applications. (a) Setup schematic of a 24-channel TSM sensor array for gas-sorption evaluation of sorbing polymeric films; (b) photo of 24 sensor crystals (including two reference sealed crystals) in a gas flow cell; (c) multi-level high-throughput materials screening strategy of sensing materials. (a) Reprinted with permission from Reference [162]. Copyright 2004 American Institute of Physics.

[161, 162] as shown in Figure 3.8c. In the primary (discovery) screen, materials were exposed to a single analyte concentration. In the secondary (focused) screen, the best materials subset was exposed to analytes and interferences. Finally, in the tertiary screen, remaining materials were tested under conditions mimicking long-term application. While all the screens were valuable, the tertiary screen provided the most intriguing data because aging of base polymers and copolymers is difficult or impossible to model [6]. From the tertiary screening, the decrease in materials response to the nonpolar analyte vapors and the increase in response to a polar interference vapor were quantified.

For detailed evaluation of diversity of the fabricated materials, the principal components analysis (PCA) tools [163] were applied (Figure 3.9). PCA is a multivariate data analysis tool that projects the data set onto a subspace of lower dimensionality with removed colinearity. PCA achieves this objective by explaining the variance of the data matrix in terms of the weighted sums of the original variables with no significant loss of information. These weighted sums of the original variables

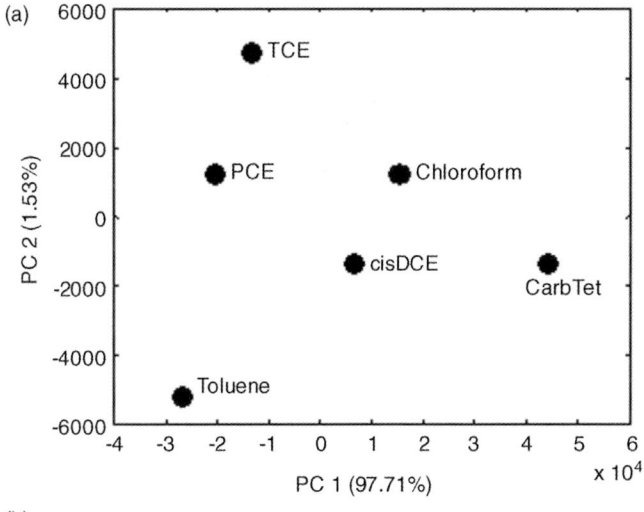

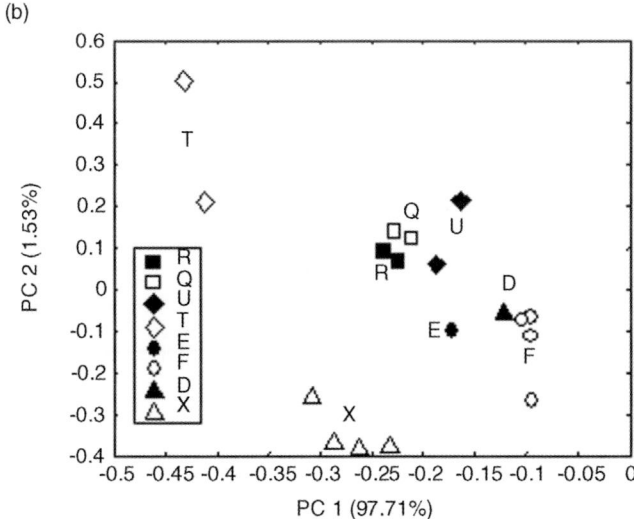

Figure 3.9 Application of PCA tools for determination of differences in the response pattern of the sensor materials toward analytes (PCE, TCE, and cis-DCE) and interferences (carbon tetrachloride, toluene, and chloroform). (a) Scores and (b) loadings plots of the first two principal components. Adapted with permission from Reference [162]. Copyright 2004 American Institute of Physics.

are called principal components (PCs). The capability for discriminating of six vapors using eight types of polymers was evaluated using a scores plot (Figure 3.9a). It demonstrated that these six vapors are well separated in the PCA space when these eight types of polymers are used for determinations. To understand what materials induce the most diversity in the response, a loadings plot was constructed (Figure 3.9b). The bigger the distance between the films of different types, the better the differences between these films. The loadings plot also demonstrates the reproducibility of the response of replicate films of the same materials. Such information served as an additional input into the materials selection for the tertiary screen. However, material selection on the basis of PCA alone does not guarantee optimal discrimination of particular vapors in the test set, because PCA measures variance, not discrimination [152]. Thus, cluster analysis tools, for example, such as those demonstrated in Figure 3.4c, can be also applied.

This 24-channel TSM sensor array system was further applied for the high-throughput screening of solvent resistance of a family of polycarbonate copolymers prepared from the reaction of Bisphenol A (BPA), hydroquinone (HQ), and resorcinol (RS) with the goal of using these copolymers as solvent-resistant supports for deposition of solvent-containing sensing formulations [164]. During the periodic exposure of the TSM crystals to polymer/solvent combinations (Figure 3.10a [83]), the mass increase of the crystal was determined, which was proportional to the amount of polymer dissolved and deposited onto the sensor from a polymer solution. The high mass sensitivity of the resonant TSM sensors (10 ng), use of only minute volume of a solvent (2 ml), and parallel operation (matching a layout of available 24 microtiter well-plates) made this system a good fit with available polymer combinatorial synthesis equipment. These parallel determinations of polymer–solvent interactions also eliminated errors associated with serial determinations. The data was further mined to construct detailed solvent-resistance maps of polycarbonate copolymers and to determine quantitative structure–property relationships (Figure 3.10b [165]). The application of this sensor-based polymer-screening system provided a lot of stimulating data that would be difficult to obtain using the conventional one-sample-at-a-time approach.

3.5.3
Electrical Energy Transduction Sensors

Sensors based on electrical energy transduction are applicable for combinatorial screening of sensing materials when these materials undergo electrically detectable changes, for example, changes in resistance or conductance during polymerization reactions and exposure to species of interest, changes in resistance due to swelling of polymers, interactions of metal oxide semiconducting surfaces with oxidizing or reducing species, and so on. Typical devices for these applications include electrochemical and electronic transducers [166–170].

The simplicity of microfabrication of electrode arrays and their subsequent application as transducer surfaces makes sensors based on electrical energy transduction among the most employed tools in combinatorial materials screening. The possibility to regulate polymerization on solid conductive surfaces by application of corresponding electrochemical potentials suggested a realization of this process in the form of multiple polymerization regions on multiple electrodes of an electronic sensor system [84, 171]. Arranging such polymerization electrodes in an array eliminated the need for dispensing systems and allowed an electrically addressable immobilization. This approach has been demonstrated on electropolymerization

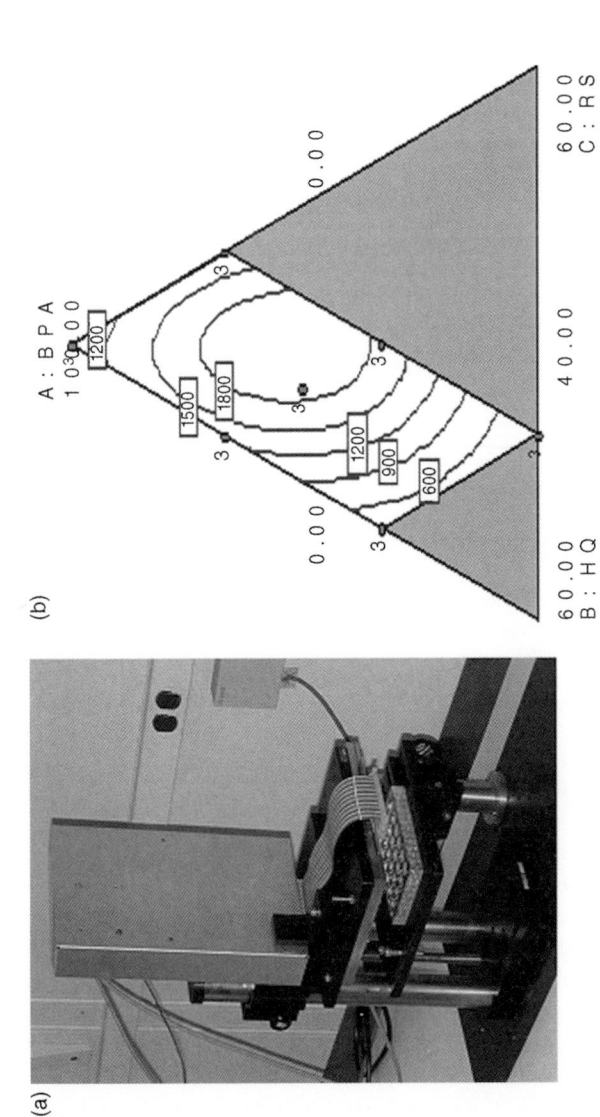

Figure 3.10 Application of the 24-channel TSM sensor array system for mapping of solvent resistance of polycarbonate copolymers. (a) General view of the screening system with a 6 × 4 microtiter well-plate positioned below the sensor array. (b) Example of property–composition mapping of solvent resistance of polycarbonate copolymers in tetrahydrofuran. Numbers in the contour lines are normalized sensor frequency shift values (Hz per mg of polymer in a well of the microtiter well-plate). (a) Reprinted with permission from Reference [83]. Copyright 2004 American Chemical Society. (b) Reprinted with permission from Reference [165]. Copyright 2006 American Chemical Society.

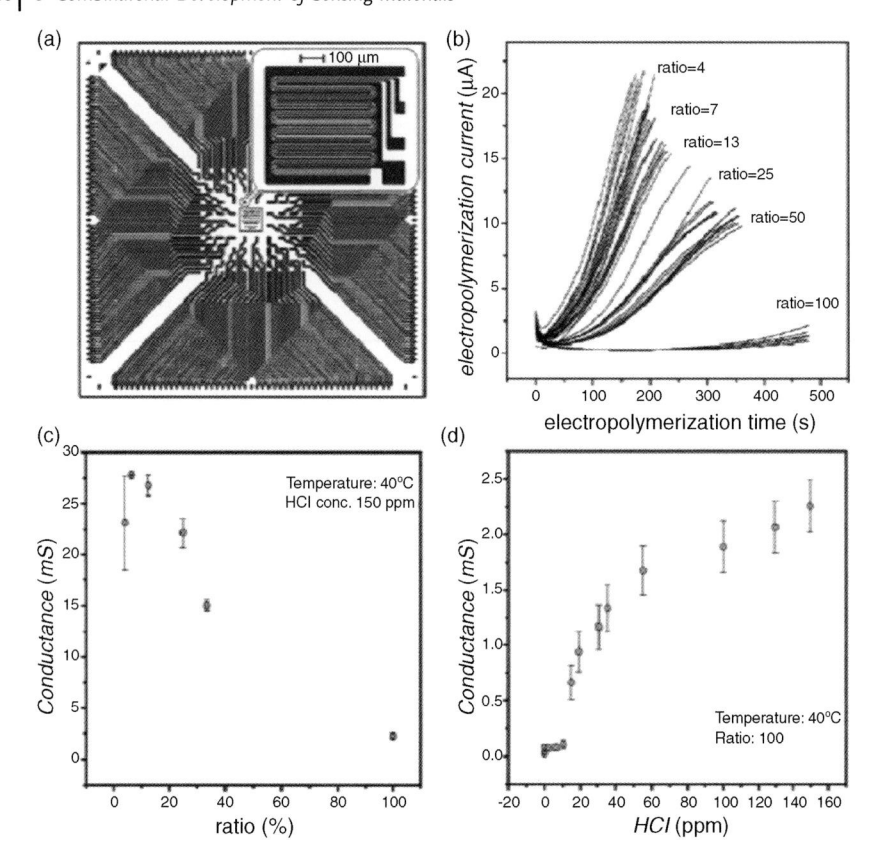

Figure 3.11 Application of a microfabricated electrode sensor array for multiple electropolymerizations and characterization of resulting conducting polymers as sensor materials. (a) Layout of the interdigital addressed electrode array. Inset: detailed structure of the single electrode for four-point measurements. (b) Current kinetics during combinatorial electropolymerization at constant potential for different ratios of non-conductive additive to conductive monomer in the polymerization mixture. (c) Dependence of conductance of one of the fabricated polymers on the ratio of non-conductive additive to conductive monomer in the polymerization mixture. (d) Dependence of conductance of one of the fabricated polymers on HCl concentration. Adapted with permission from Reference [171].

of aniline that was independently performed on different electrodes of the array [84, 171]. Thin-layer polymerization of defined mixtures of monomers was performed directly on the 96 interdigital addressed electrodes of an electrode array on an area of less than 20 mm × 20 mm (Figure 3.11a). The electrodes had an interdigital configuration designed for four-point measurements and fabricated by lithography on an oxidized silicon wafer. A computer controlled addition of analyte species provided automated investigation of the influence of different substances on the synthesized polymers. This system has been applied for combinatorial electrochemical copolymerization of mixtures of non-conductive monomer aminobenzoic acid and conductive monomer aniline at various ratios to form diverse polymers. Six groups of polymers were formed at the same conditions on 12 electrodes per group with ratios of monomers ranging from 100:1 to 4:1 to study the effects of the non-conductive aniline derivative (Figure 3.11b). Electrical characterization of synthesized polymers demonstrated that incorporation of aminobenzoic acid as a non-conductive additive in the polymer structure disturbed conductive polymer chains and led to a strong decrease of the polymer conductance (Figure 3.11c), in correspondence with predicted behavior. Exposure to HCl gas showed that polymer conductance increased with HCl concentration (Figure 3.11d).

Using this electropolymerization system, numerous copolymers have been screened. Figure 3.12 presents exemplary screening results for different binary copolymers [63]. An introduction of non-conductive monomers into polymer decreased the polymer conductance and therefore decreased the difference between

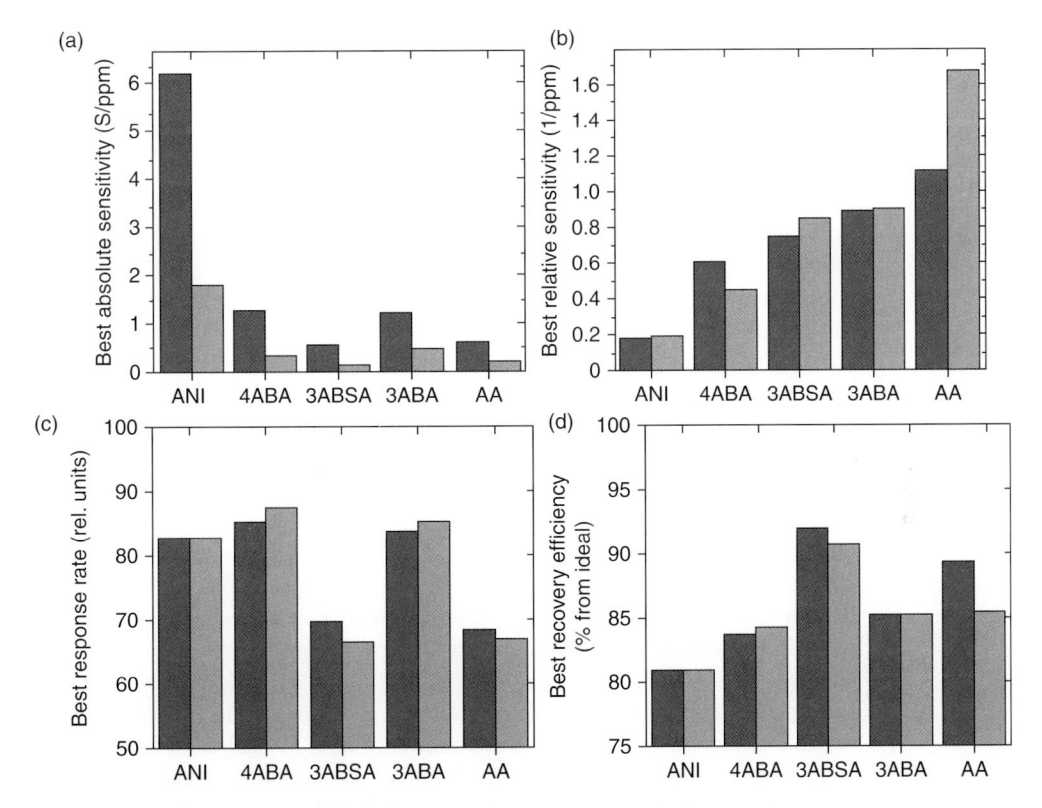

Figure 3.12 Selected results of screening of sensing materials for their response to HCl gas: (a) best absolute sensitivity, (b) best relative sensitivity, (c) best response rate, and (d) best recovery efficiency, performed by heating. Sensor materials: ANI indicates polyaniline; 4ABA, 3ABSA, 3ABA, AA indicate polymers synthesized from aniline and 4-aminobenzoic acid, 3-aminobenzenesulfonic acid, 3-aminobenzoic acid and anthranilic acid, respectively. Gray and black bars are the results obtained by two- and four-point techniques, respectively. Reprinted with permission from Reference [63]. Copyright 2008 American Chemical Society.

conductive and insulating polymer states. This caused the decrease of the absolute sensitivity (Figure 3.12a). Normalization to the polymer conductance without analyte exposure compensated this effect and demonstrated that the polymer synthesized from the mixture of aminobenzoic acid and aniline possessed the highest relative sensitivity (Figure 3.12b). This effect may be explained by the strong dependence of polymer conductance on the defect number in polymer chains. In comparison with pure polyaniline, this copolymer had better recovery efficiency but a slower response time (Figure 3.12c, d). The developed high-throughput screening system was capable of reliable ranking of sensing materials and required only ~20 min of manual interactions with the system and ~14 h of computer controlled combinatorial screening compared to ~2 weeks of laboratory work using traditional electrochemical polymer synthesis and materials evaluation [84].

Semiconducting metal oxides are another type of sensing materials that benefit from combinatorial screening technologies. Semiconducting metal oxides are typically used as gas-sensing materials that change their electrical resistance upon exposure to oxidizing or reducing gases. While, over the years, significant technological advances have been made that resulted in practical and commercially available sensors, new materials are being developed that improve further the sensing performance of these sensors. To enhance the response selectivity and stability, an accepted approach is to formulate multicomponent materials that contain additives in metal oxides. Introduction of additives into base metal oxides can change various materials properties, including concentration of charge carriers, energetic spectra of surface states, energy of adsorption and desorption, surface potential and intercrystallite barriers, phase composition, sizes of crystallites, catalytic activity of the base oxide, stabilization of a particular valence state, formation of active phases, stabilization of the catalyst against reduction, the electron exchange rate, and so on. Dopants can be added at the preparation stage (bulk dopants) that will affect the morphology, electronic properties of the base material, and its catalytic activity. However, the fundamental effects of volume dopants on base materials are not yet predictable [172]. Addition of dopants to the preformed base material (surface dopants) can lead to different dispersion and segregation effects, depending on the mutual solubility [173], and influence the overall oxidation state of the metal oxide surface [172–175].

To improve the productivity of materials evaluation by using combinatorial screening, a 36-element sensor array has been employed to evaluate various surface-dispersed catalytic additives on SnO_2 films [176, 177]. Catalysts were deposited by evaporation to nominal thicknesses of 3 nm, and then the micro-hotplates were heated to affect the formation of a discontinuous layer of catalyst particles on the SnO_2 surfaces. Figure 3.13a shows the layout of the fabricated 36-element library. The response characteristics of SnO_2 with different surface-dispersed catalytic additives are presented in Figure 3.13b. These radar plots show sensitivity results to benzene, hydrogen, methanol, and ethanol for operation at three temperatures.

To expand the capabilities of screening systems, it is attractive to characterize not only the conductance of the sensing materials with DC measurements but also their complex impedance spectra [178]. The use of complex impedance spectroscopy provides the capability to test both ion- and electron-conducting materials and to

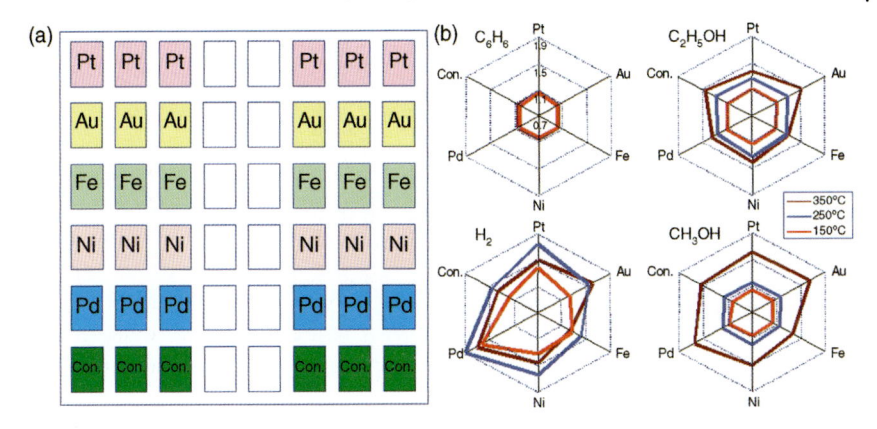

3.5 Discovery and Optimization of Sensing Materials Using Discrete Arrays | 89

Figure 3.13 Combinatorial study of effects of surface dispersion of metals into CVD-deposited SnO_2 films. (a) Layout of a 36-element library for study of the sensing characteristics of SnO_2 films with 3 nm of surface-dispersed Pt, Au, Fe, Ni, or Pd (Con. = control). Each sample was made with six replicates. (b) Radar plots of sensitivity results to benzene, hydrogen, ethanol and methanol for operation at 150, 250, and 350 °C. Reprinted from Reference [176].

study electrical properties of sensing materials that are determined by the material microstructure, such as grain boundary conductance, interfacial polarization, and polarization of the electrodes [179, 180]. A 64-multielectrode array has been designed and built for high-throughput impedance spectroscopy (10–10^7 Hz) of sensing materials (Figure 3.14a) [179]. In this system, an array of interdigital capacitors was screen-printed onto a high-temperature-resistant Al_2O_3 substrate. To ensure the high quality of determinations, parasitic effects caused by the leads and contacts have been compensated by a software-aided calibration [179]. After the system validation with doped In_2O_3 and automation of the data evaluation [180], the system was implemented for screening of various additives and matrices with the long-term goal of developing materials with improved selectivity and long-term stability. Sensing films were applied using robotic liquid-phase deposition based on optimized sol–gel synthesis procedures. Surface doping was achieved by the addition of appropriate salt solutions followed by library calcination. Screening results at 350 °C of thick films of In_2O_3 base oxide surface doped with various metals are presented as bar diagrams in Figure 3.14b [181]. It was found that some doping elements lead to changes in both the conductivity in air as well as in the gas sensing properties towards oxidizing (NO_2, NO) and reducing (H_2, CO, propene) gases. Correlations between the sensing and the electrical properties in a reference atmosphere indicated that the effect of the doping elements was due to an influence on the oxidation state of the metal oxide surface rather than to an interaction with the respective testing gases. This accelerated approach for generating reliable systematic data was further coupled to the data-mining statistical techniques that resulted in the development of (i) a model associating the sensing properties and the oxidation state of the surface layer of the metal oxide based on oxygen spillover from doping element particles to the metal oxide surface and (ii) an analytical relation for the temperature-dependent

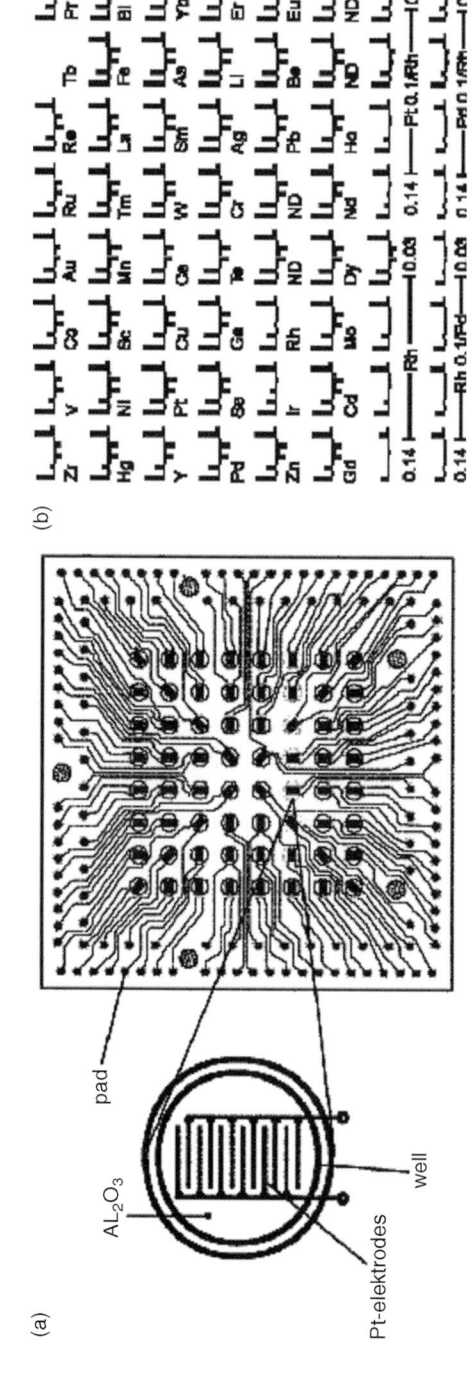

Figure 3.14 Screening of sensor metal oxide materials using complex impedance spectroscopy and a multielectrode 64-sensor array. (a) Layout of 64-sensor array. (b) Relative gas sensitivities at 350 °C of the In_2O_3 base oxide materials library surface-doped with multiple salt solutions, concentration 0.1 at.% if not denoted otherwise, ND = undoped. Sequence of test gases and their concentrations (with air in between) was H_2 (25 ppm), CO (50 ppm), NO (5 ppm), NO_2 (5 ppm), and propene (25 ppm). (a) Reprinted with permission from Reference [179]. Copyright 2002 American Chemical Society. (b) Reprinted with permission from Reference [181]. Copyright 2007 American Chemical Society.

conductivity in air and nitrogen that described the oxidation state of the metal oxide surface taking into account sorption of oxygen [181].

This high-throughput complex impedance screening system was further employed for the reliable screening of a wide variety of less explored material formulations. Polyol-mediated synthesis is an attractive method for preparation of nanoscaled metal oxide nanoparticles [182]. It requires only low annealing temperatures and provides the opportunity to tune the composition of the materials by mixing the initial components on the molecular level [183, 184]. To explore previously unknown combinations of p-type semiconducting nanocrystalline $CoTiO_3$ with different volume dopants as sensing materials, the polyol-mediated synthesis method was used to synthesize nanometer-sized $CoTiO_3$ followed by the volume-doping with Gd, Ho, K, La, Li, Na, Pb, Sb, and Sm (all at 2 at.%). The SEM-estimated primary particle size of the volume doped $CoTiO_3$ materials was in the range 30–140 nm with the smallest particle size for $CoTiO_3$:La and the largest for $CoTiO_3$:K.

The significant amount of data collected during experiments with numerous sensing materials candidates facilitated the successful efforts to develop data-mining techniques [185, 186] and a database system [187]. The developed data-mining tools based on hierarchical clustering maps (Figure 3.15) have been applied to identify several promising materials candidates such as $In_{99.5}Co_{0.5}O_x$, $W_{99}Co_{0.5}Y_{0.5}O_x$,

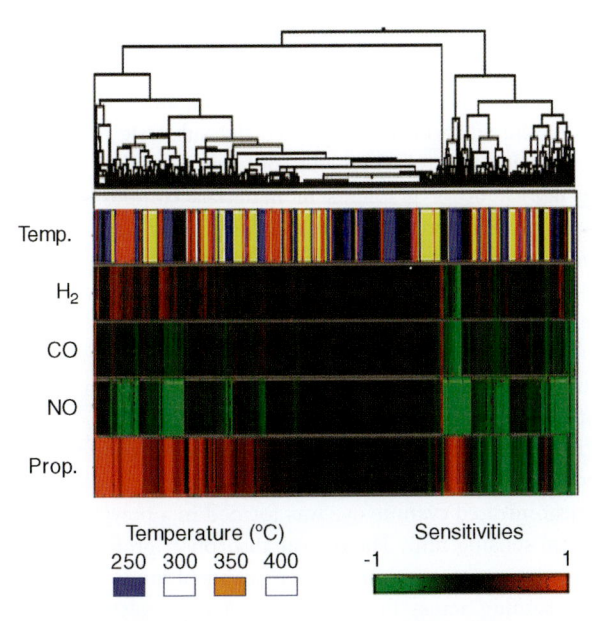

Figure 3.15 Hierarchical clustering map of 2112 responses of diverse sensing materials to H_2, CO, NO, and propene (Prop.) at four temperatures established from the high-throughput constant current measurements and processed with Spotfire data-mining software (clustering algorithm was "complete linkage" of the Euclidean distances). Reprinted with permission from Reference [188]. Copyright 2006 Molecular Diversity Preservation International.

$W_{98.3}Ta_{0.2}Y_1Mg_{0.5}O_x$, $W_{99.5}Ta_{0.5}O_x$, and $W_{99.5}Rh_{0.5}O_x$ with different gas-selectivity patterns [188].

3.6
Optimization of Sensing Materials Using Gradient Arrays

Sensor material optimization can be performed using gradient sensor materials. Spatial gradients in sensing materials can be generated by varying the nature and concentration of starting components, processing conditions, thickness, and some others factors. Once a gradient sensor array is formed, it is important to estimate the possibilities to adequately measure the variation of properties along the gradient. These can be intrinsic (thickness, chemical composition, morphology, etc.) and performance (response magnitude, selectivity, stability, immunity to poisoning, etc.) properties.

3.6.1
Variable Concentration of Reagents

Optimization of concentrations of formulation components can require significant effort because of the nonlinear relationship between additive concentration and sensor response [92, 189–195]. For detailed optimization of formulated sensor materials, concentration-gradient sensor material libraries have been employed [196]. The one, two-, and three-component composition gradients were made by flow-coating individual liquid formulations onto a flat substrate and allowing them to merge under diffusion control when still containing solvents [197]. This method combines the fabrication of gradients of materials composition with recording the materials response before and after analyte exposure and taking the ratio or difference of responses. These gradient films were applied for optimization of sensor material formulations for analysis of ionic and gaseous species [98, 197]. A very low reagent concentration in the film is expected to produce only a small signal change. The small signal change is also expected when the reagent concentration is too high. Thus, the optimal reagent concentration will depend on the analyte concentration and activity of the immobilized reagent.

Concentration optimization of a colorimetric reagent was performed in a polymer film for detection of trace concentrations of chlorine in water. A concentration gradient of a near-infrared cyanine dye was formed in a poly(2-hydroxyethyl methacrylate) hydrogel sensing film. The optical absorption profile $A_0(x)$ was obtained before analyte exposure to map the reagent concentration gradient in the film. A subsequent scanning across the gradient after the analyte exposure (1 ppm of chlorine) resulted in the determination of the optical response profile $A_E(x)$. The difference in responses, $\Delta A(x) = A_0(x) - A_E(x)$, revealed the spatial location of the optimal concentration of the reagent that produced the largest signal change (Figure 3.16a). Sensing films with the optimized concentration of the cyanine dye for chlorine determinations in industrial water were further screen-printed as a part of

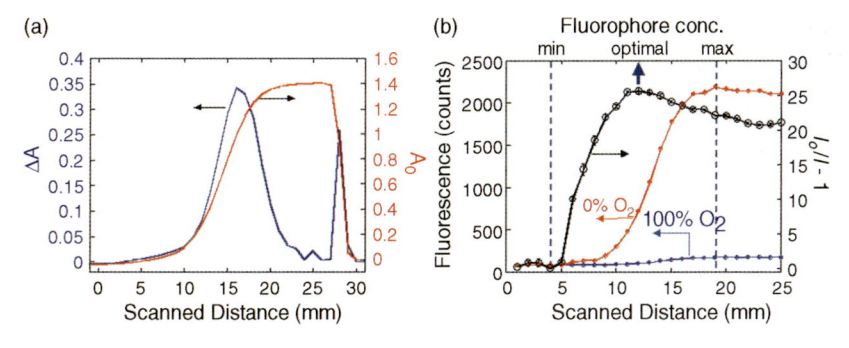

Figure 3.16 Optimization of formulated sensing materials using sensing films with gradient reagent concentration along the film length. (a) Concentration optimization of a colorimetric chlorine-responsive reagent in a formulated polymeric poly(2-hydroxyethyl methacrylate) hydrogel sensing film for detection of ions in water; exposure, 1 ppm of chlorine; (b) concentration optimization of an oxygen-responsive Pt octaethylporphyrin fluorophore in a polystyrene sensing film for detection of oxygen in air.

sensing arrays [198] onto conventional optical disks. Quantitative readout of changes in film absorbance was performed in a conventional optical disk drive in a recently developed lab-on-a-disk system [90, 198, 199].

This concentration-optimization method was also applied to optimize sensor material formulations for analysis of gaseous species [196]. Figure 3.16b shows optimization of concentration of Pt octaethylporphyrin in a polystyrene film for detection of oxygen by fluorescence quenching. This data demonstrates the simplicity, yet tremendous value, of such determinations for the rapid assessment of sensor film formulations. It shows if the optimal concentration has been reached or exceeded depending on the nonlinearity and decrease of the sensor response at the highest tested additive concentration. Unlike traditional concentration optimization approaches [192, 195], the new method provides a more dense evaluation mesh and opens opportunities for time-affordable optimization of concentration of multiple formulations components with tertiary and higher gradients [196].

3.6.2
Variable Thickness of Sensing Films

The effect of thickness of sensing films on the stability of the response in water to ionic species has also been evaluated using gradient-thickness sensing films [98]. Sensor reagent stability in a polymer matrix upon water exposure is one of the key requirements. For deposition of gradient sensor regions, several sensor coatings were flow-coated onto a 2.5-mm thick polycarbonate sheet. Typical coating dimensions were 1–1.5 cm wide and 10–15 cm long. To produce thickness gradients, the coatings were positioned vertically until the solvent evaporation in air at room temperature. The coating thickness was further evaluated using optical absorbance or profilometry. Figure 3.17a shows an example of a gradient sensor coating array. The gradient thickness of sensing films was determined from the absorbance of the

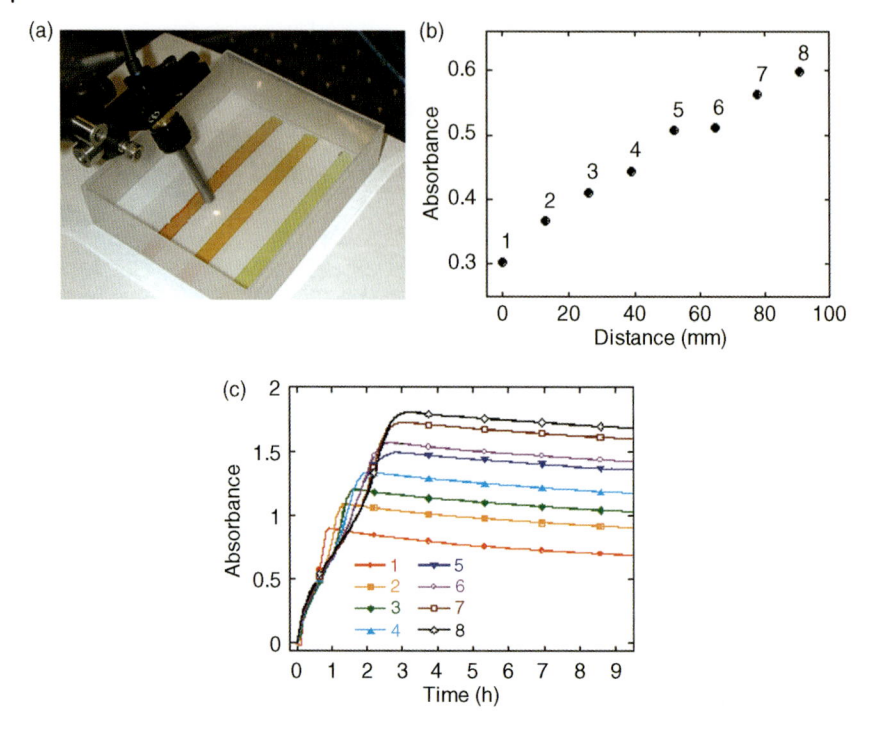

Figure 3.17 Application of gradient-thickness sensor film arrays for evaluation of reagent leaching kinetics. (a) Three gradient-thickness sensor film arrays with different loadings of an analyte-sensitive indicator; (b) film thickness as a function of film length; (c) reagent-leaching kinetics at pH 10. Reprinted with permission from Reference [98]. Copyright 2005 American Institute of Physics.

film-incorporated bromothymol blue reagent (Figure 3.17b). When these arrays were further exposed to a pH10 buffer (Figure 3.17c), an "activation" period was observed before leaching of the reagent from the polymer matrix as detected from the absorbance decrease. This activation period was roughly proportional to the film thickness. However, the leaching rate was independent of the film thickness, as indicated by the same slopes of the response curves at 3–9.5 h exposure time.

3.6.3
Variable 2D Composition

Sensors based on the change in work function of the catalytic metal gate (Pd, Pt, Rh, Ir) due to chemical reactions on the metal surface [200, 201] are attractive for detection of different gases (e.g., hydrogen sulfide, ethylene, ethanol, different amines, and others). The chemical reaction mechanisms in these sensors depend on the specific gas molecules. Optimization approaches of materials for these sensors involve several degrees of freedom [202]. To simplify screening of the desired material compositions and to reduce a common problem of batch-to-batch differences of hundreds of individually made sensors for materials development, the scanning light pulse technique (SLPT) has been developed [203–205]. In SLPT, a focused light beam is scanned over a large area semitransparent catalytic metal–insulator–semiconductor structure and the photocurrent generated in the semiconductor depletion region is measured to create a 2D response pattern of the sensing film (also known as "a chemical image").

These chemical images have been used to optimize properties such as chemical sensitivity, selectivity, and stability [103]. When combined with surface-characterization methods, this information also has led to increased knowledge of gas response phenomena. It was suggested that a 2D gradient made from two types of metal films as a double layer structure should provide new capabilities for sensor materials optimization, unavailable from thickness gradients of single metal films [206]. To make a 2D gradient, the first metal film was evaporated on the insulator with the linear thickness variation in one dimension by moving a shutter with a constant speed in front of the substrate during evaporation. On top of the first gradient thickness film, a second metal film was evaporated with a linear thickness variation perpendicular to the first film. As validation of the 2D array deposition, the response of devices with 1D thickness gradients of Pd, Pt, and Ir films to several gases has been studied with SLPT, demonstrating results similar to those of corresponding discrete components [103].

The 2D gradients have been used for studies and optimization of the two-metal structures [103, 206] and for determination of the effects of the insulator surface properties on the magnitude of sensing response [207]. Two-dimensional gradients of Pd/Rh film compositions were also studied to identify materials compositions for the most stable performance [103]. The Pd/Rh film compositions were tested for their response stability to 1000 ppm of hydrogen upon aging for 24 h at 400 °C while exposed to 250 ppm of hydrogen (Figure 3.18a, b). This accelerated aging experiment of the 2D gradient film surface demonstrated the existence of two most stable local regions. One region was a "valley" of a stable response shown as a dark color in Figure 3.18c. Another region was a thicker part of the two-component film with a ∼20-nm thick Rh film and a ∼23-nm thick Pd film. This new knowledge inspired new questions of position stability of the "valley" and the possibility to improve sensor stability by an initial annealing process.

3.6.4
Variable Operation Temperature and Diffusion Layer Thickness

At present, in conductometric sensors, semiconducting metal oxides are used as gas-sensing materials that change their electrical resistance upon exposure to oxidizing or reducing gases. While, over the years, significant technological advances have been made that resulted in practical and commercially available sensors, new materials are being developed that improve further sensing performance of these sensors. Realizing the opportunities that arise with the temperature dependence of the sensor response, temperature gradient-based sensors that utilize a single metal-oxide thin film segmented by electrodes have been developed [102, 208–213]. In addition to the

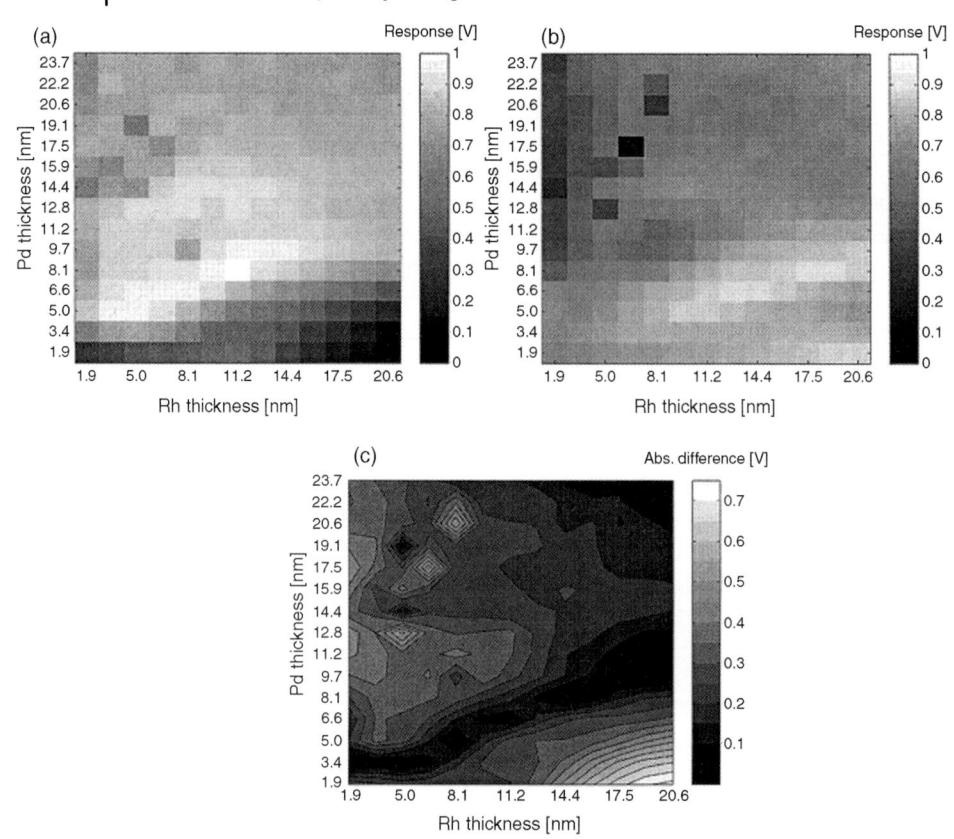

Figure 3.18 Results of the accelerated aging of a 2D combinatorial library of Rh/Pd film. Chemical response images to 1000 ppm of hydrogen (a) before and (b) after the accelerated aging; (c) differential response after and before the accelerated aging; the most stable regions have the darkest color. Reprinted with permission from Reference [103]. Copyright 2005 The Institute of Electrical and Electronics Engineers, Inc.

spatial temperature gradient heater, one of designs of the sensor chip also had a SiO_2 or Al_2O_3 membrane with a gradient thickness from 2 to 50 nm (Figure 3.19a) [214]. Such ceramic membrane provided an additional response selectivity [215] through the thickness-dependent gas transport.

To fabricate such a temperature- and membrane-gradient sensor, a gas-sensitive SnO_2:Pt film (Pt content of 0.8 at.%) was deposited onto a thermally oxidized Si wafer by RF magnetron sputtering using a shadow mask. Next, Pt strip electrodes and two meander-shaped thermoresistors were sputtered on the same side of the substrate as the SnO_2 film, under a shadow mask for structuring the films. The arrangement of the electrodes subdivided the monolithic SnO_2 film into 38 sensor segments on an area of $4 \times 8\,mm^2$. Finally, Pt heaters were deposited onto the backside of the substrate to operate the chip with A 50 °C temperature gradient from 310 to 360 °C (Figure 3.19b) [102]. The application of a temperature gradient increased the gas

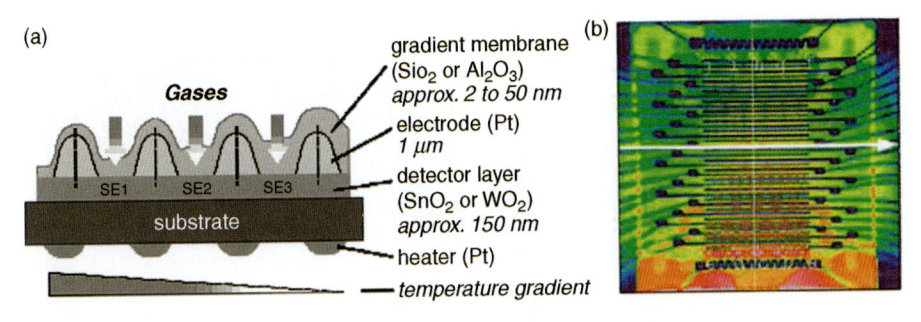

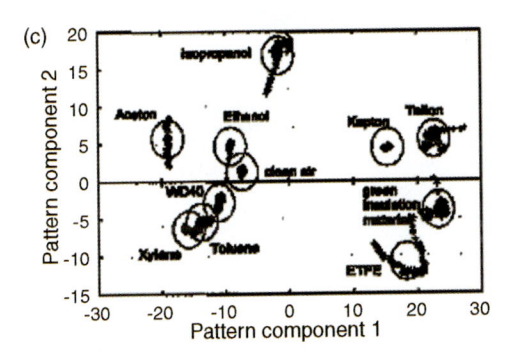

Figure 3.19 Double gradient sensor microarray for selective gas detection. (a) Sensor schematic illustrating a single metal-oxide thin film segmented by electrodes and arranged on a temperature gradient heater. The sensing film is further covered with a gradient thickness ceramic membrane. (b) False color thermal infrared image of the heated gradient sensor array with a temperature gradient of 6.7 °C mm^{-1}. The white arrow depicts the airflow direction. (c) Results of the linear discrimination analysis of the signal patterns in practical tests to detect gaseous precursors of smoldering fires induced by overheated cable insulation (ETFE: ethylene tetrafluorine ethylene). (a) Used with kind permission from Goschnick. (b) Reprinted with permission from Reference [102]. Copyright 2004 Molecular Diversity Preservation International. (c) Reprinted with permission from Reference [211]. Copyright 2002 The Institute of Electrical and Electronics Engineers, Inc.

discrimination power of the sensor by 35%. The sensor with a SiO$_2$ gradient-thickness membrane was employed for detection of gaseous precursors of smoldering fires induced by overheated cable insulation (Figure 3.19c) [211].

3.7
Emerging Wireless Technologies for Combinatorial Screening of Sensing Materials

At present, advances in wireless electronic technologies promise to add new attractive capabilities for combinatorial screening of sensing materials. Wireless proximity chemical, biological, and physical sensors provide several key advantages over wired sensors such as (i) non-contact and non-contamination measurements, (ii) operation though packaging, and (iii) rapid reading of multiple sensors with a single reader.

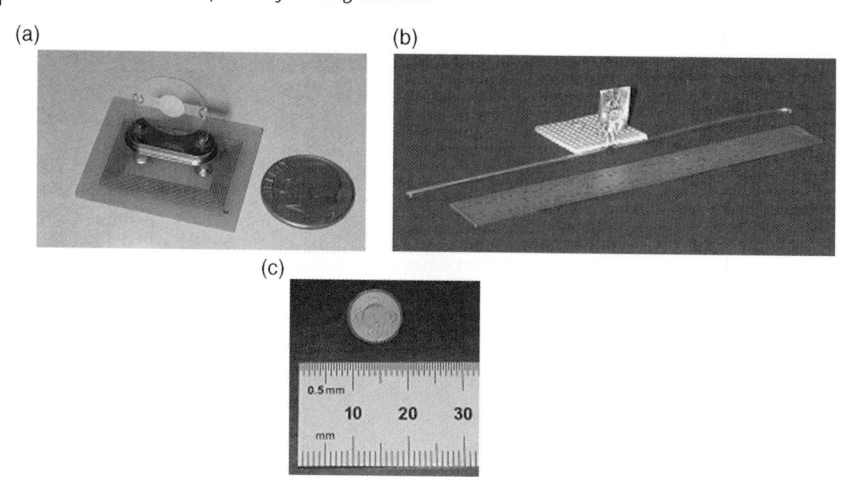

Figure 3.20 Examples of proximity wireless sensors applicable for combinatorial screening of sensing materials. (a) TSM sensor resonating at 10 MHz, (b) SAW sensor resonating at 915 MHz, and (c) RFID sensor resonating at 13.56 MHz.

Several proximity-sensing approaches based on thickness shear mode (TSM) and surface-acoustic wave (SAW) sensors connected to antennas as well as passive radio frequency identification (RFID) sensors have been developed (Figure 3.20) with possible applications for combinatorial screening of sensing materials.

To eliminate the direct wiring of individual TSM sensors and to permit materials evaluation in environments where wiring is not desirable or adds a prohibitively complex design, a wireless TSM sensor array system [216] has been developed where each sensor resonator was coupled to a receiver antenna coil and an array of these coils was scanned with a transmitter coil (Figure 3.21a). Using this sensor wireless system, sensing materials can be screened for their gas-sorption properties, analyte-binding in liquids, and for the changes in chemical and physical properties upon weathering and aging tests. The applicability of the wireless sensor materials screening approach has been demonstrated for the rapid evaluation of the effects of conditioning of polymeric sensing films at different temperatures on the vapor-response patterns. One set of high-throughput screening experiments studied Nafion film-aging effects on the selectivity pattern. Evaluation of this and many other polymeric sensing materials lacks detailed studies on the change of the chemical selectivity patterns as a function of temperature conditioning and aging. Conditioning of Nafion-coated resonators was performed at 22, 90, and 125 °C for 12 h. Temperature-conditioned sensing films were exposed to water (H_2O), ethanol (EtOH), and acetonitrile (ACN) vapors, all at concentrations (partial pressures) ranging from 0 to 0.1 of the saturated vapor pressure P_o (Figure 3.21b–d). Conditioning of sensing films at 125 °C compared to room temperature conditioning provided (i) an improvement in the linearity in response to EtOH and ACN vapors, (ii) an increase in relative response to ACN, and (iii) a ten-fold increase of the contribution to principal component #2. The latter point signifies an improvement

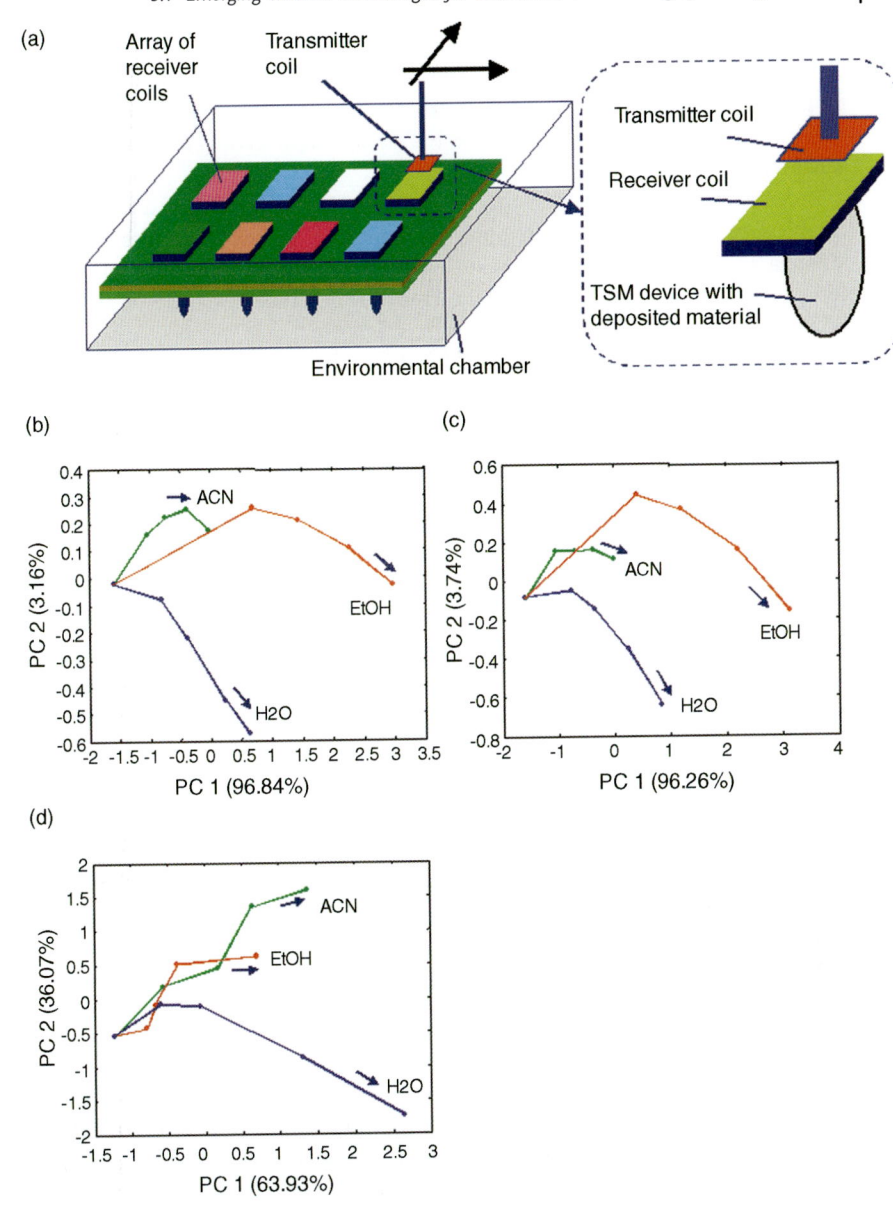

Figure 3.21 Concept for wireless high-throughput screening of materials properties using thickness shear mode resonators: (a) Configuration of a wireless proximity resonant sensor array system for high-throughput screening of sensing materials with a single transmitter coil that scans across an array of receiver coils attached to resonant sensors. (b)–(d) Evaluation of selectivity of Nafion sensing films to several vapors after conditioning at different temperatures: (b) 22, (c) 90, and (d) 125 °C. Vapors: H_2O (water), EtOH (ethanol), and ACN (acetonitrile). Concentrations of vapors are 0, 0.02, 0.04, 0.07, and $0.10P/P_o$. Arrows indicate the increase of concentrations of each vapor. Reprinted with permission from Reference [216]. Copyright 2007 American Institute of Physics.

in the discrimination ability between different vapors upon conditioning of the sensing material at 125 °C. This new knowledge will be critical in designing sensors for practical applications where the need exists to preserve sensor response selectivity over a long exploitation time or when there is a temperature cycling for an accelerated sensor-film recovery after vapor exposure.

Recently, ubiquitous and cost-effective passive RFID tags have been adapted for chemical sensing [217, 218]. By applying a sensing material onto the resonant antenna of the RFID tag and measuring the complex impedance of the RFID resonant antenna it was possible to correlate impedance response to chemical properties of interest. When a sensing film is deposited onto the resonant antenna (Figure 3.22a), the analyte-induced changes in the dielectric and dimensional properties of this sensing film affect the complex impedance of the antenna circuit through the changes in film resistance and capacitance between the antenna turns. Such changes provide selectivity in response of an individual RFID sensor and provide the opportunity to replace a whole array of conventional sensors with a single RFID sensor [217]. For this selective analyte quantitation using individual RFID sensors, complex impedance spectra of the resonant antenna are measured. Several parameters from the measured real and imaginary portions of the complex impedance are further calculated. These parameters include F_p and Z_p (the frequency and magnitude of the maximum of the real part of the complex impedance, respectively) and F_1 and F_2 (the resonant and antiresonant frequencies of the imaginary part of the complex impedance, respectively). By applying multivariate analysis of the full complex impedance spectra or the calculated parameters, quantitation of analytes and rejection of interferences is performed with individual RFID sensors.

Because temperature effects are important at all stages of sensor fabrication, testing, and end-use, understanding of temperature effects can provide ability to build robust temperature-corrected transfer functions of sensor performance in order to preserve response sensitivity, response selectivity, and response baseline stability. RFID sensors have been applied for the combinatorial screening of sensing materials to evaluate combined effects of plasticizers in polymeric formulated films and annealing temperature [219]. As a model system, a 6 × 8 array of polymer-coated RFID sensors was constructed as shown in Figure 3.22b. A solid polymer electrolyte Nafion was formulated with five types of phthalate plasticizers such as dimethyl phthalate (15), butyl benzyl phthalate (16), di-(2-ethylhexyl) phthalate (17), dicapryl phthalate (18), and diisotridecyl phthalate (19).

3.7 Emerging Wireless Technologies for Combinatorial Screening of Sensing Materials | 101

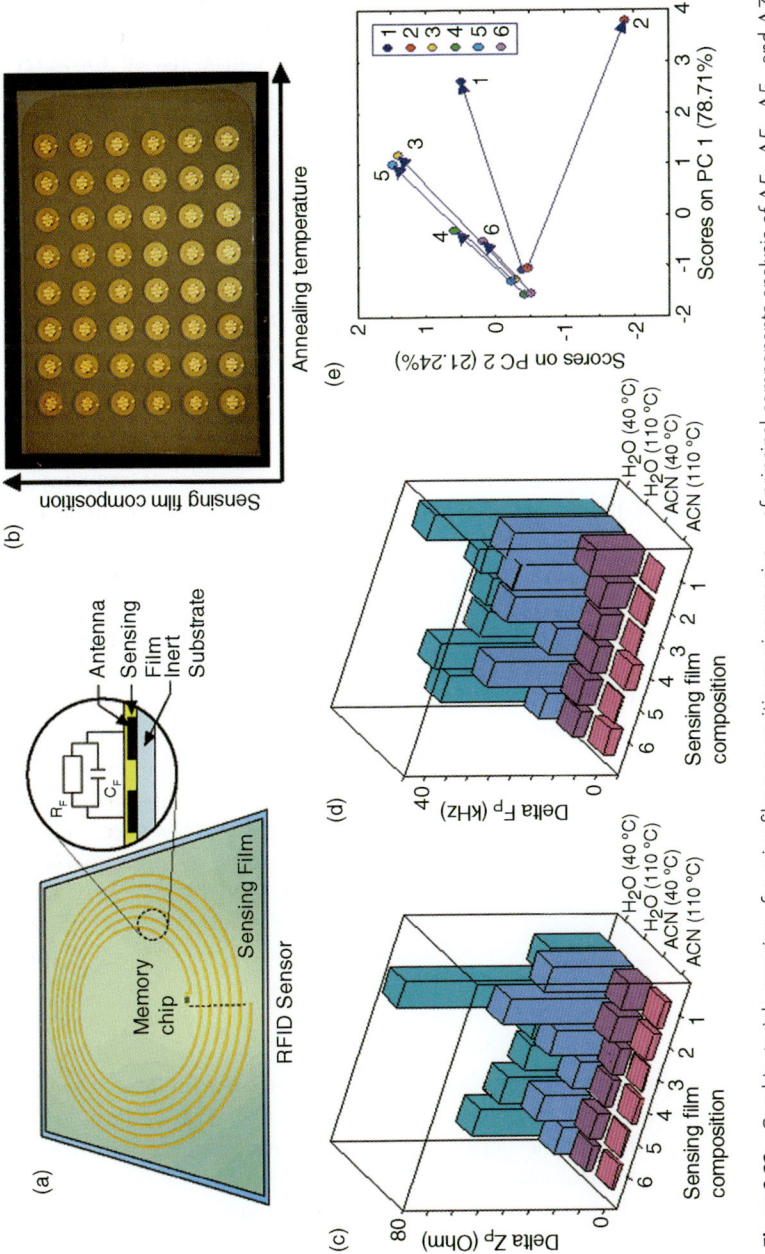

Figure 3.22 Combinatorial screening of sensing film compositions using passive RFID sensors. (a) Strategy for adaptation of conventional passive RFID tags for chemical sensing by deposition of a sensing film onto the resonant circuit of the RFID antenna. Inset: analyte-induced changes in the film material affect film resistance (R_F) and capacitance (C_F) between the antenna turns. (b) Photo of an array of 48 RFID sensors prepared for temperature-gradient evaluations of response of Nafion/phthalates compositions. (c and d) Examples of RFID sensors response to water (H_2O) and acetonitrile (ACN) vapors after annealing at different temperatures (40 and 110 °C); (c) ΔZ_P response and (d) ΔF_P response. (e) Results of principal components analysis of ΔF_1, ΔF_2, ΔF_P, and ΔZ_P responses of RFID sensors with sensing films 1–6 to H_2O and ACN vapors upon annealing at 110 °C. Arrows illustrate the H_2O–ACN Euclidean distances and the response direction of sensing films 1–6. Arrows begin at the response of the sensing film to ACN and end at the response of the sensing film to H_2O. Nafion sensing film compositions: 1, control without plasticizer; 2, dimethyl phthalate; 3, butyl benzyl phthalate; 4, di-(2-ethylhexyl) phthalate; 5, dicapryl phthalate; 6, diisotridecyl phthalate. Reprinted with permission from Reference [219]. Copyright 2009 American Chemical Society.

These sensing film formulations and control sensing films without a phthalate plasticizer were deposited onto RFID sensors, exposed to eight temperatures ranging from 40 to 140 °C using a gradient temperature heater, and evaluated for their response stability and gas-selectivity response patterns. Interrogation of RFID sensors in the array was carried out with a single transmitter (pick-up antenna) coil positioned on an X-Y translation stage and connected to a network analyzer.

To evaluate temperature effects on sensor response selectivity, the 6 × 8 array of temperature-annealed sensing films was exposed to H_2O and ACN vapors. Acetonitrile was selected as a simulant for blood chemical warfare agents (CWAs) [220] while water vapor was selected as an interference. In these experiments, the partial pressures of H_2O and ACN vapors were 0.4 of the saturated vapor pressure P_0. Nafion sensing films were used previously for humidity [221, 222] and organic vapors [223] detection. Conductance and dielectric properties of Nafion have been shown to be vapor dependent [224, 225]. Figure 3.22c and d shows representative ΔZ_p and ΔF_p responses, respectively, to H_2O and ACN upon annealing at two temperatures, 40 and 110 °C. The patterns of ΔF_1, ΔF_2, ΔF_p, and ΔZ_p responses of sensing films to H_2O and ACN vapors upon temperature annealing were further examined using PCA (Figure 3.22e). Arrows illustrate the H_2O–ACN Euclidean distances and the response direction of sensing films 1–6. Arrows begin at the response of the sensing film to ACN and end at the response of the sensing film to H_2O. The length of the arrows (the H_2O–ACN Euclidean distances) indicates response diversity of sensing films 1–6 to H_2O and ACN. The larger the Euclidean distance, the better the sensing material is in its response diversity to H_2O and ACN. From these screening experiments, it was found that different plasticizers affect the response diversity to a different extent; however, Nafion sensing films formulated with the dimethyl phthalate plasticizer improve response diversity of the sensing films to H_2O and ACN. Overall, this study demonstrated that this RFID-based sensing approach permits rapid, cost-effective combinatorial screening of dielectric properties of sensing materials. As pointed out earlier [93], in general, the increase of the level of environmental stress may be problematic if the correlation with traditional test methods is lost. To avoid this situation, it will be critical to plan the detailed accelerated aging high-throughput experiments with positive and negative controls.

3.8
Summary and Outlook

Combinatorial technologies in materials science have been successfully accepted by research groups in academia and governmental laboratories that have overcome the entry barrier of dealing with new emerging aspects in materials research such as automation and robotics, computer programming, informatics, and materials data mining. The main driving forces for combinatorial materials science in industry include broader and more detailed explored materials and process parameters space and faster time to market. Industrial research laboratories working on new catalysts and inorganic luminescent materials were among the first adopters of combinatorial methodologies in industry. The classical example of an effort by Mittasch who spent 10 years (over 1900–1909) conducting 6500 screening experiments with 2500 catalyst candidates to find a catalyst for industrial ammonia synthesis [226] will never happen again because of the availability and affordability of modern tools for high-throughput synthesis and characterization.

In the area of sensing materials, reported examples of significant screening efforts are less dramatic, yet also breathtaking. For example, over a decade ago, Cammann, Shulga, and coworkers [227] reported an "extensive systematic study" of more than 500 compositions to optimize vapor sensing polymeric materials. Walt and coworkers [19] reported screening of over 100 polymer candidates in a search for "their ability to serve as sensing matrices" for solvatochromic reagents. Seitz and coworkers [130] have investigated the influence of multicomponent compositions on the properties of pH-swellable polymers by designing $3 \times 3 \times 3 \times 2$ factorial experiments. Clearly, combinatorial technologies have been introduced at the right time to make the search for new materials more *intellectually rewarding*. Naturally, numerous academic groups that were involved in the development of new sensing materials turned to combinatorial methodologies to speed up knowledge discovery [79, 85, 92, 179, 205, 228, 229].

From numerous results achieved using combinatorial and high-throughput methods, the most successful have been in the areas of molecular imprinting, polymeric compositions, catalytic metals for field-effect devices, and metal oxides for conductometric sensors. In those materials, the desired selectivity and sensitivity have been achieved by the exploration of multidimensional chemical composition and process parameters space at a previously unavailable level of detail at a fraction of time required for conventional one-at-a-time experiments. These new tools provided the opportunity for the more challenging, yet more rewarding explorations that previously were too time-consuming to pursue.

Future advances in combinatorial development of sensing materials will be related to several key remaining unmet needs that prevent researchers having a complete combinatorial workflow and to "analyze in a day what is made in a day" [69]. At present, data management of the combinatorial workflow is, perhaps, the weakest link. However, over the last several years, there have been a growing number of reports on data mining in sensing materials [165, 188, 230, 231]. "Searching for a needle in the haystack" was popular in the early days of combinatorial materials science [21, 232, 233]. At present, it has been realized that screening of the whole materials and process parameters space is still too costly and time prohibitive even with the availability of existing tools. Instead, designing the high-throughput experiments to discover relevant descriptors will become more attractive [234].

A modern combinatorial scientist is acquiring skills as diverse as experimental planning, automated synthesis, basics of high-throughput materials characterization, chemometrics, and data mining. These new skills can now be obtained through the growing network of practitioners and through the new generation of scientists educated across the world in combinatorial methodologies. Combinatorial and high-throughput experimentation was able to bring together several previously disjoined disciplines and to combine valuable complementary attributes from each of them into a new scientific approach.

Acknowledgments

We gratefully acknowledge GE components for support of our combinatorial sensor research.

References

1. Njagi, J., Warner, J., and Andreescu, S. (2007) *J. Chem. Educ.*, **84**, 1180–1182.
2. Shtoyko, T., Zudans, I., Seliskar, C.J., Heineman, W.R., and Richardson, J.N. (2004) *J. Chem. Educ.*, **81**, 1617–1619.
3. Honeybourne, C.L. (2000) *J. Chem. Educ.*, **77**, 338–344.
4. Newnham, R.E. (1988) *Cryst. Rev.*, **1**, 253–280.
5. Akporiaye, D.E. (1998) *Angew. Chem. Int. Ed.*, **37**, 2456–2457.
6. Ulmer, C.W. II, Smith, D.A., Sumpter, B.G., and Noid, D.I. (1998) *Comput. Theor. Polym. Sci.*, **8**, 311–321.
7. Suman, M., Freddi, M., Massera, C., Ugozzoli, F., and Dalcanale, E. (2003) *J. Am. Chem. Soc.*, **125**, 12068–12069.
8. Lavigne, J.J. and Anslyn, E.V. (2001) *Angew. Chem. Int. Ed.*, **40**, 3119–3130.
9. Hatchett, D.W. and Josowicz, M. (2008) *Chem. Rev.*, **108**, 746–769.
10. Schultz, P.G. (2003) *Appl. Catal., A*, **254**, 3–4.
11. McKusick, B.C., Heckert, R.E., Cairns, T.L., Coffman, D.D., and Mower, H.F. (1958) *J. Am. Chem. Soc.*, **80**, 2806–2815.
12. Bühlmann, P., Pretsch, E., and Bakker, E. (1998) *Chem. Rev.*, **98**, 1593–1687.
13. Steinle, E.D., Amemiya, S., Bühlmann, P., and Meyerhoff, M.E. (2000) *Anal. Chem.*, **72**, 5766–5773.
14. Pedersen, C.J. (1967) *J. Am. Chem. Soc.*, **89**, 7017–7036.
15. Hu, Y., Tan, O.K., Pan, J.S., and Yao, X. (2004) *J. Phys. Chem. B*, **108**, 11214–11218.
16. Svetlicic, V., Schmidt, A.J., and Miller, L.L. (1998) *Chem. Mater.*, **10**, 3305–3307.
17. Martin, P.D., Wilson, T.D., Wilson, I.D., and Jones, G.R. (2001) *Analyst*, **126**, 757–759.
18. Potyrailo, R.A. and Sivavec, T.M. (2004) *Anal. Chem.*, **76**, 7023–7027.
19. Walt, D.R., Dickinson, T., White, J., Kauer, J., Johnson, S., Engelhardt, H., Sutter, J., and Jurs, P. (1998) *Biosens. Bioelectron.*, **13**, 697–699.
20. Eberhart, M.E. and Clougherty, D.P. (2004) *Nat. Mater.*, **3**, 659–661.
21. Jandeleit, B., Schaefer, D.J., Powers, T.S., Turner, H.W., and Weinberg, W.H. (1999) *Angew. Chem. Int. Ed.*, **38**, 2494–2532.
22. Maier, W., Kirsten, G., Orschel, M., Weiß, P.-A., Holzwarth, A., and Klein, J. (2002) Combinatorial chemistry of materials, polymers, and catalysts, in *Combinatorial Approaches to Materials Development*, vol. 814 (ed. R. Malhotra), American Chemical Society, Washington, DC, pp. 1–21.
23. Takeuchi, I., Newsam, J.M., Wille, L.T., Koinuma, H., and Amis, E.J. (eds) (2002) *Combinatorial and Artificial Intelligence Methods in Materials Science*, vol. 700, Materials Research Society, Warrendale, PA.
24. Xiang, X.-D. and Takeuchi, I. (eds) (2003) *Combinatorial Materials Synthesis*, Marcel Dekker, New York.
25. Potyrailo, R.A. and Amis, E.J. (eds) (2003) *High Throughput Analysis: A Tool for Combinatorial Materials Science*,

Kluwer Academic/Plenum Publishers, New York.
26 Koinuma, H. and Takeuchi, I. (2004) *Nat. Mater.*, **3**, 429–438.
27 Potyrailo, R.A., Karim, A., Wang, Q., and Chikyow, T. (eds) (2004) *Combinatorial and Artificial Intelligence Methods in Materials Science II*, vol. 804, Materials Research Society, Warrendale, PA.
28 Potyrailo, R.A. and Takeuchi, I. (eds) (2005) Special feature on combinatorial and high-throughput materials research. *Meas. Sci. Technol.*, **16**, 316.
29 Potyrailo, R.A. and Maier, W.F. (eds) (2006) *Combinatorial and High-Throughput Discovery and Optimization of Catalysts and Materials*, CRC Press, Boca Raton, FL.
30 Birina, G.A., and Boitsov, K.A. (1974) *Zavod. Lab. (in Russian)*, **40**, 855–857.
31 Kennedy, K., Stefansky, T., Davy, G., Zackay, V.F., and Parker, E.R. (1965) *J. Appl. Phys.*, **36**, 3808–3810.
32 Hoffmann, R. (2001) *Angew. Chem. Int. Ed.*, **40**, 3337–3340.
33 Hoogenboom, R., Meier, M.A.R., and Schubert, U.S. (2003) *Macromol. Rapid Commun.*, **24**, 15–32.
34 Anderson, F.W. and Moser, J.H. (1958) *Anal. Chem.*, **30**, 879–881.
35 Eash, M.A. and Gohlke, R.S. (1962) *Anal. Chem.*, **34**, 713.
36 Hanak, J.J. (1970) *J. Mater. Sci.*, **5**, 964–971.
37 Xiang, X.-D., Sun, X., Briceño, G., Lou, Y., Wang, K.-A., Chang, H., Wallace-Freedman, W.G., Chen, S.-W., and Schultz, P.G. (1995) *Science*, **268**, 1738–1740.
38 Chang, H., Gao, C., Takeuchi, I., Yoo, Y., Wang, J., Schultz, P.G., Xiang, X.-D., Sharma, R.P., Downes, M., and Venkatesan, T. (1998) *Appl. Phys. Lett.*, **72**, 2185–2187.
39 Briceño, G., Chang, H., Sun, X., Schultz, P.G., and Xiang, X.-D. (1995) *Science*, **270**, 273–275.
40 Danielson, E., Devenney, M., Giaquinta, D.M., Golden, J.H., Haushalter, R.C., McFarland, E.W., Poojary, D.M., Reaves, C.M., Weinberg, W.H., and Wu, X.D. (1998) *Science*, **279**, 837–839.
41 Wong, D.W. and Robertson, G.H. (1999) *Adv. Exp. Med. Biol.*, **464**, 91–105.
42 Zhao, J.-C. (2001) *Adv. Eng. Mater.*, **3**, 143–147.
43 Olk, C.H. (2005) *Meas. Sci. Technol.*, **16**, 14–20.
44 Zou, L., Savvate'ev, V., Booher, J., Kim, C.-H., and Shinar, J. (2001) *Appl. Phys. Lett.*, **79**, 2282–2284.
45 Takeuchi, I., Famodu, O.O., Read, J.C., Aronova, M.A., Chang, K.-S., Craciunescu, C., Lofland, S.E., Wuttig, M., Wellstood, F.C., Knauss, L., and Orozco, A. (2003) *Nat. Mater.*, **2**, 180–184.
46 Cui, J., Chu, Y.S., Famodu, O.O., Furuya, Y., Hattrick-Simpers, J., James, R.D., Ludwig, A., Thienhaus, S., Wuttig, M., Zhang, Z., and Takeuchi, I. (2006) *Nat. Mater.*, **5**, 286–290.
47 Holzwarth, A., Schmidt, H.-W., and Maier, W. (1998) *Angew. Chem. Int. Ed.*, **37**, 2644–2647.
48 Cooper, A.C., McAlexander, L.H., Lee, D.-H., Torres, M.T., and Crabtree, R.H. (1998) *J. Am. Chem. Soc.*, **120**, 9971–9972.
49 Lemmon, J.P., Wroczynski, R.J., Whisenhunt, D.W. Jr. and Flanagan, W.P. (2001) *Polym. Prepr.*, **42**, 630–631.
50 Reddington, E., Sapienza, A., Gurau, B., Viswanathan, R., Sarangapani, S., Smotkin, E.S., and Mallouk, T.E. (1998) *Science*, **280**, 1735–1737.
51 Greeley, J., Jaramillo, T.F., Bonde, J., Chorkendorff, I., and Nørskov, J.K. (2006) *Nat. Mater.*, **5**, 909–913.
52 Brocchini, S., James, K., Tangpasuthadol, V., and Kohn, J. (1997) *J. Am. Chem. Soc.*, **119**, 4553–4554.
53 Lai, R., Kang, B.S., and Gavalas, G.R. (2001) *Angew. Chem., Int. Ed.*, **40**, 408–411.
54 Ramirez, A.G. and Saha, R. (2004) *Appl. Phys. Lett.*, **85**, 5215–5217.
55 Jiang, R., Rong, C., and Chu, D. (2005) *J. Comb. Chem.*, **7**, 272–278.
56 Lemmon, J.P., Manivannan, V., Jordan, T., Hassib, L., Siclovan, O., Othon, M., and Pilliod, M. (2004) High throughput screening of materials for solid oxide fuel cells, in *Combinatorial and Artificial Intelligence Methods in Materials Science II. MRS Symposium Proceedings*, vol. 804 (eds R.A. Potyrailo, A. Karim, Q. Wang,

and T. Chikyow), Materials Research Society, Warrendale, PA, pp. 27–32.
57 Hänsel, H., Zettl, H., Krausch, G., Schmitz, C., Kisselev, R., Thelakkat, M., and Schmidt, H.-W. (2002) *Appl. Phys. Lett.*, **81**, 2106–2108.
58 Chisholm, B.J., Potyrailo, R.A., Cawse, J.N., Shaffer, R.E., Brennan, M.J., Moison, C., Whisenhunt, D.W., Flanagan, W.P., Olson, D.R., Akhave, J.R., Saunders, D.L., Mehrabi, A., and Licon, M. (2002) *Prog. Org. Coat.*, **45**, 313–321.
59 Wicks, D.A., and Bach, H. (2002) The coming revolution for coatings science: high throughput screening, Proceedings of The 29th International Waterborne, High-Solids, and Powder Coating Symposium, New Orleans, LA, 29, 1–24.
60 Grunlan, J.C., Mehrabi, A.R., Chavira, A.T., Nugent, A.B., and Saunders, D.L. (2003) *J. Comb. Chem.*, **5**, 362–368.
61 Stafslien, S.J., Bahr, J.A., Feser, J.M., Weisz, J.C., Chisholm, B.J., Ready, T.E., and Boudjouk, P. (2006) *J. Comb. Chem.*, **8**, 156–162.
62 Ekin, A. and Webster, D.C. (2007) *J. Comb. Chem.*, **9**, 178–188.
63 Potyrailo, R.A. and Mirsky, V.M. (2008) *Chem. Rev.*, **108**, 770–813.
64 Potyrailo, R.A. and Mirsky, V.M. (2009) Introduction to combinatorial methods for chemical and biological sensors, in *Combinatorial Methods for Chemical and Biological Sensors* (eds R.A. Potyrailo and V.M. Mirsky), Springer, New York, pp. 3–24.
65 Potyrailo, R.A., Chisholm, B.J., Olson, D.R., Brennan, M.J., and Molaison, C.A. (2002) *Anal. Chem.*, **74**, 5105–5111.
66 Potyrailo, R.A., Chisholm, B.J., Morris, W.G., Cawse, J.N., Flanagan, W.P., Hassib, L., Molaison, C.A., Ezbiansky, K., Medford, G., and Reitz, H. (2003) *J. Comb. Chem.*, **5**, 472–478.
67 MacLean, D., Baldwin, J.J., Ivanov, V.T., Kato, Y., Shaw, A., Schneider, P., and Gordon, E.M. (2000) *J. Comb. Chem.*, **2**, 562–578.
68 Potyrailo, R.A. and Takeuchi, I. (2005) *Meas. Sci. Technol.*, **16**, 1–4.
69 Cohan, P.E. (2002) Combinatorial materials science applied - mini case studies, lessons and strategies, in *2002 COMBI - The 4th Annual International Symposium on Combinatorial Approaches for New Materials Discovery*, Knowledge Foundation, Arlington, VA.
70 Department of Defence, Prepared by the Deputy Under Secretary of Defense for Science and Technology (DUSD(S&T)) (2005) https://acc.dau.mil/CommunityBrowser.aspx?id=18545 (accessed on 22 September, 2010).
71 Newman, J.D. and Turner, A.P.F. (2005) *Biosens. Bioelectron.*, **20**, 2435–2453.
72 Pickup, J.C. and Alcock, S. (1991) *Biosens. Bioelectron.*, **6**, 639–646.
73 Meyerhoff, M.E. (1993) *Trends Anal. Chem.*, **12**, 257–266.
74 Potyrailo, R.A., Morris, W.G., and Monk, D. (2008) New sensing concepts for process analytical technology. Invited Symposium "Spectroscopic and Sensing Technologies in Pharmaceutical Industry," at the Annual Meeting of the Federation of Analytical Chemistry and Spectroscopy Societies, Sept. 28–Oct. 2, Grand Sierra Resort, Reno, NV, paper 430.
75 Clark, K.J.R. and Furey, J. (2006) *BioProcess Int.*, **4** (6), S16–S20
76 Lemmo, A.V., Fisher, J.T., Geysen, H.M., and Rose, D.J. (1997) *Anal. Chem.*, **69**, 543–551.
77 Calvert, P. (2001) *Chem. Mater.*, **13**, 3299–3305.
78 de Gans, B.-J. and Schubert, U.S. (2004) *Langmuir*, **20**, 7789–7793.
79 Apostolidis, A., Klimant, I., Andrzejewski, D., and Wolfbeis, O.S. (2004) *J. Comb. Chem.*, **6**, 325–331.
80 Hassib, L. and Potyrailo, R.A. (2004) *Polym. Prepr.*, **45**, 211–212.
81 Scheidtmann, J., Frantzen, A., Frenzer, G., and Maier, W.F. (2005) *Meas. Sci. Technol.*, **16**, 119–127.
82 Schena, M. (2003) *Microarray Analysis*, John Wiley & Sons, Inc., Hoboken, NJ.
83 Potyrailo, R.A., Morris, W.G., Wroczynski, R.J., and McCloskey, P.J. (2004) *J. Comb. Chem.*, **6**, 869–873.
84 Mirsky, V.M. and Kulikov, V. (2003) Combinatorial electropolymerization: concept, equipment and applications, in *High Throughput Analysis: A Tool for Combinatorial Materials Science*

(eds R.A. Potyrailo and E.J. Amis), Kluwer Academic/Plenum Publishers, New York, Ch. 20, pp. 431–446.
85. Mirsky, V.M., Kulikov, V., Hao, Q., and Wolfbeis, O.S. (2004) *Macromol. Rapid Commun.*, **25**, 253–258.
86. Taylor, C.J. and Semancik, S. (2002) *Chem. Mater.*, **14**, 1671–1677.
87. Aronova, M.A., Chang, K.S., Takeuchi, I., Jabs, H., Westerheim, D., Gonzalez-Martin, A., Kim, J., and Lewis, B. (2003) *Appl. Phys. Lett.*, **83**, 1255–1257.
88. Cawse, J.N., Olson, D., Chisholm, B.J., Brennan, M., Sun, T., Flanagan, W., Akhave, J., Mehrabi, A., and Saunders, D. (2003) *Prog. Org. Coat.*, **47**, 128–135.
89. Amis, E.J. (2004) *Nat. Mater.*, **3**, 83–85.
90. Potyrailo, R.A., Morris, W.G., Leach, A.M., Hassib, L., Krishnan, K., Surman, C., Wroczynski, R., Boyette, S., Xiao, C., Shrikhande, P., Agree, A., and Cecconie, T. (2007) *Appl. Opt.*, **46**, 7007–7017.
91. Yoon, J., Jung, Y.-S., and Kim, J.-M. (2009) *Adv. Funct. Mater.*, **19**, 209–214.
92. Dickinson, T.A., Walt, D.R., White, J., and Kauer, J.S. (1997) *Anal. Chem.*, **69**, 3413–3418.
93. Potyrailo, R.A., Wroczynski, R.J., Pickett, J.E., and Rubinsztajn, M. (2003) *Macromol. Rapid Commun.*, **24**, 123–130.
94. Potyrailo, R.A., Szumlas, A.W., Danielson, T.L., Johnson, M., and Hieftje, G.M. (2005) *Meas. Sci. Technol.*, **16**, 235–241.
95. Potyrailo, R.A. and Wroczynski, R.J. (2005) *Rev. Sci. Instrum.*, **76**, 062222.
96. Potyrailo, R.A., Olson, D.R., Brennan, M.J., Akhave, J.R., Licon, M.A., Mehrabi, A.R., Saunders, D.L., and Chisholm, B.J. (2003) Systems and methods for the deposition and curing of coating compositions. US Patent 6,544,334 B1.
97. Potyrailo, R.A. (2004) Polymeric Materials Science and Engineering. *Polym. Prepr.*, **90**, 797–798.
98. Potyrailo, R.A. and Hassib, L. (2005) *Rev. Sci. Instrum.*, **76**, 062225.
99. Potyrailo, R.A., Ding, Z., Butts, M.D., Genovese, S.E., and Deng, T. (2008) *IEEE Sens. J.*, **8**, 815–822.
100. Bhat, R.R., Tomlinson, M.R., and Genzer, J. (2004) *Macromol. Rapid Commun.*, **25**, 270–274.
101. Turcu, F., Hartwich, G., Schäfer, D., and Schuhmann, W. (2005) *Macromol. Rapid Commun.*, **26**, 325–330.
102. Sysoev, V.V., Kiselev, I., Frietsch, M., and Goschnick, J. (2004) *Sensors*, **4**, 37–46.
103. Klingvall, R., Lundström, I., Löfdahl, M., and Eriksson, M. (2005) *IEEE Sens. J.*, **5**, 995–1003.
104. Baker, B.E., Kline, N.J., Treado, P.J., and Natan, M.J. (1996) *J. Am. Chem. Soc.*, **118**, 8721–8722.
105. Göpel, W. (1998) *Sens. Actuators B*, **52**, 125–142.
106. Alivisatos, A.P. (2004) *Nat. Biotechnol.*, **22**, 47–52.
107. Medintz, I.L., Uyeda, H.T., Goldman, E.R., and Mattoussi, H. (2005) *Nat. Mater.*, **4**, 435–446.
108. Ko, M.C. and Meyer, G.J. (1999) Photoluminescence of inorganic semiconductors for chemical sensor applications, in *Optoelectronic Properties of Inorganic Compounds* (eds D.M. Roundhill and J.P Fackler Jr.), Plenum Press, New York, pp. 269–315.
109. Lisensky, G.C., Meyer, G.J., and Ellis, A.B. (1988) *Anal. Chem.*, **60**, 2531–2534.
110. Seker, F., Meeker, K., Kuech, T.F., and Ellis, A.B. (2000) *Chem. Rev.*, **100**, 2505–2536.
111. Nazzal, A.Y., Qu, L., Peng, X., and Xiao, M. (2003) *Nano Lett.*, **3**, 819–822.
112. Vassiltsova, O.V., Zhao, Z., Petrukhina, M.A., and Carpenter, M.A. (2007) *Sens. Actuators B*, **123**, 522–529.
113. Potyrailo, R.A. and Leach, A.M. (2005) Gas sensor materials based on semiconductor nanocrystal/polymer composite films. Proceedings of TRANSDUCERS'05, The 13th International Conference on Solid-State Sensors, Actuators and Microsystems, Seoul, Korea, June 5–9, 2005, pp. 1292–1295.
114. Potyrailo, R.A. and Leach, A.M. (2006) *Appl. Phys. Lett.*, **88**, 134110.
115. Leach, A.M. and Potyrailo, R.A. (2006) Gas sensor materials based on semiconductor nanocrystal/polymer composite films, in *Combinatorial Methods and Informatics in Materials Science. MRS Symposium Proceedings*, vol. 894 (eds Q. Wang, R.A. Potyrailo,

M. Fasolka, T. Chikyow, U.S. Schubert, and A. Korkin), Materials Research Society, Warrendale, PA, pp. 237–243.
116 Potyrailo, R.A. and Leach, A.M. (2006) Multi-size semiconductor nanocrystal/polymer composite films as selective gas sensor materials. Asia-Pacific Conference of Transducer and Micro-Nano Technology APCOT 2006, June 25–28 Singapore, Abstracts Book, 121–1121
117 Otto, M. (1999) *Chemometrics: Statistics and Computer Application in Analytical Chemistry*, Wiley-VCH Verlag GmbH, Weinheim.
118 Potyrailo, R.A. (2001) Combinatorial screening, in *Encyclopedia of Materials: Science and Technology*, vol. 2 (eds K.H.J. Buschow, R.W. Cahn, M.C. Flemings, B. Ilschner, E.J. Kramer, and S. Mahajan), Elsevier, Amsterdam, pp. 1329–1343.
119 Mills, A., Lepre, A., and Wild, L. (1998) *Anal. Chim. Acta*, **362**, 193–202.
120 Bedlek-Anslow, J.M., Hubner, J.P., Carroll, B.F., and Schanze, K.S. (2000) *Langmuir*, **16**, 9137–9141.
121 Wang, J., Musameh, M., and Lin, Y. (2003) *J. Am. Chem. Soc.*, **125**, 2408–2409.
122 Potyrailo, R.A. (2003) Devices and methods for simultaneous measurement of transmission of vapors through a plurality of sheet materials. US Patent 6,567,753 B2.
123 Potyrailo, R.A. and Brennan, M.J. (2004) Method and apparatus for characterizing the barrier properties of members of combinatorial libraries. US Patent 6,684,683(B2).
124 Potyrailo, R.A. (2003) Combinatorial development and accelerated performance testing of polymers. Presented at the Second Dutch Polymer Institute workshop 'High Throughput Experimentation/Combinatorial Material Research', May 15–16 2003, Eindhoven University of Technology: Eindhoven, The Netherlands.
125 Chojnacki, P., Werner, T., and Wolfbeis, O.S. (2004) *Microchim. Acta*, **147**, 87–92.
126 Amao, Y. (2003) *Microchim. Acta*, **143**, 1–12.
127 Hartmann, P. and Trettnak, W. (1996) *Anal. Chem.*, **68**, 2615–2620.
128 Draxler, S., Lippitsch, M.E., Klimant, I., Kraus, H., and Wolfbeis, O.S. (1995) *J. Phys. Chem.*, **99**, 3162–3167.
129 Mohr, G.J. and Wolfbeis, O.S. (1996) *Sens. Actuators B*, **37**, 103–109.
130 Conway, V.L., Hassen, K.P., Zhang, L., Seitz, W.R., and Gross, T.S. (1997) *Sens. Actuators B*, **45**, 1–9.
131 Kolytcheva, N.V., Müller, H., and Marstalerz, J. (1999) *Sens. Actuators B*, **58**, 456–463.
132 Preininger, C., Mohr, G., Klimant, I., and Wolfbeis, O.S. (1996) *Anal. Chim. Acta*, **334**, 113–123.
133 Peper, S., Ceresa, A., Qin, Y., and Bakker, E. (2003) *Anal. Chim. Acta*, **500**, 127–136.
134 Legin, A., Makarychev-Mikhailov, S., Goryacheva, O., Kirsanov, D., and Vlasov, Y. (2002) *Anal. Chim. Acta*, **457**, 297–303.
135 Penco, M., Sartore, L., Bignotti, F., Sciucca, S.D., Ferrari, V., Crescini, P., and D'Antone, S. (2004) *J. Appl. Polym. Sci.*, **91**, 1816–1821.
136 Leclerc, M. (1999) *Adv. Mater.*, **11**, 1491–1498.
137 McQuade, D.T., Pullen, A.E., and Swager, T.M. (2000) *Chem. Rev.*, **100**, 2537–2574.
138 Janata, J. and Josowicz, M. (2002) *Nat. Mater.*, **2**, 19–24.
139 Dai, L., Soundarrajan, P., and Kim, T. (2002) *Pure Appl. Chem.*, **74**, 1753–1772.
140 Bobacka, J., Ivaska, A., and Lewenstam, A. (2003) *Electroanalysis*, **15**, 366–374.
141 Bidan, G. (1992) *Sens. Actuators B*, **6**, 45–56.
142 Albert, K.J., Lewis, N.S., Schauer, C.L., Sotzing, G.A., Stitzel, S.E., Vaid, T.P., and Walt, D.R. (2000) *Chem. Rev.*, **100**, 2595–2626.
143 Gomez-Romero, P. (2001) *Adv. Mater.*, **13**, 163–174.
144 Gill, I. and Ballesteros, A. (1998) *J. Am. Chem. Soc.*, **120**, 8587–8598.
145 Gill, I. (2001) *Chem. Mater.*, **13**, 3404–3421.
146 Barbero, C., Acevedo, D.F., Salavagione, H.J., and Miras, M.C. (2003) Synthesis properties and applications of functionalized conductive polymers. Jornadas Sam/Conamet/Simposio Materia 2003 C–12. Available at http://www.materiales-sam.org.ar/sitio/

biblioteca/bariloche/Conferencias/C12.PDF.
147 Lim, S.-H., Raorane, D., Satyanarayana, S., and Majumdar, A. (2006) *Sens. Actuators B*, **119**, 466–474.
148 Raorane, D., Lim, S.-H., and Majumdar, A. (2008) *Nano Lett.*, **8**, 2229–2235.
149 Hierlemann, A., Weimar, U., Kraus, G., Schweizer-Berberich, M., and Göpel, W. (1995) *Sens. Actuators B*, **26**, 126–134.
150 Wohltjen, H. (2006) A journey: from sensor ideas to sensor products. Plenary talk at the 11th International Meeting on Chemical Sensors, University of Brescia, Italy, July 16–19, 2006.
151 Grate, J.W., Abraham, H., and McGill, R.A. (1997) Sorbent polymer materials for chemical sensors and arrays, in *Handbook of Biosensors and Electronic Noses. Medicine, Food, and the Environment* (ed. E. Kress-Rogers), CRC Press, Boca Raton, FL, pp. 593–612.
152 Grate, J.W. (2000) *Chem. Rev.*, **100**, 2627–2648.
153 Grate, J.W. and Abraham, M.H. (1991) *Sens. Actuators B*, **3**, 85–111.
154 Abraham, M.H. (1993) *Chem. Soc. Rev.*, **22**, 73–83.
155 Maranas, C.D. (1996) *Ind. Eng. Chem. Res.*, **35**, 3403–3414.
156 Wise, B.M., Gallagher, N.B., and Grate, J.W. (2003) *J. Chemometrics*, **17**, 463–469.
157 Belmares, M., Blanco, M., Goddard, W.A., III, Ross, R.B., Caldwell, G., Chou, S.-H., Pham, J., Olofson, P.M., and Thomas, C. (2004) *J. Comput. Chem.*, **25**, 1814–1826.
158 Potyrailo, R.A., May, R.J., and Sivavec, T.M. (2004) *Sensor Lett.*, **2**, 31–36.
159 Potyrailo, R.A., Sivavec, T.M., and Bracco, A.A. (1999) Field evaluation of acoustic wave chemical sensors for monitoring of organic solvents in groundwater, in *Internal Standardization and Calibration Architectures for Chemical Sensors* (eds R.E. Schaffer and R.A. Potyrailo), Proceedings SPIE, vol. 3856, SPIE, Bellingham, WA, pp. 140–147.
160 Sivavec, T.M. and Potyrailo, R.A. (2002) Polymer coatings for chemical sensors. US Patent 6,357,278 B1.
161 Potyrailo, R.A., Morris, W.G., and Wroczynski, R.J. (2003) Acoustic-wave sensors for high-throughput screening of materials, in *High Throughput Analysis: A Tool for Combinatorial Materials Science* (eds R.A. Potyrailo, and E.J. Amis), Kluwer Academic/Plenum Publishers, New York, Ch. 11.
162 Potyrailo, R.A., Morris, W.G., and Wroczynski, R.J. (2004) *Rev. Sci. Instrum.*, **75**, 2177–2186.
163 Beebe, K.R., Pell, R.J., and Seasholtz, M.B. (1998) *Chemometrics: A Practical Guide*, John Wiley & Sons, Inc., New York.
164 Potyrailo, R.A., McCloskey, P.J., Ramesh, N., and Surman, C.M. (2005) Sensor devices containing co-polymer substrates for analysis of chemical and biological species in water and air. US Patent Application 2005133697.
165 Potyrailo, R.A., McCloskey, P.J., Wroczynski, R.J., and Morris, W.G. (2006) *Anal. Chem.*, **78**, 3090–3096.
166 Zemel, J.N. (1990) *Rev. Sci. Instrum.*, **61**, 1579–1606.
167 Suzuki, H. (2000) *Electroanalysis*, **12**, 703–715.
168 Wang, J. (2002) *Talanta*, **56**, 223–231.
169 Hagleitner, C., Hierlemann, A., Brand, O., and Baltes, H. (2002) CMOS single chip gas detection systems - Part I, in *Sensors Update*, Vol. 11 (eds H. Baltes, W. Göpel, and J. Hesse), Wiley-VCH Verlag, GmbH, Weinheim, pp. 101–155.
170 Hagleitner, C., Hierlemann, A., Brand, O., and Baltes, H. (2003) CMOS single chip gas detection systems - Part II, in *Sensors Update*, Vol. 12 (eds H. Baltes, W. Göpel, and J. Hesse), Wiley-VCH Verlag, GmbH, Weinheim, pp. 51–120.
171 Kulikov, V. and Mirsky, V.M. (2004) *Meas. Sci. Technol.*, **15**, 49–54.
172 Siemons, M., Koplin, T.J., and Simon, U. (2007) *Appl. Surf. Sci.*, **254**, 669–676.
173 Franke, M.E., Koplin, T.J., and Simon, U. (2006) *Small*, **2**, 36–50.
174 Korotcenkov, G. (2005) *Sens. Actuators B*, **107**, 209–232.
175 Barsan, N., Koziej, D., and Weimar, U. (2007) *Sens. Actuators B*, **121**, 18–35.
176 Semancik, S. (2002) Correlation of chemisorption and electronic effects for metal oxide interfaces: transducing principles for temperature programmed gas microsensors. Final Technical Report Project Number: EMSP 65421, Grant

Number: 07-98ER62709, US Department of Energy. Available from http://www.osti.gov/bridge/.

177 Semancik, S. (2003) Temperature-dependent materials research with micromachined array platforms, in *Combinatorial Materials Synthesis* (eds X.-D. Xiang and I. Takeuchi), Marcel Dekker, New York, pp. 263–295.

178 Barsoukov, E. and Macdonald, J.R. (eds) (2005) *Impedance Spectroscopy: Theory, Experiment, and Applications*, 2nd edn, John Wiley & Sons, Inc., Hoboken, NJ.

179 Simon, U., Sanders, D., Jockel, J., Heppel, C., and Brinz, T. (2002) *J. Comb. Chem.*, **4**, 511–515.

180 Simon, U., Sanders, D., Jockel, J., and Brinz, T. (2005) *J. Comb. Chem.*, **7**, 682–687.

181 Sanders, D. and Simon, U. (2007) *J. Comb. Chem.*, **9**, 53–61.

182 Feldmann, C. (2001) *Script. Mater.*, **44**, 2193–2196.

183 Siemons, M. and Simon, U. (2006) *Sens. Actuators B*, **120**, 110–118.

184 Siemons, M. and Simon, U. (2007) *Sens. Actuators B*, **126**, 595–603.

185 Sieg, S., Stutz, B., Schmidt, T., Hamprecht, F., and Maier, W.F. (2006) *J. Mol. Mod.*, **12**, 611–619.

186 Sieg, S.C., Suh, C., Schmidt, T., Stukowski, M., Rajan, K., and Maier, W.F. (2007) *QSAR Comb. Sci.*, **26**, 528–535.

187 Frantzen, A., Sanders, D., Scheidtmann, J., Simon, U., and Maier, W.F. (2005) *QSAR Comb. Sci.*, **24**, 22–28.

188 Frenzer, G., Frantzen, A., Sanders, D., Simon, U., and Maier, W.F. (2006) *Sensors*, **6**, 1568–1586.

189 Collaudin, A.B. and Blum, L.J. (1997) *Sens. Actuators B*, **38–39**, 189–194.

190 Mills, A. (1998) *Sens. Actuators B*, **51**, 60–68.

191 Eaton, K. (2002) *Sens. Actuators B*, **85**, 42–51.

192 Papkovsky, D.B., Ponomarev, G.V., Trettnak, W., and O'Leary, P. (1995) *Anal. Chem.*, **67**, 4112–4117.

193 Levitsky, I., Krivoshlykov, S.G., and Grate, J.W. (2001) *Anal. Chem.*, **73**, 3441–3448.

194 Florescu, M. and Katerkamp, A. (2004) *Sens. Actuators B*, **97**, 39–44.

195 Basu, B.J., Thirumurugan, A., Dinesh, A.R., Anandan, C., and Rajam, K.S. (2005) *Sens. Actuators B*, **104**, 15–22.

196 Potyrailo, R.A. and Hassib, L. (2004) Development of polymeric sensor materials using combinatorial approaches. Presented at the MACRO 2004 - World Polymer Congress, the 40th IUPAC International Symposium on Macromolecules, July 4–9 IUPAC: Paris, France.

197 Potyrailo, R.A. and Hassib, L. (2005) Development of gas and liquid optical sensor materials using combinatorial screening techniques. Proceedings of TRANSDUCERS'05, The 13th International Conference on Solid-State Sensors, Actuators and Microsystems, Seoul, Korea, June 5–9, 2005 pp. 2099–2102.

198 Potyrailo, R.A., Morris, W.G., Leach, A.M., Sivavec, T.M., Wisnudel, M.B., Krishnan, K., Surman, C., Hassib, L., Wroczynski, R., Boyette, S., Xiao, C., Agree, A., and Cecconie, T. (2007) *Am. Lab.*, **42** (6) 32–35.

199 Potyrailo, R.A., Morris, W.G., Leach, A.M., Sivavec, T.M., Wisnudel, M.B., and Boyette, S. (2006) *Anal. Chem.*, **78**, 5893–5899.

200 Lundström, I., Shivaraman, S., Svensson, C., and Lundkvist, L. (1975) *Appl. Phys. Lett.*, **26**, 55–57.

201 Lundström, I., Svensson, C., Spetz, A., Sundgren, H., and Winquist, F. (1993) *Sens. Actuators B*, **13–14**, 16–23.

202 Eriksson, M., Klingvall, R., and Lundström, I. (2006) A combinatorial method for optimization of materials for gas sensitive field-effect devices, in *Combinatorial and High-Throughput Discovery and Optimization of Catalysts and Materials* (eds R.A. Potyrailo and W.F. Maier), CRC Press, Boca Raton, FL, pp. 85–95.

203 Lundström, I., Erlandsson, R., Frykman, U., Hedborg, E., Spetz, A., Sundgren, H., Welin, S., and Winquist, F. (1991) *Nature*, **352**, 47–50.

204 Löfdahl, M., Eriksson, M., and Lundström, I. (2000) *Sens. Actuators B*, **70**, 77–82.

205 Lundström, I., Sundgren, H., Winquist, F., Eriksson, M., Krantz-Rülcker, C., and Lloyd-Spetz, A. (2007) *Sens. Actuators B*, **121**, 247–262.

206 Klingvall, R., Lundstrom, I., Lofdahl, M., and Eriksson, M. (2003) *Proc. IEEE Sensors*, **2**, 1114–1115.

207 Eriksson, M., Salomonsson, A., Lundström, I., Briand, D., and Åbom, A.E. (2005) *J. Appl. Phys.*, **98**, 034903.

208 Goschnick, J., Frietsch, M., and Schneider, T. (1998) *Surf. Coat. Technol.*, **108–109**, 292–296.

209 Koronczi, I., Ziegler, K., Kruger, U., and Goschnick, J. (2002) *IEEE Sens. J.*, **2**, 254–259.

210 Arnold, C., Andlauer, W., Häringer, D., Körber, R., and Goschnick, J. (2002) *Proc. IEEE Sens.*, **1**, 426–429.

211 Arnold, C., Harms, M., and Goschnick, J. (2002) *IEEE Sens. J.*, **2**, 179–188.

212 Goschnick, J., Koronczi, I., Frietsch, M., and Kiselev, I. (2005) *Sens. Actuators B*, **106**, 182–186.

213 Schneider, T., Betsarkis, K., Trouillet, V., and Goschnick, J. (2004) *Proc. IEEE Sens.*, **1**, 196–197.

214 Frietsch, M., Koronczi, I., and Goschnick, J. (2004) Electronic noses as tools for online odour management in the environment. Environmental Odour Management. Nov. 17–19, Cologne, Lecture 26.

215 Goschnick, J. (2001) *Microelectron. Eng.*, **57–58**, 693–704.

216 Potyrailo, R.A. and Morris, W.G. (2007) *Rev. Sci. Instrum.*, **78**, 072214.

217 Potyrailo, R.A. and Morris, W.G. (2007) *Anal. Chem.*, **79**, 45–51.

218 Potyrailo, R.A., Morris, W.G., Sivavec, T., Tomlinson, H.W., Klensmeden, S., and Lindh, K. (2009) RFID sensors based on ubiquitous passive 13.56-MHz RFID tags and complex impedance detection. *Wireless Commun. Mobile Comput.*, **9**, 1318. doi: 10.1002/wcm.711

219 Potyrailo, R.A., Surman, C., and Morris, W.G. (2009) Combinatorial screening of polymeric sensing materials using RFID sensors: combined effects of plasticizers and temperature. *J. Comb. Chem.*, **11** (4), 598–603. doi: 10.1021/cc900001n

220 Lee, W.S., Lee, S.C., Lee, S.J., Lee, D.D., Huh, J.S., Jun, H.K., and Kim, J.C. (2005) *Sens. Actuators B*, **108**, 148–153.

221 Feng, C.-D., Sun, S.-L., Wang, H., Segre, C.U., and Stetter, J.R. (1997) *Sens. Actuators B*, **40**, 217–222.

222 Tailoka, F., Fray, D.J., and Kumar, R.V. (2003) *Solid State Ionics*, **161**, 267–277.

223 Sun, L.-X. and Okada, T. (2001) *J. Membr. Sci.*, **183**, 213–221.

224 Cappadonia, M., Erning, J.W., Niaki, S.M.S., and Stimming, U. (1995) *Solid State Ionics*, **77**, 65–69.

225 Wintersgill, M.C. and Fontanella, J.J. (1998) *Electrochim. Acta*, **43**, 1533–1538.

226 Ertl, G. (1990) *Angew. Chem. Int. Ed.*, **29**, 1219–1227.

227 Buhlmann, K., Schlatt, B., Cammann, K., and Shulga, A. (1998) *Sens. Actuators B*, **49**, 156–165.

228 Cho, E.J., Tao, Z., Tang, Y., Tehan, E.C., Bright, F.V., Hicks, W.L. Jr., Gardella, J.A.. Jr. and Hard, R. (2002) *Appl. Spectrosc.*, **56**, 1385–1389.

229 Frantzen, A., Scheidtmann, J., Frenzer, G., Maier, W.F., Jockel, J., Brinz, T., Sanders, D., and Simon, U. (2004) *Angew. Chem. Int. Ed.*, **43**, 752–754.

230 Villoslada, F.N. and Takeuchi, T. (2005) *Bull. Chem. Soc. Jpn.*, **78**, 1354–1361.

231 Mijangos, I., Navarro-Villoslada, F., Guerreiro, A., Piletska, E., Chianella, I., Karim, K., Turner, A., and Piletsky, S. (2006) *Biosens. Bioelectron.*, **22**, 381–387.

232 Potyrailo, R.A., Olson, D.R., Chisholm, B.J., Brennan, M.J., Lemmon, J.P., Cawse, J.N., Flanagan, W.P., Shaffer, R.E., and Leib, T.K. (2001) High throughput analysis of polymer materials and coatings. Presented at the Invited Symposium "Analytical Tools For High Throughput Chemical Analysis and Combinatorial Materials Science", Pittsburgh Conference on Analytical Chemistry and Applied Spectroscopy,

March 4–9, New Orleans, Louisiana, 2001.
233 Jansen, M. (2002) *Angew. Chem. Int. Ed.*, **41**, 3746–3766.
234 Potyrailo, R.A. (2007) High-throughput experimentation in early 21st century: searching for materials descriptors, not for a needle in the haystack. Presented at the 6th DPI Workshop on Combinatorial and High-Throughput Approaches in Polymer Science, September 10–11, 2007, Darmstadt, Germany.

4
Fluorescent Cyclodextrins as Chemosensors for Molecule Detection in Water
Hiroshi Ikeda

4.1
Introduction

The development of new chemical sensing methods for the recognition of molecules or ions in water is an important theme in the field of supramolecular chemistry [1–4]. Extensive efforts have been made to construct chemosensors that selectively interact with analytes to produce detectable changes in signals (e.g., fluorescence, absorption or circular dichroism). Several chromophore-appended cyclodextrins (CDs) have been prepared for this purpose [5–8] and used to detect colorless neutral molecules in water via changes in the intensity of their fluorescence, absorption, or circular dichroism. CDs are torus-shaped cyclic oligosaccharides consisting of six, seven, and eight D-glucopyranose units for α-CD, β-CD, and γ-CD, respectively (Figure 4.1) [9–11]. In aqueous solution, CDs can accommodate various organic compounds in their central cavities. The hydrophobic interaction, the van der Waals force interaction, and dipole–dipole interactions between the CD and guest are usually considered the major driving forces of inclusion phenomena, and the stability of the inclusion complex is affected by the fit (shape and size) of the guest to the CD cavity. Various values for the internal diameter of CDs have been reported using different estimation methods (Table 4.1) [12–14]. Recently, Müller and Wenz calculated the minimum internal diameters of CDs by semi-empirical AM1 and PM3 calculation and showed that the stability of an inclusion complex of an α-CD with a guest is strongly correlated with the cross-sectional diameter of the α-CD and the guest [12]. Many types of chemosensors can be constructed using CDs that recognize the shape and size of analytes. This chapter discusses the molecule-sensing abilities of modified CDs, CD–peptide conjugates, and CD–protein conjugates.

(a)

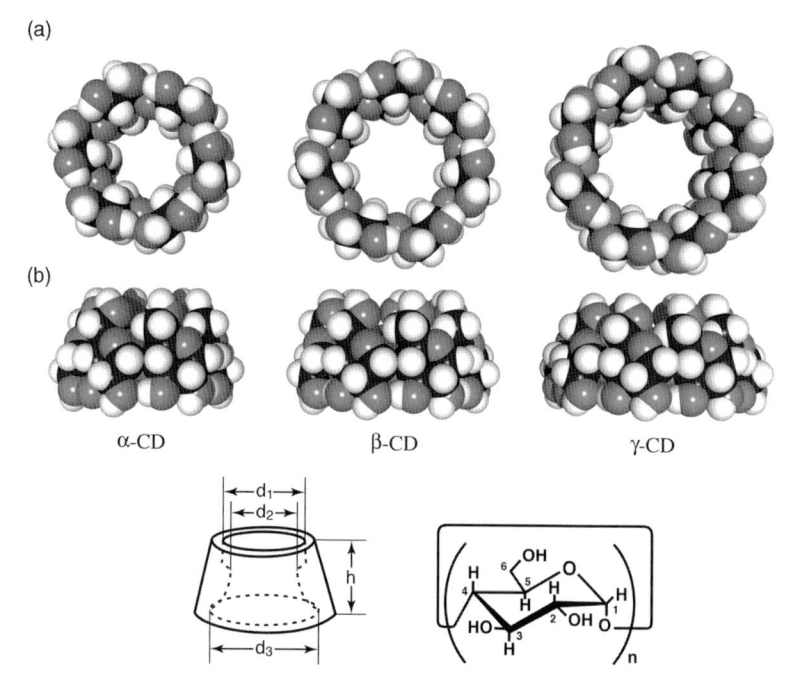

(b)

α-CD   β-CD   γ-CD

Figure 4.1 Structures of cyclodextrins: (a) view from the secondary hydroxy side, (b) side view; for α-CD, β-CD, and γ-CD, $n = 6$, 7, and 8 (D-glucopyranose units), respectively.

4.2
Pyrene-Appended Cyclodextrins

4.2.1
Pyrene-Appended γ-Cyclodextrins as Molecule Sensors

The pyrene molecule forms an excimer by excited-state face-to-face interaction. Herkstroeter has observed a remarkable increase in the excimer emission of pyrene in the presence of γ-CD and proposed the formation of a 2:2 γ-CD/pyrene

Table 4.1 Properties of cyclodextrins.

	n	d_1 (A)[a]	d_2 (A)[a]	d_3 (A)[a]	h (A)[a]	d_1 (A)[b]	d_3 (A)[b]	d_2 (A) AM1[c]	d_2 (A) PM3[c]
α-CD	6	5.6	4.2	8.8	7.8	4.7	5.2	4.4	4.4
β-CD	7	6.8	5.6	10.8	7.8	6.0	6.4	5.8	6.5
γ-CD	8	8.0	6.8	12.0	7.8	7.5	8.3	7.4	8.1

a) Reference [13].
b) Reference [14].
c) Reference [12].

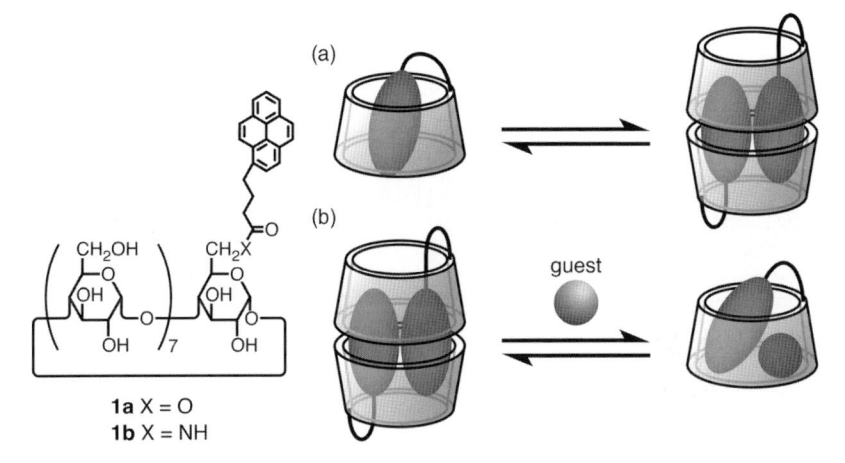

Figure 4.2 Pyrene-appended γ-cyclodextrins (**1a** and **1b**) as molecule sensors: (a) conformational equilibrium in aqueous solution; (b) guest-induced conformational change. Adapted from Reference [17].

complex [15]. Ueno has prepared pyrene-appended γ-CDs, **1a** and **1b**, which formed an association dimer in aqueous solution (Figure 4.2a) and exhibited excimer emission around 470 nm (Figure 4.3a) [16, 17]. This excimer emission decreased remarkably upon addition of a guest, whereas the monomer emission at 378 and 397 nm increased (Figure 4.3a). Analysis of the guest-induced circular dichroism and absorption variations substantiated the fact that this fluorescence variation occurs upon conversion from the association dimer into a 1:1 host–guest complex (Figure 4.2b). Since the guest-induced fluorescence variation is greatly affected by the shape and size of the guest species, it is possible that this system can be used as a sensory system of molecular recognition for detecting various organic compounds (Figure 4.3b, c). The molecular recognition abilities were evaluated using the sensitivity parameter $\Delta I/I_0$, where $\Delta I = I - I_0$, with I and I_0 being the excimer emission intensities in the presence and absence of guests. Because of these findings, chemosensors based on CDs have been improved.

4.2.2
Pyrene-Appended γ-Cyclodextrin as a Bicarbonate Sensor

Suzuki has prepared a new pyrene-appended γ-CD (**2**) with a triamine linker (Figure 4.4) [18]. This derivative formed an association dimer in aqueous solution and exhibited typical pyrene fluorescence around 370–400 nm together with strong excimer-like fluorescence centering at 475 nm. A new fluorescence band appeared around 390–460 nm upon addition of $NaHCO_3$. None of the anions except bicarbonate induced the new fluorescence band, and metal cations (Zn^{2+}, Na^+, or K^+) showed no ability to cause the new fluorescence band. This selective fluorescence change can act as a chemosensor for bicarbonate, which is a physiologically

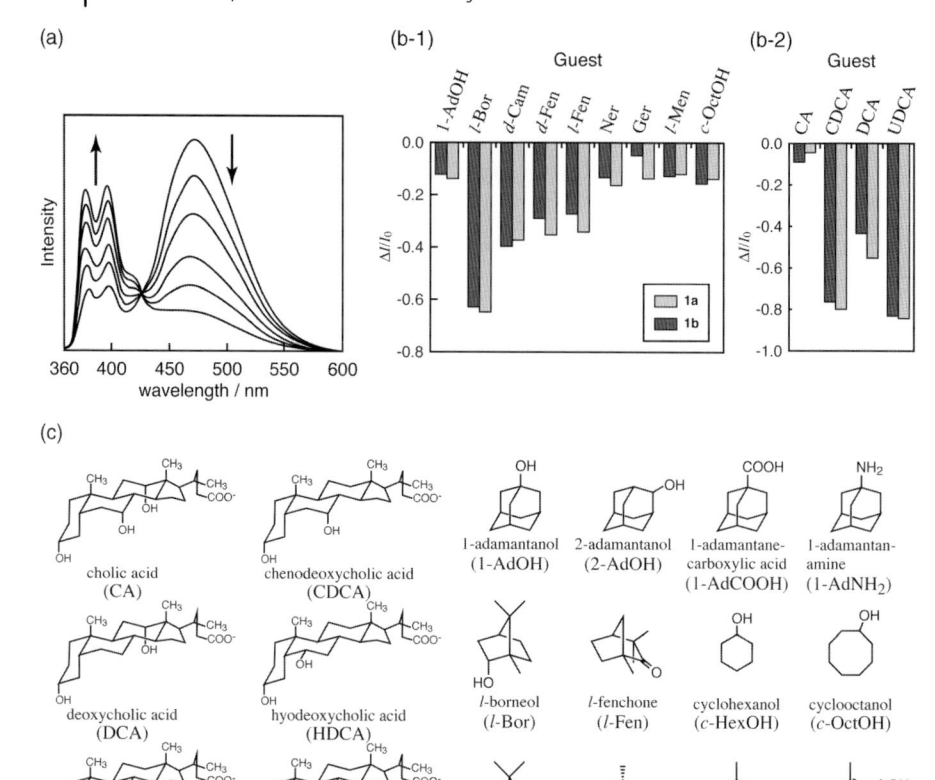

Figure 4.3 (a) Fluorescence spectra of **1b** in the presence of various concentrations of *l*-borneol; [**1b**] = 2.85 × 10^{-5} M (10% DMSO), [*l*-borneol] = 0, 0.333, 0.667, 1.33, 2.00 and 3.33 × 10^{-3} M. (b) Sensitivity parameters ($\Delta I/I_0$) of **1a** and **1b** ($\Delta I = I - I_0$, with I and I_0 being the excimer emission intensities in the presence and absence of guests, respectively); (b-1) [**1b**] = 3 × 10^{-5} M, [guest] = 2 × 10^{-3} M, and (b-2) [**1b**] = 3 × 10^{-5} M, [guest] = 2 × 10^{-4} M. (c) Structures of guests. Adapted from References [16] and [17].

important anion that plays a vital role not only in maintaining the pH of biological fluids but also in signal transduction in intracellular events.

4.2.3
Bis-Pyrene-Appended γ-Cyclodextrins

The sensor system based on compound **1** has a defect from a practical point of view, because the formation of association dimers depends on the concentration of the γ-CD derivatives, and the concentration of the γ-CD derivatives has to be set at a constant value to obtain comparable results for various guests. Aoyagi has prepared

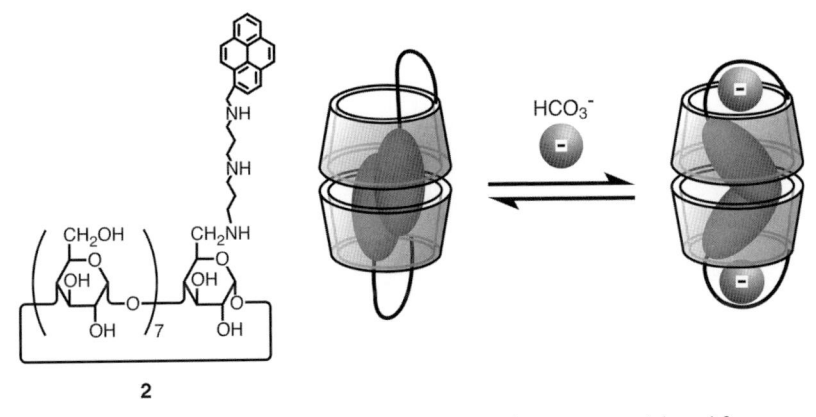

Figure 4.4 Pyrene-appended γ-cyclodextrin (**2**) as a bicarbonate sensor. Adapted from Reference [18].

γ-CD derivatives **3–6** bearing two pyrene units to overcome this defect and thereby construct more effective sensory systems (Figure 4.5) [19]. Because there are four regioisomers, each of whose two pyrenes are located on different glucose units, it is interesting to see how the regioisomers respond differently to the guest species. Compounds **3–6** have a methylene spacer between the γ-CD and the pyrene unit to improve the flexibility of the appending unit. All derivatives showed weak normal emission with peaks around 380 and 400 nm, and extremely strong excimer emission with the peak around 490 nm. The excimer emission arose not from intermolecular pyrene–pyrene interaction but from an intramolecular interaction under the experimental conditions. The peak intensity of the excimer emission decreased upon addition of lithocholic acid, while the monomer emission peaks were enhanced. The isoemissive point was observed at 445 nm, indicating the equilibrium between the self-inclusion complex and the 1 : 1 host–guest complex. This observation suggests that the two pyrene rings, which are included in the cavity and form the excimer, are excluded from the cavity upon guest accommodation and tend to be separated from each other (Figure 4.6a). The guest selectivities of **3–6** were evaluated using the sensitivity parameters $\Delta I/I_0$ (Figure 4.6b). The sensitivity parameters of each host are

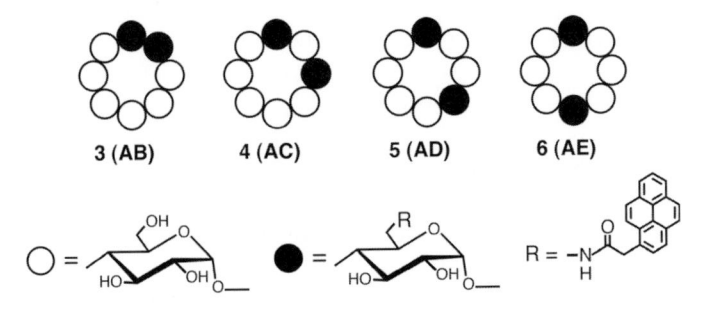

Figure 4.5 Bis-pyrene-appended γ-cyclodextrins (**3–6**).

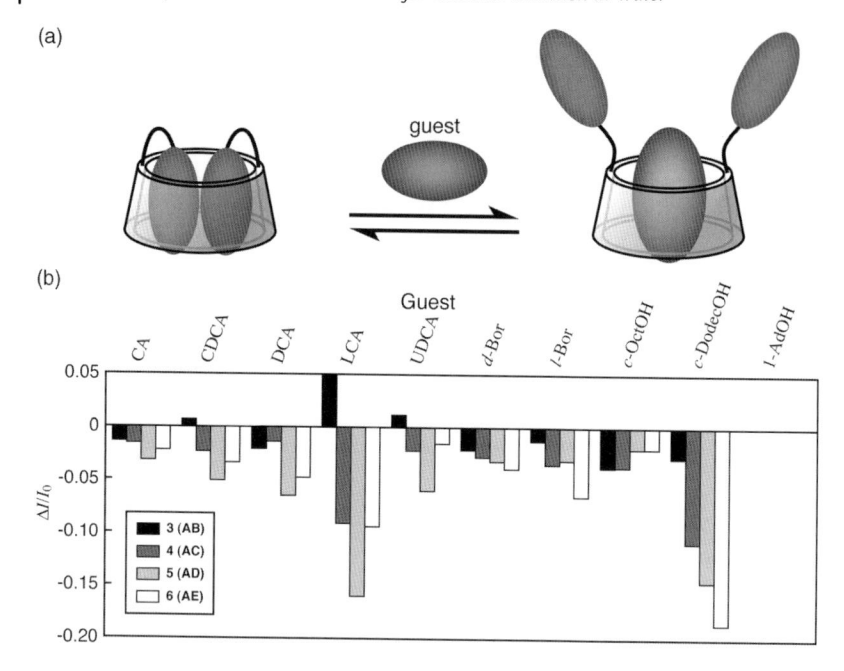

Figure 4.6 (a) Guest-induced conformational change of bis-pyrene-appended γ-cyclodextrins; (b) sensitivity parameters ($\Delta I/I_0$) of **3–6** ($\Delta I = I - I_0$, with I and I_0 being the excimer emission intensities in the presence and absence of guests, respectively); [**3–6**] = 1.5×10^{-7} M, [LCA] = [CA] = [DCA] = [UDCA] = [CDCA] = 1.0×10^{-5} M, [d-Bor] = [l-Bor] = [1-AdOH] = [c-OctOH] = [c-DodecOH] = 1.0×10^{-4} M. Adapted from Reference [19].

different and combined use of these hosts could produce a pattern recognition system for molecule recognition.

4.3
Fluorophore–Amino Acid–CD Triad Systems

4.3.1
Dansyl-Leucine-Appended CDs

Ikeda *et al.* have prepared fluorophore–leucine–CD triad systems **7–10** as chemosensors for molecule detection (Figure 4.7) [20]. These chemosensors have a leucine unit as a spacer between the CD cavity and the dansyl (DNS) unit which acts as a fluorophore. The hydrophobic side chain of the amino acid is expected to increase binding affinity due to a hydrophobic cap effect. The chirality of the leucine unit is also expected to affect the binding affinity for guests.

Conformational changes of dansyl-leucine-appended CDs upon addition of guests were first investigated by fluorescence decay and NMR techniques. While the

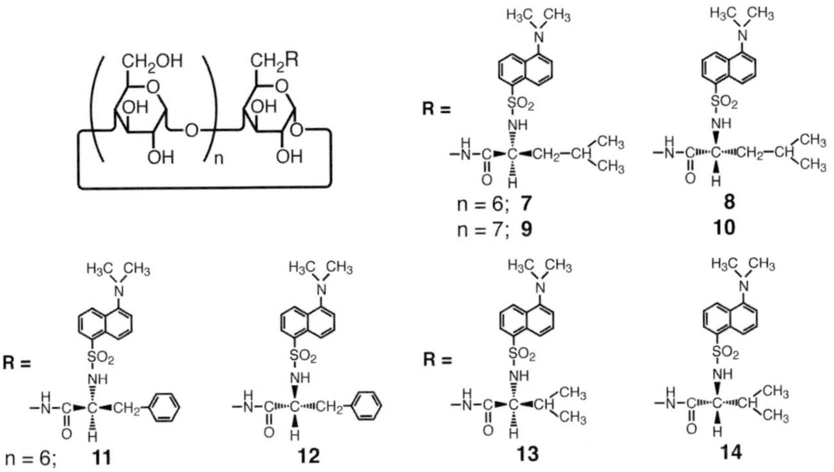

4.3 Fluorophore–Amino Acid–CD Triad Systems

Figure 4.7 Dansyl-amino acid-appended cyclodextrins (**7–14**).

conformational interconversion occurs too rapidly to be followed by NMR spectroscopy, analysis of the fluorescence decay of the fluorophore pendant can provide useful information with respect to conformational features. This is because the fluorescence lifetimes of many fluorophores are of a measurable magnitude (nanoseconds) for each conformation. The DNS unit is sensitive to hydrophobicity around it and has different lifetimes when located inside the cavity as opposed to in bulk water solution. All of the dansyl unit fluorescence decays were analyzed by a simple double-exponential function, and the results are summarized in Table 4.2. This means that there are two kinds of observable conformational isomers and that these two species are in equilibrium (Figure 4.8). The longer and shorter lived species are the ones with the dansyl unit inside and outside the cavity, respectively.

Table 4.2 Fluorescence lifetimes of **7–10** alone and in the presence of guest in aqueous solution.[a].

Host	Guest	τ_1 (ns)	A_1	τ_2 (ns)	A_2	χ^2
7	—	5.7	0.33	17.7	0.67	1.32
8	—	6.9	0.23	17.8	0.77	1.38
9	—	7.3	0.46	13.1	0.54	1.13
10	—	7.9	0.78	13.5	0.22	1.06
7	1-AdOH	5.5	0.99	19.0	0.01	1.38
8	1-AdOH	5.9	0.97	17.9	0.03	1.07
9	l-Bor	5.5	0.57	13.7	0.43	1.23
10	l-Bor	5.7	0.86	13.2	0.14	1.03
9	c-HexOH	7.6	0.40	14.8	0.60	1.22
10	c-HexOH	7.3	0.45	14.1	0.55	0.95

a) [a][**7–10**] = 2×10^{-5} M, [1-AdOH] = 5×10^{-3} M, [l-Bor] = 3×10^{-3} M, [c-HexOH] = 5×10^{-2} M. Decay curves were fitted to Eq. (4.1); χ^2 is the parameter for the goodness of the fit.

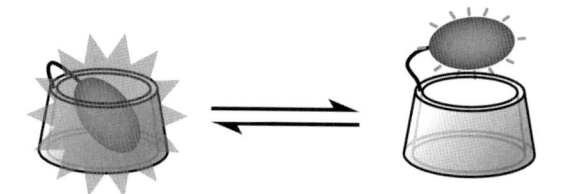

Figure 4.8 Conformational equilibrium of chromophore-appended cyclodextrin in aqueous solution. Adapted from Reference [20].

The fluorescence decay is expressed by Eq. (4.1):

$$D(t) = A_1 \exp(-t/\tau_1) + A_2 \exp(-t/\tau_2) \tag{4.1}$$

where A_i is a pre-exponential factor contributing to the signal at zero time and τ_i is the lifetime of the ith component.

Alternatively, A_1/A_2 can be expressed by Eq. (4.2):

$$A_1/A_2 = (C_1 \varepsilon_1 \Phi_1/\tau_1)(C_2 \varepsilon_2 \Phi_2/\tau_2) = (\varepsilon_1 C_1/\tau_{01})/(\varepsilon_2 C_2/\tau_{02}) \tag{4.2}$$

where

C_i is the concentration of species i,
ε_i is the molar extinction coefficient,
Φ_i is the quantum efficiency,
τ_{0i} is the intrinsic fluorescence lifetime.

If:

$$\varepsilon_1 = \varepsilon_2 \text{ and } \tau_{01} = \tau_{02} \tag{4.3}$$

then:

$$A_1/A_2 = C_1/C_2 \tag{4.4}$$

Since the fluorescence of the self-inclusion state and of the non-self-inclusion state arises from the same unit, a reasonable approximation of the ratio of the intrinsic lifetimes would be unity [21–25]. Therefore, the equilibrium can be quantified from parameters obtained directly from the analysis of fluorescence decay curves. This estimation is useful for discussion of the equilibrium semi-quantitatively, although it is reported that τ_0 of the dansyl unit depends slightly on solvent polarity and, therefore, Eq. (4.4) is only a rough approximation [23].

The fractions of longer lifetime species for β-CD (**7, 8**) derivatives are larger than those of γ-CD derivatives (**9, 10**) (Table 4.2). Thus, the self-inclusion states of the β-CD derivatives are more stable than those of the γ-CD derivatives, and the cavity of γ-CD is too large to form a stable self-inclusion complex of the DNS unit. The stability of the self-inclusion state is affected by the chirality of the leucine unit. The self-inclusion states of **8**, which have the D-Leu unit, are more stable than those of **7**, which have the L-Leu unit, and the self-inclusion states of **9**, which have the L-Leu unit, are more stable than those of **10**, which have the D-Leu unit.

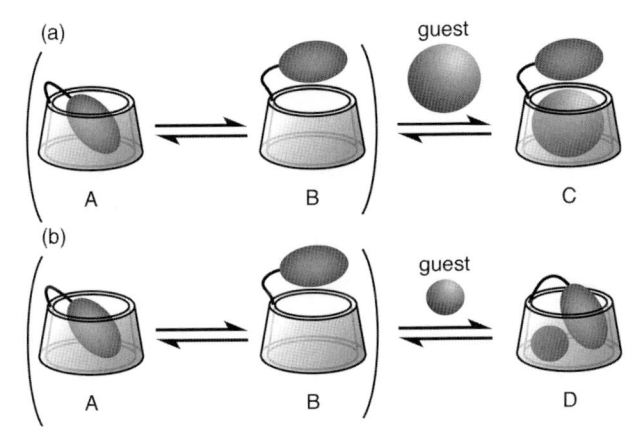

Figure 4.9 Conformational equilibrium in aqueous solution and guest-induced conformational change of chromophore-appended cyclodextrins; (a) larger guest and (b) smaller guest. Adapted from Reference [20].

The fractions of the shorter lifetime species of **7** and **8** increase upon addition of 1-adamantanol, while the fractions of the longer lifetime species of the hosts decrease (Table 4.2). This indicates that the induced-fit conformational change of the chromophore-appended CD occurs in association with guest accommodation, excluding the chromophore, from the inside to the outside of the CD cavity (Figure 4.9a). The fact that the fluorescence decay curves of the hosts are still double-exponential indicates that the fluorescence lifetimes of species B and C in Figure 4.9a are similar and not resolved by this lifetime measurement system.

In contrast, the fractions of the longer lifetime component of **9** and **10** increase upon addition of cyclohexanol but decrease upon addition of *l*-borneol (Table 4.2). This result suggests that the dansyl unit of **9** and **10** is co-included with cyclohexanol in the cavity (Figure 4.9b), whereas it is excluded by *l*-borneol (Figure 4.9a). Since the cavity of γ-CD is too large to form a stable intramolecular complex with the dansyl unit, a small guest such as cyclohexanol may act as a spacer to form stable complexes [20]. However, *l*-borneol is too large to act as a spacer, and accommodation of the guest displaces the chromophore from the inside to the outside of the CD cavity.

The NMR spectra of **7** and **8** provide information about their structures in the self-inclusion state, because the DNS units of **7** and **8** are located inside the cavity in the major conformational state. The ^{1}H NMR spectra of **7** and **8** were assigned by the combined use of various 1D and 2D NMR techniques. Their structures were estimated from NOE data and the degree of the anisotropic ring current effect from the DNS unit in the ^{1}H resonances of the CD protons [26]. The DNS unit of **8** is located deeper within the cavity than that of **7** (Figure 4.10). This deeper inclusion into its cavity makes **8** the more stable self-inclusion complex, and this higher stability is also reflected in its lower binding abilities.

Figure 4.11 shows the sensitivity parameters expressed as $\Delta I/I_0$ (where $\Delta I = I - I_0$, with I and I_0 being the fluorescence intensities in the presence and absence of the guest, respectively). Compounds **7** and **8** exhibit similar trends in guest dependency

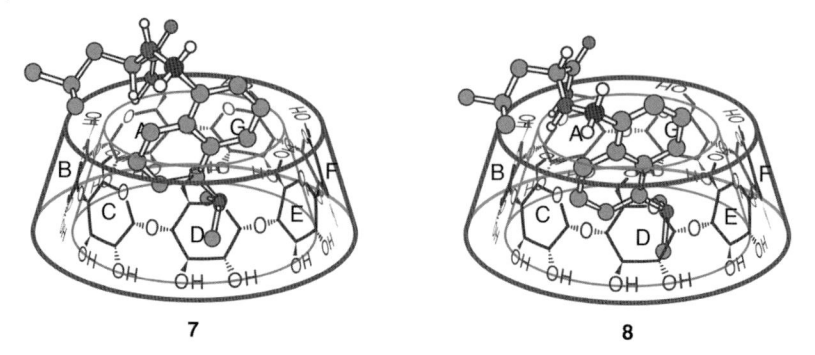

Figure 4.10 Estimated structures of **7** and **8**. Adapted from Reference [20].

of the $\Delta I/I_0$ value, but the $\Delta I/I_0$ value of **7** is larger than that of **8** in all cases, because the binding constants of **7** are over twice as large as those of **8** for most of the guests (Figure 4.11a).

The guest dependencies of the sensitivity parameters ($\Delta I/I_0$) of **9** and **10** are similar, although the $\Delta I/I_0$ value of **10** is larger than that of **9** (Figure 4.11b). Notably, this chirality dependence is opposite to that of the corresponding **7** and **8**. The $\Delta I/I_0$ values of **9** and **10** for Ger, *l*-Men, and *c*-HexOH are positive, which suggests that these guests are co-included with the DNS unit in the cavity (Figure 4.9b).

4.3.2
Dansyl-Valine-Appended and Dansyl-Phenylalanine-Appended CDs

There is interest in whether the tendency of the chirality of the leucine spacer unit to influence the stability of the self-inclusion state is shared by other amino acid spacer units. This was investigated by preparing new fluorophore–amino acid–CD triad

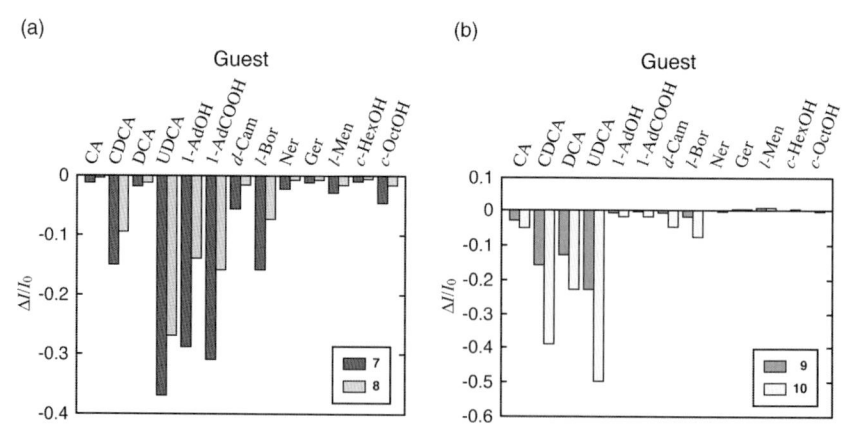

Figure 4.11 Sensitivity parameters ($\Delta I/I_0$) of (a) **7** and **8** and (b) **9** and **10** for various guests ($\Delta I = I - I_0$, with I and I_0 being the fluorescence intensities in the presence and absence of guests, respectively); (a) [**7** and **8**] = 2×10^{-6} M, [guest] = 1×10^{-5} M and (b) [**9** and **10**] = 2×10^{-6} M, [guest] = 1×10^{-4} M. Adapted from Reference [20].

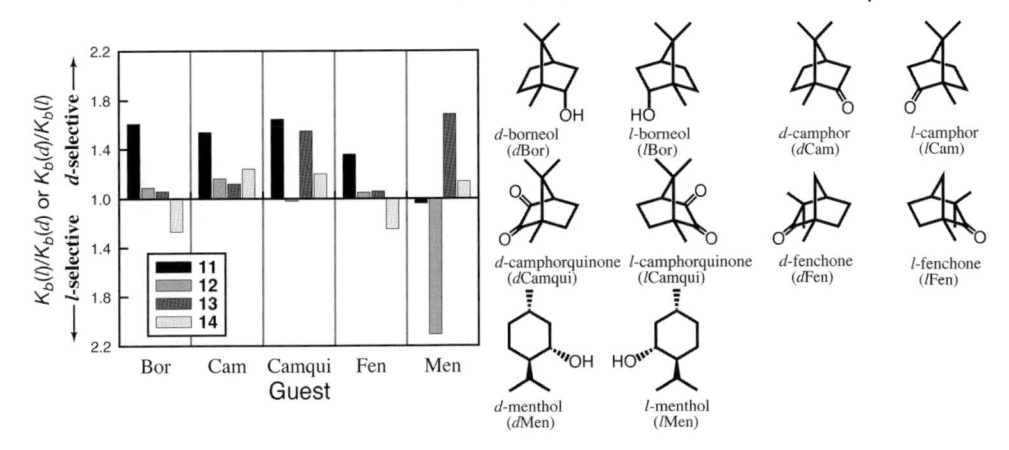

Figure 4.12 Chiral discrimination abilities of **11–14**. Adapted from Reference [27].

systems, having either phenylalanine or valine as the spacer (Figure 4.7) [27]. The fluorescence intensity of **12** is much larger than that of **11**. This influence of the spacer chirality on the fluorescence intensity is similar to the N-dansyl-leucine-appended β-CD. In contrast, the fluorescence intensity of **13** is similar to that of **14**, and both intensities are larger than that of **7** or **8**. The steric hindrance of the isopropyl side chain of valine scarcely inhibits the self-inclusion of the DNS unit into the CD cavity, whereas the benzyl unit of phenylalanine affects the self-inclusion depth of the DNS unit.

There is also interest in the ability of the chirality of the spacer to influence enantioselective recognition of chiral guests. Enantioselective chemosensors are currently of great interest to determine the enantiomeric excess of samples in drug discovery or that of products in high-throughput screening of enantioselective catalysts and biocatalysts. Figure 4.12 shows the chiral discrimination abilities of **11–14** for some norbornane and cyclohexane derivatives. Compound **11** showed good d-selectivity for borneol, camphor, camphorquinone, and fenchone, but poor selectivity for menthol. Compounds **12** and **13** showed good l-selectivity and d-selectivity, respectively, for menthol. Only menthol has a cyclohexane skeleton, whereas the other guests have the same norbornane framework. Compound **11** shows high selectivity for the norbornane derivatives but does not show selectivity for the cyclohexane derivative. Each of the four chemosensors **11–14** shows a different selectivity pattern for each of the norbornane derivatives having the same framework but a different functional group.

4.4
Molecular Recognition by Regioisomers of Dansyl-Appended CDs

4.4.1
6-O, 2-O, and 3-O-Dansyl-γ-CDs

A CD has three types of hydroxy groups: 6-OH on the primary hydroxy side, and 2-OH and 3-OH on the secondary hydroxy side. It is of interest to note how the regioisomers of the modified CDs differ in molecule recognition. In organic solvents, the primary hydroxy groups of CDs are more reactive than the secondary hydroxy groups and can be selectively modified by a sulfonation reagent. Under alkaline conditions, the secondary hydroxy group can be sulfonated, but the reaction yield is not high because the sulfonated product is easily decomposed through epoxide formation. Wang found a new method for sulfonation of CDs in which a solution of γ-CD in a pH 10 carbonate buffer was poured into a vigorously stirred solution of dansyl chloride in DMF [28]. The reaction was stopped after a few seconds by addition of hydrochloric acid. The mono-dansylated γ-CDs at the C2 or C3 position (**16** or **17**) was produced in high yields, and no C6 dansylated product was found. Compounds **16** and **17** were separated by HPLC. The C6 mono-dansylated γ-CD (**15**) was prepared by the reaction of γ-CD with dansyl chloride in pyridine. The molecular recognition abilities of **15–17** were evaluated using sensitivity parameters ($\Delta I/I_0$) (Figure 4.13a). These three hosts show

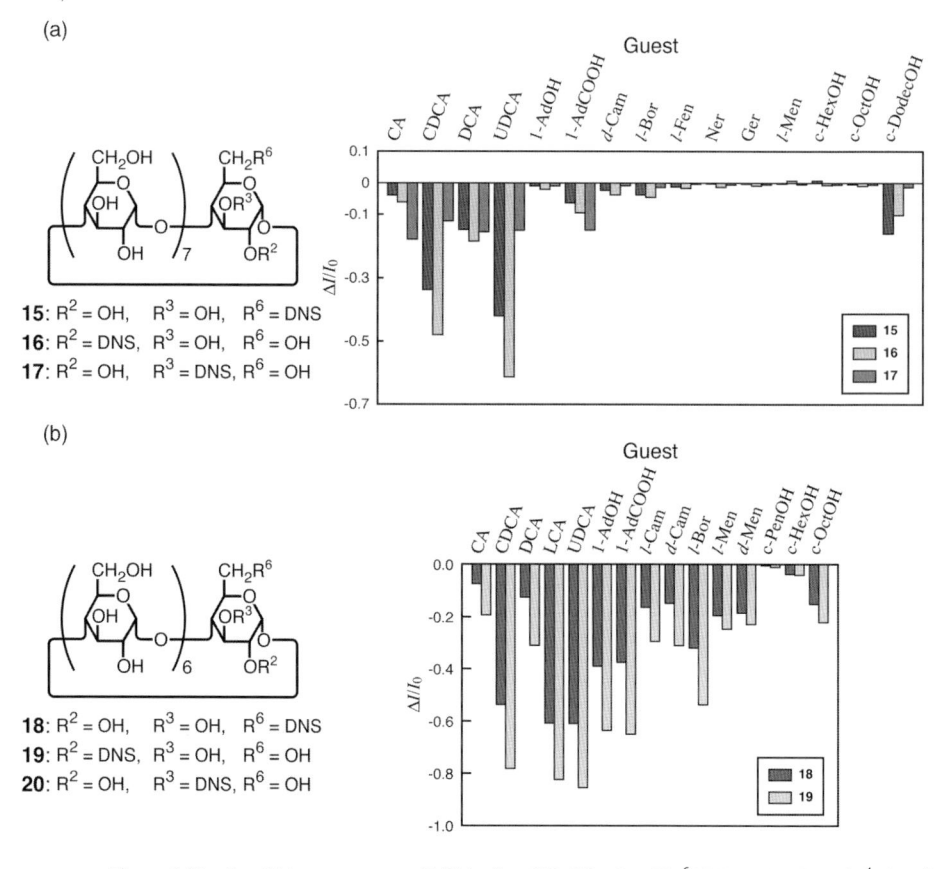

Figure 4.13 Sensitivity parameters ($\Delta I/I_0$) of (a) dansyl-appended γ-cyclodextrins (**15–17**) and (b) dansyl-appended β-cyclodextrins (**18** and **19**) for various guests; (a) [**15–17**] = 5 × 10^{-6} M, [guest] = 1 × 10^{-4} M and (b) [**18** and **19**] = 1 × 10^{-5} M, [guest] = 1 × 10^{-4} M. Adapted from References [28] and [29].

remarkably different sensitivities for guests. The sensitivity of **16** was higher than that of **15** in most cases, and the sensitivity of **17** was the lowest. This difference results from the difference in location of the DNS unit in the CD cavity. Only the fluorescence intensity of **15** increased upon addition of small guests (geraniol, nerol, cyclooctanol, and cyclohexanol). This suggests that small guests can be co-included only with the DNS unit of **15** in the cavity, because the presence of a single methylene group between the DNS unit and the CD framework causes flexibility in the DNS unit of **15**.

4.4.2
6-O, 2-O, and 3-O-Dansyl-β-CDs

The regioisomers of mono-dansylated β-CDs were also prepared using a method similar to that for the γ-CD derivatives [29]. The fluorescence intensity of **20** was much stronger than that of **18** or **19**. The fluorescence decay analysis indicates that the DNS unit of **20** is completely accommodated inside the CD cavity. The decrease in fluorescence intensity of **20** upon addition of a guest was very slow, and exclusion of the DNS unit required a dramatic deformation of the CD framework. Compound **20** seems to be unsuitable for use as a fluorescence sensor. The fluorescence decay analysis suggests that **18** and **19** are in a dynamic equilibrium between two conformations in aqueous solution, wherein the DNS unit is either inside or outside the cavity as the longer and shorter lifetime species, respectively. The primary conformation of the DNS unit of **18** is located in a more hydrophilic environment, while that of **19** is located in a more hydrophobic environment. This conformational difference causes the difference in variation in fluorescence when guests are added. Sensitivity parameters for **18** and **19** are shown in Figure 4.13b. The fluorescence variations of **19** are larger than those of **18** for all guests added, whereas they show a similar sensitivity pattern for guests.

4.5
Turn-On Fluorescent Chemosensors

The "turn-off" fluorescent chemosensor system is effective for detecting molecules, as shown in the previous section, but it has some defects. First, self-inclusion of the chromophore can inhibit accommodation of the guest. Second, the affinities of both β-CD and γ-CD for bile acid derivatives are greater than their affinities for adamantane derivatives. Therefore, adamantanol cannot be detected in a mixed solution of a bile acid and adamantanol using 'turn-off' fluorescent chemosensors. Finally, the detection of a guest is accompanied by a decrease in the fluorescence intensity in the "turn-off" fluorescent chemosensor system, whereas an increase in the emission intensity in response to a guest would be a more effective approach for chemical sensing systems. Therefore, a new "turn-on" fluorescent chemosensor that overcomes these defects is desired.

4-Amino-7-nitrobenz-2-oxa-1,3-diazole (NBDamine) was selected as a fluorophore for a new type of chemosensor, and new chemosensors, **21** and **22**, were prepared

n = 6; **21**
n = 7; **22**

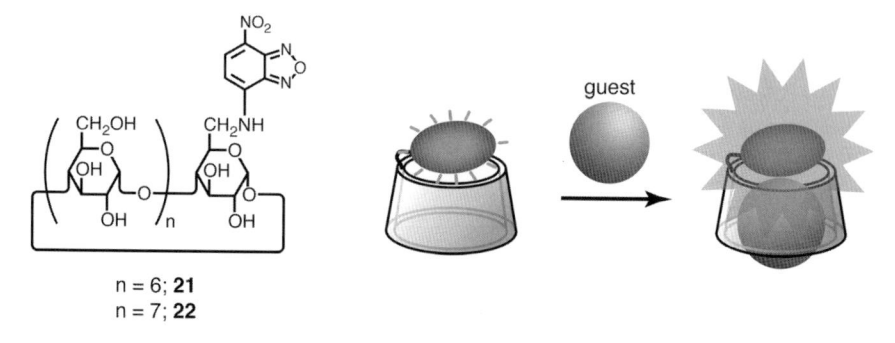

Figure 4.14 NBDamine-appended cyclodextrins (**21** and **22**). Adapted from Reference [30].

(Figure 4.14) [30]. NBDamine displays the interesting property of fluorescing weakly in water and strongly in organic solvents, membranes, or hydrophobic environments. The fluorescence intensity of **21** increased upon addition of 1-adamantanol (1-AdOH), indicating an increase in hydrophobicity near the NBDamine unit induced by accommodation of the guest. Figure 4.15a shows the sensitivity parameters of **21** for various guests; **21** is quite sensitive to the adamantane and borneol derivatives, which have a comparatively spherical shape that fits the β-CD cavity. The response of **21** to these guests is a large increase in the fluorescence intensity. Notably, **21** is not sensitive to bile acids, although bile acids are strongly bound by the native β-CD [31, 32]. This phenomenon can be explained by the equation in Figure 4.16. The bile acid derivative can be accommodated inside the β-CD cavity, but the NBD unit moves away from the entrance of the CD cavity after making the inclusion complex with the bile acid derivative. The NBD is still located in a hydrophilic environment in the complex with the bile acid and its fluorescence intensity does not increase.

The sensitivity parameters of **22** for various guests are also shown in Figure 4.15a. Compound **22** is relatively sensitive to each bile acid but has no response to the

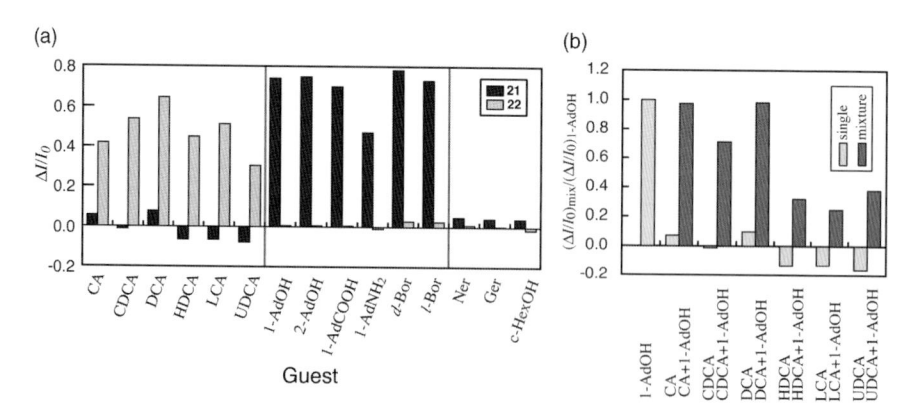

Figure 4.15 (a) Sensitivity parameters ($\Delta I/I_0$) of **21** and **22** for various guests; [**21** and **22**] = 5×10^{-6} M, [guest] = 1×10^{-5} M. (b) Relative sensitivity parameters (($\Delta I/I_0)_{mix}/(\Delta I/I_0)_{1\text{-AdOH}}$) of **21** ($5 \times 10^{-6}$ M) for mixtures of guests (each at 1×10^{-5} M). Adapted from Reference [30].

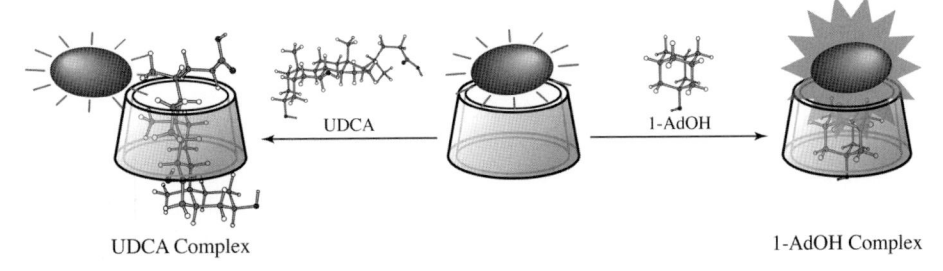

Figure 4.16 Guest-induced conformational changes of NBDamine-appended cyclodextrins for UDCA (ursodeoxycholic acid) complex and 1-AdOH complex.

adamantane derivatives. The responses of **22** to guests are also increases in the fluorescence intensity. Considering the differing guest-response patterns of **21** and **22**, discrimination between the bile acids and adamantane derivatives at any concentration by the combined use of **21** and **22** is now a simple matter.

Notably, 1-AdOH can be detected even in the presence of a bile acid by using **21** [30]. This is the first example of adamantanol being detected in the presence of a bile acid by a CD-based chemosensor. The relative sensitivity parameters $[(\Delta I/I_0)_{mix}/(\Delta I/I_0)_{1\text{-AdOH}}]$ of **21** (5×10^{-6} M) for a solution containing both a bile acid (1×10^{-5} M) and 1-AdOH (1×10^{-5} M) are positive and are comparable to that for the addition of 1-AdOH alone (Figure 4.15b).

4.6
Effect of Protein Environment on Molecule Sensing

Nakamura *et al.* have prepared a monensin-capped dansyl-appended β-CD (**23**) as a fluorescent chemosensor for molecule recognition (Figure 4.17) [33, 34]. Monensin is an antibacterial hydrophobic substance and can bind the sodium cation, forming a ring-like conformation with a sodium ion at its center. Monensin that is introduced to

Figure 4.17 Monensin-appended (**23**) and biotin-appended cyclodextrins (**24–26**).

the primary hydroxyl side of the CD can act as a hydrophobic cap that is responsive to sodium ion. The guest-dependent fluorescence variation of **23** is enhanced by the addition of alkali metal cations in the order $Na^+ > K^+ > Li^+ > Cs^+$, which is consistent with the expected order for binding of alkali metal cations by monensin itself. The binding constants of **23** increase twofold for *l*-camphor, nerol, and geraniol upon addition of sodium ions, and sevenfold for cyclohexanol.

Wang has prepared a fluorescent β-CD (**24**) that has *p*-N,N-dimethylaminobenzoyl and biotin units as a fluorophore and a protein (avidin)-binding site, respectively (Figure 4.17) [35]. The binding constant of **24** for UDCA increases 8.9-fold when avidin was added. The fluorescence intensity of **24** in the presence of avidin increases upon addition of guests, while it decreases when these same guests are added in the absence of avidin. In the presence of avidin, accommodation of a guest into the CD cavity displaces the fluorophore unit from the CD cavity to the hydrophobic pocket of avidin instead of the bulk water solution. These results imply that avidin can act as a localized hydrophobic environment, which leads to highly responsive sensors.

Ikunaga has prepared biotin-appended dansyl-appended CDs **25** and **26** (Figure 4.17) [36]. The fluorescence intensities of these CD derivatives were enhanced more than three times by the addition of avidin. The binding constants of **25** and **26** for CDCA increase 1.4-fold and 2.5-fold, respectively, in the presence of avidin. The fluorescence decay of the dansyl unit of **25** was analyzed by simple double-exponential functions. The ratio of the longer lifetime component of **25** in the presence of avidin increases from 7.3% to 36.2% upon addition of HDCA. The fluorophore is displaced from the CD cavity to a very hydrophobic pocket of avidin upon addition of the guest. Avidin provides a wide hydrophobic environment at the primary hydroxyl side of β-CD, excluding the bulk water.

4.7
CD–Peptide Conjugates as Chemosensors

It is necessary to introduce multiple functional groups on CD to improve the sensitivity and selectivity of CD-based chemosensors, but synthesis of a multifunctionalized CD is not easy due to the difficulty in selective modification of the many hydroxy groups on a CD. If the rigid α-helix scaffold of a peptide was used, two different photoreactive moieties could easily be placed on both the primary hydroxy side and the secondary hydroxy side of CD.

Matsumura has prepared CD–peptide conjugates that have the DNS unit and the β-CD in the α-helix peptide side chain (Figure 4.18) [37]. A 17-residual alanine-based peptide (**27–30**) was used for the α-helix-stabilizing properties of alanine. In addition, three intramolecular Glu-Lys salt bridges were introduced into the peptide to stabilize the α-helix. Four types of peptides were prepared to examine the effect of the positions and directions of the DNS and β-CD moieties on the α-helix peptide structure and the molecular sensing abilities. The order of helix formation was **28** (75%) > **27** (61%) > **30** (53%) > **29** (44%). Compounds **27**, **28**, and **30** form a more helical structure than peptide **31** (45% helix), which has neither the β-CD nor

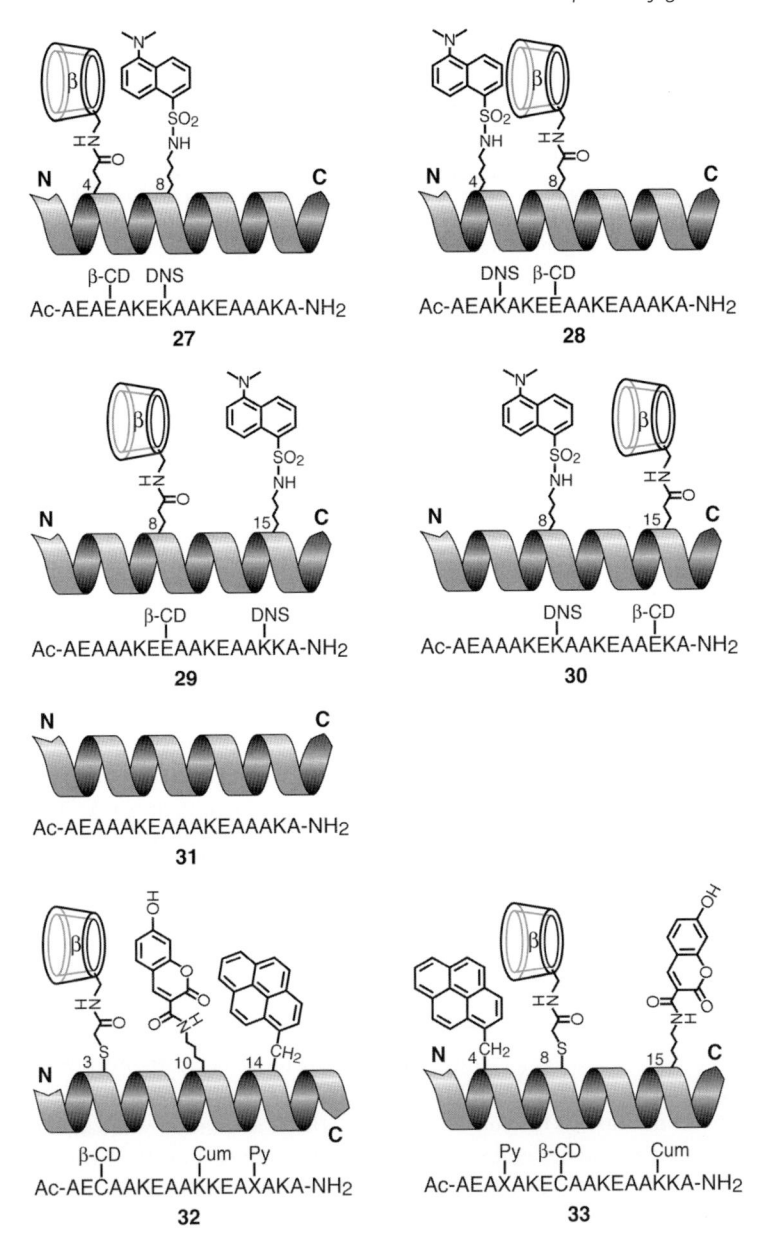

Figure 4.18 Cyclodextrin–peptide conjugates (**27–33**).

DNS unit. The helix content is reduced upon addition of 1-AdOH. The intramolecular inclusion complex of the β-CD with the DNS unit promotes higher helical structures. Fluorescence spectra studies indicate that deeper inclusion of the DNS unit into its cavity causes stabilization of the α-helix structure. Time-resolved fluorescence studies suggest that the DNS unit is in an equilibrium between two

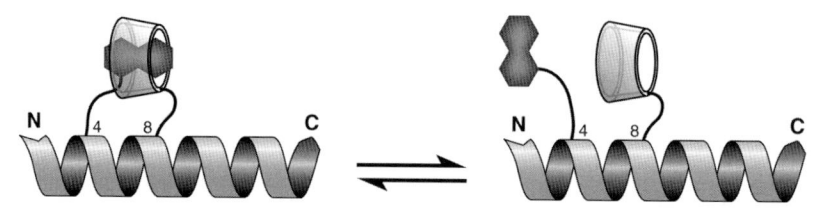

Figure 4.19 Conformational equilibrium of cyclodextrin–peptide conjugates in aqueous solution. Adapted from Reference [37].

conformations, located inside and outside the CD cavity, similar to DNS-Leu-appended β-CD (**7** or **8**) (Figure 4.19). The binding affinity to guests is related to the orientation of the β-CD unit along the α-helix peptide. The binding constants of **27** and **29** are larger than those of **28** and **30**, respectively. The binding affinities and selectivities for the bile acids as guests are similar to the chromophore-appended CD. The guest selectivity of β-CD on the α-helix is not disturbed by the peptide scaffold, and the CD–peptide conjugate can be used as a chemosensor.

It is interesting to know what happens if two chromophores are attached to both primary and secondary hydroxyl sites so as to sandwich the CD between the two chromophores. Hossain has prepared CD–peptide hybrids **32** and **33** [38, 39]. Pyrene (donor) and coumarin (acceptor) were introduced on either one (**32**) or each side (**33**) of the CD on an α-helix peptide scaffold to make a FRET system for a chemosensor. Hybrid **32** showed high fluorescence emission of the acceptor coumarin at around 448 nm upon irradiating at 340 nm, corresponding to the absorption wavelength of pyrene, and indicating that energy transfer occurs from pyrene to coumarin in **32**. The fluorescence intensity of both the acceptor coumarin and the donor pyrene decreases upon addition of a guest, hyodeoxycholic acid (HDCA). This attenuation of the fluorescence intensity is thought to be associated with the exclusion of the coumarin unit from the inside to the outside of the β-CD cavity upon accommodating a HDCA molecule (Figure 4.20a). Being excluded from the CD cavity in **32**, coumarin may come into very close contact with the pyrene unit, resulting in the fluorescence quenching.

Hybrid **33** exhibits intramolecular FRET without quenching of the two fluorophores in the absence of guests. The fluorescence intensity of the acceptor (coumarin) upon irradiating at the absorption wavelength of pyrene (340 nm) is markedly decreased with an increase in the concentration of HDCA as a guest, whereas the intensity of the donor (pyrene) emission is increased. Addition of HDCA causes coumarin to be excluded from the CD cavity, and, thus, the distance between pyrene and coumarin is increased, resulting in less efficient energy transfer (Figure 4.20b).

4.8
Immobilized Fluorescent CD on a Cellulose Membrane

Tanabe has immobilized a dansyl-appended CD on a cellulose membrane for practical use [40, 41]. Some of the glucose rings in the cellulose were opened by

4.8 Immobilized Fluorescent CD on a Cellulose Membrane | 131

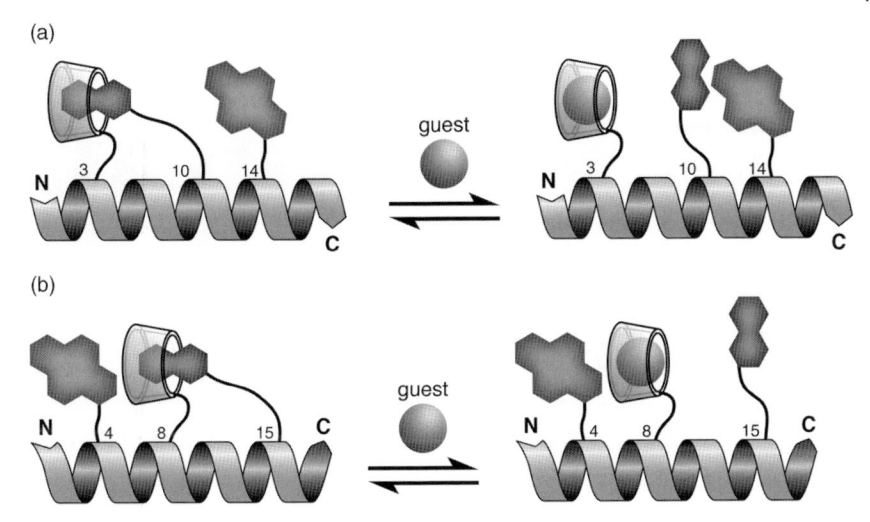

Figure 4.20 Guest-induced conformational change of cyclodextrin–peptide conjugates: (a) **32** and (b) **33**. Adapted from References [38] and [39].

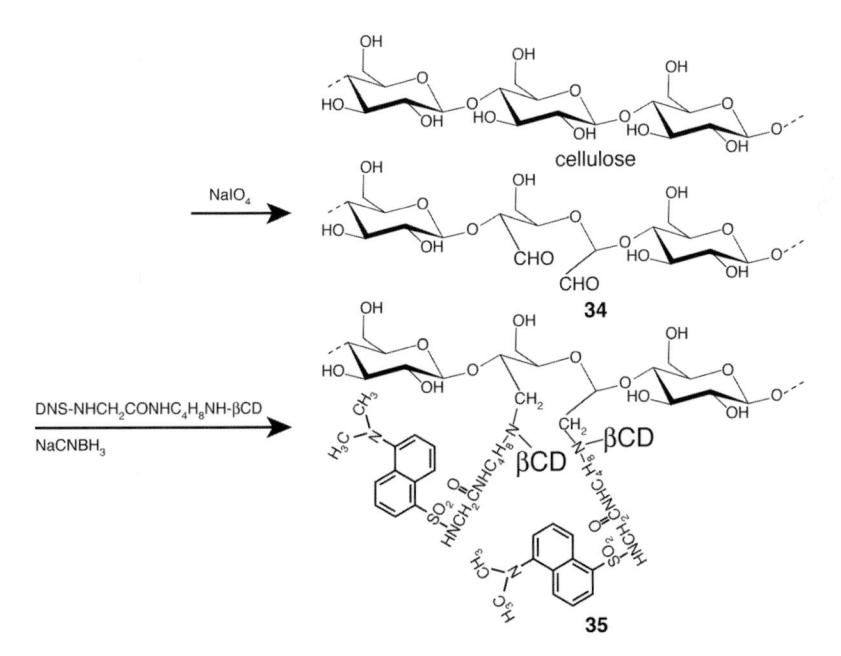

Figure 4.21 Immobilized fluorescent cyclodextrins on a cellulose membrane. Adapted from Reference [41].

(a)

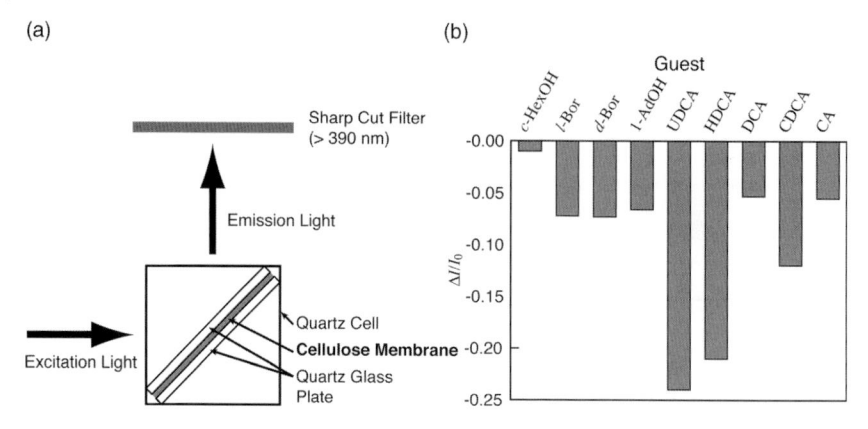

Sharp Cut Filter (> 390 nm)

Emission Light

Excitation Light

Quartz Cell
Cellulose Membrane
Quartz Glass Plate

(b)

Guest: c-HexOH, l-Bor, d-Bor, l-AdOH, UDCA, HDCA, DCA, CDCA, CA

$\Delta I/I_0$: -0.00, -0.05, -0.10, -0.15, -0.20, -0.25

Figure 4.22 (a) Schematic representation for membrane fluorescence measurements of **35**; (b) sensitivity parameters ($\Delta I/I_0$) of **35**; [guest] = 2×10^{-5} M. Adapted from Reference [41].

NaIO$_4$ oxidation to produce two aldehyde groups (Figure 4.21). The fluorescent CDs were immobilized by reaction with the aldehyde groups on the membrane. The responses of fluorescent CDs immobilized on the cellulose membrane upon addition of guests are similar to that of CD chemosensors in aqueous solution (Figure 4.22).

4.9
Conclusion

Chromophore-appended CDs are effective chemosensors for detecting molecules in water. CD chemosensors can recognize the shape and size of guests. The response pattern of CD chemosensors for guests can be changed by varying the spacer between the CD cavity and the chromophore or modifying the position of the chromophore. The cavity size is also an important factor for the response pattern. Application of supramolecular technology such as a rotaxane would improve the sensitivity and/or selectivity of the chemosensors. Many types of CD chemosensors having different responses for guests are expected to be used for new pattern recognition systems to detect molecules.

References

1. Anslyn, E.V. (2007) Supramolecular analytical chemistry. *J. Org. Chem.*, **72**, 687–699.
2. de Silva, A.P., Gunaratne, H.Q.N., Gunnlaugsson, T., Huxley, A.J.M., McCoy, C.P., Rademacher, J.T., and Rice, T.E. (1997) Signaling recognition events with fluorescent sensors and switches. *Chem. Rev.*, **97**, 1515–1566.
3. Desvergne, J.P. and Czarnik, A.W.(eds) (1997) *Chemosensors of Ion and Molecule Recognition*, NATOASI Series C492, Kluwer Academic, Dordrecht.

4 Czarnik, A.W.(ed.) (1993) *Fluorescent Chemosensors for Ion and Molecule Recognition*, American Chemical Society, Washington, DC.

5 Ikeda, H. and Ueno, A. (2006) in *Cyclodextrin Materials Photochemistry, Photophysics, and Photobiology* (ed. A. Douhal), Elsevier, Paris, ch 12, pp. 267–283.

6 Ueno, A. and Ikeda, H. (2001) in *Molecular and Supramolecular Photochemistry*, vol. 8 (eds V. Ramamurthy and K.S. Schanze), Marcel Dekker, New York, ch 8 pp. 461–503.

7 Ueno, A. (1996) Fluorescent cyclodextrins for molecule sensing. *Supramol. Sci.*, **3**, 31–36.

8 Ueno, A. (1993) Fluorescent sensors and color-change indicators for molecules. *Adv. Mater.*, **5**, 132–134.

9 Dodziuk, H.(ed.) (2006) *Cyclodextrins and their Complexes -Chemistry, Analytical Methods, Applications*, Wiley-VCH Verlag GmbH, Weinheim.

10 Szejtli, J. and Osa, T.(eds) (1996) *Comprehensive Supramolecular Chemistry*, vol. **3**, Pergamon, Oxford.

11 Szejtli, J. (1988) *Cyclodextrin Technology*, Kluwer, Dordrecht.

12 Müller, A. and Wenz, G. (2007) Thickness recognition of bolaamphiphiles by α-cyclodextrin. *Chem. Eur. J.*, **13**, 2218–2223.

13 Hall, K.S., Nissan, R.A., Quintana, R.L., and Hollins, R.A. (1988) Inclusion complexes of diisopropylfluorophosphate with cyclodextrins. *J. Catal.*, **112**, 464–468.

14 Saenger, W. (1980) Cyclodextrin inclusion compounds in research and industry. *Angew. Chem.*, **92**, 343–361;Saenger, W. (1980) *Angew. Chem. Int. Ed. Engl.*, **19**, 344–362.

15 Herkstroeter, W.G., Martic, P.A., Evans, T.R., and Farid, S. (1986) Cyclodextrin inclusion complexes of 1-pyrenebutyrate: the role of coinclusion of amphiphiles. *J. Am. Chem. Soc.*, **108**, 3275–3280.

16 Ueno, A., Suzuki, I., and Osa, T. (1989) Association dimers, excimers, and inclusion complexes of pyrene-appended γ-cyclodextrins. *J. Am. Chem. Soc.*, **111**, 6391–6397.

17 Ueno, A., Suzuki, I., and Osa, T. (1990) Host-guest sensory systems for detecting organic compounds by pyrene excimer fluorescence. *Anal. Chem.*, **62**, 2461–2466.

18 Suzuki, I., Ui, M., and Yamauchi, A. (2006) Supramolecular probe for bicarbonate exhibiting anomalous pyrene fluorescence in aqueous media. *J. Am. Chem. Soc.*, **128**, 4498–4499.

19 Aoyagi, T., Ikeda, H., and Ueno, A. (2001) Fluorescence properties, induced-fit guest binding and molecular recognition abilities of modified γ-cyclodextrins bearing two pyrene moieties. *Bull. Chem. Soc. Jpn.*, **74**, 157–164.

20 Ikeda, H., Nakamura, M., Ise, N., Oguma, N., Nakamura, A., Ikeda, T., Toda, F., and Ueno, A. (1996) Fluorescent cyclodextrins for molecule sensing: fluorescent properties, NMR characterization, and inclusion phenomena of N-dansylleucine-modified cyclodextrins. *J. Am. Chem. Soc.*, **118**, 10980–10988.

21 Hashimoto, S. and Thomas, J.K. (1985) Fluorescence study of pyrene and naphthalene in cyclodextrin-amphiphile complex systems. *J. Am. Chem. Soc.*, **107**, 4655–4662.

22 Nelson, G., Patonay, G., and Warner, I.M. (1988) Effects of selected alcohols on cyclodextrin inclusion complexes of pyrene using fluorescence lifetime measurements. *Anal. Chem.*, **60**, 274–279.

23 Li, Y.-H., Chan, L.-M., Tyer, L., Moody, R.T., Himel, C.M., and Hercules, D.M. (1975) Study of solvent effects on the fluorescence of 1-(dimethylamino)-5-naphthalenesulfonic acid and related compounds. *J. Am. Chem. Soc.*, **97**, 3118–3126.

24 Huang, J. and Bright, F.V. (1990) Unimodal Lorentzian lifetime distributions for the 2-anilinonaphthalene-6-sulfonate-β-cyclodextrin inclusion complex recovered by multifrequency phase-modulation fluorometry. *J. Phys. Chem.*, **94**, 8457–8463.

25 Dunbar, R.A. and Bright, F.V. (1994) Comparison of inter- and intramolecular

26 Ikeda, H., Nakamura, M., Ise, N., Toda, F., and Ueno, A. (1997) NMR studies of conformations of N-dansyl-L-leucine-appended and N-dansyl-D-leucine-appended β-cyclodextrin as fluorescent indicators for molecular recognition. *J. Org. Chem.*, **62**, 1411–1418.

27 Ikeda, H., Li, Q., and Ueno, A. (2006) Chiral recognition by fluorescent chemosensors based on N-dansyl-amino acid-modified cyclodextrins. *Bioorg. Med. Chem. Lett.*, **16**, 5420–5423.

28 Wang, Y., Ikeda, T., Ueno, A., and Toda, F. (1992) Syntheses and molecular recognition abilities of 6-O, 2-O, and 3-O-dansyl-γ-cyclodextrins. *Chem. Lett.*, 863–866.

29 Wang, Y., Ikeda, T., Ikeda, H., Ueno, A., and Toda, F. (1994) Dansyl-β-cyclodextrins as fluorescent sensors responsive to organic compounds. *Bull. Chem. Soc. Jpn.*, **67**, 1598–1607.

30 Ikeda, H., Murayama, T., and Ueno, A. (2005) Skeleton-selective fluorescent chemosensor based on cyclodextrin bearing a 4-amino-7-nitrobenz-2-oxa-1,3-diazole moiety. *Org. Biomol. Chem.*, **3**, 4262–4267.

31 Wallimann, P., Marti, T., Fürer, A., and Diederich, F. (1997) Steroids in molecular recognition. *Chem. Rev.*, **97**, 1567–1608.

32 Rekharsky, M.V. and Inoue, Y. (1998) Complexation thermodynamics of cyclodextrins. *Chem. Rev.*, **98**, 1875–1917.

33 Nakamura, M., Ikeda, A., Ise, N., Ikeda, H., Ikeda, T., Toda, F., and Ueno, A. (1995) Dansyl-modified β-cyclodextrin with a monensin residue as a hydrophobic, metal responsive cap. *J. Chem. Soc., Chem. Commun.*, 721–722.

34 Ueno, A., Ikeda, A., Ikeda, H., Ikeda, T., and Toda, F. (1999) Fluorescent cyclodextrins responsive to molecules and metal ions. Fluorescence properties and inclusion phenomena of N^{α}-dansyl-L-lysine-β-cyclodextrin and monensin-incorporated N^{α}-dansyl-L-lysine-β-cyclodextrin. *J. Org. Chem.*, **64**, 382–387.

35 Wang, J., Nakamura, M., Hamasaki, K., Ikeda, T., Ikeda, H., and Ueno, A. (1996) A fluorescent molecule-recognition sensor with a protein as an environmental factor. *Chem. Lett.*, 303–304.

36 Ikunaga, T., Ikeda, H., and Ueno, A. (1999) The effects of avidin on inclusion phenomena and fluorescent properties of biotin-appended dansyl-modified β-cyclodextrin. *Chem. Eur. J.*, **5**, 2698–2704.

37 Matsumura, S., Sakamoto, S., Ueno, A., and Mihara, H. (2000) Construction of α-helix peptides with β-cyclodextrin and dansyl units and their conformational and molecular sensing properties. *Chem. Eur. J.*, **6**, 1781–1788.

38 Hossain, M.A., Mihara, H., and Ueno, A. (2003) Novel peptides bearing pyrene and coumarin units with or without β-cyclodextrin in their side chains exhibit intramolecular fluorescence resonance energy transfer. *J. Am. Chem. Soc.*, **125**, 11178–11179.

39 Hossain, M.A., Mihara, H., and Ueno, A. (2003) Fluorescence resonance energy transfer in a novel cyclodextrin–peptide conjugate for detecting steroid molecules. *Bioorg. Med. Chem. Lett.*, **13**, 4305–4308.

40 Tanabe, T., Touma, K., Hamasaki, K., and Ueno, A. (2001) Fluorescent cyclodextrin immobilized on a cellulose membrane as a chemosensor system for detecting molecules. *Anal. Chem.*, **73**, 1877–1880.

41 Tanabe, T., Touma, K., Hamasaki, K., and Ueno, A. (2001) Immobilized fluorescent cyclodextrin on a cellulose membrane as a chemosensor for molecule detection. *Anal. Chem.*, **73**, 3126–3130.

5
Cyclopeptide Derived Synthetic Receptors
Stefan Kubik

5.1
Introduction

Cyclopeptides are a group of natural products of wide structural diversity that are mainly produced by lower organisms such as plants, bacteria, fungi, and lower sea animals and that usually possess high biological activity [1–3]. Several cyclopeptides are toxic or cytotoxic, for example, while others possess antibiotic activity, act as hormones, or act as ionophores.

The strong biological responses often caused by cyclopeptides are partly a consequence of the cyclic nature of these compounds. Bioavailability of a cyclic peptide is, for example, usually much higher than that of an acyclic analog because ring closure increases proteolytic stability, particularly against degradation by exoproteases. In addition, reduction of the conformational flexibility of a peptide upon ring closure diminishes the entropic penalty that has to be paid upon binding to its target, usually causing an increase in binding affinity.

Cyclopeptides can be classified systematically according to various aspects. Four structurally distinctly different types of cyclopeptides can be distinguished, for example, by considering the possible ways a linear peptidic precursor can cyclize, namely, by coupling of the terminal groups in the peptide backbone, of functional groups in two different side chains, or of a side chain functionality to the N- or the C-terminus. Alternatively, cyclopeptide classification can be based on the origin of the peptide or on the presence of special subunits in the ring. For this chapter a classification system is proposed in which cyclopeptides are divided into two major classes: peptides that act as substrates for a larger ligand and ones that act as receptors in the sense of supramolecular chemistry by binding to a smaller guest. Examples of the latter type are compounds **1–3**.

Artificial Receptors for Chemical Sensors. Edited by V.M. Mirsky and A.K. Yatsimirsky
Copyright © 2011 WILEY-VCH Verlag GmbH & Co. KGaA, Weinheim
ISBN: 978-3-527-32357-9

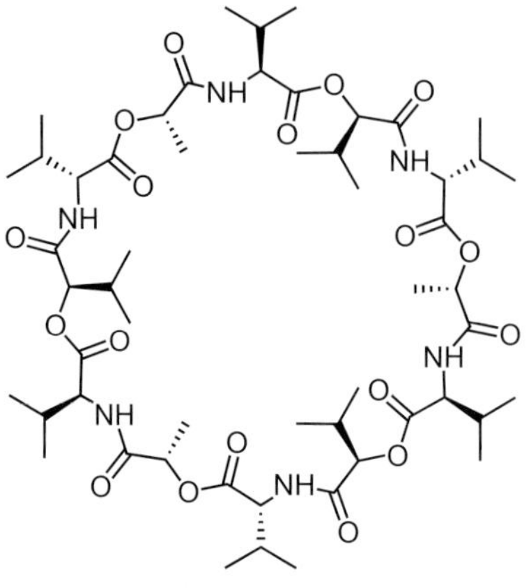

1 Valinomycin
cyclo(L-Lac-L-Val-D-Lac-D-Hyv)₃

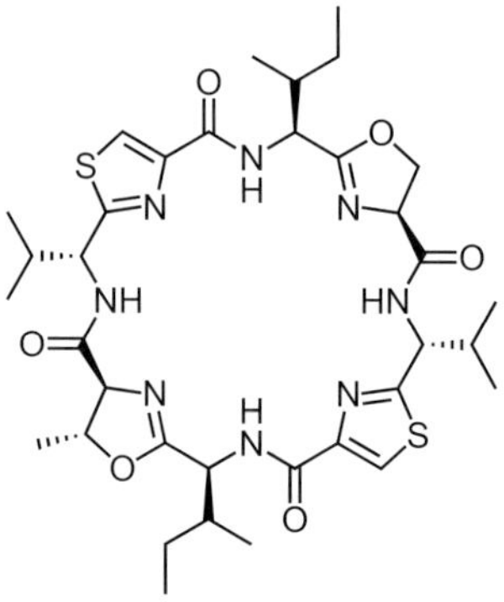

2 Patellamide A

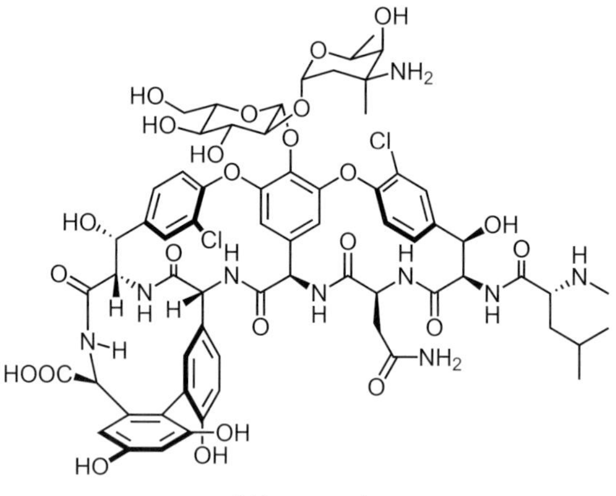

3 Vancomycin

The antibiotic valinomycin (**1**) is obtained from *Streptomyces fulvissimus* and several other *Streptomyces* strains [4]. It is made up of D- and L-valine (Val), D-hydroxyisovaleric acid (Hyv), and L-lactic acid (Lac) with ester and amide bonds alternating along the ring [5]. Biological activity of valinomycin is due to its ability to transport potassium ions across lipid membranes, which disturbs the membrane potential, ultimately leading to cell death.

The sea squirt *Lissoclinum patella* produces several families of structurally closely related cyclic peptides, of which patellamide A (**2**) is one example [6]. Besides L-isoleucine and D-valine this cyclic octapeptide contains two oxazoline and two thiazole rings that derive from, respectively, threonine and cysteine. Patellamides bind transition metals such as Cu^{2+} or Zn^{2+} within their cavity [7, 8], an ability that is supposedly the reason why *Lissoclinum patella*, the organism from which **2** has been isolated, can contain several metals, including copper, up to 10 000 times the concentration found in the local marine environment [9].

The glycopeptide antibiotic vancomycin (3) is produced by the bacterium *Amycolatopsis orientalis* and possesses high activity against Gram-positive bacteria [10–12]. The aglycon of 3 consists of a linear heptapeptide containing several non-proteinogenic subunits. Five aromatic subunits, arranged along the chain of this heptapeptide, are involved in the formation of three macrocycles, one containing a biaryl and two containing a diphenyl ether moiety. Antibiotic activity of vancomycin is due to its inhibitory effect on the mechanical stabilization of the bacterial cell wall during cell wall biosynthesis [10–12].

The identification and characterization of these and other cyclopeptide-based receptors together with the discovery of the cation binding properties of cyclic oligo(ethylene glycol)s (crown ethers) by C. J. Pedersen in 1967 [13] also triggered activity in the development of synthetic cyclopeptides with receptor properties. The appeal of such receptors lies in the fact that even the simplest cyclopeptide that has a cavity large enough for the inclusion of a guest contains an alternating arrangement of hydrogen bond donors (NH-groups) and hydrogen bond acceptors (C=O groups) along the ring, which induce affinity for, respectively, anions and cations.

Substrate binding is, however, not restricted to the peptide groups but could also involve functionalities in the side chains of the amino acids subunits. Since cyclopeptide synthesis allows these subunits to be varied over a wide range, including incorporation of non-natural amino acids, cyclopeptide derived receptors can be devised not only for charged but also for neutral substrates. Structural diversity can additionally be achieved by variation of ring size and sequence of the amino acid subunits along the ring by changing the length of the acyclic cyclopeptide precursor or the order in which it is assembled from the individual subunits, allowing receptor properties to be fine tuned [14]. Notably, a defined sequence of binding sites along the cavity of a synthetic receptor is much more difficult to achieve on the basis of conventional hosts such as crown ethers, cyclodextrins, or calixarenes, which are usually not synthesized sequentially but in a modular fashion starting from symmetrical cyclic precursors [15]. Chirality is an additional feature of cyclopeptide derived receptors that can play a role in their binding properties.

A disadvantage of the use of cyclic peptides as artificial receptors is their conformational flexibility, which is smaller than that of the corresponding acyclic precursor but often still substantial. Molecular recognition studies involving cyclopeptides therefore generally include detailed conformational analyses. In addition, cyclopeptides, especially larger ones, tend to adopt conformations in solution stabilized by intramolecular hydrogen bonds, unsuitable for the inclusion of a guest molecule [16]. Both problems can, however, be overcome to a certain extent by introduction of rigid abiotic amino acid subunits into the ring, a strategy employed in many cyclopeptide derived synthetic receptors.

In this chapter, representative examples of such receptors are presented. Receptor classification on structural grounds is difficult since cyclopeptide derived receptors are structurally much more heterogeneous than other types of receptors, for example, cyclodextrins or calixarenes, which can often be organized according to either ring size or functional groups attached to a certain position. The receptors presented here are therefore grouped by the type of substrate bound. This chapter is not meant to be comprehensive. Instead, several successful attempts in receptor design are highlighted to illustrate the potential of the approach.

5.2
Receptors for Cations

Analysis of the interaction of valinomycin (**1**) with potassium ions revealed that complex formation causes a conformational stabilization of the otherwise rather flexible valinomycin ring, which adopts a bracelet-like conformation in the complex stabilized by six hydrogen bonds between each NH group and the carbonyl group of the preceding amide moiety [17]. This arranges the six remaining ester carbonyl groups to point into the center of the cavity where they can coordinate to a metal ion in an almost perfect octahedral fashion. Cavity dimensions are optimal for K^+, but also allow the inclusion of larger metal ions such as Rb^+. The valinomycin backbone is too rigid, however, to allow tight binding of smaller ions such as Na^+ or even Li^+. Complementarity between valinomycin and its substrate is clearly visible in the crystal structure depicted in Figure 5.1 [18].

To gain insight into the interplay between conformation and binding affinity various valinomycin analogs have been synthesized. The group around Ovchinnikov has, for example, shown in a series of systematic investigations that changes in the side chains of **1** have a relatively small effect on cation affinity [17]. Inversion of the amino acid or hydroxy acid configurations induces steric strain, thereby weakening complex stability. In addition, replacement of the amide groups with esters or N-methylated amides causes a decrease in complex stability because intramolecular hydrogen bond formation is impaired. Replacement of the ester groups with amides or N-methylated amides, in contrast, has no effect on intramolecular hydrogen bonding and yields efficient ionophores.

In the valinomycin analogs described by Gisin *et al.* [19] the D-hydroxyisovaleric acid residues of **1** were replaced by D-proline, a structural change that does not affect intramolecular hydrogen bonding. Compound **4a**, in which the lactic acid residues of **1** were additionally replaced by L-proline, exhibits an about three orders of

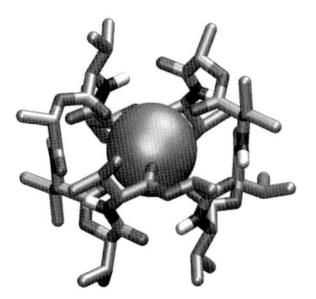

Figure 5.1 Crystal structure of the K^+ complex of **1** [18].

magnitude larger affinity for alkali picrates than **1** while the selectivity sequence, namely, $K^+ \approx Rb^+ > Cs^+ > Na^+ > Li^+$, is retained. Similar results were obtained for **4b**, showing that the side chain in the L-amino acid residues responsible for intramolecular hydrogen bond formation has no large effect on receptor properties. The large binding affinity observed for **4a** and **4b** was attributed to the higher electron-donating properties of the carbonyl groups involved in cation coordination in these peptides (amides) with respect to **1** (esters) and the reduced flexibility of the rings imposed by the introduction of proline subunits. In line with this interpretation is that reduction of the nucleophilicity of the cation coordinating carbonyl groups by replacement of the tertiary L-proline amides in **4a** and **4b** by secondary amides in **4c** causes a drop in cation affinity. Decrease of ring size not only diminishes cation affinity but also reverses cation selectivity. The sodium affinity of **4d** is, for example, slightly larger than the affinity of this receptor for potassium.

4a
cyclo(L-Pro-L-Val-D-Pro-D-Val)₃

4b
cyclo(L-Pro-L-Ala-D-Pro-D-Val)₃

4c
cyclo(L-Ala-L-Val-D-Pro-D-Val)₃

4d
cyclo(L-Pro-L-Val-D-Pro-D-Val)₂

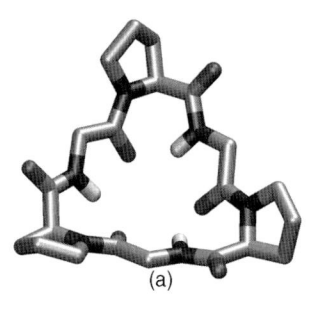

Figure 5.2 Calculated structures of conformations S (a) and S1* with included Mg^{2+} (b) of cyclopeptide **5**. Calculations were performed by the author based on the results reported in Reference [24] by using MacSpartan 04 (Wavefunction, Inc.) and the MMFF force-field.

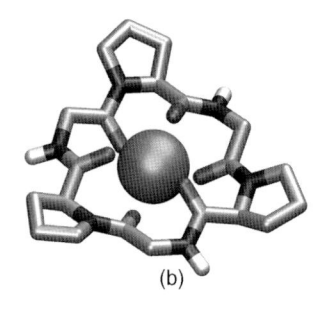

Related studies concerned with the identification of valinomycin mimics or analogs of other naturally occurring metal binding cyclopeptides such as antamanide [20] have been performed in several other groups and the more recent characterization of the binding properties of a cyclolinopeptide A analog demonstrates that the metal binding properties of cyclopeptides are still a matter of interest [21].

Design of synthetic receptors based on cyclopeptides for which there are no natural analogs was pioneered by the Blout group [22, 23]. Conformational analysis revealed, for example, that cyclohexapeptide **5**, containing alternating glycine and L-proline residues, adopts a C_3-symmetrical conformation, termed S, in dioxane and chloroform, which has all peptide bonds trans, and is stabilized by three intramolecular hydrogen bonds [24]. Addition of metal ions induces conformational reorganization, leading to conformation S_1^* with no intramolecular hydrogen bonds. Instead, all six carbonyl groups of the amino acids converge toward the center of the cavity where cation coordination takes place. To illustrate this behavior conformations S and S_1^* are depicted in Figure 5.2. Among the alkali metal ions, **5** shows selectivity for Li^+ and Na^+ over K^+ and larger ions in 80% methanol–water. Among divalent ions Ca^{2+} is bound in preference of Ba^{2+}. Complexes of higher stoichiometry were observed in acetonitrile, especially with Mg^{2+}, while in polar solvents such as water and in the absence of salts an asymmetrical conformation of **5** is preferred. A similar conformational reorganization of a cyclopeptide derived metal ion receptor has been demonstrated recently for a hexameric cyclic α-peptoid [25].

5
cyclo(L-Pro-Gly)₃

Cyclopeptide **5** has not only been shown to interact with metal ions but also with ammonium ions, a property **5** shares with crown ethers [26]. Complex formation is due to hydrogen bond formation between the cyclopeptide carbonyl groups and the acidic guest protons. Complexation of the enantiomers of chiral ammonium ions leads to the formation of diastereomeric complexes, which was demonstrated by ^{13}C NMR spectroscopy. While the hydrochloride of L-proline benzyl ester forms discreet 1:1 complexes with **5**, a complex of higher stoichiometry with two amino acids bound to one molecule of **5** was observed with the hydrochloride salt of L-valine methyl ester [27]. The latter complex structurally relates to the $(Mg^{2+})_2 \subset \mathbf{5}$ complex. A stepwise binding mechanism is proposed for its formation in which interaction of three carbonyl groups of **5** with the ammonium group of the first valine molecule induces a cyclopeptide conformation that allows binding of a second guest to the remaining three carbonyl groups on the opposite side of the ring. Protonated primary amines are required for this to occur, as shown by the inability of proline to cause binding of a second guest to **5**.

To test whether aromatic subunits induce affinity of cyclopeptides for quaternary ammonium ions as in calixarenes [28] we synthesized cyclic hexapeptide **6** [29]. Conformational analysis revealed that this peptide indeed preferentially adopts conformations in chloroform related to the *cone* conformation of calixarenes. Interconversion of these conformations is, however, associated with the inversion of the orientation of the aromatic subunits with respect to the ring plane and preorganization of **6** for cation binding is therefore not optimal. Weak interactions between **6** and, for example, n-butyltrimethylammonium (BTMA$^+$) iodide could still be detected in 0.2% d_6-DMSO/CDCl$_3$ ($K_a = 300$ M^{-1}) [29].

Reduction of conformational flexibility of **6** by replacing the glutamic acid subunits with proline caused an approximately fourfold improvement in BTMA$^+$ affinity ($K_a = 1260$ M^{-1}) [30], and cation affinity was further increased by restricting the residual conformational mobility in **7** at the bonds flanking the nitrogen and the carbon atom of the secondary amide groups by introduction of aromatic substituents that form intramolecular hydrogen bonds to the neighboring amide

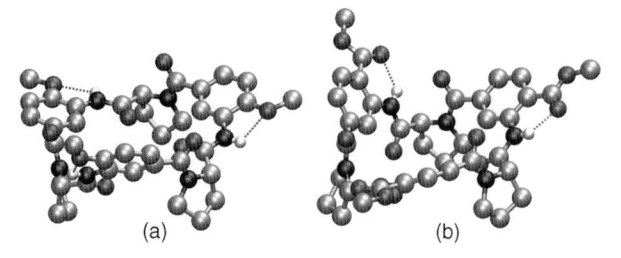

Figure 5.3 Crystal structures of cyclopeptides **8a** [31] (a) and **8b** [32] (b). Non-acidic protons are omitted for clarity.

NH groups. Peptides **8a** and **8b**, for example, bind BTMA$^+$ picrate significantly stronger in 0.2% d_6-DMSO/CDCl$_3$ than **6** (BTMA$^+$ ⊂ **8b**: $K_a = 2700$ M^{-1}; BTMA$^+$ ⊂ **8b**: $K_a = 10\,800$ M^{-1}). Interaction of the aromatic substituents with the aromatic NH groups in these peptides is nicely visible in the crystal structures depicted in Figure 5.3 [31, 32].

6 R = CH$_2$CH$_2$COOCH(CH$_3$)$_2$
cyclo[L-Glu(O*i*Pr)-Aba]$_3$

7 cyclo(L-Pro-Aba)$_3$

8a R = OCH$_3$
8b R = COOCH$_3$

Which type of interaction is responsible for the complexation of quaternary ammonium ions by cyclopeptides **6–8** is an open question. The shielding of guest protons observed in the NMR upon complex formation hints toward cation–π interactions. The unusual high binding constant of the BTMA$^+$ complex of **8b** and the fact that binding strength does not correlate with the electronic effects aromatic substituents exert on the electrostatic potentials of the 3-aminobenzoic acid derived subunits [32] indicate that cation–π interactions are unlikely to be the only driving force of complex formation, however. Additionally, hydrogen bonds between cyclopeptide carbonyl groups and C–H protons in α-position of the quaternary nitrogen atom in the guest, which can be quite strong [33], could at least partly contribute to complex stability.

Chirality of these peptides allows differentiation of chiral quaternary ammonium cations [34]. Interestingly, enantioselectivity is not only affected by the structure of the cyclopeptide but also by the absolute configuration of the BTMA$^+$ counterion if a salt is used in the binding studies that contains a chiral anion [35]. This is an indication that both components of the salt are bound in halogenated solvents as a close ion pair (even if only one component of the ion pair is bound specifically by the receptor, vide infra) and that structural aspects of the intact ion pair affect complex stability.

Kojima, Miyake, and coworkers have used N,N'-ethylene-bridged dipeptide subunits to restrict the conformational mobility of a cyclopeptide and to prevent the formation of intramolecular hydrogen bonds [36]. They showed that cyclic octapeptides **9a** and **9b** bind divalent alkaline earth as well as transition metal ions with ionic radii of approximately 0.95–1.0 Å [37, 38]. This behavior differs from that of other cyclic octapeptides but is similar to that of 18-membered cyclic hexapeptides, suggesting that the cavity dimensions of the 24-membered ring of **9a** are similar to those of cyclohexapeptides containing no N,N'-ethylene-bridged dipeptide subunits.

9a R = CH$_2$CH(CH$_3$)$_2$
9b R = CH(CH$_3$)$_2$

10 R = CH$_2$CH(CH$_3$)$_2$

^{13}C NMR spectroscopy has also allowed the differentiation of the enantiomers of chiral ammonium ions [39]. In this context, cyclopeptide **10** proved to possess superior properties over the smaller 24-membered derivative. With **10**, not only ^{13}C NMR differentiation but also selective transport through liquid membranes of the enantiomers of chiral α-amino acids could be achieved [40].

Affinity for transition metals, specifically Cu^{2+} and Zn^{2+}, is a feature shared by cyclopeptides containing oxazoline, thiazole, or thiazoline subunits, most of which are natural products isolated from the sea squirt *Lissoclinum patella*. Examples of this structurally diverse family of cyclopeptides are patellamide A (**2**) and cyclopeptides **11–13**. The metal binding properties of these peptides have been reviewed recently in detail [41] and only a summary of relevant aspects is therefore presented here.

11 Ascidiacyclamide   **12** Lissoclinamide 3   **13** Westiellamide

Solid evidence for the metal binding properties of azole-based cyclopeptides was provided by the crystal structure of the dicopper(II) complex of ascidiacyclamide (**11**) [42]. In this complex, two copper ions separated by a bridging carbonate anion are included into the cavity of **11**, each metal ion coordinating to two nitrogens of adjacent heterocyclic subunits. A third coordination site is provided by a deprotonated amide NH situated between the azole units involved in metal binding. Figure 5.4 shows that **11** adopts a saddle-shaped conformation to accommodate the two copper ions.

Azole-based cyclopeptide have been shown to adopt several conformations in the absence of metal ions, termed type I–IV according to a systematic classification proposed by Ishida [43], mainly differing in macrocycle symmetry. Yet, there is spectroscopic evidence, particularly by CD spectroscopy, that, independent of initial conformation, coordination of Zn^{2+} or Cu^{2+} causes a conformational reorganization and stabilization of a similar saddle-shaped structure as found in the dicopper(II) complex of **11** [44]. Titration experiments allowed quantification of binding strength, showing that binding of the first metal ion is usually associated with a K_a of 2×10^4 to $2 \times 10^5 \, M^{-1}$ and binding of the second ion with a K_a between 20 and 230 M^{-1}. Similar binding studies have also been performed with lissoclinamides, for example, **12** [45].

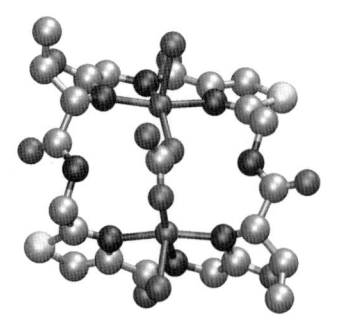

Figure 5.4 Crystal structure of the dicopper(II) complex of **11** [42]. Hydrogen atoms and valine and isoleucine side chains are omitted for clarity.

Synthetic analogs of patellamide A (**2**), cyclopeptides **14a–d** lacking the oxazoline rings, have been prepared by the Comba group to probe the importance of these subunits in metal coordination [46, 47].

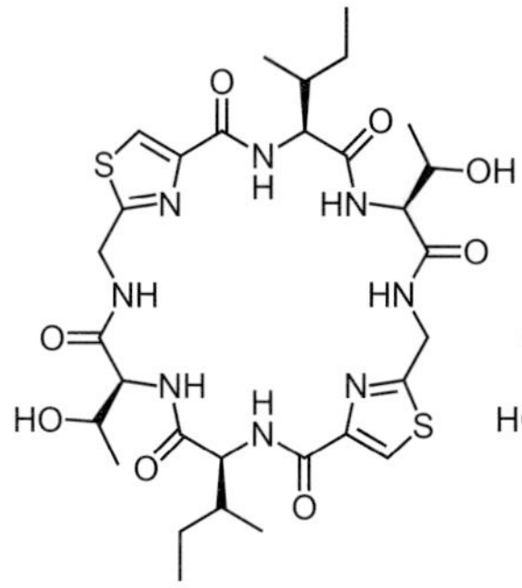

14a PatJ^1
cyclo[L-Ile-L-Thr-(Gly)Thz]$_2$

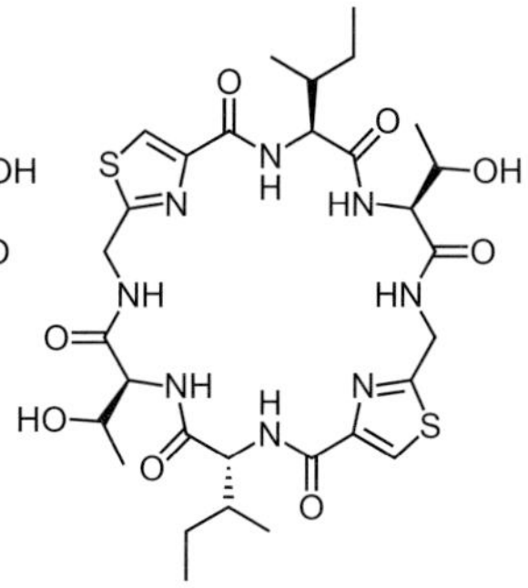

14b PatJ^2
cyclo[L-Ile-L-Thr-(Gly)Thz-
D-Ile-L-Thr-(Gly)Thz]

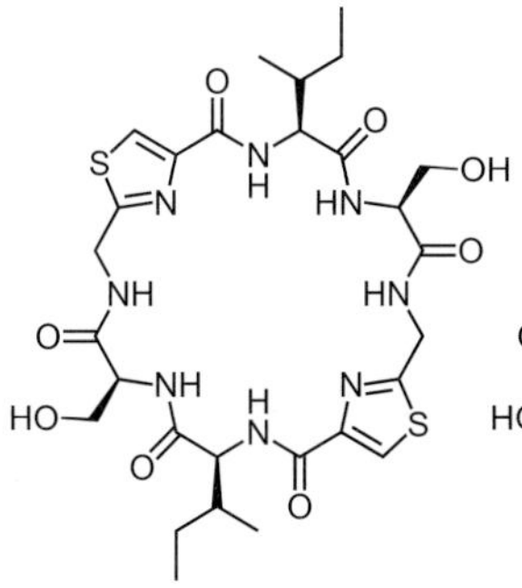

14c PatL
cyclo[L-Ile-L-Ser-(Gly)Thz-
L-Ile-L-Ser-(Gly)Thz]

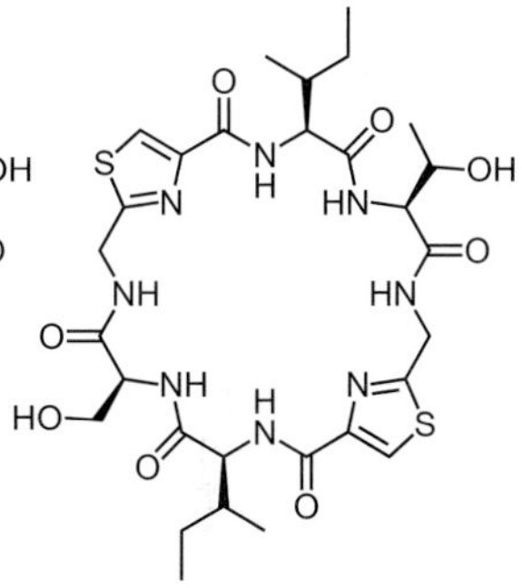

14d PatN
cyclo[L-Ile-L-Thr-(Gly)Thz-
L-Ile-L-Ser-(Gly)Thz]

In the crystal, **14a** adopts a conformation with two almost coparallel thiazole rings, similar to patellamide D, in which the thiazole-N and amide-N donors are well preorganized for transition metal binding. In solution, formation of several mono- and dinuclear Cu^{2+} species was observed, with dinuclear complexes being the more stable ones. Two types of dicopper complexes were analyzed structurally, demonstrating that the cyclopeptides in these complexes adopt a conformation similar to that of the oxazoline containing analogs [46]. Thus, metal complexation is not significantly affected by the absence of the oxazoline residues although slight but significant structural differences were found in the dinuclear Cu^{2+} complexes depending on cyclopeptide structure.

The Imanishi group was the first to study the effect of side chain functional groups on the binding properties of cyclopeptides by using derivatives **15–17**.

Comparison of the Ba^{2+} affinity of peptides **15a** and **15b** revealed little difference in binding strength or cyclopeptide conformation [48]. It was thus concluded that the carboxyl groups in the side chains of **15b** do not participate in cation binding, most probably because the tetrapeptide cavity is too small to allow both carboxylate groups to interact with the metal. To test whether larger and more flexible cyclic skeletons are better suited, cyclic hexapeptides **16a–c** were synthesized. Since only **16c** showed affinity for Ca^{2+} and Ba^{2+}, while no binding of these ions could be detected by **16a,b**, these investigations allowed no conclusive evaluation of the effect of the peripheral carboxyl groups in **16c** on ion affinity [49]. Comparison of the cation affinity of cyclopeptides **17a** and **17b**, however, revealed a significantly higher cation affinity of **17b** containing carbonyl groups in two side chains, indicating that a cooperative participation of the side chains in **17b** in cation coordination is very likely [50].

15a R = COOCH$_3$; *cyclo*[L-Pro-L-Glu(OMe)]$_2$
15b R = COOH; *cyclo*[L-Pro-L-Glu]$_2$

16a R = COOH; *cyclo*[L-Pro-L-Asp-L-Phe]$_2$
16b R = CH$_2$CH$_2$COOH; *cyclo*[L-Pro-L-Aad-L-Phe]$_2$
17a R = CH(CH$_3$)$_2$; *cyclo*[L-Pro-L-Leu-L-Phe]$_2$
17b R = SCH$_2$NHCOCH$_3$; *cyclo*[L-Pro-L-Cys(Acm)-L-Phe]$_2$

16c *cyclo*[L-Pro-D-Asp-D-Phe]$_2$

A structurally innovative cyclopeptide with transition metal binding ligands in the side chains has been described recently by Haberhauer. This synthetic peptide (**18a**) is structurally related to westiellamide (**13**) but contains three N-alkylated imidazole residues as heterocyclic subunits [51]. The X-ray structure of this compound showed that the substituents on the imidazole moieties converge and are arranged in a triple helix-like fashion, suggesting that coordination of all three 2,2′-bipyridyl groups to a metal ion would give one of the two enantiomeric octahedral complexes in excess, possibly exclusively. Indeed, upon reaction of **18a** with Ru²⁺ or Os²⁺ salts only one diastereomer of the corresponding complexes [**18a**·Ru]²⁺ and [**18a**·Os]²⁺ could be isolated. A crystal structure showed that the octahedral metal complexes possess the Λ-configuration in these compounds, a finding that agrees well with predictions made on the basis of DFT calculations that indicated that the Λ-complex should be significantly more stable than the Δ-complex.

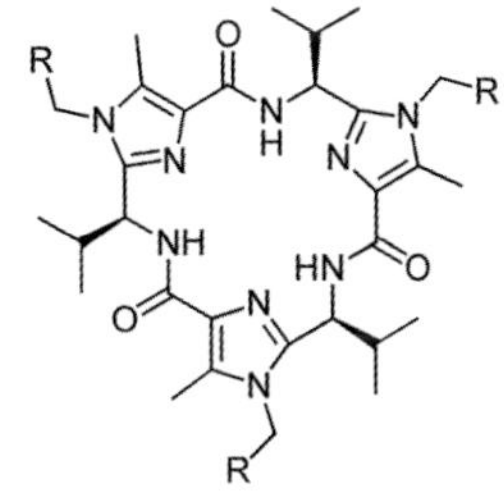

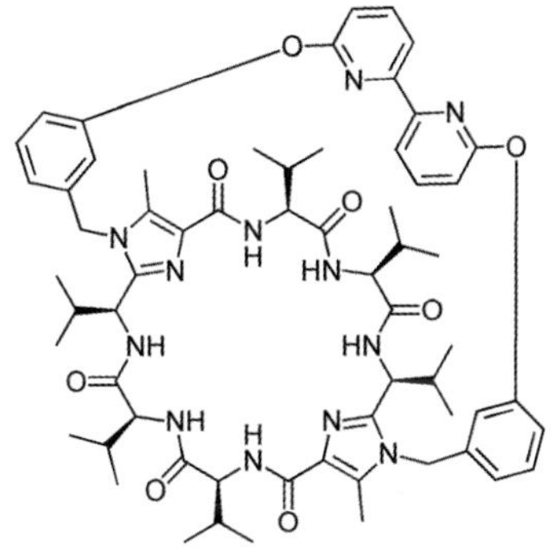

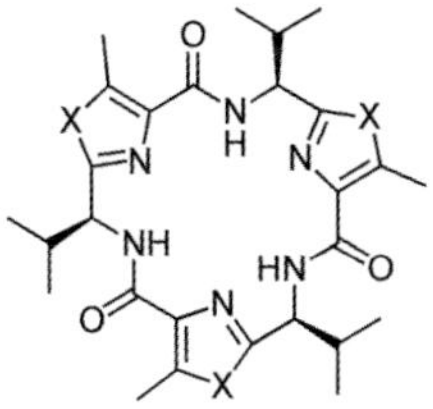

18a R = [pyridine-pyridine group attached via N]

18b R = [HN-C(=O)- pyridine-pyridine group]

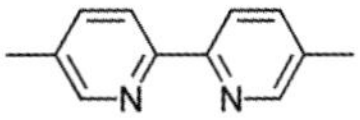

19

20a X = NCH₃
20b X = O
20c X = S

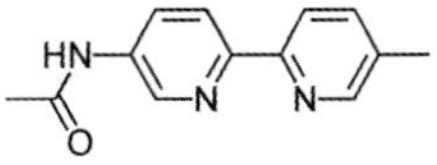

This concept was extended to the synthesis of kinetically labile metal complexes by using peptide **18b** [52]. Hydrogen bonds between the NH groups in the substituents and C=O groups in the cyclopeptide ring produce a preferential M-helical arrangement of the substituents around the cyclic scaffold, a stabilizing interaction that is reminiscent of a β-turn in an acyclic peptide. Without exception, coordination of **18b** to various metal ions, including Ni²⁺, Cu²⁺, Zn²⁺, Mn²⁺, and Fe²⁺, leads to the formation of the Λ-configured metal complexes, clearly demonstrating the strong chiral induction of the cyclopeptide.

Another interesting system introduced by Haberhauer is molecular hinge **19** [53]. The uncomplexed form of **19** represents the open state of the hinge and exhibits planar chirality. The 2,2′-bipyridine moiety in this form adopts the M conformation because the P conformation is strongly destabilized by the chiral peptidic scaffold. Copper(II) binding induces rotation around the central bond in the bipyridine moiety and stabilizes a ligand conformation with *syn*-oriented nitrogen atoms. This closing of the hinge destroys its planar chirality, which can nicely be followed by CD spectroscopy. The hinge can be opened again by addition of cyclam, which causes decomplexation of [**19**·Cu²⁺], but since only the M conformation can be reached from the closed state the resulting molecular motion is unidirectional, a characteristic feature of molecular machines. The interaction of westiellamide (**13**) and analogs **20a–c** with Cu²⁺ has been studied in a collaboration between the Haberhauer and the Comba groups [54].

Few groups have developed chemosensors for ions on the basis of cyclopeptides. The Ghadiri group has constructed a sensor by incorporation of a self-assembling cyclic octapeptide **21** containing alternating D-leucine and L-tryptophan residues into organosulfur self-assembled monolayers on gold films [55]. Such peptides have previously been shown by the same group to self-assemble into tubular structures [56–59]. Tubes of **21** incorporated into a monolayer of dodecanethiol or dodecyl sulfide on gold and oriented perpendicularly to the gold surface form channels whose diameter is defined by the inner diameter of the cyclopeptide ring. A monolayer without channels is an electrical isolator, as demonstrated by the fact that no response was observed in cyclic voltammograms if the solution over the monolayer contained redox-active ions such as $[Fe(CN)_6]^{3-}$, $[Mo(CN)_8]^{4-}$, or $[Ru(NH_3)_6]^{3+}$. Monolayers containing peptide tubes, on the other hand, allow the electrochemical detection of $[Fe(CN)_6]^{3-}$ and $[Ru(NH_3)_6]^{3+}$ but not of $[Mo(CN)_8]^{4-}$ because the latter ion cannot diffuse though the channels. Selectivity of ion detection therefore depends on cyclopeptide size.

21
cyclo(L-Trp-D-Leu)$_4$

22a R = COOH; *cyclo*[L-Trp-(D-Glu-L-Glu)$_2$-D-Leu-L-Leu-D-Leu]
22c R = CH$_2$CH$_2$NH$_2$; *cyclo*[L-Trp-(D-Lys-L-Lys)$_2$-D-Leu-L-Leu-D-Leu]

22b *cyclo*[D-Trp-L-Glu-D-Leu-L-Cys-D-Leu-L-Glu-D-Leu-L-Cys]

In another approach, cyclopeptides **22a–c**, which also contain alternating D- and L-amino acids, have been described as optical sensors. Peptide **22a** forms complexes with divalent transition metal ions such as Cu^{2+}, Zn^{2+}, Cd^{2+}, Hg^{2+}, and Pb^{2+} and with the trivalent group III metal Al^{3+} but not with alkali or earth alkaline metal ions in water [60]. Highest affinity, amounting to a K_a of $2.2 \times 10^6 \, M^{-1}$, is observed for Hg^{2+}. The interaction with Pb^{2+}, Hg^{2+}, and Cu^{2+} modulates the fluorescence emission properties of the tryptophan chromophore in **22a** even in the presence of an excess of other metal ions, allowing this cyclopeptide to be used as an optical sensor for such analytes in water. Cysteine residues in **22b** improve affinity and fluorescence response for Hg^{2+} in water ($Hg^{2+} \subset $ **22b**: $K_a = 7.6 \times 10^7 \, M^{-1}$) [61]. Peptide **22c** exhibits no significant affinity for metal ions in water, but in the zwitterionic micellar detergent dodecylphosphocholine weak but specific metal ion binding is observed [62]. No results about the effect of metal complexation on the fluorescence of **22c** were reported.

5.3
Receptors for Ion Pairs

We have seen earlier that complexation of an ion pair is often unavoidable in nonpolar solvents even if the receptor interacts only with one component of the salt since individual ions cannot be solvated in such media (vide supra). This situation can change in a more polar environment. Ion pair receptors should thus have proper binding sites for a cation *and* an anion [63]. Depending on the distance of the binding sites, the ion pair may be bound as a contact ion pair or a separated ion pair. In the first case binding of one ion usually reinforces binding of the second since electrostatic interactions between the two oppositely charged substrates remain intact in the ternary complex [64–66].

Owing to the presence of C=O and NH groups along a cyclopeptide ring such compounds should, in principle, have the potential ability to act as ion pair receptors. The only peptides for which this property has been demonstrated so far, however, are to the best of my knowledge compounds **6** and **7**. Indications for the ability of these peptides to bind both components of an ion pair came from characterization of the counterion dependence of cation complex stability. Thus, while the cation of BuNMe$_3$I binds to **6** with an association constant of $300 \, M^{-1}$ in chloroform, the cation complex stability is about 10^4 times higher if the corresponding tosylate salt is used [29]. This drastic increase in cation complex stability was explained by coordination of the tosylate anion to the NH groups of **6**, which (i) stabilizes a cyclopeptide conformation well suited for cation binding and (ii) allows the cation to interact with the simultaneously bound anion in the ternary complex that is formed. The more weakly coordinating iodide, on the other hand, does not interact with **6**. Molecular modeling showed that the optimal conformation for anion complexation is one with the maximum number of NH groups converging toward the center of the peptide ring [29].

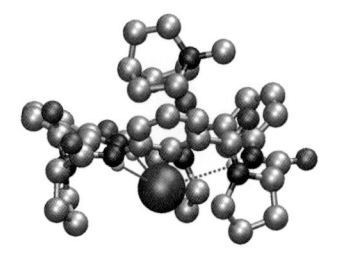

Figure 5.5 Crystal structure of the N-methylquinuclidinium iodide complex of **7**. Non-acidic hydrogen atoms are omitted for clarity.

Similar effects were observed for the proline-containing cyclopeptide **7**, but since the aromatic NH groups in this peptide are not involved in intramolecular hydrogen bonds to the aromatic C=O groups as in **6** they are free to interact also with weakly coordinating anions, allowing interaction with tosylate *or* with iodide [30]. The tosylate salt is, however, more strongly bound. Complexation of the anion leads to a significant increase in cation complex stability in comparison to complexes that contain no simultaneously bound anion. The crystal structure shown in Figure 5.5 of the complex between N-methylquinuclidinium iodide and **6** demonstrates the complexation of the ion pair that has been detected in solution and is retained the solid state [30].

5.4
Receptors for Anions

Anion coordination chemistry has become an area in supramolecular chemistry of considerable importance because of the potential applications synthetic anion receptors might have in, for example, medicinal chemistry or environmental monitoring [67–75]. The structural diversity of anion receptors is immense but there are only a limited number of receptors deriving from cyclopeptides. Complex formation of these receptors usually relies on hydrogen bond formation that involves the peptide NH groups arranged along the ring and only in a few cases functional groups in amino acid side chains additionally contribute to binding.

The group around Ishida has studied the ability of cyclopeptides containing an alternating sequence of natural α-amino acids and aromatic amino acids along the ring [76, 77] to serve as serine protease mimics [78] and for the construction of ion channels [79, 80]. They also showed that cyclic hexapeptide **23**, in which 3-aminobenzoic acid subunits alternate with L-alanine, strongly interacts with phosphate esters in DMSO [81]. Replacement of alanine in **23** with serine causes a slight reduction of complex stability. A much larger drop was observed for acyclic and larger analogs of **23**, for example, the corresponding octapeptide, demonstrating that the hexapeptide has the correct size and shape for phosphate ester recognition.

23 R = CH$_3$
cyclo(L-Ala-Aba)$_3$

During our investigations on the interactions of cyclopeptides **6** and **7** with cations we also came across the anion binding ability of these compounds (vide supra). For both compounds we showed that to allow for optimal interactions with anions the peptide must adopt a conformation with all NH groups pointing toward the cavity center [30], and by stabilizing a conformation with diverging NH groups in **8a** and **8b** we could prevent anion binding [32]. We were therefore curious whether the reversed effect could also be achieved, namely, an increase in anion affinity by the stabilization of a peptide conformation with converging NH groups.

Our approach to induce such a conformation consisted in replacing the 3-aminobenzoic acids in **7** with 6-aminopicolinic acid subunits. NH groups in the 2-position of pyridine rings are known to preferentially adopt a conformation with their dipole moment arranged in an antiparallel fashion to that of the free electron pair on the ring nitrogen. This bias should cause a convergent arrangement of all NH groups in cyclopeptide **24a**, which was indeed found [82]. Although the improved preorganization of **24a** for anion binding with respect to **7** does cause an increase in anion affinity, this increase is only moderate. Stability of the tosylate complex in DMSO increases from $<10\,M^{-1}$ for **7** to $4500\,M^{-1}$ for **24a**, for example. In chloroform, investigation of anion affinity of **24a** is complicated by the fact that the peptide adopts a non-symmetrical conformation stabilized by an intramolecular hydrogen bond [83].

24a R = H cyclo(L-Pro-Apa)$_3$
24b R = OH cyclo(L-Hyp-Apa)$_3$

Since affinity of neutral receptors that rely on hydrogen bonding for anion complexation generally decreases with increasing polarity of the solvent [84] we were surprised to notice strong interactions of **24a** with anions such as sulfate and halides in highly competitive protic solvent mixtures, for example, 80% D$_2$O–CD$_3$OD [82]. The reason for this unusual property of the cyclopeptide turned out to be a change in complex stoichiometry from 1 : 1 in DMSO to 2 : 1 in aqueous solvents. In the latter complexes, a completely desolvated anion is bound by six hydrogen bonds in a cavity between two peptide rings as illustrated by the crystal structure of the iodide complex of **24a** in Figure 5.6a.

Quantitative evaluation of halide and sulfate affinity of **24a** showed that K_{11}, the stability constant of the 1 : 1 complex between the cyclopeptide and the anion, is significantly smaller than K_{21}, the binding constant representing the formation of the 2 : 1 complex from the 1 : 1 complex [85]. This means that once formed, the 1 : 1 complexes of **24a** have a strong tendency to bind a second cyclopeptide molecule, making complex formation a highly cooperative process.

Cyclopeptide **24b**, which contains hydroxyproline instead of proline subunits, also interacts with anions in aqueous media, but only forms 1 : 1 complexes whose stability is similar to the 1 : 1 complexes of **24a** [85]. The different stoichiometries of the anion complexes of **24a** and **24b** were attributed to the fact that desolvation of the hydroxyproline subunits required for an aggregation of two molecules of **24b** is energetically more difficult than that of the proline units in **24a**. Notably, cyclopeptide **24b** forms weakly stable anion complexes even in 100% D$_2$O [85].

In subsequent work, we converted the 2 : 1 complexes formed by **24a** into 1 : 1 complexes by covalently linking two peptide units together via adipic acid [86]. The corresponding bis(cyclopeptide) **25a** exhibits high anion affinity in 50% H$_2$O–CH$_3$OH; the stability of the sulfate complex approaches a log K_a of 5, for example. Halides are bound somewhat weaker, with the association constants of the complexes increasing in the order Cl$^-$ < Br$^-$ < I$^-$, thus correlating with the

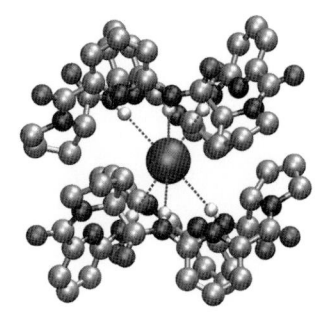

Figure 5.6 Crystal structure of the iodide complex of **24a**. Non-acidic hydrogen atoms are omitted for clarity.

size of the anion. Comparison of the binding equilibria on the basis of the association constants determined for **24a** and **25a** showed that anion complexation of the bis(cyclopeptide) is significantly more efficient than that of the monotopic analog [83].

25d X = —NH—C(O)—CH₂—(1,3-C₆H₄)—CH₂—C(O)—NH—

25e X = —NH—C(O)—[4-(Me₂N)-C₆H₃]—[3-(Me₂N)-C₆H₃]—C(O)—NH—

To systematically evaluate the influence of linker structure on anion affinity the receptor properties of bis(cyclopeptides) **25b** and **25c**, whose linkers become progressively more rigid, were compared to those of **24a** [87]. The linker of the fourth bis(cyclopeptide) included in this study, compound **25d**, was identified by using *de novo* structure-based design methods as implemented in the HostDesigner software [88–90]. These investigations showed that while the differences in the affinity of receptors **25a–d** toward a given anion are not very pronounced, there are profound differences in the thermodynamics of anion complexation. In general, a decrease in conformational rigidity of the linker improves the entropic advantage of complex formation, but not necessarily the overall complex stability because this advantage is largely compensated by a less favorable enthalpic contribution. The best receptor in this series is **25d**, whose linker was identified by computational methods.

Slight structural modification of **25c** gave the chemosensor **25e** [91]. Fluorescence of this receptor is quenched upon sulfate complexation but not in the presence of any of eight other anions tested. Bis(cyclopeptide) **25e** thus allows the quantitative and qualitative optical detection of sulfate even in the presence of a large excess of chloride ions. Interestingly, the sulfate affinity of **25e** is significantly lower than that of **25c**, which was attributed to different electronic effects of the substituents in the linkers of **25c** and **25e** on the preferred conformation of these bis(cyclopeptides).

In another approach we improved the binding properties of **25a** by using dynamic combinatorial chemistry [92]. This approach furnished bis(cyclopeptides) **26a** and **26b**, which proved to bind sulfate and iodide about one order of magnitude stronger than **25a**. Sulfate affinity of **26b**, for example, approaches an impressive log K_a of 7 in 33% H_2O–CH_3CN. Anion binding of **26a** and **26b** is also stronger than that of receptor **25d**, which was identified by computational methods [87].

The anion affinity of rigid macrocyclic triamide **27**, which consists of only non-natural amino acid subunits, namely 3′-amino-3-biphenylcarboxylic acid, has been studied in the Hamilton group [93, 94]. Three NH groups point toward the center of the approximately 5 Å wide cavity of this receptor, inducing affinity toward tetrahedral anions. Interaction in 2% d_6-DMSO/$CDCl_3$ with Bu_4NOTs leads to a 1:1 complex while inorganic anions such as halides, nitrate, hydrogen sulfate, and dihydrogen phosphate anions give rise to more complex binding equilibria due to the formation of complexes containing two receptor molecules and one anion. In DMSO, the formation of complexes with higher stoichiometries was not observed.

27 **28**

While the stabilities of the halide and nitrate complexes of this receptor are significantly smaller in d_6-DMSO, high affinity for tetrahedral anions is retained. Binding selectivity of **27** was correlated with the size and shape of the receptor cavity as well as with the arrangement of the NH groups. Thus, iodide binds more strongly to **27** than chloride despite the fact that the charge density of a chloride ion is higher because the fit of iodide inside the receptor cavity is better, allowing stronger interactions with the converging NH groups.

On the basis of **27**, fluorescent receptor **28** was developed [95]. The aminocoumarin fluorophore was integrated in **28** such that it is directly involved in hydrogen bonding to the anion. While anion binding properties of **28** proved to be similar as those of **27**, with high selectivity for tetrahedral ions, complex formation also causes an increase in fluorescence intensity and a redshift of the coumarin emission, an effect that was attributed to the stabilization of the fluorophore's excited state relative to the ground state on anion binding.

An anion binding cyclopeptide containing only acyclic non-natural amino is the cyclic hexapseudopeptide composed of D,L-α-aminoxy acids **29** described by Yang et al. [96, 97]. This compound adopts a C_3 symmetric bracelet-like conformation in CDCl$_3$ stabilized by intramolecular C=O···H−N hydrogen bonds (Figure 5.7). All α-protons of the aminoxy acid residues point inward and the side chains outward, with those of the D- and the L-aminoxy acid residues residing on opposite sides of the ring plane. Since all C=O and N−H groups are involved in hydrogen bonding, prediction of whether **29** binds ions at all and, if it does, which type of ion is preferred was not straightforward. Binding studies using cations and anions were therefore carried out that revealed that **29** selectively interacts with anions in CDCl$_3$. Structural investigations showed that for binding to occur a conformational reorganization leads to a structure in which all N−H groups converge toward the center of the macrocycle. To illustrate this behavior the calculated structure of the chloride complex of **29** is depicted in Figure 5.7b.

29

Several synthetic analogs of vancomycin (**3**) have been devised. Receptor **30** in which the AB and the CD rings of the vancomycin skeleton were deleted, leaving only the DE ring framework, that is, the carboxylate binding pocket intact has, for example, been developed by the Hamilton group [98]. Binding of carboxylates by **30** involves a combination of hydrogen bonding from the amide and ammonium functionalities, thus resembling the binding mode of vancomycin. A series of related receptors (**31a–d**) has been described by Pieters [99, 100]. These compounds bind the tetrabutylammonium salt of Ac-D-Ala in $CDCl_3$. Binding affinity is almost independent of the nature of the residue R; however, the leucine containing receptor **32** exhibited an approximately fourfold weaker affinity to Ac-D-Ala than **31b**, showing that variation of the amino acid in the center of the carboxylate binding pocket has an influence on complex stability [100].

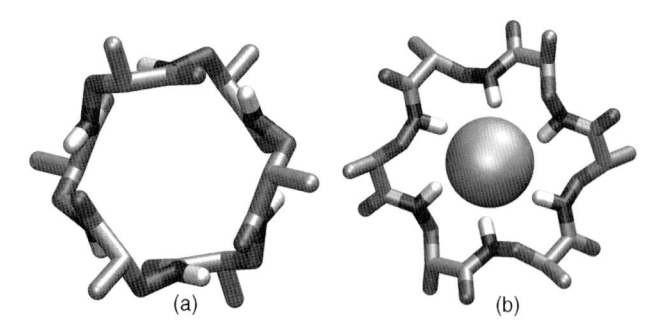

Figure 5.7 Calculated structures of the preferred conformation of **29** in the absence of anions (a) and in the chloride complex (b). Non-acidic hydrogen atoms are omitted and amino acid side chains are replaced by methyl groups for clarity. The coordinates of the structures were taken from Reference [96].

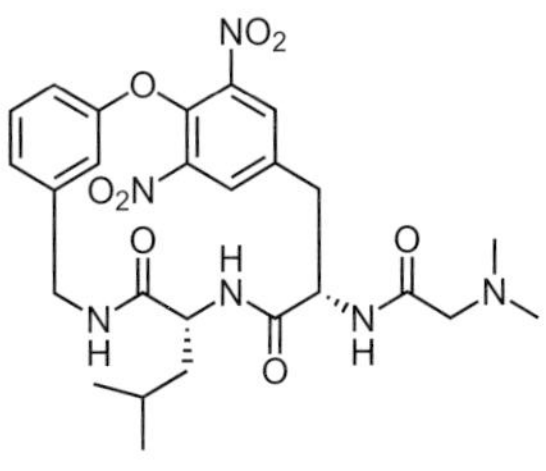

30

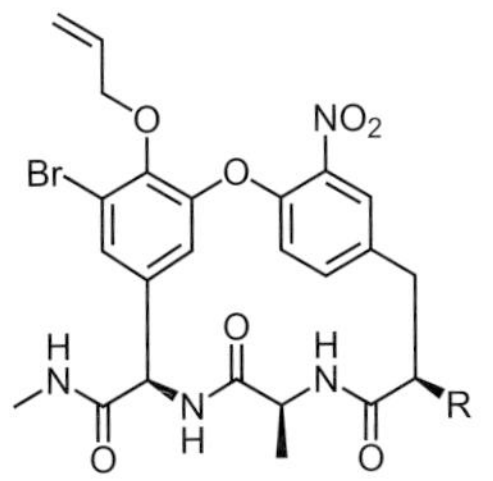

31a R = NHBoc
31b R = NH$_3^{\oplus}$ TFA$^{\ominus}$
31c R = —N(H)C(O)N(H)^{t}Bu
31d R = —N(H)C(O)CH(CH$_3$)NHBoc

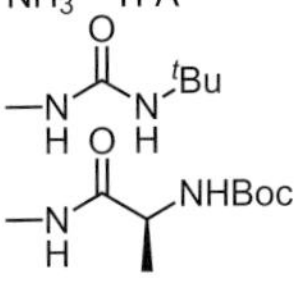

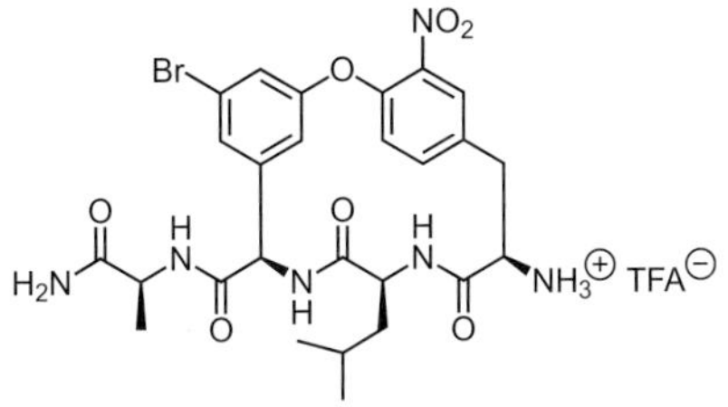

32

A different approach for devising cyclopeptide derived anion receptors than the ones described so far has been pursued by the Jolliffe group. They used a C_2 symmetrical cyclic octapeptide as a scaffold that structurally relates to ascidiacyclamide (**11**) [101]. Two pendant dipicolylamino groups complexed to Zn(II) serve as binding sites for phosphate anions. An indicator displacement assay was established on the basis of **33** by assembling this receptor with a coumarin derivative whose fluorescence was quenched upon complex formation. Of 15 anions tested, only pyrophosphate and to a lesser extent ATP, ADP, and citrate were able to displace the indicator from its complex with **33**. The large selectivity for pyrophosphate is due to the distance of the primary binding sites in **33**, which allows optimal contacts with this anion. In addition, NMR spectroscopic investigations indicated that the amide groups along the peptidic scaffold of **33** could provide secondary binding sites for the pyrophosphate anion.

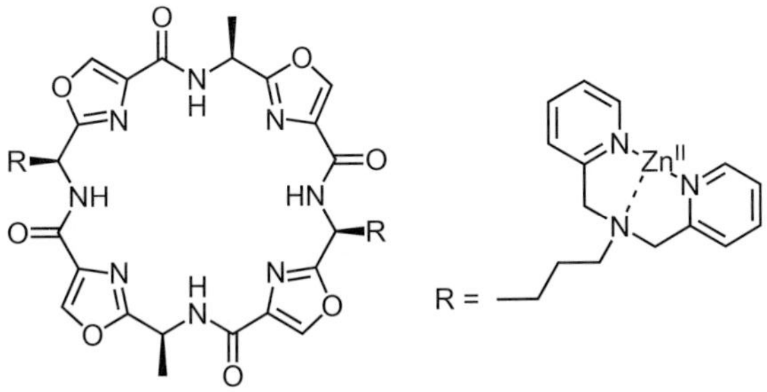

33

5.5
Receptors for Neutral Substrates

The NH and C=O groups along a peptide ring are rarely appropriately arranged to allow for strong hydrogen bonding interactions with polar neutral guest molecules and the polar nature of cyclopeptides is unsuitable for binding to hydrophobic compounds. Complexation of neutral guests therefore usually involves the side chains of a cyclic peptide. In this context, cyclopeptides **34a–d** containing hydrophobic amino acids have been synthesized, in a collaboration between the Jung and the Gauglitz groups, as receptors for volatile organic compounds such as tetrachloroethene, anisole, ethylacetate, or *n*-octane [102]. Monolayers of these peptides were prepared on a silica surface by reaction of the amino groups in the lysine subunit(s) with epoxy groups on the surface. Subsequently, interactions of the immobilized cyclopeptides and the analytes were studied by reflectometric interference spectroscopy. These results showed that monolayers containing cyclopeptides are generally much more sensitive toward the analytes than ones containing a linear analog, indicating that specific interactions between the cyclopeptides and the analytes are responsible for the observed effects. Moreover, cyclopeptides with only one point of attachment (**34a,b**) are more effective than ones attached to the surface via three lysine residues (**34c,d**), and interactions of peptides containing *p*-chlorophenyl side chains (**34b,d**) are stronger than those with the *p*-nitrophenyl side chains (**34a,c**). The authors conclude that exploiting the receptor properties of cyclopeptides in such sensors for the selective detection of an analyte in solution is a promising approach, particularly since the possibility to synthesize cyclopeptides in a combinatorial fashion should allow the selection of efficient binders from a large pool of different receptor candidates. Successful application of cyclohexapeptide libraries as chiral selectors in capillary electrophoresis has also been demonstrated [103].

34a R = NO$_2$; *cyclo*[L-Lys-(L-Phe(pNO$_2$)-L-Ala)$_2$-L-Phe(pNO$_2$)]
34b R = Cl; *cyclo*[L-Lys-(L-Phe(pCl)-L-Ala)$_2$-L-Phe(pCl)]

34c R = NO$_2$; *cyclo*[L-Lys-L-Phe(pNO$_2$)]$_3$
34d R = Cl; *cyclo*[L-Lys-L-Phe(pCl)]$_3$

In a similar approach the same team of authors showed that immobilization of cyclopeptides **34c** or **35a,b** on silica furnishes sensors for the selective detection of amino acids in aqueous solution [104]. Besides cyclopeptide structure, selectivity also depends on medium effects as a monolayer containing cyclopeptide **35a** shows highest sensitivity for arginine in neutral, buffer-free water while in phosphate buffered saline at pH 7.4 glutamate is the preferred analyte. Enantiomeric differentiation could also be achieved this way.

35a R = pC$_6$H$_4$NO$_2$; *cyclo*[L-Lys-L-Arg-(L-Lys-L-Phe(pNO$_2$))$_2$]

35b R = —O-[glucose] *cyclo*[L-Lys-L-Arg-L-Lys-L-Phe(pNO$_2$)-L-Lys-L-Ser(βGlc)]

Cyclopeptides **36a,b** have been immobilized on gold via the cysteine SH groups. Interactions between these peptides and amino acids were sensed by recording the frequency change of a quartz crystal microbalance induced by the presence of different amino acids in the supernatant aqueous solution [105]. In general, the sensitivity of the quartz microbalance is higher when coated with **36a**. This peptide induces highest sensitivity toward the positively charged amino acids lysine and arginine and possesses good enantiomeric discrimination ability for D- and L-arginine. Distinct differences in selectivity and sensitivity were observed for monolayers containing **36b**, demonstrating the dependence of sensor properties on cyclopeptide structure.

36a R = (CH$_2$)$_3$NH$_2$; *cyclo*[L-Lys-L-Cys-L-Trp-L-Cys-L-Lys-L-Cys]$_3$
36b R = C$_6$H$_5$; *cyclo*[L-Lys-L-Cys-L-Trp-L-Cys-L-Phe-L-Cys]$_3$

Haberhauer has shown that cyclopeptide **18a** with pendant 2,2′-bipyridine residue binds phenols in 10% acetonitrile–chloroform [106]. Highest affinity was observed for 1,3,5-trihydroxybenzene, a guest that best matches the symmetry of the host, while 1,2,3-trihydroxybenzene or dihydroxybenzenes are bound much less efficiently. Binding is due to hydrogen bonding interactions between the hydroxy groups of the guest and the nitrogen atoms in the biphenyl residues of the host. Interestingly, a cage-like receptor in which two cyclopeptide rings are bridged via three 4,4'-disubstituted 2,2'-bipyridine subunits possesses only weak affinity for 1,3,5-trihydroxybenzene although it should be much better preorganized for binding. This result was attributed to the fact that the inner cavity of the cage is slightly too small to allow for efficient interactions with the guest. While **18a** can easily adapt its cavity dimensions to the steric requirements of 1,3,5-trihydroxybenzene, this is much more difficult for the cage in which the mutual arrangement of the bipyridine linkers is fixed.

My own group has investigated the binding of monosaccharides by cyclopeptides **37a–c**, which contain substituents with carboxylate groups in the 5-position of the aromatic subunits [107]. All three cyclopeptides possess modest affinities for different methyl glycosides in 4% d_4-MeOD–CDCl$_3$. With regard to binding selectivity, certain trends were visible. Thus, peptide **37b** generally forms more stable complexes than the other two receptors, demonstrating that the orientation of the carboxylate groups in the aspartic acid residues of **37b** is better suited for monosaccharide binding. Furthermore, hexopyranosides are better bound than a pentofuranoside, which could be due to the larger ring size of the pyranosides but also to the additional hydroxy group in hexoses with which the peptides can interact. Both results indicate that binding is due to specific interactions between the guests and the hosts. Receptor **37b** also possesses affinity for arginine derivatives in DMSO, methanol, and even in water [108].

5.6
Conclusion

In 1981 E. R. Blout published a short review entitled "Cyclic peptides – past, present, and future" [23]. In the section on the future of cyclopeptide chemistry he made several predictions of trends this field might see, most of which have now become reality. NMR spectroscopy has, for example, developed into an indispensable tool for conformational analyses in solution, a huge number of new naturally cyclic peptides has been identified since 1981, and stabilizing bioactive conformations of linear peptides by ring closure has furnished potent leads for medicinal applications [109]. Two of Blout's predictions involve the use of cyclic peptides in molecular recognition. Specifically, he speculated that one day cyclopeptides will be available that possess three-dimensional cavities where ions or other guest molecules can be included and that conformationally restrained cyclopeptides will serve as mimics for the recognition sites of proteins. This chapter shows that the last almost 30 years have seen remarkable developments also in these areas. Cyclopeptides may not be regarded as conventional macrocyclic host molecules in supramolecular chemistry yet – crown ethers, cyclodextrins, and calixarenes play a much larger role in this respect – but major advantages of cyclopeptides are their wide structural (and conformational) variability and their close relationship to the ligands used by Nature, allowing the design of truly biomimetic receptors.

References

1 Pomilio, A.B., Battista, M.E., and Vitale, A.A. (2006) Naturally-occurring cyclopeptides: structures and bioactivity. *Curr. Org. Chem.*, **10**, 2075–2121.

2 Tan, N.-H. and Zhou, J. (2006) Plant cyclopeptides. *Chem. Rev.*, **106**, 840–895.

3 Liskamp, R.M.J., Rijkers, D.T.S., and Bakker, S.E. (2008) Bioactive macrocyclic peptides and peptide mimics, in *Modern Supramolecular Chemistry: Strategies for Macrocyclic Synthesis* (eds F. Diederich, P.J. Stang, and R.R. Tykwinski), Wiley-VCH Verlag GmbH, Weinheim, pp. 1–27.

4 Brockmann, H. and Schmidt-Kastner, G. (1955) Valinomycin I, XXVII. Mitteilung über Antibiotica aus Actinomyceten. *Chem. Ber.*, **88**, 57–61.

5 Shemyakin, M.M., Aldanova, N.A., Vinogradova, E.I., and Feigina, M.Y. (1963) The structure and total synthesis of valinomycin. *Tetrahedron Lett.*, **4**, 1921–1925.

6 Davidson, B.S. (1993) Ascidians: producers of amino acid derived metabolites. *Chem. Rev.*, **93**, 1771–1791.

7 Hawkins, C.J. (1988) Unusual chelates isolated from Ascidiacea. *Pure Appl. Chem.*, **60**, 1267–1270.

8 Morris, L.A., Jaspars, M., Kettenes-van den Bosch, J.J., Versluis, K., Heck, A.J.R., Kelly, S.M., and Price, N.C. (2001) Metal binding of *Lissoclinum patella* metabolites. Part 1: Patellamides A, C and ulithiacyclamide A. *Tetrahedron*, **57**, 3185–3197.

9 van den Brenk, A.L., Fairlie, D.P., Hanson, G.R., Gahan, L.R., Hawkins, C.J., and Jones, A. (1994) Binding of copper(II) to the cyclic octapeptide patellamide D. *Inorg. Chem.*, **33**, 2280–2289.

10 Rao, A.V.R., Gurjar, M.K., Reddy, K.L., and Rao, A.S. (1995) Studies directed toward the synthesis of vancomycin and related cyclic peptides. *Chem. Rev.*, **95**, 2135–2167.

11 Nicolaou, K.C., Boddy, C.N.C., Bräse, S., and Winssinger, N. (1999) Chemistry, biology, and medicine of the glycopeptide antibiotics. *Angew. Chem. Int. Ed.*, **38**, 2097–2152.

12 Williams, D.H. and Bardsley, B. (1999) The vancomycin group of antibiotics and the fight against resistant bacteria. *Angew. Chem. Int. Ed.*, **38**, 1173–1193.

13 Pedersen, C.J. (1967) Cyclic polyethers and their complexes with metal salts. *J. Am. Chem. Soc.*, **89**, 7017–7036.

14 Lambert, J.N., Mitchell, J.P., and Roberts, K.D. (2001) The synthesis of cyclic peptides. *J. Chem. Soc., Perkin Trans. 1*, 471–484.

15 Higler, I., Timmerman, P., Verboom, W., and Reinhoudt, D.N. (1998) The modular approach in supramolecular chemistry. *Eur. J. Org. Chem.*, 2689–2702.

16 Karle, I.L. (1981) X-ray analysis conformation of peptides in the crystalline state, in *The Peptides - Analysis, Synthesis, Biology*, vol. 4 (eds E. Gross and J. Meienhofer), Academic Press, New York, pp. 1–54.

17 Ovchinnikov, Y.A. and Ivanov, V.T. (1975) Conformational states and biological activity of cyclic peptides. *Tetrahedron*, **31**, 2177–2209.

18 Pletnev, V.Z., Tsygannik, I.N., Fonarev, Y.D., Mikhailova, I.Y., Kulikov, Y.V., Ivanov, V.T., Langs, D.A., and Duax, W.L. (1995) Crystal and molecular structure of K$^+$ complex of *meso*-valinomycin, *cyclo*[-(D-Val-L-Hyi-L-Val-D-Hyi)$_3$-]·KAuCl$_4$. *Bioorg. Khim.*, **21**, 828–833.

19 Gisin, B.F., Ting-Beall, H.P., Davis, D.G., Grell, E., and Tostesone, D.C. (1978) Selective ion binding and membrane activity of synthetic cyclopeptides. *Biochim. Biophys. Acta*, **509**, 201–217.

20 Hollósi, M. and Wieland, T. (1977) Ion binding properties in acetonitrile of cyclopeptides built up from proline and glycine residues. *Int. J. Peptide Protein Res.*, **10**, 329–341.

21 Saviano, G., Rossi, F., Benedetti, E., Pedone, C., Mierke, D.F., Maione, A., Zanotti, G., Tancredi, T., and Saviano, M. (2001) Structural consequences of metal complexation of *cyclo*[Pro-Phe-Phe-Ala-Xaa]$_2$ decapeptides. *Chem. Eur. J.*, **7**, 1176–1183.

22 Deber, C.M., Madison, V., and Blout, E.R. (1976) Why cyclic peptides? Complementary approaches to conformations. *Acc. Chem. Res.*, **9**, 106–113.

23 Blout, E.R. (1981) Cyclic-peptides: past, present, and future. *Biopolymers*, **20**, 1901–1912.

24 Madison, V., Atreyi, M., Deber, C.M., and Blout, E.R. (1974) Cyclic peptides. IX. Conformations of a synthetic ion-binding cyclic peptide, *cyclo*(Pro-Gly)$_3$, from circular dichroism and proton and carbon-13 nuclear magnetic resonance. *J. Am. Chem. Soc.*, **96**, 6725–6734.

25 Maulucci, N., Izzo, I., Bifulco, G., Aliberti, A., De Cola, C., Comegna, D., Gaeta, C., Napolitano, A., Pizza, C., Tedesco, C., Flot, D., and De Riccardis, F. (2008) Synthesis, structures, and properties of nine-, twelve-, and eighteen-membered *N*-benzyloxyethyl cyclic α-peptoids. *Chem. Commun.*, 3927–3929.

26 Deber, C.M. and Blout, E.R. (1974) Amino acid cyclic peptide complexes. *J. Am. Chem. Soc.*, **96**, 7566–7568.

27 Bartman, B., Deber, C.M., and Blout, E.R. (1977) Cyclic peptides. 16. Carbon-13 NMR relaxation studies of complexes between *cyclo*(L-Pro-Gly)$_3$ and amino acids. Conformational aspects of stepwise binding. *J. Am. Chem. Soc.*, **99**, 1028–1033.

28 Ma, J.C. and Dougherty, D.A. (1997) The cation-π interaction. *Chem. Rev.*, **97**, 1303–1324.

29 Kubik, S. (1999) Large increase in cation binding affinity of artificial cyclopeptide receptors by an allosteric effect. *J. Am. Chem. Soc.*, **121**, 5846–5855.

30 Kubik, S. and Goddard, R. (1999) A new cyclic pseudopeptide composed of (L)-proline and 3-aminobenzoic acid subunits as a ditopic receptor for the simultaneous complexation of cations and anions. *J. Org. Chem.*, **64**, 9475–9486.

31 Kubik, S. and Goddard, R. (2000) Intramolecular conformational control in a cyclic peptide composed of alternating (L)-proline and substituted 3-aminobenzoic acid subunits. *Chem. Commun.*, 633–634.

32 Kubik, S. and Goddard, R. (2001) Fine tuning of the cation affinity of artificial receptors based on cyclic peptides by

intramolecular conformational control. *Eur. J. Org. Chem.*, 311–322.

33 Cannizzaro, C.E. and Houk, K.N. (2002) Magnitudes and chemical consequences of R_3N^+-C-H···O=C hydrogen bonding. *J. Am. Chem. Soc.*, **124**, 7163–7169.

34 Heinrichs, G., Vial, L., Lacour, J., and Kubik, S. (2003) Enantioselective recognition of a quaternary ammonium ion by C_3 symmetric cyclic hexapeptides. *Chem. Commun.*, 1252–1253.

35 Heinrichs, G., Kubik, S., Lacour, J., and Vial, L. (2005) Matched/mismatched interaction of a cyclic hexapeptide with ion pairs containing chiral cations and chiral anions. *J. Org. Chem.*, **70**, 4498–4501.

36 Miyake, H. and Kojima, Y. (1996) Macrocyclic pseudopeptides containing N,N'-ethylene-bridged-dipeptide units: synthesis, binding properties toward metal and organic ammonium cations, and conformations. The first step in designing artificial metalloproteins. *Coord. Chem. Rev.*, **148**, 301–314.

37 Kojima, Y., Ikeda, Y., Miyake, H., Iwadou, I., Hirotsu, K., Shibata, K., Yamashita, T., Ohsuka, A., and Sugihara, A. (1991) Macrocyclic peptides VI. Complex formations and conformations of an ionophorous cyclic octapeptide containing N,N'-ethylene-bridged (S)-leucyl-(S)-leucine and glycine in acetonitrile. *Polym. J.*, **23**, 1359–1363.

38 Kojima, Y., Miyake, H., Ikeda, Y., Shibata, K., Yamashita, T., Ohsuka, A., and Sugihara, A. (1992) Macrocyclic peptides VII. Solution conformation and cation-binding properties of an ionophorous cyclic octapeptide containing N,N'-ethylene-bridged (S)-valyl-(S)-valine and glycine. *Polym. J.*, **24**, 591–595.

39 Miyake, H., Kojima, Y., Yamashita, T., and Ohsuka, A. (1993) Macrocyclic peptides, 8. Enantiomeric differentiations of various (R)- and (S)- ammonium and (R)- and (S)-α-amino acid ester salts by macrocyclic pseudopeptides. *Makromol. Chem.*, **194**, 1925–1933.

40 Miyake, H., Yamashita, T., Kojima, Y., and Tsukube, H. (1995) Enantioselective transport of amino acid ester salts by macrocyclic pseudopeptides containing N,N'-ethylene-bridged-dipeptide units. *Tetrahedron Lett.*, **36**, 7669–7672.

41 Bertram, A. and Pattenden, G. (2007) Marine metabolites: metal binding and metal complexes of azole-based cyclic peptides of marine origin. *Nat. Prod. Rep.*, **24**, 18–30.

42 van den Brenk, A.L., Byriel, K.A., Fairlie, D.P., Gahan, L.R., Hanson, G.R., Hawkins, C.J., Jones, A., Kennard, C.H.L., Moubaraki, B., and Murray, K.S. (1994) Crystal structure and electrospray ionization mass spectrometry, electron paramagnetic resonance, and magnetic susceptibility study of $[Cu_2(ascidH_2)(1,2-\mu\text{-}CO_3)(H_2O)_2]\cdot 2H_2O$, the bis(copper(II)) complex of ascidiacyclamide (ascidH$_4$), a cyclic peptide isolated from the ascidian *Lissoclinum patella*. *Inorg. Chem.*, **33**, 3549–3557.

43 Ishida, T., In, Y., Shinozaki, F., Doi, M., Yamamoto, D., Hamada, Y., Shioiri, T., Kamigauchi, M., and Sugiura, M. (1995) Solution conformations of patellamides B and C, cytotoxic cyclic hexapeptides from marine tunicate, determined by NMR spectroscopy and molecular dynamics. *J. Org. Chem.*, **60**, 3944–3952.

44 Freeman, D.J., Pattenden, G., Drake, A.F., and Siligardi, G. (1998) Marine metabolites and metal ion chelation. Circular dichroism studies of metal binding to *Lissoclinum* cyclopeptides, *J. Chem. Soc., Perkin Trans. 2*, 129–135.

45 Morris, L.A., Milne, B.F., Jaspars, M., Kettenes-van den Bosch, J.J., Versluis, K., Heck, A.J.R., Kelly, S.M., and Price, N.C. (2001) Metal binding of *Lissoclinum patella* metabolites. Part 2: lissoclinamides 9 and 10. *Tetrahedron*, **57**, 3199–3207.

46 Comba, P., Cusack, R., Fairlie, D.P., Gahan, L.R., Hanson, G.R., Kazmaier, U., and Ramlow, A. (1998) The solution structure of a copper(II) compound of a new cyclic octapeptide by EPR spectroscopy and force field calculations. *Inorg. Chem.*, **37**, 6721–6727.

47 Bernhardt, P.V., Comba, P., Fairlie, D.P., Gahan, L.R., Hanson, G.R., and Lötzbeyer, L. (2002) Synthesis and structural properties of patellamide a

48 Fusaoka, Y., Ozeki, E., Kimura, S., and Imanishi, Y. (1989) Synthesis and interaction with metal ions of cyclic oligopeptides bearing carboxyl groups. *Int. J. Peptide Protein Res.*, **34**, 104–110.

49 Ozeki, E., Miyazu, T., Kimura, S., and Imanishi, Y. (1989) Cyclic hexapeptides bearing carboxyl groups. Interaction with metal ions and lipid membrane. *Int. J. Peptide Protein Res.*, **34**, 97–103.

50 Ozeki, E., Kimura, S., and Imanishi, Y. (1989) Conformation and complexation with metal ions of cyclic hexapeptides: *cyclo*(L-Leu-L-Phe-L-Pro)$_2$ and *cyclo*[L-Cys(Acm)-L-Phe-L-Pro]$_2$. *Int. J. Peptide Protein Res.*, **34**, 111–117.

51 Haberhauer, G., Oeser, T., and Rominger, F. (2005) A widely applicable concept for predictable induction of preferred configuration in C_3-symmetric systems. *Chem. Commun.*, 2799–2801.

52 Haberhauer, G. (2008) Control of helicity in C_3-symmetric systems by peptide-like β-turns. *Tetrahedron Lett.*, **49**, 2421–2424.

53 Haberhauer, G. (2008) Control of planar chirality: the construction of a copper-ion-controlled chiral molecular hinge. *Angew. Chem. Int. Ed.*, **47**, 3635–3638.

54 Comba, P., Gahan, L.R., Haberhauer, G., Hanson, G.R., Noble, C.J., Seibold, B., and van den Brenk, A.L. (2008) Copper(II) coordination chemistry of westiellamide and its imidazole, oxazole, and thiazole analogues. *Chem. Eur. J.*, **14**, 4393–4403.

55 Motesharei, K. and Ghadiri, M.R. (1997) Diffusion-limited size-selective ion sensing based on SAM-supported peptide nanotubes. *J. Am. Chem. Soc.*, **119**, 11306–11312; Motesharei, K. and Ghadiri, M.R. (1998) *J. Am. Chem. Soc.*, **120**, 1347.

56 Ghadiri, M.R., Granja, J.R., Milligan, R.A., McRee, D.E., and Khazanovich, N. (1993) Self-assembling organic nanotubes based on a cyclic peptide architecture. *Nature*, **366**, 324–327.

57 Ghadiri, M.R. (1995) Self-assembled nanoscale tubular ensembles. *Adv. Mater.*, **7**, 675–677.

58 Hartgerink, J.D., Granja, J.R., Milligan, R.A., and Ghadiri, M.R. (1996) Self-assembling peptide nanotubes. *J. Am. Chem. Soc.*, **118**, 43–50.

59 Bong, D.T., Clark, T.D., Granja, J.R., and Ghadiri, M.R. (2001) Self-assembling organic nanotubes. *Angew. Chem. Int. Ed.*, **40**, 988–1011.

60 Ngu-Schwemlein, M., Butko, P., Cook, B., and Whigham, T. (2006) Interactions of an acidic cyclooctapeptide with metal ions: microcalorimetric and fluorescence analyses. *J. Peptide Res.*, **66** (Suppl. 1), 72–81.

61 Ngu-Schwemlein, M., Gilbert, W., Askew, K., and Schwemlein, S. (2008) Thermodynamics and fluorescence studies of the interactions of cyclooctapeptides with Hg^{2+}, Pb^{2+}, and Cd^{2+}. *Bioorg. Med. Chem.*, **16**, 5778–5787.

62 Gates, W.D., Rostas, J., Kakati, B., and Ngu-Schwemlein, M. (2005) Amphipathic cyclooctapeptides: interactions with detergent micelles and metal ions. *J. Mol. Struct.*, **733**, 5–11.

63 Kirkovits, G.J., Shriver, J.A., Gale, P.A., and Sessler, J.L. (2001) Synthetic ditopic receptors. *J. Inclusion Phenom. Macrocyclic Chem.*, **41**, 69–75.

64 Deetz, M.J., Shang, M., and Smith, B.D. (2000) A macrobicyclic receptor with versatile recognition properties: simultaneous binding of an ion pair and selective complexation of dimethyl sulfoxide. *J. Am. Chem. Soc.*, **122**, 6201–6207.

65 Mahoney, J.M., Beatty, A.M., and Smith, B.D. (2001) Selective recognition of an alkali halide contact ion-pair. *J. Am. Chem. Soc.*, **123**, 5847–5848.

66 Mahoney, J.M., Davis, J.P., Beatty, A.M., and Smith, B.D. (2003) Molecular recognition of alkylammonium contact ion-pairs using a ditopic receptor. *J. Org. Chem.*, **68**, 9819–9820.

67 Bianchi, A., Bowman-James, K., and García-España, E. (1997) *Supramolecular*

Chemistry of Anions, Wiley-VCH Verlag GmbH, New York.
68 Schmidtchen, F.P. and Berger, M. (1997) Artificial organic host molecules for anions. *Chem. Rev.*, **97**, 1609–1646.
69 Snowden, T.S. and Anslyn, E.V. (1999) Anion recognition: synthetic receptors for anions and their application in sensors. *Curr. Opin. Chem. Biol.*, **3**, 740–746.
70 Beer, P.D. and Gale, P.A. (2001) Anion recognition and sensing: the state of the art and future perspectives. *Angew. Chem. Int. Ed.*, **40**, 487–516.
71 Suksai, C. and Tuntulani, T. (2003) Chromogenic anion sensors. *Chem. Soc. Rev.*, **32**, 192–202.
72 Gale, P.A. (2003) Special issue on anion coordination chemistry: celebrating the 35th anniversary of Park and Simmon's publication. *Coord. Chem. Rev.*, **240**, 1–226.
73 Kubik, S., Reyheller, C., and Stüwe, S. (2005) Recognition of anions by synthetic receptors in aqueous solution. *J. Inclusion Phenom. Macrocyclic Chem.*, **52**, 137–187.
74 Sessler, J.L., Gale, P.A., and Cho, W.-S. (2006) *Anion Receptor Chemistry*, RSC Publishing, Cambridge.
75 Gale, P.A. (2006) Special issue on anion coordination chemistry. *Coord. Chem. Rev.*, **250**, 2918–3244.
76 Ishida, H. and Inoue, Y. (1999) Peptides that contain unnatural amino acids: toward artificial proteins. *Rev. Heteroatom Chem.*, **19**, 79–142.
77 Ishida, H., Kyakuno, M., and Oishi, S. (2004) Molecular design of functional peptides by utilizing unnatural amino acids: toward artificial and photofunctional protein. *Biopolymers*, **76**, 69–82.
78 Ishida, H., Donowaki, K., Suga, M., Shimose, K., and Ohkubo, K. (1995) Serine proteases mimics: hydrolytic activity of cyclic peptides which include a non-natural amino acid. *Tetrahedron Lett.*, **36**, 8987–8990.
79 Ishida, H., Qi, Z., Sokabe, M., Donowaki, K., and Inoue, Y. (2001) Molecular design and synthesis of artificial ion channels based on cyclic peptides containing unnatural amino acids. *J. Org. Chem.*, **66**, 2978–2989.
80 Qi, Z., Sokabe, M., Donowaki, K., and Ishida, H. (1999) Structure-function study on a *de novo* synthetic hydrophobic ion channel. *Biophys. J.*, **76**, 631–641.
81 Ishida, H., Suga, M., Donowaki, K., and Ohkubo, K. (1995) Highly effective binding of phosphomonoester with neutral cyclic peptides which include a non-natural amino acid. *J. Org. Chem.*, **60**, 5374–5375.
82 Kubik, S., Goddard, R., Kirchner, R., Nolting, D., and Seidel, J. (2001) A cyclic hexapeptide containing L-proline and 6-aminopicolinic acid subunits binds anions in water. *Angew. Chem. Int. Ed.*, **40**, 2648–2651.
83 Kubik, S., Goddard, R., Otto, S., Pohl, S., Reyheller, C., and Stüwe, S. (2005) Optimization of the binding properties of a synthetic anion receptor using rational and combinatorial strategies. *Biosens. Bioelectron.*, **20**, 2364–2375.
84 Jorgensen, W.L. (1989) Interactions between amides in solution and the thermodynamics of weak binding. *J. Am. Chem. Soc.*, **111**, 3770–3771.
85 Kubik, S. and Goddard, R. (2002) Conformation and anion binding properties of cyclic hexapeptides containing L-4-hydroxyproline and 6-aminopicolinic acid subunits. *Proc. Natl. Acad. Sci. USA*, **99**, 5127–5132.
86 Kubik, S., Kirchner, R., Nolting, D., and Seidel, J. (2002) A molecular oyster: a neutral anion receptor containing two cyclopeptide subunits with a remarkable sulfate affinity in aqueous solution. *J. Am. Chem. Soc.*, **124**, 12752–12760.
87 Reyheller, C., Hay, B.P., and Kubik, S. (2007) Influence of linker structure on the anion binding affinity of biscyclopeptides. *New J. Chem.*, **31**, 2095–2102.
88 Hay, B.P. and Firman, T.K. (2002) HostDesigner: A program for the *de novo* structure-based design of molecular receptors with binding sites that

complement metal ion guests. *Inorg. Chem.*, **41**, 5502–5512.

89 Hay, B.P., Oliferenko, A.A., Uddin, J., Zhang, C. and Firman, T.K. (2005) Search for improved host architectures: application of de novo structure-based design and high-throughput screening methods to identify optimal building blocks for multidentate ethers. *J. Am. Chem. Soc.*, **127**, 17043–17053.

90 Bryantsev, V.S. and Hay, B.P. (2006) De novo structure-based design of bisurea hosts for tetrahedral oxoanion guests. *J. Am. Chem. Soc.*, **128**, 2035–2042.

91 Reyheller, C. and Kubik, S. (2007) Selective sensing of sulfate in aqueous solution using a fluorescent bis(cyclopeptide). *Org. Lett.*, **9**, 5271–5274.

92 Otto, S., Kubik, S. (2003) Optimization of a neutral receptor that binds inorganic anions in aqueous solution using dynamic combinatorial chemistry. *J. Am. Chem. Soc.*, **125**, 7804–7805.

93 Choi, K. and Hamilton, A.D. (2001) Selective anion binding by a macrocycle with convergent hydrogen bonding functionality. *J. Am. Chem. Soc.*, **123**, 2456–2457.

94 Choi, K. and Hamilton, A.D. (2003) Rigid macrocyclic triamides as anion receptors: anion-dependent binding stoichiometries and ^{1}H chemical shift changes. *J. Am. Chem. Soc.*, **125**, 10241–10249.

95 Choi, K. and Hamilton, A.D. (2001) A dual channel fluorescence chemosensor for anions involving intermolecular excited state proton transfer. *Angew. Chem. Int. Ed.*, **40**, 3912–3915.

96 Yang, D., Qu, J., Li, W., Zhang, Y.-H., Ren, Y., Wang, D.-P., and Wu, Y.-D. (2002) Cyclic hexapeptide of D,L-α-aminoxy acids as a selective receptor for chloride ion. *J. Am. Chem. Soc.*, **124**, 12410–12411.

97 Li, X., Wu, Y.-D., and Yang, D. (2008) α-Aminoxy acids: new possibilities from foldamers to anion receptors and channels. *Acc. Chem. Res.*, **41** 1428–1438.

98 Pant, N. and Hamilton, A.D. (1988) Carboxylic acid complexation by a synthetic analog of the "carboxylate-binding pocket" of vancomycin. *J. Am. Chem. Soc.*, **110**, 2002–2003.

99 Pieters, R.J. (2000) Synthesis and binding studies of carboxylate binding pocket analogs of vancomycin. *Tetrahedron Lett.*, **41**, 7541–7545.

100 Arnusch, C.J. and Pieters, R.J. (2003) Solid phase synthesis of vancomycin mimics. *Eur. J. Org. Chem.*, 3131–3138.

101 McDonough, M.J., Reynolds, A.J., Lee, W.Y.G., and Jolliffe, K.A. (2006) Selective recognition of pyrophosphate in water using a backbone modified cyclic peptide receptor. *Chem. Commun.*, 2971–2973.

102 Leipert, D., Rathgeb, F., Herold, M., Mack, J., Gauglitz, G., and Jung, G. (1999) Interaction between volatile organic compounds and cyclopeptides detected with reflectrometric interference spectroscopy. *Anal. Chim. Acta*, **392**, 213–221.

103 Jung, G., Hofstetter, H., Feiertag, S., Stoll, D., Hofstetter, O., Wiesmüller, K.-H., and Schurig, V. (1996) Cyclopeptide libraries as new chiral selectors in capillary electrophoresis. *Angew. Chem. Int. Ed. Engl.*, **35**, 2148–2150.

104 Leipert, D., Nopper, D., Bauser, M., Gauglitz, G., and Jung, G. (1998) Investigation of the molecular recognition of amino acids by cyclopeptides with reflectrometric interference spectroscopy. *Angew. Chem. Int. Ed.*, **37**, 3308–3311.

105 Weiß, T., Leipert, D., Kaspar, M., Jung, G., and Göpel, W. (1999) Monolayers of cyclopeptides: a new concept for molecular recognition and enantiomeric discrimination. *Adv. Mater.*, **11**, 331–335.

106 Haberhauer, G., Oeser, T., and Rominger, F. (2005) Molecular scaffold for the construction of three-armed and cage-like receptors. *Chem. Eur. J.*, **11**, 6718–6726.

107 Bitta, J. and Kubik, S. (2001) Cyclic hexapeptides with free carboxylate groups as new receptors for monosaccharides. *Org. Lett.*, **3**, 2637–2640.

108 Bitta, J. and Kubik, S. (2001) Complexation of arginine with a cyclopeptide in polar solvents and water. *J. Supramol. Chem.*, **1**, 293–297.

109 Haubner, R., Finsinger, D., and Kessler, H. (1997) Stereoisomeric peptide libraries and peptidomimetics for designing selective inhibitors of the αvβ3 integrin for a new cancer therapy *Angew. Chem. Int. Ed. Engl.*, **36**, 1374–1389.

6
Boronic Acid-Based Receptors and Chemosensors

Xiaochuan Yang, Yunfeng Cheng, Shan Jin, and Binghe Wang

6.1
Introduction

Boronic acid is a commonly used moiety in the design of binders and sensors for carbohydrates and other nucleophile/Lewis base-containing compounds. Such applications largely depend on the unique electronic structure of the boron atom in a boronic acid. Specifically, the boron atom of a boronic acid in its neutral form has an open shell (six valence electrons) and, therefore, is electron-deficient. Consequently, it has a high tendency to react with a Lewis base to reach the stable octet form. Because of this unique electronic property, boronic acids are able to form tight and reversible complexes with 1, 2- and 1, 3-substituted Lewis base donors such as hydroxyl, amino, and carboxylate groups [1–4]. Boronic acids are also known to bind with simple Lewis bases such as fluoride [5–8], cyanide ions [9, 10], and hydroxyl groups [11–14]. All these properties have made boronic acid an important functional group for designing receptors and sensors for carbohydrates [15–20], α-hydroxyacids [21–24], α-amino alcohols [4, 25, 26], cyanide [9, 10], and fluoride [5–8, 27]. However, for two reasons, this chapter will only cover boronic acid-based sensors for diol-containing compounds such as carbohydrates and catechol. First, the principles that will be discussed in the design and selection of sensors/binders for carbohydrates are generally applicable to other classes of organic compounds and anions. Secondly, carbohydrate and catechol sensing represents an area of enormous biological importance. This is due to the need for sensors for catechol-based neurotransmitters and the tremendous development in the glycobiology and glycomics areas, which requires a large number of tools that allow high affinity and high specificity recognitions. For example, many cell surface glycan structures change when cells are transformed from benign to malignant cells; [28–33] many viruses and bacteria use carbohydrates as a key recognition moiety for infection and toxicity; [34–41] and the glycosylation patterns for certain glycoproteins change depending on whether the glycoproteins come from normal or cancerous origin [42–47]. Binders and sensors for such glycans are very useful as potential diagnostic tools and therapeutic agents. Because the boronic acid-based binders of carbohydrates are similar in function to

Artificial Receptors for Chemical Sensors. Edited by V.M. Mirsky and A.K. Yatsimirsky
Copyright © 2011 WILEY-VCH Verlag GmbH & Co. KGaA, Weinheim
ISBN: 978-3-527-32357-9

lectins, we have named these compounds boronolectins [48], which include small molecule boronolectins (SMBL) [15, 49–54], nucleic acid-based boronolectins (NABL) [55, 56], and peptide-based boronolectins (PBL) [57–62].

In the boronic acid-based carbohydrate sensing field, there have been hundreds of publications from a large number of active laboratories including those of Strongin, Shinkai, Czarnik, Anslyn, Lavigne, Heagy, Duggan, Hall, Asher, James, Wang, Lakowicz, Geddes, Norrild, Hindsgaul, Singaram, Tao, Lowe, Eggert, Smith, Okano, Taylor, Takeuchi, Ishihara, Katterle, Sporzynski, Freund, Mattiasson, Yoon, Scheller, Tuncel, Anzai, Houston, and so on. It would not be feasible or necessary to comprehensively review this field. Instead, the focus will be on key issues and approaches to consider in designing and selecting boronic acid sensors/binders for carbohydrates and catechols. For detailed descriptions of the boronic acid sensing field, readers are referred to several outstanding reviews on boronic acid-based sensor development [48, 63–71], boronic acid reporters that change fluorescent properties upon binding [48, 66], factors that affect the binding equilibrium [2, 3], and other issues to consider when designing boronic acid-based carbohydrate sensors [48]. In addition, there has also been a review [48] and several in-depth studies on the details of the structures of carbohydrate–boronic acid complexes [18, 72–75], which are very worthwhile reading materials for those who are new to this field.

To start the discussion, a brief description of the salient features of the boronic acid–diol complexation reaction is presented. The acidity of a boronic acid **1** is different from that of a conventional Brønsted acid, that is, ionization by releasing a proton. In a protic solvent, the acidity of a boronic acid arises from its reaction with a solvent molecule to form an anionic tetrahedral boronate **2** with the release of one proton (Scheme 6.1). The deprotonation of the boronic acid hydroxyl group actually has a higher pK_a than the boron open-shell reaction with water or alcohol and is very difficult, especially after the boron is converted to its tetrahedral anionic state **2**. Thus the acidity of a boronic acid results from its Lewis acidity *per se*. In an aprotic solvent, boronic acids act purely as strong Lewis acids. Therefore, a boronic acid can react with

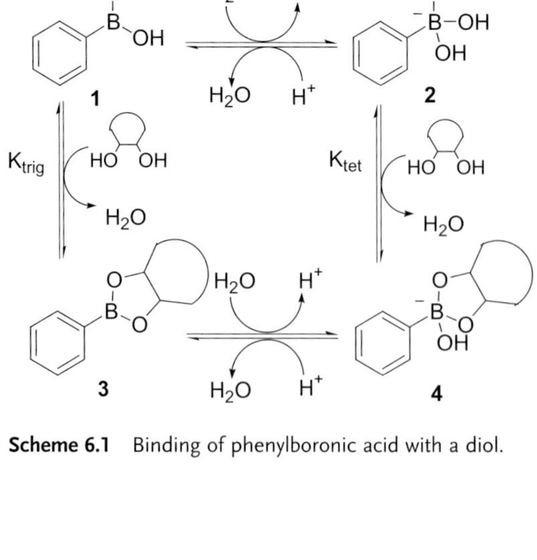

Scheme 6.1 Binding of phenylboronic acid with a diol.

a variety of Lewis bases such as hydroxyl, sulfhydryl, and amino groups as well as fluoride and cyanide. For this reason, boronic acids have also been used as Lewis acid catalysts and chelating agents for reaction controls [76]. Following the interaction with a Lewis base-containing molecule such as a diol, the boronic acid is transformed to its corresponding boronic ester **3**, which is also an acid because of the existence of the boron open shell. The boronic ester can react with a protic solvent (e.g., water) molecule, release a proton, and generate the boronate ester species **4** in much the same way as a boronic acid (Scheme 6.1).

Not all boronic acids have the same affinity toward a given diol and not all diols have the same affinity for a given boronic acid. The intrinsic preference in binding is a very important issue to consider for the design of boronic acid-based carbohydrate sensors. Although it is not always true, boronic acids with low pK_a values tend to have high binding affinities [1–3, 48]. Several factors can affect the intrinsic affinities of diols for a boronic acid. Low pK_a values, small O–C–C–O dihedral angles, and restricted rotations around the C–C bond of the diol part all favor binding [3]. High pH is commonly believed to favor boronic acid binding to diols. This is true most of the time, but not always. Optimal binding depends on the interplay of the pK_a values of the boronic acid and diol, the solution pH, and the nature and concentration of the buffer solution [3, 48]. The optimal pH for binding is often between the pK_a values of the boronic acid and diols. Thus, the optimal binding pH can be lower than physiological pH with low pK_a diols such as catechol compounds and boronic acids [3, 48]. For example, the optimal pH for the binding between alizarin red S and a phenyl boronic acid is close to or below 7.4 [3, 77]. Other factors such as solvent, buffer, temperature and steric hindrance can also affect the binding between a diol and a boronic acid [2, 3, 48].

Looking at individual reactions and binding between either the trigonal boronic acid (**1**, Scheme 6.1) or the tetrahedral boronate (**2**, Scheme 6.1) with a diol, K_{trig} represents the equilibrium constant between the trigonal form **1** of boronic acid and the boronic ester **3**; K_{tet} describes the equilibrium between the tetrahedral boronate form **2** and boronate ester **4**. The binding between a phenylboronic acid and a diol can also be described in Scheme 6.2, which represents the overall equilibrium between

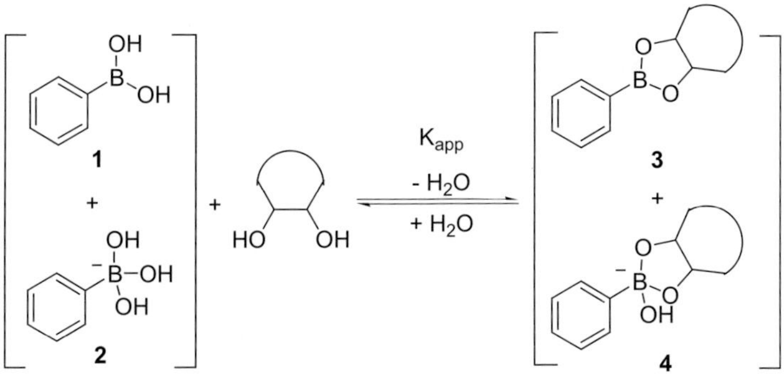

Scheme 6.2 Overall binding equilibrium of phenylboronic acid with a diol.

boronic acid/boronate **1**, **2** and boronic and boronate esters **3**, **4** [48]. K_{app} is used to describe the overall binding constant (Scheme 6.2), which of course is an apparent binding constant, but is directly related to the equilibrium between a sensor and a diol. In the literature, binding constants commonly refer to the apparent overall binding constant, K_{app}, which does not take into consideration the ionization states of either the complexed or the free form of a boronic acid [2, 3].

The strong and reversible interactions between a boronic acid moiety and a Lewis base/nucleophile in aqueous solution make it an ideal functional group for binder/receptor and chemosensor design. However, the strong individual functional group interactions of boronic acids alone are not sufficient for the design of specific chemosensors and receptors for complex molecules such as carbohydrates. One of the most important contemporary challenges in using boronic acids for the design of carbohydrate sensors is the selection of the appropriate scaffold to position the boronic acid and other functional groups in appropriate places and orientations in order to afford the necessary specificity and affinity. This chapter will be focused on this aspect. Other factors such as boronic acid-based reporter design and the modulation of their spectroscopic properties have been summarized in excellent reviews [48, 66]. For scaffold selection and design, one can consider three approaches: (i) *de novo* design, (ii) a combinatorial library approach, and (iii) template-directed synthesis. The following discussions will be divided into these three sections.

6.2
De Novo Design

In any sensor/binder effort, one of the most challenging aspects is the proper design of the three-dimensional scaffold, which allows the desired functional group arrangement and orientation for complementary interactions with the analyte. In this regard, carbohydrate chemosensors are no exception. In a *de novo* design effort, it is essential that the conformation of the analyte is well understood. This is hard to achieve with most saccharides except the simplest monosaccharides such as glucose. Though there have been extensive and impressive efforts [78–84], molecular modeling studies of saccharides lag behind similar studies of proteins and peptides for several reasons. First, there has been much more effort expended for protein and peptide conformational studies than for carbohydrates. Therefore, collectively, there are more data available for peptides and proteins. Secondly, there have been many more experimental studies of protein and peptide structure than for saccharides, which allows ready validation and homology modeling. Third, peptide and protein structures with polyamide as the backbone are more rigid than saccharides. The conformational flexibility of saccharides adds to the complexity and difficulty in conformational studies. Nevertheless, there have been some computational chemistry-guided carbohydrate sensor design efforts. The most prominent example is the work of Drueckhammer [85], which took a vector-based approach for the design of boronic acid-based sensors for glucose in its pyranose form (Scheme 6.3) instead of the furanose form, which is the binding form with many other diboronic acid

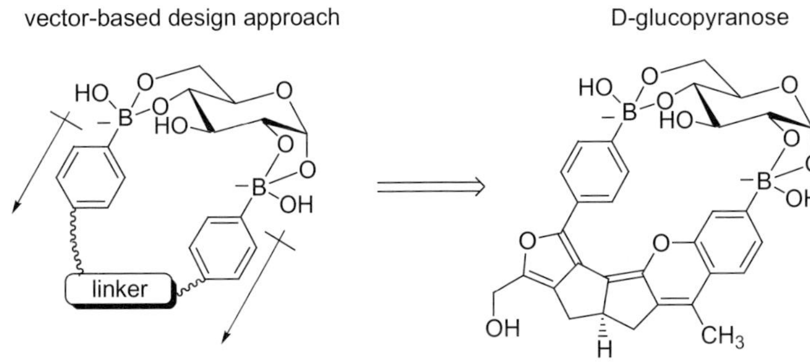

Scheme 6.3 Vector-based approach for the design of boronic acid-based sensor **5** for glucose in its pyranose form.

compounds reported in the literature [18, 72, 86–91]. The work started with the virtual "reaction" of glucopyranose with two phenylboronic acid moieties at the 1,2 and 4,6 positions. Then the system was subjected to quantum mechanics geometry optimization. The two phenylboronic acid moieties, after optimization, were considered as vectors to allow the computational search, using the CAVEAT program, of rigid scaffolds that allow the proper arrangements and orientation of these two boronic acid vectors for optimal complementarity. Compound **5** was an example candidate sensor for D-glucopyranose. As designed, **5** was able to bind D-glucose with high affinity, and binding resulted in a decrease in fluorescence intensity. Indeed, the binding was in the pyranose form, as demonstrated by ^{1}H NMR. The selectivity of the sensor for D-glucose versus other saccharides including D-galactose, D-mannose, and D-fructose was also studied. The sensor showed at least a 400-fold higher affinity for glucose ($K_d = 2.5 \times 10^{-5}$ M) than for the two other saccharides, D-galactose ($K_d = 1.0 \times 10^{-2}$ M) and D-mannose ($K_d = 1.6 \times 10^{-2}$ M). The high selectivity for D-glucose was attributed to the proper scaffold of the sensor, which positioned the two boronic acid moieties complementary to the diol pairs of glucose. Compound **5** represents a very successful example of using *de novo* computational design for the construction of carbohydrate sensors. Very importantly, the same general approach should be applicable to the design of sensors for other saccharides and organic molecules.

Using a similar approach, the Wang laboratory designed a boronic acid-based sensor for dopamine [92]. Dopamine has three important structural components (Figure 6.1), which should allow for strong interactions in an aqueous environment: (i) a catechol unit, which should form an ester with the boronic acid functional group [93–97]; (ii) a central aromatic ring, which should allow strong hydrophobic interactions; and (iii) an amino group, which, upon protonation, should afford strong interactions with an anionic functional group. The crystal structure of dopamine was used as the starting point for the design. Vectors were built similarly to that described by Drueckhammer (Scheme 6.3). Calculation of the dopamine–boronic acid complex was conducted using *ab initio*/6–31G optimization in the PCM water model

6 Boronic Acid-Based Receptors and Chemosensors

Figure 6.1 Structures of dopamine and reference compounds.

integrated in the Gaussian 03 program [98]. Based on optimized vector arrangements, three structural features were identified as important for the design of dopamine receptors (Figure 6.2a). A 2D virtual library of compounds containing these three structural components, a boronic acid, a carboxylic acid and a hydrophobic linker, was developed with the Iilib diverse 1.02 program [99, 100] and then converted into 3D structures using the CORINA program [101]. Finally, ISIS/BASE was used for a database search to give candidate structures. Figure 6.2a and b shows the calculation results with defined "ideal" distances and orientation of receptors for optimal binding as well as the receptors designed (Figure 6.2c).

Several structurally analogous sensors were synthesized, which differed in their X-, Y- substitutions on the aryl ring (Figure 6.2c). These receptors showed significant

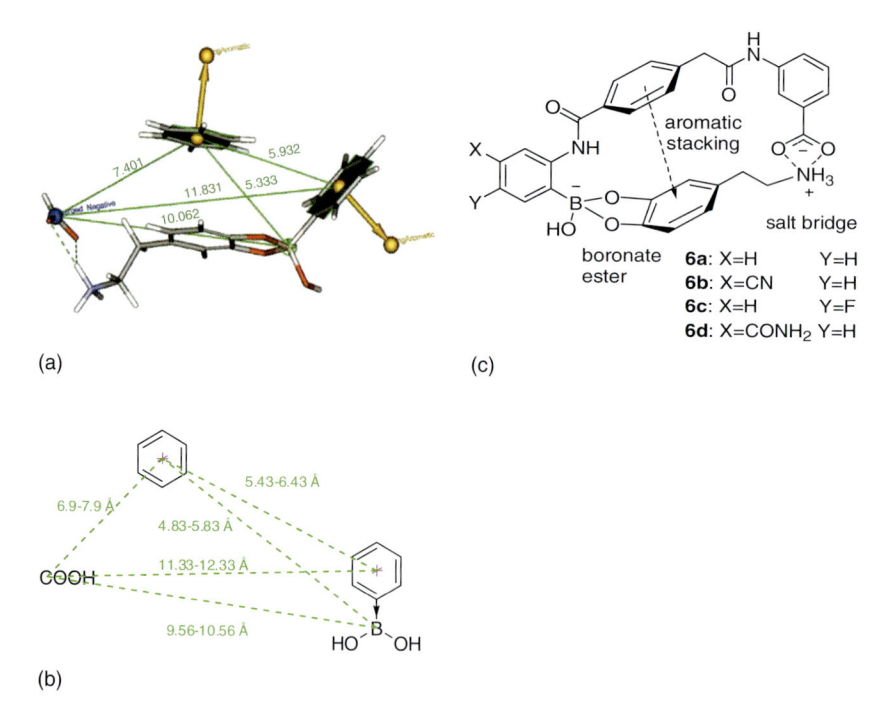

Figure 6.2 (a) Calculated distances among the three binding parts, (b) proposed ideal vector arrangement of the three binding components of the target receptors based on computational results, and (c) proposed receptor molecules. (Figures adapted and modified from Shan et al. [92].).

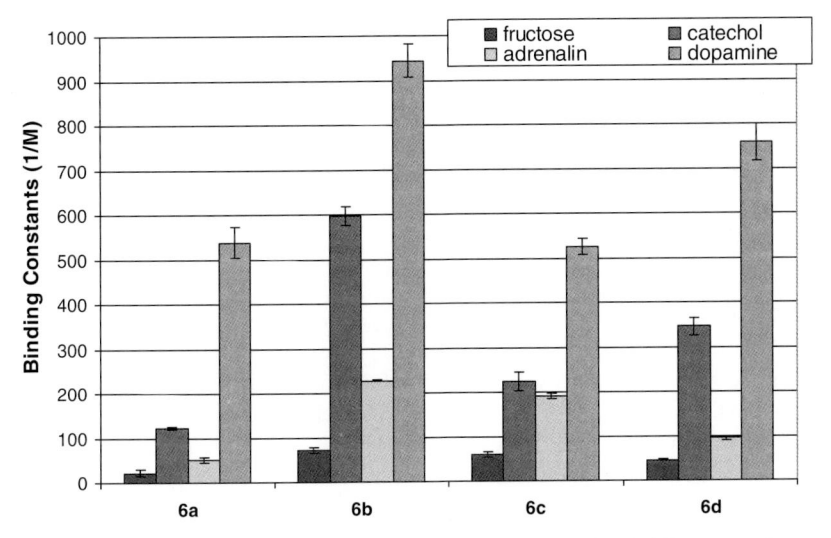

Figure 6.3 Comparison of the apparent binding constants of **6a–d** with different catechols in phosphate buffer at pH 7.4 (modified from Shan et al. [92]).

binding with and selectivity for dopamine (Figure 6.3). The apparent binding constants with dopamine were in the range 520–940 M^{-1}. In comparison, the binding constants with catechol and epinephrine ranged from 120 to 600 and 50 to 230 M^{-1}, respectively. Compound **6b** (X = CN, Y = H) showed the highest affinity (940 M^{-1}), but only less than twofold selectivity over catechol and about fourfold selectivity over epinephrine. Compounds **6a** and **6c** showed the lowest affinity for dopamine, but compound **6a** showed the highest selectivity over catechol (fourfold) and epinephrine (about tenfold). One very unique feature of these receptors is the significant selectivity of **6a–d** for dopamine over epinephrine though these two structures differ only by one hydroxyl group on the side chain. Such results indicated the significance of using a hydrophobic linker, which could help to bias against epinephrine due to the presence of an extra hydrophilic hydroxyl group on the side chain of epinephrine. When a linear alkyl chain was used in the sensors reported by Glass [97] and Yoon [102, 103], the selectivity over epinephrine was non-existent or very small. Though the dopamine receptor compounds tested only showed modest affinity and selectivity, which were far from being applicable for analysis in real life, the study further validated the vector-based approach that Druekhammer developed. In *de novo* design, the initial "lead" structure scaffold is very hard to come up with. The vector-based approach described at least serves as a very good starting point for generating "lead" structures for further optimization.

In addition to these two *de novo* design approaches, there have been other designs. Though these designs were not entirely based on *de novo* computational studies, they were certainly based on the structural features of the analytes and receptors in accomplishing complementary interactions. One such example is the work of Norrild and coworkers [89]. In this study, they designed a bispyridiniumboronic acid sensor **7**

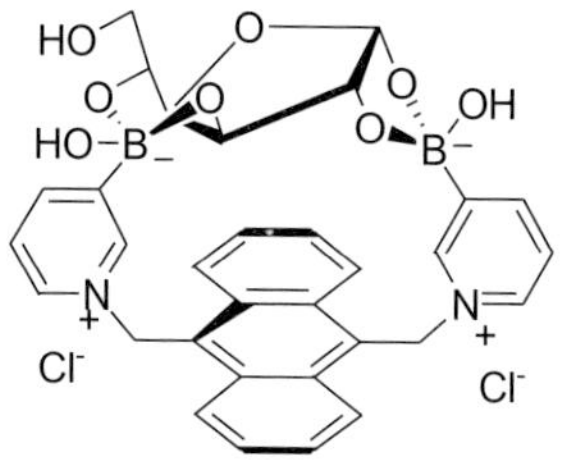

Figure 6.4 Bispyridiniumboronic acid sensor **7** for D-glucose.

for glucose sensing (Figure 6.4). The introduction of two positive charges was designed to (i) lower the pK_a of the boronic acid, which is known to improve binding and (ii) to improve water solubility. The anthracene linker seemed to provide the proper spacing and orientation of the two boronic acid moieties for glucose binding. The anthracene group was also chosen as a fluorophore. The fluorescence intensity increased upon addition of D-glucose in a concentration-dependent fashion with the formation of the sensor–glucose complex in the furanose form, as determined by ^{1}H and ^{13}C NMR in D$_2$O, pH 7.4.

The second example was a uronic acid sensor from the Shinkai laboratory [104]. It was based on the idea of functional group interactions at two points: (i) diol–boronic acid and (ii) anion/carboxylate–metal interactions. Specifically, boronic acid was used for binding to the diol moiety; and 1,10-phenanthroline chelated with Zn(II) was used for the coordination with carboxylate in uronic acids (Scheme 6.4). The association constants (log K_{ass}) for uronic acids in water-MeOH (1:2 v/v) buffer (pH 8.0) were determined to be 3.4 (for D-glucuronic acid) and 3.1 (for D-galacturonic acid) in the presence of Zn(II). On the other hand, without Zn(II), the binding constant for D-glucuronic acid was too weak to be determined and log K_{ass} for D-galacturonic acid was decreased to 1.9. In contrast, D-fructose gave a high binding affinity both in the presence (log K_{ass} = 2.4), and in the absence of Zn(II) (log K_{ass} = 2.5). Such results supported that uronic acids were bound to two binding sites simultaneously.

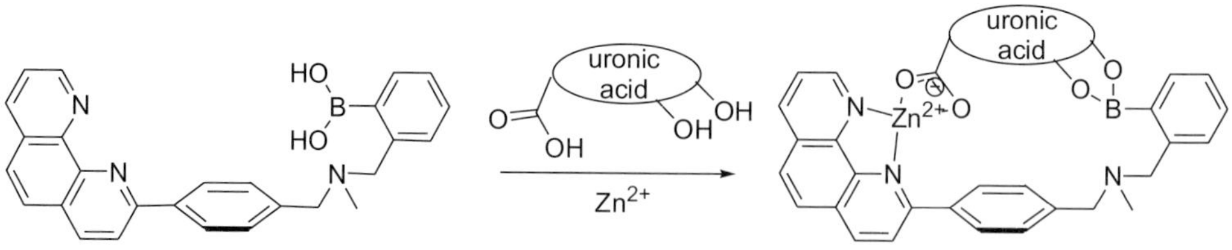

Scheme 6.4 Boronic acid sensor **8** for uronic acid in presence of Zn(II).

Another chemosensor **9** for dopamine and norepinephrine was designed and synthesized by the Glass laboratory [97]. The design of sensor **9** was also based on complementary interactions with the catechol and primary amine moieties. In this case, imine formation was used for interaction with the primary amine and a boronic acid moiety was used for interaction with the catechol moiety. It was proposed that iminium ion formation between the coumarin aldehyde and dopamine would generate a chromophoric response due to internal hydrogen bond formation (Scheme 6.5). In addition, the selectivity and affinity of **9** towards catecholamines were further enhanced by cyclic boronate ester formation with the catechol diols. All three catecholamines, dopamine, epinephrine and norepinephrine, quenched the fluorescence of **9** to 39–77% of its original intensity. All three catecholamines bond to the sensor with high affinity ($K_a = 3400 \text{ M}^{-1}$ for dopamine, 6500 M^{-1} for norepinephrine, 5000 M^{-1} for epinephrine). For the control analytes tested, all of them showed much lower affinity than the cachecholamines ($K_a = 250 \text{ M}^{-1}$ for tyramine, and less than 7 M^{-1} for L-glutamic acid, L-lysine, D-glucosamine, D-glucose).

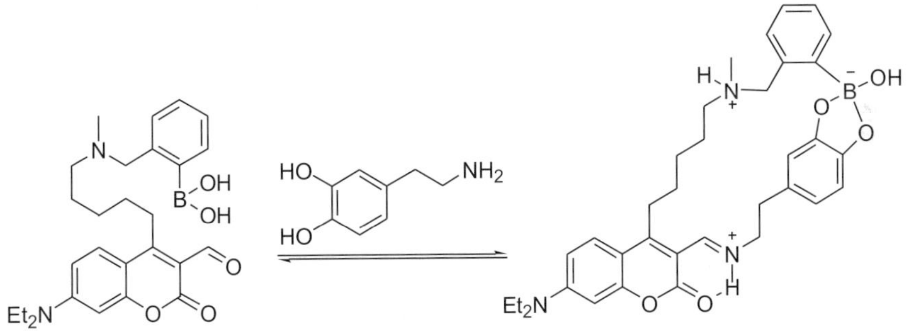

Scheme 6.5 Iminium ion and internal hydrogen bond formation in **9**-dopamine binding.

Overall, *de novo* design of boronic acid-based sensors is still in its infancy. However, promising results have been obtained. One would expect to see increasing work in this area.

6.3
Combinatorial Approaches

Although some success has been achieved through *de novo* design, this approach is time-consuming and the level of success varies depending on the accuracy of the computational design. Given that the current state of the art does not allow for complete *in silico* modeling of saccharide structures, large scale *de novo* design of sensors, especially those for oligosaccharides, has still a long way to go. One method of constructing sensors that does not require a prior detailed understanding of the structural features of the analyte is combinatorial chemistry [105–110]. In such an approach, a large library of compounds can be created and the "binder" selected using various approaches. In this combinatorial approach, one requirement is the availability of methods to create "polymeric" boronic acids in order to generate the desired diversity. In this sense, nucleic acid- [55, 56] and peptide-based [57–62] poly (boronic acids) have been successfully prepared with promising results.

With glycoproteins, their glycosylation patterns quite often change in cancer. For example, prostate specific antigen (PSA) is a prostate cancer biomarker and a glycoprotein. For PSA from cancer cells, the glycosylation patterns are known to be different from that of normal tissue [111–121]. Tools that allow detection and differentiation of the various glycoforms of a given glycoprotein will be very useful. However, until very recently, there had not been a general method for the preparation of "binders" that can achieve this. Based on the general aptamer selection method developed about 18 years ago by the laboratories of Szostak, Joyce, and Gold [122–124], the Wang laboratory are working to develop a methodology of using boronic acid-modified DNA to select aptamers that can recognize an intact glycoprotein with the ability to differentiate glycosylation variations. This was achieved through the incorporation of a boronic acid modified thymidine moiety (B-TTP, Figure 6.5) into the DNA sequences [55]. Because of the intrinsic affinity of boronic acids for carbohydrates, the hypothesis was that the incorporation of the boronic acid moiety would allow for the aptamer selection process to gravitate toward the "sweet spot" (glycosylation site) and, therefore, allow distinction of glycosylation variations. In other words, this method of using boronic acid-modified DNA allows the selection to go after the "sweet spot." The boronic acid moiety was functionalized with a terminal azido group and attached to the linear linker at 5-position of thymidine triphosphate via a Cu(I)-catalyzed Huisgen cyclization reaction [125–127]. It was demonstrated that B-TTP can be successfully recognized by DNA polymerase as substrate and incorporated into an elongating DNA sequence. Both MALDI-MS and gel electrophoresis results support completion of full length DNA polymerization using B-TTP in place of natural TTP in the reaction. Further experiment showed that DNA polymerization and amplification using boronic acid-labeled DNA as template was achieved with no noticeable difference from using natural DNA template. Thus, it is not surprising that PCR amplification of DNA using B-TTP was also attained. Among all the observed results, it should be specifically noticed that, in one particular experiment, the boronic acid-modified DNA gel-shift band was separated from the natural DNA band with the same length and composition in the catechol-modified PAGE result. Since it is commonly known that catechol is capable of forming a tight complex with boronic acid [2, 3, 92, 128], the gel-shift result serves as a good

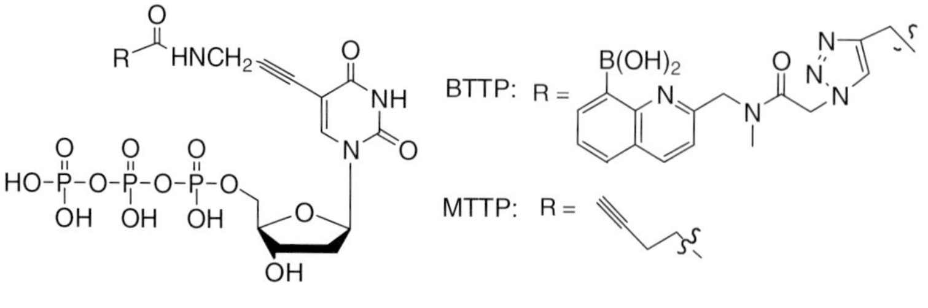

Figure 6.5 Boronic acid-modified TTP (B-TTP) for DNA aptamers selection.

indication that the diol-binding ability of boronic acid was well retained after being tethered to the DNA backbone. Thus, all these preliminary results are good proof-of-concept evidence supporting the feasibility of the proposed idea of developing boronic acid-based aptamers, or nucleic acid-based boronlectins (NABLs) in the future.

Independently, and prior to the boronic acid-modified DNA aptamer work in the Wang laboratory, the Anslyn laboratory [56] used unmodified RNA aptamers to tune the selectivity of a boronic acid-based receptor with great success. For example, bis-boronic acid **10** (Figure 6.5) by itself was known to bind citrate ($K_d = 5.5 \times 10^{-6} \, M^{-1}$) slightly better than tartrate ($K_d = 7.1 \times 10^{-6} \, M^{-1}$). The immobilized bis-boronic acid receptor **11** on glyoxal agarose beads (Figure 6.5) was subjected to an RNA pool (N50) SELEX (Systematic Evolution of Ligands by Exponential Enrichment method) in the presence of 200 μM tartrate. Negative selection was performed on the RNA pool against the immobilized tris-amine beads **12** as blank control (Figure 6.6) before each round of SELEX to increase the selection specificity. The RNA binding to the complex of receptor–tartrate became dominant in the pool from the seventh round. In radio-labeled binding assays, all seven families of the cloned RNA aptamers showed preference for binding the receptor-tartrate complex (from 12 to 25%) over the receptor alone (<7%). One specific sequence was used in the fluorescence binding test of receptor towards analytes. In the presence of this aptamer, the receptor was able to bind tartrate with $K_d = 2.1 \times 10^{-4} \, M^{-1}$ while no binding to citrate was observed ($K_d > 3 \times 10^{-3} \, M^{-1}$). Thus, the selectivity of the bis-boronic acid was tuned to at least 14-fold higher for binding tartrate than for citrate in the presence of the selected aptamer. In comparison, the receptor had a 1.3-fold higher preference for citrate over tartrate in the absence of aptamer.

In addition to aptamer-based boronic acid libraries, there have been efforts to make PBL libraries for the same purpose. In this section, it is important to mention the work from the Hall laboratory on developing a general solid-phase approach to the synthesis and isolation of functionalized boronic acids [57], which should be very useful in combinatorial library synthesis for boronic acid-based carbohydrate sensors. As shown in Scheme 6.6, N,N-diethanolaminomethyl polystyrene (DEAM-PS, **13**) can be used for the immobilization of boronic acids. These DEAM-PS-supported arylboronic acids can be used in resin-to-resin transfer reactions (RRTR) or for further reactions. Along a similar line, the Hall laboratory has established a

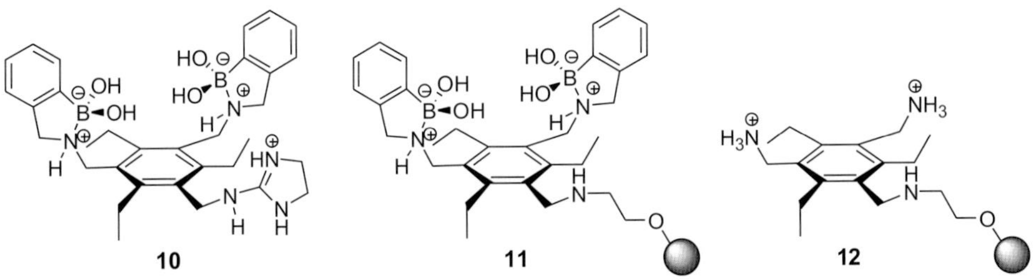

Figure 6.6 Bis-boronic acid **10**, its immobilized form **11** on glyoxal agarose beads and blank control tris-amine beads **12**.

Scheme 6.6 Immobilization and derivatization of boronic acids using N, N-diethanolaminomethyl polystyrene (DEAM-PS **13**) for combinatorial PBL library synthesis.

prototypic bead-supported split-pool library of triamine-derived triboronic acid receptors (Figure 6.7) [62]. Well controlled synthesis, bead decoding by HPLC, and preliminary screening of a combinatorial library **14a** and **14b** of 289 triamine-derived triboronic acid receptors for oligosaccharides have been accomplished. The PBLs from single beads were decoded from the library and detected by electrospray ionization mass spectrometry (ESI-MS).

Anslyn and coworkers prepared a PBL library **15** with a boronic acid side chain **16** for pattern-based carbohydrate sensing (Figure 6.8) [58]. In this work, bromopyrogallol red (BPR) was used as an indicator for saccharide binding studies. The binding response signal was recorded by a CCD camera. Linear discriminant analysis (LDA) was used and discrimination between monosaccharides and disaccharides and within various saccharide groups was achieved. This LDA data set could also be

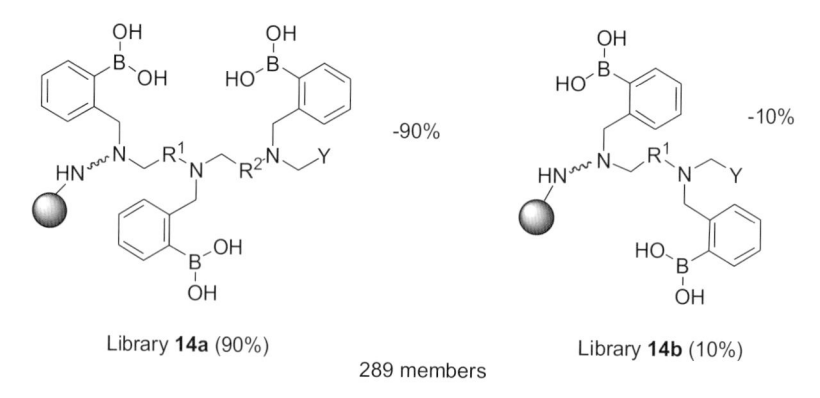

Figure 6.7 The general structures of PBL libraries **14a** and **14b** with 289 members.

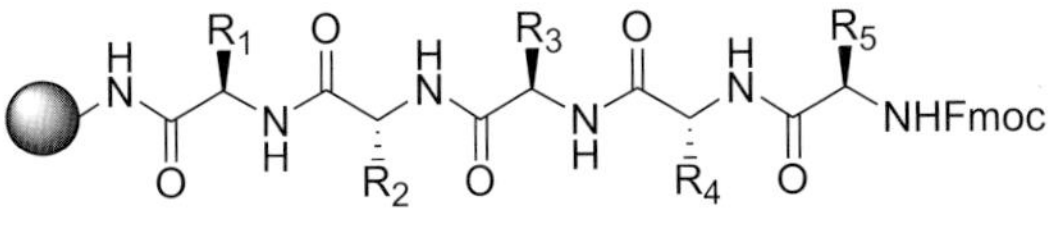

15, R$_1$-R$_5$ = side chains of natural amino acids and **16**   **16** Boronic acid modified side chain

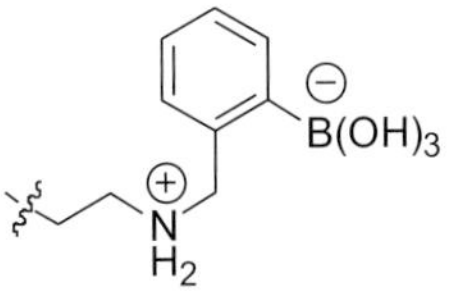

Figure 6.8 Structures of PBL library **15** and boronic acid modified side chain **16**.

used for identifying sucralose in a real world beverage sample. Their chemosensor assay system had the advantage of good water solubility and high sensitivity. This method was one of the first assays where supramolecular pattern-based sensors were used to identify a specific target in a complex beverage.

The Lavigne laboratory has also reported their PBLs for carbohydrate recognition [61]. In this study, PBLs were synthesized on beads and used to probe fluorescently labeled glycoproteins and oligosaccharides. A small 12-mer PBL library on a solid phase was synthesized using a biased split-and-pool combinatorial approach. Fluorescent microscope imaging of the beads after mixing them with various fluorescently labeled glyco-products was used to examine binding. Binding patterns using microplates allowed them to establish that binding was both PBL- and saccharide-dependent. Further expansion of such libraries could allow the selection of binding sequences for a variety of glycan-based products.

Duggan and coworkers also prepared solid-supported PBLs derived from 4-borono-L-phenylalanine and studied their affinity for a chromophoric diol, alizarin [59]. Libraries of solid-supported pentapeptide-based bisboronic acids, including a "lysine series" and an "arginine series", were prepared. The authors developed a technique for measuring the affinity between alizarin and the solid-supported PBLs. Significant variations in alizarin binding strengths, both within and between arginine and lysine series, were observed, with binding constants in the range 200–1100 M^{-1}. Such results suggest that different sequences do show preferential binding and selectivity. Though no specific "sensor" was identified, the general method developed could be very useful in the future.

6.4
Template-Directed Synthesis

Another approach to the construction of carbohydrate sensors/binders is template-directed synthesis, which also has the desired feature of not requiring prior knowledge of the carbohydrate three-dimensional structures. Among the different template-directed synthesis approaches, molecular imprinting is especially useful in creating receptor sites for organic molecules including carbohydrates. Molecular imprinting is a technique first demonstrated in the late 1940s by Dickey [129]. It involves three steps: (i) mixing the template/analyte with the functional monomers and crosslinkers at low temperature to allow complementary intermolecular interactions; (ii) polymerization of the monomers and crosslinkers under mild conditions; and (iii) extraction of the template molecules from the polymers, which leaves behind polymeric sites with memories of the template molecules in terms of size, shape, and functional group orientations (Scheme 6.7). This technique has been used for the preparation of selective recognition sites for a wide variety of molecules [130–132]. Naturally, the same method has been used for the preparation of sensors [133], including boronic acid-based carbohydrate sensors.

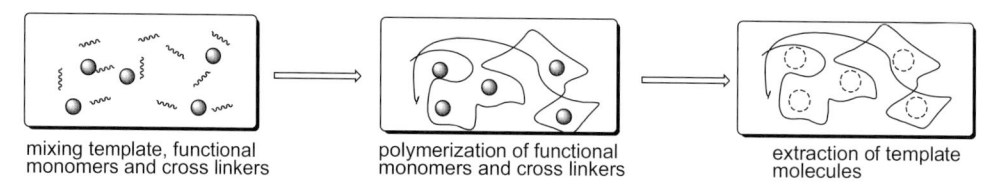

Scheme 6.7 Molecular imprinting process.

Wulff first demonstrated the use of boronic acid compounds in molecular imprinting for the construction of polymeric cavities specific for certain saccharides [134]. D-fructose and D-galactose were used as the template compounds. Specifically, these sugars were first conjugated with 2 equiv of VPB ((4-vinylphenyl)boronic acid), presumably at 2,3;4,5-OH of fructose and 1,2;3,4-OH of galactose, to yield their respective polymerizable monomers. EGDMA (ethylene glycol dimethacrylate) was used as the crosslinker. The polymers prepared were shown to have "memories" of the template, including stereochemical features. For example, the D-fructose-imprinted polymer could differentiate the D-form from the L-form by a factor of 1.47–1.63. It is interesting to note that the D-fructose- imprinted polymer could also differentiate L-galactose from D-galactose by a factor of 1.17 while D-galactose-imprinted polymer could differentiate L-fructose from D-frucose by a factor of 1.25. It was proposed that the high similarity between D-fructose and L-galactose as well as L-fructose and D-galactose in terms of their spatial hydroxyl group arrangements was responsible for the observed cross reactivities.

Using the same imprinting principle, the Wang laboratory was the first to synthesize boronic acid-based fluorescent polymers for saccharides as potential fluorescent sensors [135, 136]. In such studies, a polymerizable boronic acid monomer **17** was prepared (Figure 6.9). The anthracene-based boronic acid unit in **17** is known to increase fluorescent intensity upon sugar binding, as reported by the Shinkai laboratory [16, 49]. The monomer was first reacted with D-fructose to form a covalent complex before polymerization in the presence of EGDMA as the crosslinker. In re-binding studies, the fructose-imprinted polymer showed significant fluorescent intensity changes (over 100%) upon fructose addition. This polymer also showed preference for D-fructose over D-glucose and D-mannose, which was at least twice as large as that determined with the control boronic acid alone.

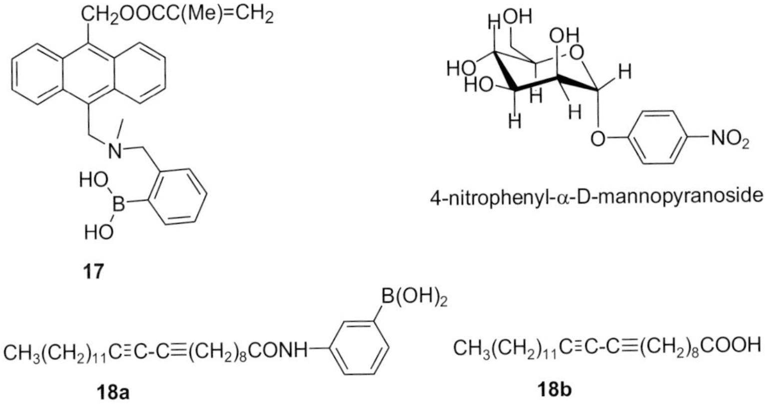

Figure 6.9 Boronic acid monomer **17**, **18a**, crosslinker **18b** and 4-nitrophenyl-α-D-mannopyranoside.

In yet another example, Kurihara and coworkers [137] conducted surface imprinting of 4-nitrophenyl α-D-mannopyranoside with a polymerizable phenylboronic acid monomer, (3-(10′,12′-penta-cosadiynamido)phenylboronic acid) **18a** and crosslinker **18b** (Figure 6.9). 4-Nitrophenyl α-D-mannopyranoside (Figure 6.9) was used as the template molecule and the 4-nitrophenoxy group can be used to quantify the binding by UV–visible spectroscopy. In re-binding studies using both imprinted and control polymers, about 75% of the binding sites of the imprinted surface were occupied by the template, while about 57% of the non-imprinted surface was covered. This implied that the imprinted monolayers had an improved affinity for the template molecule.

The molecular imprinting technique is also used in developing boronic acid-based electrochemical sensors. Usually, a thick imprinted polymer layer on the electrode surface is unfavorable since it hinders analyte diffusion and retards polymer–electrode communication. The problem can be partially overcome by utilizing thin layer polymer film electropolymerization on electrodes or on the gate surface of an ion-sensitive field effect transistor (ISFET) device through radical polymerization [138–140]. In such cases, the barrier to analyte diffusion and electrical contact is minimized. In this regard, the Willner group reported the copolymerization of acrylamide–acrylamidephenylboronic acid on TiO_2 film for imprinting AMP, GMP and CMP [140]. Charging the boronic acid polymer membrane with nucleotides controlled the gate potential of the ISFET to allow detection of each nucleotide. Similar thin layer boronic acid polymers were constructed by the same group for various nucleotides and monosaccharides on Au electrodes, piezoelectric Au quartz crystals, or on the gate surface of ISFET device [138]. Faradaic impedance spectroscopy measurement showed enhanced selectivity for the template molecules by increasing the crosslinking degree of the polymer. Most recently, the same group reported fabrication of stereoselective and enantioselective imprinted polymer for monosaccharides such as D-glucose and D-mannose [139]. The co-electropolymerization of phenol with a 3-hydroxyphenyl boronic acid–monosaccharide complex on an Au support created a thin film of boronate-conjugated polyphenol layer coated electrode (**19**, Figure 6.10). Competitive electrochemical assay was carried out by measuring the amperometric responses using ferrocene modified-monosaccharides (**20**, Figure 6.10) as the redox labels to analyze the binding affinity of each saccharide. The observed stereoselectivity and enantioselectivity indicated that a rather delicate structural contour was constructed around the saccharide template with boronic acid and other motifs accommodating the hydroxyl groups in a delicate coordination way.

Flow calorimetry was used by the Scheller group [141] to study the interactions of phenylboronic acid-based polymers imprinted with fructosyl valine (Fru-Val), fructose and pinacol, respectively. Fru-Val is known as the N-terminal glycated amino acid on glycosylated hemoglobin (HbA1c) (HbA0 is the non-glycated hemoglobin). The imprinted polymer for Fru-Val was prepared by copolymerizing the complex of Fru-Val and VPB with crosslinker TRIM (trimethylolpropane trimethacrylate). An exothermic peak signal was observed when Fru-Val flew through the imprinted polymer in a thermistor. This was followed by an almost steady state signal. The intensity of the peak signal increased in response to Fru-Val in a concentration-dependent manner (up to 5 mM). Control polymer prepared from the pinacol–VPB complex in the same way did not give such significant signal changes. The apparent imprinting factor (IF), the ratio of binding with imprinted polymer to that with the control polymer at equilibrium, was determined as 41 by measuring the signal intensity ratios. The temperature increase induced by Fru-Val interactions with imprinted polymer was 2.4-fold higher than that of fructose-imprinted polymer interactions and 270-fold higher than that of Val-imprinted polymer interactions. Thus, a selective Fru-Val MIP sensor was developed and used in calorimeter measurements.

Though the discussion of template-directed boronic acid-based sensor work has been focused on polymers, one should not feel restricted by this. It is very conceivable and should be feasible to make small molecule boronic acid-based sensors (SMBL) for carbohydrates through template-directed synthesis. Such an approach is in line with dynamic combinatorial library work, which has been proven useful [142–146]. For example, one could envision the linking of two boronic acids through a tether in a library consisting of multiple boronic acid and linker species. Such a library could give a large number of bisboronic acid end products. However, if a template carbohydrate is present, it is possible that the library can be biased to give preferentially bisboronic acids that have high affinity for the template. The accomplishment of such a goal could pave the way for the template-directed synthesis of SMBLs for carbohydrates.

In conclusion, boronic acids have very unique properties and can be used as a key functional group in the construction of sensors/binders for diol-containing compounds such as carbohydrates and catechols as well as other Lewis bases such as fluoride and cyanide. Generally speaking, there are three approaches to the construction of such binders/sensors: *de novo* design, combinatorial chemistry, and

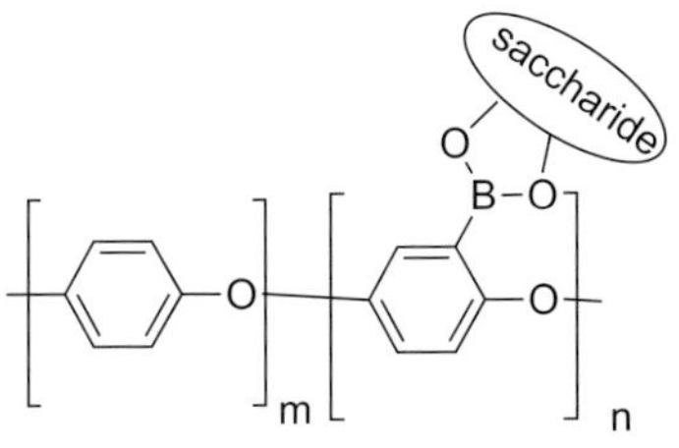

Figure 6.10 Boronic acid-containing polyphenols **19** and ferrocene-monosaccharide redox labels **20**.

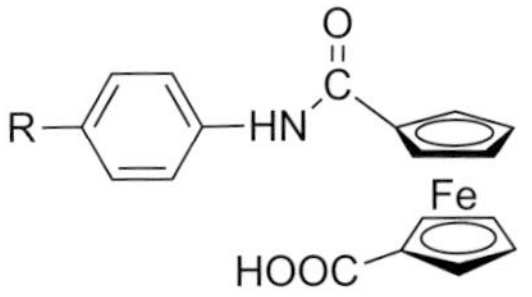

template-directed synthesis. Because of the tremendous development and interest in glycobiology and glycomics, one would expect to see increasing interests in boronic acid-based sensors/binders for carbohydrates.

Acknowledgment

Financial support of the work in the PI's laboratory by the National Institutes of Health (CA123329, CA113917, DK55062, CA88343, NO1-CO-27184, GM086925, and GM084933), Georgia Research Alliance, Georgia Cancer Coalition, and the Georgia State University Molecular Basis of Disease program is gratefully acknowledged.

References

1 Lorand, J.P. and Edwards, J.O. (1959) *J. Org. Chem.*, **24**, 769–774.
2 Springsteen, G. and Wang, B. (2002) *Tetrahedron*, **58**, 5291–5300.
3 Yan, J., Springsteen, G., Deeter, S., and Wang, B. (2004) *Tetrahedron*, **60**, 11205–11209.
4 Jabbour, A., Steinberg, D., Dembitsky, V.M., Moussaieff, A., Zaks, B., and Srebnik, M. (2004) *J. Med. Chem.*, **47**, 2409–2410.
5 DiCesare, N. and Lakowicz, J.R. (2002) *Anal. Biochem.*, **301**, 111–116.
6 Badugu, R., Lakowicz, J.R., and Geddes, C.D. (2005) *Sensor Actuat. B-Chem.*, **104**, 103–110.
7 Swamy, K.M.K., Lee, Y.J., Lee, H.N., Chun, J., Kim, Y., Kim, S.J., and Yoon, J. (2006) *J. Org. Chem.*, **71**, 8626–8628.
8 Oehlke, A., Auer, A.A., Jahre, I., Walfort, B., Ruffer, T., Zoufala, P., Lang, H., and Spange, S. (2007) *J. Org. Chem.*, **72**, 4328–4339.
9 Badugu, R., Lakowicz, J.R., and Geddes, C.D. (2004) *Anal. Chem.*, **327**, 82–90.
10 Badugu, R., Lakowicz, J.R., and Geddes, C.D. (2005) *Dyes Pigments*, **64**, 49–55.
11 Brindley, P.B., Gerrard, W., and Lappert, M.F. (1955) *J. Chem. Soc.*, 2956–2958.
12 Morrison, J.D. and Letsinger, R.L. (1964) *J. Org. Chem.*, **29**, 3405–3407.
13 Matteson, D.S. and Schaumberg, G.D. (1967) *J. Organomet. Chem.*, **8**, 359–360.
14 Sienkiewicz, P.A. and Roberts, D.C. (1980) *J. Inorg. Nucl. Chem.*, **42**, 1559–1575.
15 Yoon, J. and Czarnik, A.W. (1992) *J. Am. Chem. Soc.*, **114**, 5874–5875.

16 James, T.D., Sandanayake, K.R.A.S., Iguchi, R., and Shinkai, S. (1995) *J. Am. Chem. Soc.*, **117**, 8982–8987.

17 Eggert, H., Frederiksen, J., Morin, C., and Norrild, J.C. (1999) *J. Org. Chem.*, **64**, 3846–3852.

18 Bielecki, M., Eggert, H., and Norrild, J.C. (1999) *J. Chem. Soc., Perkin Trans. 2*, 449–455.

19 Yang, W., Fan, H., Gao, S., Gao, X., Ni, W., Karnati, V., Hooks, W.B., Carson, J., Weston, B., and Wang, B. (2004) *Chem. Biol.*, **11**, 439–448.

20 Asher, S.A., Alexeev, V.L., Goponenko, A.V., Sharma, A.C., Lednev, I.K., Wilcox, C.S., and Finegold, D.N. (2003) *J. Am. Chem. Soc.*, **125**, 3322–3329.

21 Wang, Z., Zhang, D.Q., and Zhu, D.B. (2005) *J. Org. Chem.*, **70**, 5729–5732.

22 Gray Jr., C.W. and Houston, T.A. (2002) *J. Org. Chem.*, **67**, 5426–5428.

23 Wiskur, S.L., Lavigne, J.L., Ait-Haddou, H., Lynch, V., Chiu, Y.H., Canary, J.W., and Anslyn, E.V. (2001) *Org. Lett.*, **3**, 1311–1314.

24 Gamsey, S., Miller, A., Olmstead, M.M., Beavers, C.M., Hirayama, L.C., Pradhan, S., Wessling, R.A., and Singaram, B. (2007) *J. Am. Chem. Soc.*, **129**, 1278–1286.

25 Cho, B.T. (2005) in *Boronic Acids* (ed. D.G. Hall), Wiley-VCH Verlag GmbH, Weinheim, pp. 411–439.

26 Aharoni F R, Bronsteyn, M., Jabbour, A., Zaks, B., Srebnik, M., and Steinberg, D. (2008) *Bioorg. Med. Chem.*, **16**, 1596–1604.

27 Cooper, C.R., Spencer, N., and James, T.D. (1998) *Chem. Commun.*, 1365–1366.

28 Dennis, J.W. (1992) in *Cell Surface Carbohydrates and Cell Development* (ed. M. Fukuda), CRC Press, Boca Raton, pp. 161–194.

29 Fukuda, M. (1992) in *Cell Surface Carbohydrates and Cell Development* (ed. M. Fukuda), CRC Press, Boca Raton, pp. 127–160.

30 Culig F Z, Hittmair, A., Hobisch, A., Bartsch, G., Klocker, H., Pai, L.H., and Pastan, I. (1998) *Prostate*, **36**, 162–167.

31 Fukuda, M. (1996) *Cancer Res.*, **56**, 2237–2244.

32 Meyer, T. and Hart, I.R. (1998) *Eur. J. Cancer*, **34**, 214–221.

33 Kannagi, R., Izawa, M., Koike, T., Miyazaki, K., and Kimura, N. (2004) *Cancer Sci.*, **95**, 377–384.

34 Karlsson, K.A. (1998) *Acta. Biochim. Pol.*, **45**, 429–438.

35 Liu, J., Shriver, Z., Pope, R.M., Thorp, S.C., Duncan, M.B., Copeland, R.J., Raska, C.S., Yoshida, K., Eisenberg, R.J., Cohen, G., Linhardt, R.J., and Sasisekharan, R. (2002) *J. Biol. Chem.*, **277**, 33456–33467.

36 Hartshorn, K.L., White, M.R., Mogues, T., Ligtenberg, T., Crouch, E., and Holmskov, U. (2003) *Am. J. Physiol.-Lung Cell. Mol. Physiol.*, **285**, L1066–L1076.

37 Goodman, J.L., Nelson, C.M., Klein, M.B., Hayes, S.F., and Weston, B.W. (1999) *J. Clin. Invest.*, **103**, 407–412.

38 Thomas, R.J. and Brooks, T.J. (2004) *Microb. Pathol.*, **36**, 83–92.

39 Chiba, H., Inokoshi, J., Nakashima, H., Omura, S., and Tanaka, H. (2004) *Biochem. Biophys. Res. Commun.*, **316**, 203–210.

40 Masoud, H., Martin, A., Thibault, P., Moxon, E.R., and Richards, J.C. (2003) *Biochemistry*, **42**, 4463–4475.

41 Shroll, R.M. and Straatsma, T.P. (2002) *Biopolymers.*, **65**, 395–407.

42 Miyazaki, K., Ohmori, K., Izawa, M., Koike, T., Kumamoto, K., Furukawa, K., Ando, T., Kiso, M., Yamaji, T., Hashimoto, Y., Suzuki, A., Yoshida, A., Takeuchi, M., and Kannagi, R. (2004) *Cancer Res.*, **64** 4498–4505.

43 Zhang, J., Nakayama, J., Ohyama, C., Suzuki, M., Suzuki, A., Fukuda, M., and Fukuda, M.N. (2002) *Cancer Res.*, **62**, 4194–4198.

44 Krzeslak, A., Pomorski, L., Gaj, Z., and Lipinska, A. (2003) *Cancer Lett.*, **196**, 101–107.

45 Takano, R., Muchmore, E., and Dennis, J.W. (1994) *Glycobiology*, **4**, 665–674.

46 Peracaula, R., Royle, L., Tabares, G., Mallorqui-Fernandez, G., Barrabes, S., Harvey, D.J., Dwek, R.A., Rudd, P.M., and de Llorens, R. (2003) *Glycobiology*, **13**, 227–244.

47 Fernandez-Salas, E., Peracaula, R., Frazier, M.L., and De Llorens, R. (2000) *Eur. J. Biochem.*, **267**, 1484–1494.

48 Yan, J., Fang, H., and Wang, B. (2005) *Med. Res. Rev.*, **25**, 490–520.

49 James, T.D., Sandanayake, K.R.A.S., and Shinkai, S. (1994) *Chem. Commun.*, 477–478.

50 Cao, H., McGill, T., and Heagy, M.D. (2004) *J. Org. Chem.*, **69**, 2959–2966.

51 DiCesare, N. and Lakowicz, J.R. (2002) *Tetrahedron Lett.*, **43**, 2615–2618.

52 Davis, C.J., Lewis, P.T., McCarroll, M.E., Read, M.W., Cueto, R., and Strongin, R.M. (1999) *Org. Lett.*, **1**, 331–334.

53 Arimori, S., Bosch, L.I., Ward, C.J., and James, T.D. (2001) *Tetrahedron Lett.*, **42**, 4553–4555.

54 Zhang, Y., Li, M., Chandrasekaran, S., Gao, X., Fang, X., Lee, H.-W., Hardcastle, K., Yang, J., and Wang, B. (2007) *Tetrahedron*, **63**, 3287–3292.

55 Lin, N., Yan, J., Huang, Z., Altier, C., Yan, J., Carrasco, N., Suyemoto, M., Johnson, L., Fang, H., Wang, Q., Wang, S., and Wang, B. (2007) *Nucl. Acid Res.*, **35** 1222–1229.

56 Manimala, J.C., Wiskur, S.L., Ellington, A.D., and Anslyn, E.V. (2004) *J. Am. Chem. Soc.*, **126**, 16515–16519.

57 Gravel, M., Kim, A., Mark Zak, T., Berube, C., and Hall, D.G. (2002) *J. Org. Chem.*, **67**, 3–15.

58 Edwards, N.Y., Sager, T.W., McDevitt, J.T., and Anslyn, E.V. (2007) *J. Am. Chem. Soc.*, **129**, 13575–13583.

59 Duggan, P.J. and Offermann, D.A. (2007) *Aust. J. Chem.*, **60**, 829–834.

60 Stones, D., Manku, S., Xiaosong, L., and Hall, D. (2004) *Chem. Eur. J.*, **10**, 92–100.

61 Zou, Y., Broughton, D.L., Bicker, K.L., Thompson, P.R., and Lavigne, J.J. (2007) *ChemBiochem.*, **8**, 2048–2051.

62 Manku, S. and Hall, D.G. (2007) *Aust. J. Chem.*, **60**, 824–828.

63 Wang, W., Gao, X., and Wang, B. (2002) *Curr. Org. Chem.*, **6**, 1285–1317.

64 Yang, W., Gao, X., and Wang, B. (2003) *Med. Res. Rev.*, **23**, 346–368.

65 James, T.D. and Shinkai, S. (2002) *Top. Curr. Chem.*, **218**, 159–200.

66 Cao, H.S. and Heagy, M.D. (2004) *J. Fluoresc.*, **14**, 569–584.

67 Striegler, S. (2003) *Curr. Org. Chem.*, **7**, 81–102.

68 Badugu, R., Lakowicz, J.R., and Geddes, C.D. (2005) *Curr. Anal. Chem.*, **1**, 157–170.

69 Kim, S.K., Kim, H.N., Xiaoru, Z., Lee, H.N., Lee, H.N., Soh, J.H., Swamyand, K.M.K., and Yoon, J. (2007) *Supramol. Chem.*, **19**, 221–227.

70 Pickup, J.C., Hussain, F., Evans, N.D., Rolinski, O.J., and Birch, D.J.S. (2005) *Biosens. Bioelectron.*, **20**, 2555–2565.

71 Mader, H.S. and Wolfbeis, O.S. (2008) *Microchim. Acta*, **162**, 1–34.

72 Norrild, J.C. and Eggert, H. (1995) *J. Am. Chem. Soc.*, **117**, 1479–1484.

73 Norrild, J.C. and Eggert, H. (1996) *J. Chem. Soc. Perkin Trans. 2*, 2583–2588.

74 Norrild, J.C. (2001) *J. Chem. Soc. Perkin Trans. 2*, 719–726.

75 Norrild, J.C. and Sotofte, I. (2001) *J. Chem. Soc. Perkin Trans. 2*, 727–732.

76 Hall, D.G. (ed.) (2005) *Boronic Acids: Preparation and Applications in Organic Synthesis and Medicine*, Wiley-VCH Verlag GmbH.

77 Springsteen, G. and Wang, B.H. (2001) *Chem. Commun.*, 1608–1609.

78 Woods, R.J., Dwek, R.A., Edge, C.J., and Fraser-Reid, B. (1995) *J. Phy. Chem.*, **99**, 3832–3846.

79 Bras, N.F., Cerqueira, N.M.F.S.A., Fernandes, P.A., and Ramos, M.J. (2008) *Int. J. Quantum. Chem.*, **108**, 2030–2040.

80 DeMarco, M.L. and Woods, R.J. (2008) *Glycobiology*, **18**, 426–440.

81 Neumann, D., Lehr, C.M., Lenhof, H.P., and Kohlbacher, O. (2004) *Adv. Drug Deliver. Rev.*, **56**, 437–457.

82 Parrill, A.L., Mamuya, N., Dolata, D.P., and Gervay, J. (1997) *Glycoconjugate J.*, **14**, 523–529.

83 Peters, T. and Pinto, B.M. (1996) *Curr. Opin. Struc. Biol.*, **6**, 710–720.

84 Woods, R.J. (1998) *Glycoconjugate J.*, **15**, 209–216.

85 Yang, W., He, H., and Drueckhammer, D.G. (2001) *Angew. Chem. Int. Edit.*, **40**, 1714–1718.

86 Alavi, A and Axford, J.S. (eds) (1995) *Glycoimmunology*, vol. **376**. Plenum Press, New York.

87 Alexeev, V.L., Sharma, A.C., Goponenko, A.V., Das, S., Lednev, I.K., Wilcox, C.S., Finegold, D.N., and Asher, S.A. (2004) *Anal. Chem.*, **75**, 2316–2323.
88 Cooper, C.R. and James, T.D. (1998) *Chem. Lett.*, 883–884.
89 Eggert, H., Frederiksen, J., Morin, C., and Norrild, J.C. (1999) *J. Org. Chem.*, **64**, 3846–3852.
90 Kaur, G., Fang, H., Gao, X., Li, H., and Wang, B. (2006) *Tetrahedron*, **62**, 2583–2589.
91 Nicholls, M.P. and Paul, P.K.C. (2004) *Org. Biomol. Chem.*, **2**, 1434–1441.
92 Jin, S., Li, M., Zhu, C., Tran, V., and Wang, B. (2008) *ChemBiochem.*, **9**, 1431–1438.
93 Springsteen, G. and Wang, B. (2001) *Chem. Commun.*, 1608–1609.
94 Stack, D.E., Hill, A.L., Diffendaffer, C.B., and Burns, N.M. (2002) *Org. Lett.*, **4**, 4487–4490.
95 Zhu, L., Shabbir, S.H., Gray, M., Lynch, V.M., Sorey, S., and Anslyn, E.V. (2006) *J. Am. Chem. Soc.*, **128**, 1222–1223.
96 Sarson, L.D., Ueda, K., Takeuchi, R., and Shinkai, S. (1996) *Chem. Comm.*, **5**, 619–620.
97 Secor, K.E. and Glass, T.E. (2004) *Org. Lett.*, **6**, 3727–3730.
98 Frisch, M.J., Trucks, G.W., Schlegel, H.B., Scuseria, G.E., Robb, M.A., Cheeseman, J.R., Montgomery Jr., J.A., Vreven, T., Kudin, K.N., Burant, J.C., Millam, J.M., Iyengar, S.S., Tomasi, J., Barone, V., Mennucci, B., Cossi, M., Scalmani, G., Rega, N., Petersson, G.A., Nakatsuji, H., Hada, M., Ehara, M., Toyota, K., Fukuda, R., Hasegawa, J., Ishida, M., Nakajima, T., Honda, Y., Kitao, O., Nakai, H., Klene, M., Li, X., Knox, J.E., Hratchian, H.P., Cross, J.B., Bakken, V., Adamo, C., Jaramillo, J., Gomperts, R., Stratmann, R.E., Yazyev, O., Austin, A.J., Cammi, R., Pomelli, C., Ochterski, J.W., Ayala, P.Y., Morokuma, K., Voth, G.A., Salvador, P., Dannenberg, J.J., Zakrzewski, V.G., Dapprich, S., Daniels, A.D., Strain, M.C., Farkas, O., Malick, D.K., Rabuck, A.D., Raghavachari, K., Foresman, J.B., Ortiz, J.V., Cui, Q., Baboul, A.G., Clifford, S., Cioslowski, J., Stefanov, B.B., Liu, G., Liashenko, A., Piskorz, P., Komaromi, I., Martin, R.L., Fox, D.J., Keith, T., Al-Laham, M.A., Peng, C.Y., Nanayakkara, A., Challacombe, M., Gill, P.M.W., Johnson, B., Chen, W., Wong, M.W., Gonzalez, C., and Pople, J.A. (2004) Gaussian 03, Revision C.02. Gaussian, Inc.: Wallingford, CT.
99 Langer, T. and Wolber, G. (2004) *Pure Appl. Chem.*, **76**, 991–996.
100 Dornhofer, A., Biely, M., Wolber, G., and Langer, T. (2006) A novel 2D depiction method using breadth-first ordering and an adapted 2D force field. In: Esin Aki Sener and Ismail Yalcin, editors, QSAR and Molecular Modelling in Rational Design of Bioactive Molecules, Computer Aided Drug Design & Development Society, Ankara, Turkey, pp. 421–422.
101 Gasteiger, J., Rudolph, C., and Sadowski, J. (1990) *Tetrahedron Comput. Methodol.*, **3**, 537–547.
102 Jang, Y.J., Jun, J.H., Swamy, K.M.K., Nakamura, K., Koh, H.S., Yoon, Y.J., and Yoon, J. (2005) *Bull. Korean Chem. Soc.*, **26**, 2041–2043.
103 Yoon, J.Y. and Czamik, A.W. (1993) *Bioorg. Med. Chem.*, **1**, 267–271.
104 Takeuchi, M., Yamamoto, M., and Shinkai, S. (1997) *Chem. Commun.*, 1731–1732.
105 Diaz-Garcia, M.E., Pina-Luis, G., and Rivero, I.A. (2006) *Trac-Trend Anal. Chem.*, **25**, 112–121.
106 Potyrailo, R.A. (2004) *Macromol. Rapid Commun.*, **25**, 78–94.
107 Hermkens, P.H.H., Ottenheijm, H.C.J., and Rees, D. (1996) *Tetrahedron*, **52**, 4527–4554.
108 Potyrailo, R.A. and Mirsky, V.M. (2009) *Combinatorial Methods for Chemical and Biological Sensors*, Springer, New York.
109 Lam, K.S., Liu, R.W., Miyamoto, S., Lehman, A.L., and Tuscano, J.M. (2003) *Acc. Chem. Res.*, **36**, 370–377.
110 Brown, B.B., Wagner, D.S., and Geysen, H.M. (1995) *Mol. Divers.*, **1**, 4–12.
111 Basu, P.S., Majhi, R., and Batabyal, S.K. (2003) *Clin. Biochem.*, **36**, 373–376.

112 Belanger, A., van Halbeek, H., Graves, H.C., Grandbois, K., Stamey, T.A., Huang, L., Poppe, I., and Labrie, F. (1995) *Prostate*, **27**, 187–197.

113 Dudkin, V.Y., Miller, J.S., and Danishefsky, S.J. (2004) *J. Am. Chem. Soc.*, **126**, 736–738.

114 Huber, P.R., Schmid, H.P., Mattarelli, G., Strittmatter, B., van Steenbrugge, G.J., and Maurer, A. (1995) *Prostate*, **27**, 212–219.

115 Huizen, I.V., Wu, G., Moussa, M., Chin, J.L., Fenster, A., Lacefield, J.C., Sakai, H., Greenberg, N.M., and Xuan, J.W. (2005) *Clin. Cancer Res.*, **11**, 7911–7919.

116 Jankovic, M.M. and Kosanovic, M.M. (2005) *Clin. Biochem.*, **38**, 58–65.

117 Kanoh, Y., Mashiko, T., Danbara, M., Takayama, Y., Ohtani, S., Egawa, S., Baba, S., and Akahoshi, T. (2004) *Anticancer Res.*, **24**, 3135–3139.

118 Ohyama, C., Hosono, M., Nitta, K., Oh-eda, M., Yoshikawa, K., Habuchi, T., Arai, Y., and Fukuda, M. (2004) *Glycobiology*, **14**, 671–679.

119 Tabares, G., Jung, K., Reiche, J., Stephan, C., Lein, M., Peracaula, R., de Llorens, R., and Hoesel, W. (2007) *Clin. Biochem.*, **40**, 343–350.

120 Tabares, G., Radcliffe, C.M., Barrabes, S., Ramirez, M., Aleixandre, R.N., Hoesel, W., Dwek, R.A., Rudd, P.M., Peracaula, R., and de Llorens, R. (2006) *Glycobiology*, **16**, 132–145.

121 Villoutreix, B.O., Getzoff, E.D., and Griffin, J.H. (1994) *Protein Sci.*, **3**, 2033–2044.

122 Ellington, A.D. and Szostak, J.W. (1990) *Nature*, **346**, 818–822.

123 Robertson, D.L. and Joyce, G.F. (1990) *Nature*, **344**, 467–468.

124 Tuerk, C. and Gold, L. (1990) *Science*, **249**, 505–510.

125 Huisgen, R. (1963) *Angew. Chem., Int. Edit.*, **2**, 565–598.

126 Kolb, H.C., Finn, M.G., and Sharpless, K.B. (2001) *Angew. Chem. Int. Edit. Engl.*, **40**, 2004–2021.

127 Tornoe, C.W., Christensen, C., and Meldal, M. (2002) *J. Org. Chem.*, **67**, 3057–3064.

128 Yang, W., Gao, X., Springsteen, G., and Wang, B. (2002) *Tetrahedron Lett.*, **43**, 6339–6342.

129 Dickey, F.H. (1949) *Proc. Natl. Acad. Sci.*, **35**, 227–229.

130 Yan, M. and Ramstrom, O. (2005) *Molecularly Imprinted Materials: Science and Technology*, Marcel Dekker.

131 Zhang, H.Q., Ye, L., and Mosbach, K. (2006) *J. Mol. Recognit.*, **19**, 248–259.

132 Shea, K. (1994) *Trends Polym. Sci.*, **2**, 166.

133 Wang, W., Gao, S., and Wang, B. (2005) in *Molecularly Imprinted Materials: Science and Technology* (eds M. Yan and O. Ramstrom), Marcel Dekker, pp. 701–726.

134 Wulff, G. and Schauhoff, S. (1991) *J. Org. Chem.*, **56**, 395–400.

135 Gao, S., Wang, W., and Wang, B. (2001) *Bioorg. Chem.*, **29**, 308–320.

136 Wang, W., Gao, S., and Wang, B. (1999) *Org. Lett.*, **1**, 1209–1212.

137 Miyahara, T. and Kurihara, K. (2000) *Chem. Lett.*, 1356–1357.

138 Sallacan, N., Zayats, M., Bourenko, T., Kharitonov, A.B., and Willner, I. (2002) *Anal. Chem.*, **74**, 702–712.

139 Granot, E., Tel-Vered, R., Lioubashevski, O., and Willner, I. (2008) *Adv. Funct. Mater.*, **18**, 478–484.

140 Zayats, M., Lahav, M., Kharitonov, A.B., and Willner, I. (2002) *Tetrahedron*, **58**, 815–824.

141 Rajkumar, R., Katterle, M., Warsinke, A., Moehwald, H., and Scheller, F.W. (2008) *Biosens Bioelectron.*, **23**, 1195–1199.

142 Lehn, J.-M. (1999) *Chem. Eur. J.*, **5**, 2455–2463.

143 Linton, B. and Hamilton, A.D. (1999) *Curr. Opin. Chem. Biol.*, **3**, 307–312.

144 Otto, S., Furlan, R.L.E., and Sanders, J.K.M. (2002) *Drug Discov. Today*, **7**, 117–125.

145 Eliseev, A.V. and Lehn, J.M. (1999) *Combinatorial Chem. Biol.*, **243**, 159–172.

146 Alonso, H., Bliznyuk, A.A., and Gready, J.E. (2006) *Med. Res. Rev.*, **26**, 531–568.

7
Artificial Receptor Compounds for Chiral Recognition

Thomas J. Wenzel and Ngoc H. Pham

7.1
Introduction

Chiral receptor compounds have been used extensively for chiral recognition studies. This chapter describes receptor compounds that form host–guest complexes through noncovalent interactions. Furthermore, only receptor compounds that have a well-defined cavity or motif that specifically interacts with a guest through multiple points of attraction are included. Discussion is restricted to receptor compounds used for the analysis of chiral compounds in gas or solution phase studies.

The most extensive applications of receptor compounds in chiral recognition involve the use of chromatographic [1, 2] or NMR spectroscopic [3] methods. Ultraviolet–visible absorption spectroscopy is often used to determine whether a receptor compound exhibits enantiodifferentiation with substrates. Mass spectrometric methods are increasingly being developed and used in studies of enantiomeric recognition [4].

Cyclodextrins, crown ethers, calixarenes, and calix[4]resorcinarenes are broad and widely studied classes of receptor compounds. Studies on these systems are too extensive to describe exhaustively, so only illustrative examples are presented. Similarly, a range of specific enantioselective receptor compounds has been described. The number of reports of such systems is so large that only examples designed to illustrate the range of receptor compounds are presented.

7.2
Cyclodextrins

Cyclodextrins are a series of cyclic oligosaccharides, the most common of which contain six (α), seven (β), or eight (γ) glucose units (Figure 7.1). Cyclodextrins have a basket-shaped cavity in which the narrower opening is ringed with primary hydroxy groups at the 6-position and the wider opening is ringed with secondary hydroxy groups at the 2- and 3-positions of each glucose ring. Because the α-, β-, and

---

Artificial Receptors for Chemical Sensors. Edited by V.M. Mirsky and A.K. Yatsimirsky
Copyright © 2011 WILEY-VCH Verlag GmbH & Co. KGaA, Weinheim
ISBN: 978-3-527-32357-9

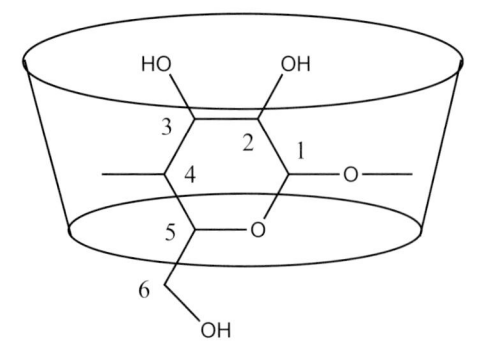

Figure 7.1 Representation of cyclodextrin superimposed with one D-glucose subunit.

γ-cyclodextrins have different sizes, they can accommodate different guest molecules. Enantiomeric recognition often depends on a complementary match between the size of the substrate and size of the cavity. The native, underivatized cyclodextrins are natural compounds. However, derivatization at the 2-, 3-, and/or 6-hydroxy groups has been carried out to prepare an array of chiral cyclodextrins with different properties. Derivatization affects both the solubility and enantiomeric recognition properties of the cyclodextrins. An important factor with certain cyclodextrin derivatives involves the degree of substitution (DS). Sometimes the cyclodextrin is fully derivatized, whereas in other cases substituent groups are added in an indiscriminate fashion and a certain DS is achieved. Cyclodextrins are probably the most frequently studied family of receptor compounds and they have been widely used in chiral recognition applications.

7.2.1
Alkylated Cyclodextrins

Per-O-alkylated cyclodextrins are prepared by the reaction of a large excess of an alkylhalide with cyclodextrin in the presence of a suitable base. Per-O-methylation of cyclodextrin has been achieved using sodium hydroxide, potassium hydroxide [5], or sodium hydride [6] with methyl iodide in methyl sulfoxide. Per-O-methyl cyclodextrins have been coated with moderately polar siloxanes as capillary gas chromatographic phases and used for the separation of a wide range of volatile enantiomeric compounds, including unfunctionalized hydrocarbons [6]. Heptakis(2,3,6-tri-O-methyl)-β-cyclodextrin is now commercially available.

Hexakis(2,3,6-tri-O-methyl)-α-cyclodextrin and heptakis(2,3,6-tri-O-methyl)-β-cyclodextrin are water- and organic-soluble and have also been used as chiral NMR solvating agents. For example, hexakis(2,3,6-tri-O-methyl)-α-cyclodextrin caused chiral recognition in the ^{1}H NMR spectra of metal complexes such as Ru(bipy)$_3^{2+}$ (bipy = 2,2′-bipyridine) and Ru(phen)$_3^{2+}$ (phen = 1,10-phenanthroline) [7]. Heptakis(2,3,6-tri-O-methyl)-β-cyclodextrin is an especially effective chiral NMR solvating agent for trisubstituted allenes. The allene hydrogen resonance was a useful one to employ in determining enantiomeric purity. The allene resonance of the *(S)*-isomer was consistently deshielded relative to that of the (R)-isomer in the presence of the hexakis(2,3,6-tri-O-methyl)-α-cyclodextrin [8].

Hexakis(2,3,6-tri-O-pentyl)-α-cyclodextrin and heptakis(2,3,6-tri-O-pentyl)-β-cyclodextrin have been evaluated as capillary gas chromatographic phases for the separation of enantiomers [9–11]. Reaction of the cyclodextrin with an excess of 1-bromopentane in methyl sulfoxide in the presence of sodium hydroxide resulted in alkylation of the 2 and 6-positions. Alkylation of the less reactive 3-position was achieved either by reaction of the 2,6-di-O-pentyl cyclodextrin with excess 1-bromopentane in tetrahydrofuran (THF) in the presence of sodium hydride [10, 11] or by addition of more sodium hydroxide and 1-bromopentane with continuing reflux for five days [9].

Per-substituted 2,6-di-O-pentyl-α, β- or γ-cyclodextrins, which are readily obtained by alkylation using 1-bromopentane and sodium hydroxide in methyl sulfoxide, have also been used as chiral gas chromatographic phases [12]. Similarly, per-substituted (2,6-di-O-methyl)-β-cyclodextrin has been prepared followed by reaction of 1-bromopentane and sodium hydride to alkylate the 3-position [13]. The heptakis(2,6-di-O-methyl-3-O-pentyl)-β-cyclodextrin was employed as a capillary gas chromatographic phase.

Substitution of alkyl groups at other positions is achieved either by selective blockage of certain positions followed by alkylation and deprotection or by differences in the reactivity of the 2-, 3-, and 6-hydroxy groups in the presence of appropriate bases. For example, sodium hydride selectively reacts at the 2-position of cyclodextrin. A reaction of one equivalent of β-cyclodextrin with seven equivalents of sodium hydride in the presence of methyl iodide produces the heptakis(2-O-methyl)-β-cyclodextrin [14]. The heptakis(2,3-di-O-pentyl-6-O-methyl)-β-cyclodextrin has been obtained by a multistep process. The first step involved blocking the 6-position by reaction with tert-butyldimethylsilyl chloride (TBDMS) and imidazole in dimethylformamide (DMF) [15]. Reaction of the blocked cyclodextrin with n-pentyl iodide and sodium hydride in DMF afforded the corresponding heptakis(2,3-di-O-pentyl)-β-cyclodextrin. After removal of the TBDMS using tetrabutylammonium fluoride (TBAF) in THF, the 6-position was methylated by reaction with methyl iodide and sodium hydride in THF.

Heptakis(2-O-methyl-3,6-di-O-pentyl)-β-cyclodextrin has been prepared using a multistep procedure. The 6- and 2-positions were first protected using TBDMS and allyl bromide, respectively. Deprotection of the 6-position enabled pentylation of the 3- and 6-positions. After de-allylation, methylation of the 2-position was performed. The cyclodextrin derivative phase was then evaluated as a stationary phase for gas chromatographic separations [16].

Schemes to selectively attach permethylated β-cyclodextrin to silica gel at the 2-, 3-, or 6-positions have been described [17]. Starting with established procedures for making the heptakis(2,3-di-O-methyl)- or heptakis(2,6-di-O-methyl)-β-cyclodextrin derivative, it was possible to prepare the corresponding mono-oct-7-enyl derivative at either the 6- or 3-position using one equivalent of sodium hydride and 8-bromo-1-octene in DMF. The remaining hydroxy groups were then methylated using established methods. The alkene group was then utilized to covalently link the cyclodextrin derivative to silica gel. Using a known 2,3,6-tri-O-methylated derivative with a single underivatized hydroxy group at the 2-position [18], the corresponding 2-bound cyclodextrin was prepared by attachment of the oct-7-enyl group at the remaining 2-hydroxy group [17]. Of these, the 2-O-bonded phase was judged best as a gas chromatographic stationary phase for the separation of enantiomeric compounds.

7.2.2
Acylated and Mixed Acylated/Alkylated Cyclodextrins

Per-O-acetylated derivatives of cyclodextrins are readily prepared by reaction with excess acetic anhydride in pyridine [19, 20]. A per-O-acetylated β-cyclodextrin immobilized on silica gel has been evaluated as a stationary phase in liquid chromatography [19]. However, mixed alkylated-acylated cyclodextrin derivatives have been more commonly used as chromatographic phases.

Hexakis(2,6-di-O-pentyl)-α- or heptakis(2,6-di-O-pentyl)-β-cyclodextrin has been acetylated at the 3-position using acetic anhydride, triethylamine, and 4-N-dimethylaminopyridine (DMAP) [10, 11]. The resulting cyclodextrins were used as stationary phases in capillary gas chromatography. Using analogous synthetic procedures, cyclodextrin derivatives with substitution patterns such as 3-O-trifluoroacetyl-2,6-di-O-pentyl (trifluoroacetic anhydride) [12, 21], 3-O-butyryl-2,6-di-O-pentyl (butyric anhydride) [16], and 3-O-trifluoroacetyl-2,6-di-O-methyl [6] were also evaluated as chiral gas chromatographic stationary phases. A particularly interesting observation was a temperature-dependent reversal in elution order for the methyl esters of certain phenoxypropionic acid herbicides on a gas chromatographic phase consisting of octakis(3-O-butyryl-2,6-di-O-pentyl)-γ-cyclodextrin [16].

The 6-O-acetyl-2,3-di-O-pentyl derivatives of α-, β-, and γ-cyclodextrin have been prepared and evaluated as chiral gas chromatographic phases [15]. This compound was not as effective as the corresponding 3-O-acetyl-2,6-di-O-pentyl or 6-O-methyl-2,3-di-O-pentyl derivatives.

The octakis(3-O-butyryl-2,6-di-O-pentyl)-γ-cyclodextrin has also been exploited for enantiorecognition using thickness shear mode resonators [22], surface acoustic wave devices [23], Fourier-transform infrared reflectance spectra [23], and capacitive microsensors [24]. In these techniques, the cyclodextrin derivative is coated into a polymeric thin film such as poly(dimethylsiloxane) on a sensor surface. In the presence of enantiomers, differences in signals are noted for the two enantiomers. Differences in the vibrating frequency of a quartz crystal are observed for a pair of enantiomers when mass is deposited onto the surface. In the capacitance sensor, enantiomers of methyl propionate caused antipode signals on adsorption into the cyclodextrin layer [24].

7.2.3
Carbamoylated Cyclodextrins

Various carbamoylated cyclodextrins have been used for chiral recognition in chromatographic and NMR applications. The derivatives are prepared by reacting the cyclodextrin and appropriate isocyanate under reflux for several hours in pyridine [25, 26]. In many cases, steric hindrance created by attachment of the carbamoyl groups limits the extent of the reaction and unreacted hydroxy groups remain [25]. Liquid chromatographic applications include the use of a perphenyl carbamoylated derivative of β-cyclodextrin [19], an indiscriminately substituted 2,6-dimethylphenyl [25], 3,5-dimethylphenyl [26, 27], or (R)- or (S)-1-(1-naphthyl)ethyl [25] carbamoylated cyclodextrin derivatives as bonded stationary phases. Derivatives of persubstituted (2,6-di-O-pentyl)-α-, β-, or γ-cyclodextrin with either a n-propyl, isopropyl, or phenyl carbamoyl group at the 3-position were used as gas chromatographic stationary phases for free alcohols, amines, and epoxides [28].

Exhaustively 3,5-dimethylphenyl carbamoylated, partially 3,5-dimethylphenyl carbamoylated (2,3-positions), and 6-TBDMS/3,5-dimethylphenyl carbamoylated (2,3-positions) derivatives of β-cyclodextrin have been compared for their effectiveness as chloroform-soluble chiral NMR solvating agents. The per-carbamoyl derivative was generally the most effective. The ^{1}H NMR spectra of N-(3,5-dinitrobenzoyl) derivatives of amino acid methyl esters and amines exhibited enantiomeric discrimination. However, association of the substrate with the cyclodextrin involved dipole–dipole and π–π interactions at the external carbamoyl groups instead of insertion into the cavity [29].

7.2.4
2-Hydroxypropylether Cyclodextrins

The reaction of either (R)- or (S)-propylene oxide with α-, β-, or γ-cyclodextrin in aqueous sodium hydroxide solution produces indiscriminately substituted 2-hydroxypropylether derivatives. The degrees of substitution range from 6.2 for α-cyclodextrin to 7.0 for γ-cyclodextrin. The hydroxypropyl derivative was bonded to silica gel and used as a liquid chromatographic phase. The permethylated derivatives of 2-hydroxypropylether α-, β-, and γ-cyclodextrins were evaluated as capillary gas chromatographic stationary phases. The 2-hydroxypropylether cyclodextrin phases could be used to separate compounds with a wide variety of functional groups [30].

7.2.5
Tert-butyldimethylsilyl Chloride-Substituted Cyclodextrins

The reaction of cyclodextrins with tert-butyldimethylsilyl chloride in DMF in the presence of imidazole leads to selective derivatization at the 6-position [31, 32]. Using established procedures, it is then possible to alkylate or acylate the 2- and 3-positions of the cyclodextrin [31]. Removal of the TBDMS group using TBAF in oxolane or boron trifluoride etherate in dichloromethane opens the 6-position for further derivatization [31]. Alternatively, the 6-TBDMS derivative can be selectively alkylated at the 2-position using methyl iodide in the presence of barium oxide and barium hydroxide in DMF [32].

Heptakis(2,3-di-O-acetyl-6-O-TBDMS)-β-cyclodextrin [33], heptakis(2,3-di-O-methyl-6-O-TBDMS)-β-cyclodextrin [34], and heptakis(2,3-di-O-pentyl-6-O-

TBDMS)-β-cyclodextrin [35] have been evaluated as gas chromatographic stationary phases.

7.2.6
Anionic Cyclodextrins

Several negatively charged cyclodextrins have been employed in the enantiomeric discrimination of cationic substrates. The negatively charged derivatives are especially effective in capillary electrophoretic applications [36]. One example is an indiscriminately substituted sulfobutyl ether β-cyclodextrin (SBE-CD) that is prepared by the reaction of cyclodextrin with butane sultone in an aqueous solution of sodium hydroxide [37]. The anionic SBE-CD derivative has been used as an eluent phase additive in reversed-phase liquid chromatography [38], capillary electrophoresis [39], and electrokinetic chromatography [40].

Indiscriminately substituted sulfated cyclodextrins are prepared by reaction with chlorosulfonic acid in pyridine or using a mixture of triethylamine and sulfur trioxide in DMF [41]. It is usually possible to obtain a relatively high DS (9–14 substituent groups) for the sulfated derivatives. Sulfated cyclodextrins have been used as eluent additives in capillary electrophoresis [36] or as bonded phases in liquid chromatography [42]. The heptakis(2,3-di-O-acetyl-6-sulfato)-β-cyclodextrin [43] or octakis(2,3-di-O-acetyl-6-sulfato)-γ-cyclodextrin [44] were used as additives in capillary electrophoresis. Synthesis of the mixed acetyl-sulfate cyclodextrins first involved blocking the 6-position with TBDMS, acetylation of the 2,3-positions, deprotection with boron trifluoride etherate, and reaction with pyridine and sulfur trioxide in DMF.

A commercially available sulfated β-cyclodextrin with a DS of 9 has been used as a chiral NMR solvating agent for water-soluble cationic substrates such as pheniramine (1), carbinoxamine (2), doxylamine (3), and propranolol (4) and was more effective than native β-cyclodextrin [45]. The addition of paramagnetic lanthanide ions such as ytterbium(III) or dysprosium(III) to mixtures of substrates and sulfated β-cyclodextrin frequently led to enhancements in enantiomeric discrimination in the ^{1}H NMR spectrum. The lanthanide ion bound at the sulfate groups of the cyclodextrin and caused changes in chemical shifts in the NMR spectra of substrates that were often different for the two enantiomers.

$H_3C-\underset{\underset{\text{(2-pyridyl)}}{|}}{\overset{\overset{\text{Ph}}{|}}{C}}-OCH_2CH_2\overset{+}{N}H(CH_3)_2$ $\underset{\text{(1-naphthyl)}}{OCH_2\underset{\underset{}{|}}{\overset{\overset{OH}{|}}{C}}HCH_2\overset{+}{N}HCH(CH_3)_2}$

3 **4**

Carboxymethylation is another common way of preparing anionic cyclodextrin derivatives. Carboxymethylated cyclodextrins have often been used to affect separations in capillary electrophoresis [36]. Indiscriminate carboxymethylation is achieved by reacting sodium iodoacetate with cyclodextrin in an aqueous solution of sodium hydroxide [46]. While substitution of carboxymethyl groups occurs at the 2-, 3-, and 6-positions, addition at the 2-position predominates [46]. Procedures for the selective incorporation of carboxymethyl groups at the 2- (activation with sodium hydride) or 6-position (activation with pyridine) have been described [47].

Indiscriminately substituted carboxymethylated α-, β-, and γ-cyclodextrins were especially effective chiral NMR solvating agents for cationic substrates [45, 47, 48]. The enhanced effectiveness of the indiscriminately substituted cyclodextrins in NMR applications relative to the 2- and 6-substituted derivatives was probably because of its considerably higher degree of substitution. Enantiomeric discrimination in the NMR spectrum with the carboxymethylated cyclodextrins was much larger than observed with the native cyclodextrins. Similar to studies with sulfated cyclodextrins described earlier, paramagnetic praseodymium(III) or ytterbium(III) bound at the carboxymethyl groups of the cyclodextrin derivatives and often enhanced the enantiomeric discrimination in the ^{1}H NMR spectra of substrates [47–49]. In several cases, resonances that did not exhibit enantiomeric discrimination in the presence of only the carboxymethylated cyclodextrin split into two signals when Yb(III) or Pr(III) was added [47, 48]. In addition, the enantiomeric discrimination with the lanthanide ion was often so large that much lower concentrations of the carboxymethylated cyclodextrin could be used [49].

Analogous hexakis(6-carboxymethylthio-6-deoxy)-α-cyclodextrin and heptakis(6-carboxymethylthio-6-deoxy)-β-cyclodextrin derivatives have been used to separate metal chelate complexes (M(phen)$_3$$^+$) [M = Ru(III), Rh(III), Fe(II), Co(II), and Zn(II); phen = 1,10-phenanthroline] by capillary electrophoresis or produce enantiomeric discrimination in the ^{1}H NMR spectra [7, 50].

7.2.7
Cationic Cyclodextrins

Several cationic cyclodextrins have been prepared and used for enantiomeric discrimination [36]. Reaction of (2,3-epoxypropyl)trimethylammonium chloride with cyclodextrin in an aqueous solution of sodium hydroxide produces an indiscriminately substituted O-2-hydroxypropyltrimethylammonium derivative [51]. Because of the quaternary amine, the O-2-hydroxypropyltrimethylammonium cyclodextrin derivative has a fixed positive charge irrespective of pH. It has been used as a mobile phase additive in capillary electrophoresis [52, 53] and reversed-phase liquid chromatography [54] to enhance the separation of neutral and anionic compounds.

Reaction of a per-substituted 6-bromo-6-deoxy-cyclodextrin derivative with amines such as ethanolamine [55] or methoxyethylamine [56] produced the corresponding per-6-amino-6-deoxy derivative. The per-substituted 6-bromo-6-deoxy-cyclodextrin was prepared by reaction of cyclodextrin with triphenylphosphine and bromine in DMF [55]. The amino cyclodextrins are positively charged at acidic pH and facilitate the analysis of anionic species in capillary electrophoresis [55, 56]. A mono-(6-butylammonium-6-deoxy)-β-cyclodextrin was prepared by reaction of n-butylamine with mono-(6-tosyl)-β-cyclodextrin in DMF [57]. The tosyl derivative was prepared by reaction of cyclodextrin with p-toluenesulfonyl chloride in pyridine. Protonation of the monobutylamine derivative at acidic pH provided a cationic cyclodextrin that was used as a mobile phase additive in capillary electrophoresis to separate carboxylate species [57].

Either mono or per-substituted 6-amino-6-deoxy cyclodextrins have been prepared by reaction of ammonia with the corresponding 6-tosyl-6-deoxy derivative [20] or by reducing the corresponding 6-azido-6-deoxy derivative with triphenylphosphine and concentrated ammonium hydroxide in dioxane–methanol [58]. Enantiomeric discrimination of carboxylate species by ^{1}H NMR spectroscopy [59], anionic species by capillary electrophoresis [60], or various species by measured thermodynamic data [61] shows the utility of the protonated 6-amino-6-deoxy cyclodextrin derivatives over native cyclodextrins for distinguishing anionic species.

Heptakis(6-amino-6-deoxy)-β-cyclodextrin caused non-equivalence in the ^{1}H NMR spectra of several N-acetylated amino acids. Native β-cyclodextrin did not cause enantiomeric discrimination and the heptakis(6-amino-6-deoxy)-β-cyclodextrin was more effective than the corresponding mono(6-amino-6-deoxy)-β-cyclodextrin [62].

7.2.8
Miscellaneous Cyclodextrins

Cyclodextrin derivatives with a mono-diethylenetriaminepentaacetic acid (DTPA) moiety attached at either the 2- or 6-position have been prepared and examined with paramagnetic Dy(III) as water-soluble chiral NMR shift reagents [63, 64]. An ethylenediamino [63] or amino group [64] was selectively attached at the 2- or 6-position by reaction of ethylenediamine or ammonia with the corresponding mono (O-tosyl) derivative. Reaction of the ethylenediamino or amino cyclodextrin with DTPA dianhydride afforded the corresponding cyclodextrin-DTPA amide derivative. Dysprosium(III) bound at the DTPA moiety and caused perturbations in chemical shifts in the ^{1}H NMR spectra of substrates bound in the cyclodextrin cavity. The results were much better for the 2-ethylenediamino derivative than the 6-ethylenediamino derivative, presumably because Dy(III) was positioned closer to the secondary face, which is usually the location where enantiomeric discrimination occurs [63]. For the amine derivative, the presence of the DTPA moiety at the 2-position seemed to block access to the cavity and it was not as effective as either the 6-amino or 2-ethylenediamino derivative [64].

Mixed cyclodextrin–crown ether systems, one example of which is **5**, have been prepared, attached to silica gel, and used as chiral liquid chromatographic stationary phases. The cyclodextrin was anchored to silica gel at one of the C2 positions and then an aza crown moiety was attached to the cyclodextrin through one of the primary hydroxy groups. Compounds associated through either the cyclodextrin or crown functionality, thereby increasing the versatility and selectivity of these chromatographic phases [65].

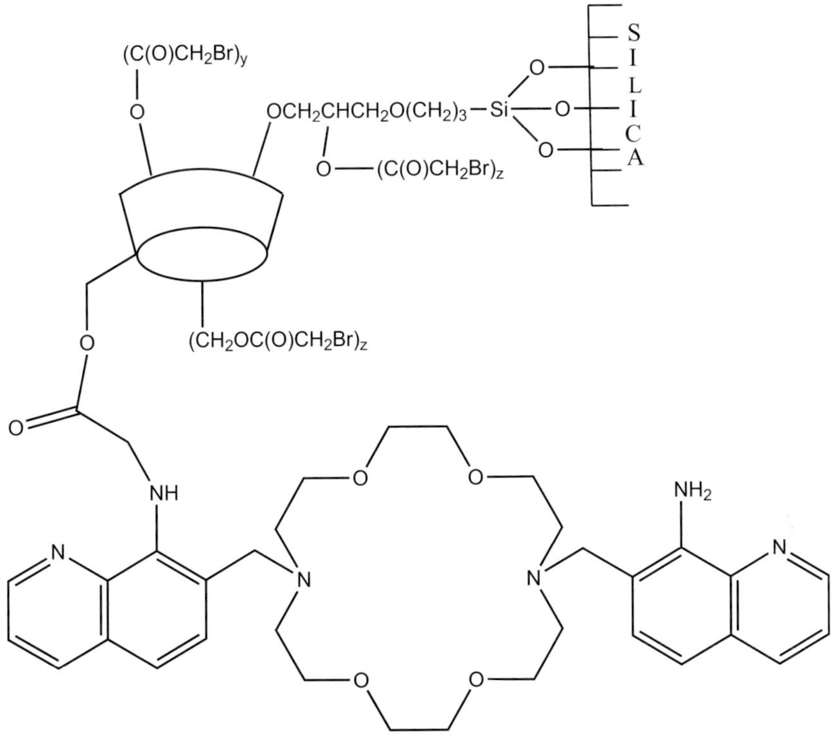

$(x + y + z \sim 3.2)$

5

A comprehensive review of the application of commercially available, cyclodextrin-based liquid chromatographic phases has been published [1]. In addition to the use of native cyclodextrins, the full range of applications of (2,3-di-O-methylated)-β-cyclodextrin, acetylated β-cyclodextrin, hydroxypropylether β-cyclodextrin, 1-(1-naphthyl)ethylcarbamoylated β-cyclodextrin, 3,5-dimethylphenylcarbamoylated β-cyclodextrin, carboxymethylated β-cyclodextrin, and permethylated cyclodextrins are described. A comprehensive review of the use of cyclodextrins as chiral NMR solvating agents was published recently [3].

7.3
Crown Ethers

Crown ethers are notable for their binding to protonated primary amines. Chiral crown ethers have been widely used for the enantiomeric recognition of chiral primary amines. In particular, the three hydrogen atoms in the primary ammonium ion are ideally aligned to form hydrogen bonds to three of the oxygen atoms in 18-crown-6-ethers (Figure 7.2a). Secondary ammonium ions generally do not associate well with crown ethers because they can only form two hydrogen bonds and the second substituent group of the ammonium salt often hinders association. One notable exception is (18-crown-6)-2,3,11,12-tetracarboxylic acid (**6**). When a neutral secondary amine is mixed with **6**, a neutralization reaction occurs to form the protonated secondary ammonium salt and carboxylate ion of **6**. Favorable association occurs through the formation of two hydrogen bonds and an ion pair (Figure 7.2b) [66–69].

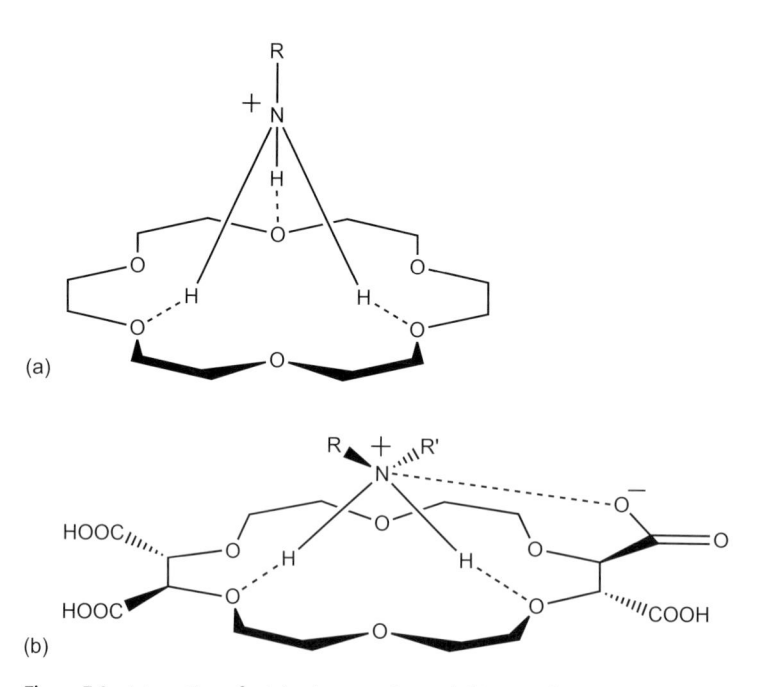

Figure 7.2 Interaction of a (a) primary amine and (b) secondary amine with (18-crown-6)-2,3,11,12-tetracarboxylic acid. Primary amines are studied as protonated species. Secondary amines are studied as neutral species and undergo a neutralization reaction with the crown ether.

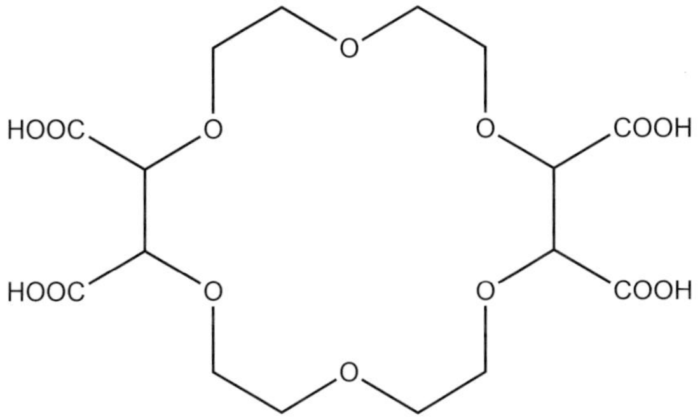

6

The preparation of crown ethers often involves the use of a chiral diol that is then reacted with an appropriate bridging ethylene glycol unit to complete the macrocycle. Many crown ethers use two identical chiral units that are bridged by appropriate ethylene glycol linkers. Several comprehensive reviews have been published that describe the use of crown ethers as bonded liquid chromatographic stationary phases [2], chiral NMR solvating agents [3], and for the analysis of enantiomeric amines [70].

7.3.1
1,1'-Binaphthalene-Based Crown Ethers

The 1,1'-binaphthalene-2,2'dihydroxy moiety has been an important chiral unit used in the formation of crown ethers. Compound **7** was the first chiral crown ether employed for enantiomeric recognition [71]. Initial studies explored the distribution of racemic amine salts between aqueous and chloroform layers. Liquid chromatographic separation of amino acid esters was achieved either by adding **7** to the mobile phase [72] or covalently bonding it to silica gel [73]. Enantiomeric discrimination was also observed in the ^{1}H NMR spectrum of the salt of 1-phenylethylamine [71]. Among the 1,1'-binaphthalene-based crown ethers, **8** was especially effective for its enantioselectivity [74, 75]. Dynamically coated liquid chromatographic phases of **8** are commercially available and used by many investigators, and a method for bonding **8** to silica gel for liquid chromatographic applications has been described [76].

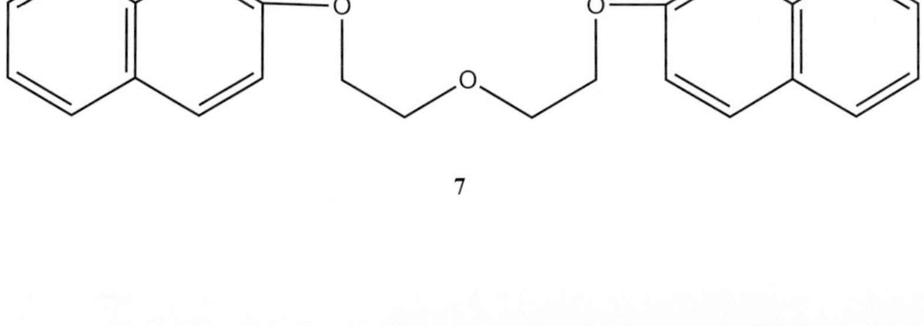

7

8

7.3.2
Carbohydrate-Based Crown Ethers

Carbohydrates have been used as chiral moieties in the preparation of a large family of crown ethers. A comprehensive review of the early work on these systems has been published [77].

The crown ether with a 1,2:5,6-diisopropylidene-D-mannitol and *tert*-butyl-substituted phenyl group (**9**) caused enantiomeric discrimination in extraction and liquid chromatographic studies [78]. The crown ether was coated onto a C_{18} silica gel bonded phase for the liquid chromatographic separations. In a subsequent study, **9** was found to be more effective than **8** as an organic-soluble chiral NMR solvating agent [79]. Enantiomeric discrimination of primary ammonium salts with **9** was greater in methanol-d_4 and acetonitrile-d_3 than in chloroform-*d*. It was also shown that achiral organic-soluble paramagnetic lanthanide tetrakis-β-diketonate anions of the form Ln(fod)$_4^-$ (fod = 6,6,7,7,8,8,8-heptafluoro-2,2-dimethyl-3,5-octanedione) could be added to solutions of **9** and substrates in chloroform-*d* and acetonitrile-d_3 to enhance the enantiomeric discrimination. The tetrakis chelate complex was formed in solution by adding Eu(fod)$_3$ and Ag(fod) to the sample. In the presence of the ammonium chloride salt, silver chloride precipitated out of solution. Ammonium ions in the bulk solution formed an ion pair with the Ln(fod)$_4^-$. The two substrate enantiomers have different association constants with the crown ether. Therefore, the enantiomer with the smaller association constant with **9**, which has a higher proportion in the bulk solution, exhibited larger changes in chemical shifts in the NMR spectrum.

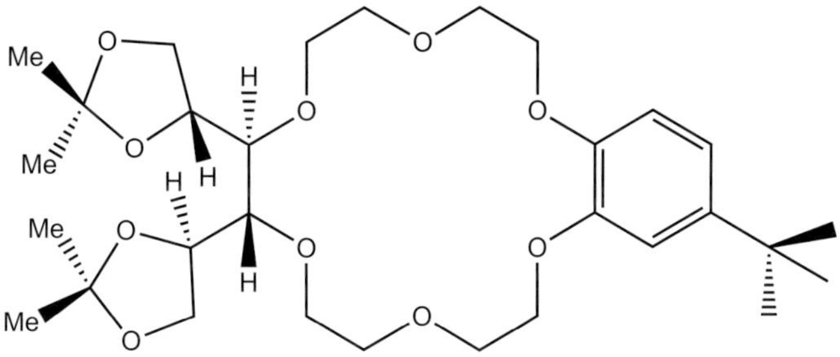

9

In a comparison with ten other glycoside-derived crown ethers, **10** was the most effective in causing enantiomeric discrimination in the methine resonance of phenylglycine methyl ester hydrochloride [80]. A crown ether derived from β-galactopyranoside (**11**) has been examined as a chloroform-soluble chiral NMR solvating agent [80, 81]. Only modest enantiomeric discrimination was observed in the ^{1}H NMR spectra of amino acid ester hydrochlorides. The corresponding diol derivative (**12**) was soluble in acetonitrile-d_3 and methanol-d_4 and was far more effective as a chiral NMR solvating agent than **11**. Of particular significance was the observation that ytterbium(III) could be added as its nitrate salt to solutions of **12** and substrates in acetonitrile-d_3 and often enhanced the enantiomeric discrimination in the NMR spectrum. In this case, the Yb(III) bound to **12** in a chelate manner through the two hydroxy groups and caused changes in chemical shifts in the NMR spectrum of the substrate bound to the crown ether [81].

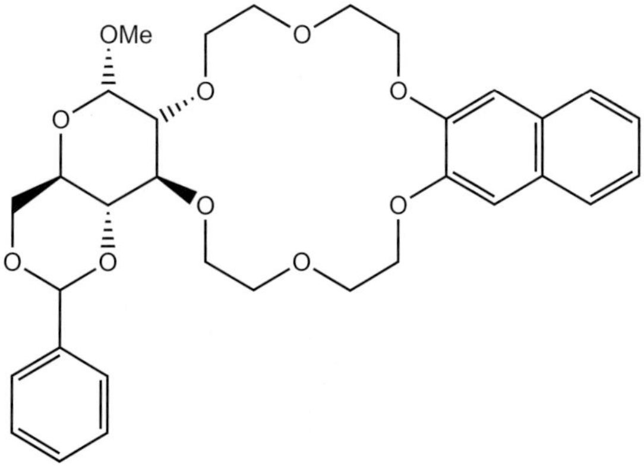

10

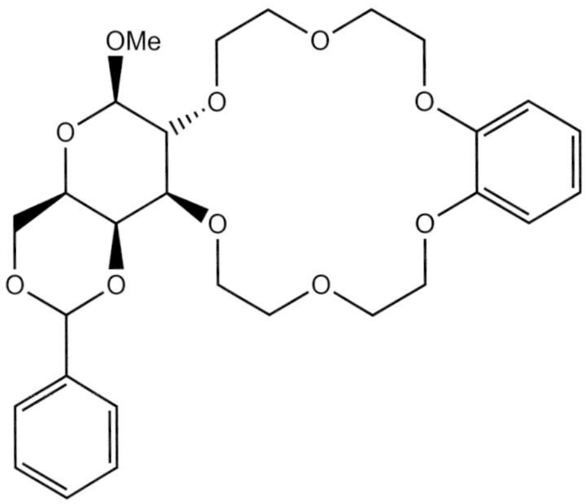

11

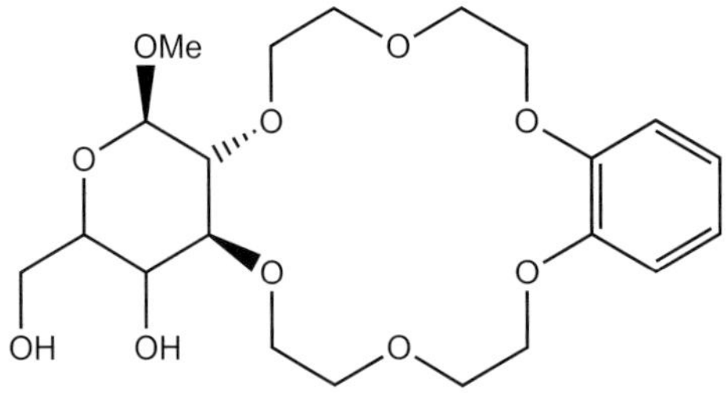

12

7.3.3
Tartaric Acid-Based Crown Ethers

The (18-crown-6)-2,3,11,12-tetracarboxylic acid **6**, which is derived from tartaric acid and was first synthesized in 1975 [82], is unquestionably the most versatile and useful crown ether for chiral recognition. The initial preparation of **6** involved an alkylation of the dithallium alcoholate of (N,N,N'N'-tetramethyl)tartramide in DMF by an aliphatic diiodide or dibromide to produce the crown ether as a tetraamide derivative [82, 83]. Heating the tetraamide in hydrochloric acid produced the tetracarboxylic acid form of the crown. An improved preparation using (N,N,N'N'-tetramethyl)tartramide, a glycol ditosylate, and sodium hydride has been reported [84]. Both enantiomers of **6** are now commercially available.

Refluxing **6** in acetyl chloride yielded the corresponding dianhydride (**13**) [83]. Reaction of the dianhydride with amines produced *syn-* and *anti-*diamide derivatives that were easily separated by column chromatography [85]. However, if the reaction was run in the presence of triethylamine, only the *syn-*isomer formed. Reaction of the methyl tetraamide derivative with amino acid methyl ester hydrochloride salts and triethylamine in methylene chloride produced the amino acid-containing tetraamide derivatives [86]. Thermodynamic values for the association of the methyl esters of phenylglycine and the glycine-phenylalanine dipeptide with various tetraamide derivatives of **6** have been measured and show enantiomeric discrimination [87]. A membrane electrode system that incorporated tetraamides of **6** showed enantiomeric selectively toward 1-phenylethylamine and the methyl ester of phenylalanine [88].

13

Of more significance has been the utilization of **6** in liquid chromatographic separations and as a chiral NMR solvating agent. One method of immobilizing **6** on silica gel involved reacting it with a 3-aminopropylsilanized silica gel in the presence of 2-ethoxy-1-ethoxycarbonyl-1,2-dihydroquinone (EEDQ) to form an amide linkage [89]. It was believed that a mixture of monoamide- and diamide-linked crown ether occurred. A *syn*-diamide-linked phase was obtained by reacting the dianhydride of TCA with an aminopropyl silica gel [90]. A comparative study using dynamically coated and bonded phases of **8**, the mixed monoamide/diamide phase, and the *syn*-diamide phase was undertaken with various amines [91]. In an effort to characterize the separation mechanism of these stationary phases, the effect of column temperature, inorganic modifier, organic modifier, and acidic modifier was explored. Lipophilic and hydrophilic effects were also examined. Whereas the effect of temperature and inorganic modifier was equivalent for all of the crown ethers, the effect of organic and acidic modifier varied with the different crown ethers.

It has also been shown that liquid chromatographic stationary phases with **6** can be used to separate secondary amines [92]. The significance of the two hydrogen bonds and ion pairing interaction (Figure 7.2b) in explaining the unusual ability of **6** among 18-crown-6 ethers to associate with secondary ammonium salts was proposed in this study. A comprehensive review of the utilization of **6** in liquid chromatographic applications has been published [93].

The utility of **6** as a chiral NMR solvating agent for primary amines has been demonstrated [94–97]. Enantiomeric discrimination was observed in methanol-d_4, acetonitrile-d_3, or deuterium oxide [94, 97], although the best results were usually obtained in methanol-d_4. The amine can be added as either a protonated salt or in its neutral form [94, 96]. In the latter case, a neutralization reaction between the amine and **6** produces the protonated ammonium ion that is needed for association. Analysis of amino acids in deuterium oxide with the (+)-enantiomer of **6** always shielded the methine resonance of the D-enantiomer more than that of the L-enantiomer. Addition of ytterbium(III) as its nitrate salt to solutions of ammonium salts and neutral **6** caused enhancements in the enantiomeric discrimination [94, 96]. The Yb(III) bound to the carboxylic acid groups of the crown and perturbed the chemical shifts of the resonances of the substrate in its bound form.

The utility of **6** as an especially effective chiral NMR solvating agent for secondary amines has been shown [66–69]. It is necessary to add the neutral form of the secondary amine to **6**. A neutralization reaction produced the ammonium ion and monocarboxylate of **6**, thereby enabling the formation of two hydrogen bonds and an ion pairing interaction needed for favorable association of the two species. The utility of **6** as a chiral NMR solvating agent has been demonstrated for pyrrolidines [67], piperidines [68], piperazines [68], and prochiral amines [69]. A small degree of enantiomeric discrimination was even observed in the ^{1}H NMR spectra of tertiary amines with **6** [69]. Better results were obtained by examining the ^{13}C NMR spectra of the tertiary amines.

7.3.4
Crowns Ethers with Phenol Moieties

A wide variety of crown ethers that incorporate phenol units have been prepared (**14** and **15** are two examples) and evaluated for their chiral recognition properties. General procedures used to prepare these crown ethers involve coupling an appropriate aromatic dibromide and diol [98] or ditosylate and diol [99]. Chromophoric crown ethers are obtained by incorporation of a 2,4-dinitrophenylhydrazine unit onto a quinone intermediate in the synthesis.

Crown ethers with the chromophoric dinitroazo aromatic unit exhibit distinct differences in the UV-vis absorption spectra in the presence of enantiomeric pairs, thereby providing a convenient colorimetric indicator for the analysis of chiral primary amines [100]. Neutral primary amines can be added to the phenol-containing crown ethers since the phenolic hydrogen atom will protonate the amine to form the ammonium needed for favorable association [101].

An interesting observation with several of phenol-containing crown ethers is a temperature-dependent reversal in the enantiomeric discrimination [102], which is observed using either UV-vis or NMR spectroscopic measurements. The temperature dependence follows a van't Hoff relationship and it was subsequently shown that chemical shifts in the NMR spectra measured as a function of temperature could be used to assign absolute configuration of the amine [103].

The phenol-containing crown ethers were also used in the development of fast atom bombardment-mass spectrometry (FAB-MS) methods designed to show the enantioselective complexation of chiral ammonium ion guests by the hosts. One method involved comparing the relative peak intensities of a target host–guest complex ion to that of an internal standard host–guest ion. The association constants measured using the FAB-MS method were comparable to those obtained using NMR spectroscopy [99]. A second method used a racemic mixture in which one of the enantiomers was deuterium labeled [104]. Examination of the relative peak heights of the host–guest complexes allowed a determination of the comparative binding preference of the enantiomers. A comparison of FAB and electrospray ionization mass spectrometric methods for the analysis of deuterium-labeled amino acid esters found that FAB was more effective [105]. A procedure for determining the enantiomeric purity of amines using FAB-MS involved the use of a racemic mixture of the (R)- and (S)-isomers of the crown ether, one of which was isotopically labeled [106]. Relative peak heights of the complexes were used to determine enantiomeric purity.

In some cases, the phenol-containing crown ethers also cause enantiomeric discrimination in the ^{1}H NMR spectra of substrates. For example, the NMR spectrum of methionine exhibited significant enantiomeric discrimination in the presence of the adamantyl-substituted crown ether (14) [104].

Pseudo-crown ethers with larger rings, such as pseudo-24-crown-8 (16) and 27-crown-9 (17), have been shown by ^{1}H NMR titration experiments to enantioselectively bind secondary amines. The larger crown macrocycle reduced steric hindrance, thereby facilitating stronger complex formation of the secondary amine [107].

16

17

A compound with two phenol-containing crown ethers attached to a phenolphthalein moiety (**18**) is an effective colorimetric reagent for the enantiomeric analysis of amino acid derivatives and β-amino alcohols. The phenolphthalein-based crown ether also showed an inversion of enantiomeric recognition for some substrates as a function of temperature [108].

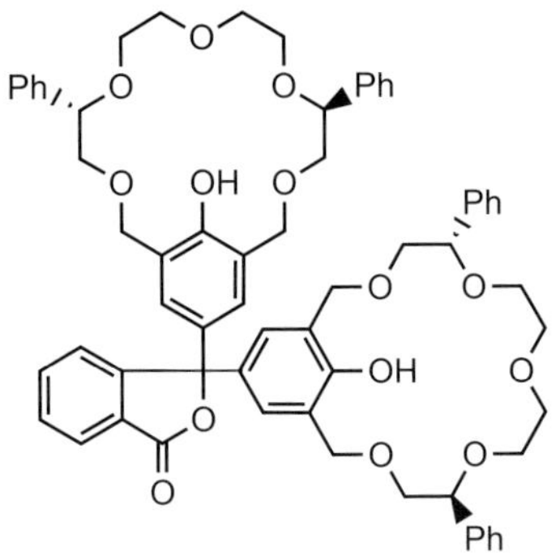

18

7.3.5
Crown Ethers with Pyridine Moieties

Chiral crown ethers containing pyridine functionalities in the macrocycle have been prepared and examined for their enantioselective properties. In a typical preparation (Scheme 7.1), an appropriately substituted tetraethylene glycol reacts with a 2,6-

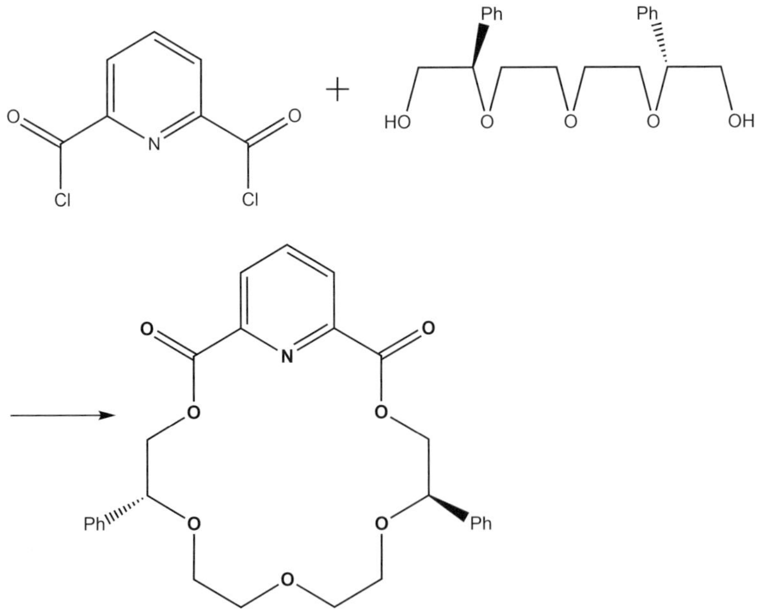

19

Scheme 7.1

pyridinedicarbonyl chloride to yield the pyridine-containing crown ether [109]. The diol with a phenyl ring was prepared from *(R)-* and *(S)-*mandelic acid and represented the most challenging aspect of the synthesis. Thermodynamic data measured using ^{1}H NMR spectroscopy indicated enantiomeric discrimination of amino acid methyl esters and 1-(1-naphthyl)ethylamine with the pyridine-containing crown ethers [110]. Chiral recognition was influenced by the R-substituent group of the crown ether and those with methyl, phenyl, or *tert*-butyl groups were most effective [111].

Compound **19** was used in the development of a FAB-MS method for determining association constants between the crown ether and ammonium ions that was reportedly better than those described above in Section 7.3.4 on phenol-containing crown ethers. The method uses a mixture of an achiral crown ether as a reference host with the chiral crown ether and substrate. The relative peak intensities of the different species can be used to determine the association constants [112].

Utilization of a 4-allyloxy-2,6-pyridinedimethyl tosylate in the preparation of the crown ether facilitated its attachment to silica gel and its use for liquid chromatographic separations [113]. A similar crown ether containing a dimethylacridino unit **(20)** was attached to silica gel and used as a liquid chromatographic phase [114].

20

7.4
Calixarenes

The reaction of phenol with formaldehyde in the presence of hydrochloric acid furnishes a cavity compound in which the phenol rings are bridged by methylene groups **(21)**. The most common of the calixarenes have four phenol rings in the cavity, although varying the reaction conditions leads to analogues with higher numbers of rings [115]. An asymmetric addition of a substituent group to the calixarene, for example, adding a methyl substituent to only one of the phenol rings in a *p-tert*-butylcalix[4]arene, leads to a pair of enantiomers that are inherently chiral. While inherently chiral calixarenes have been the focus of many studies, they tend not to be especially practical for chiral recognition applications. A more practical approach is to attach an optically pure chiral group to the calixarene. Strategies for adding chiral substituent groups through reaction at the phenol oxygen atoms (usually referred to as the lower rim) or the position para to the phenol group (usually referred to as the upper rim) have been described. A comprehensive review of chirality in calixarenes and calixarene assemblies has been published [116].

21

Optically pure amino acids are often coupled to calixarenes. Scheme 7.2 illustrates a common scheme for upper-rim addition of amino acids [117]. The phenol group is converted into a propoxy ether to hold the calixarene in a cone conformation. Amino acids can then be N-terminally coupled to the carboxylic acid moiety using standard peptide coupling reagents. An alternative means of attaching amino acids to the upper rim through the C-terminal end involves the use of the amino-substituted calixarene [117].

Scheme 7.3 illustrates a common strategy for lower-rim attachment of amino acids [117]. The phenol groups are alkylated with ethyl bromoacetate and then saponified to the carboxylic acid. Amino acids can then be added at their N-terminal end using standard peptide coupling procedures. A thorough review of the use of peptide calixarenes in molecular recognition has been published [117].

p-tert-Butylcalix[4]arenes with amino acids such as alanine and valine on the lower rim have been used to separate 1,1′-binaphthalene-2,2′-dihydroxy, 1,1′-binaphthalene-2,2′-diamine, and binaphthyldiyl hydrogen phosphate by capillary electrophoresis [118]. Similar p-tert-butylcalix[4]arenes with alanine, valine, leucine, and proline

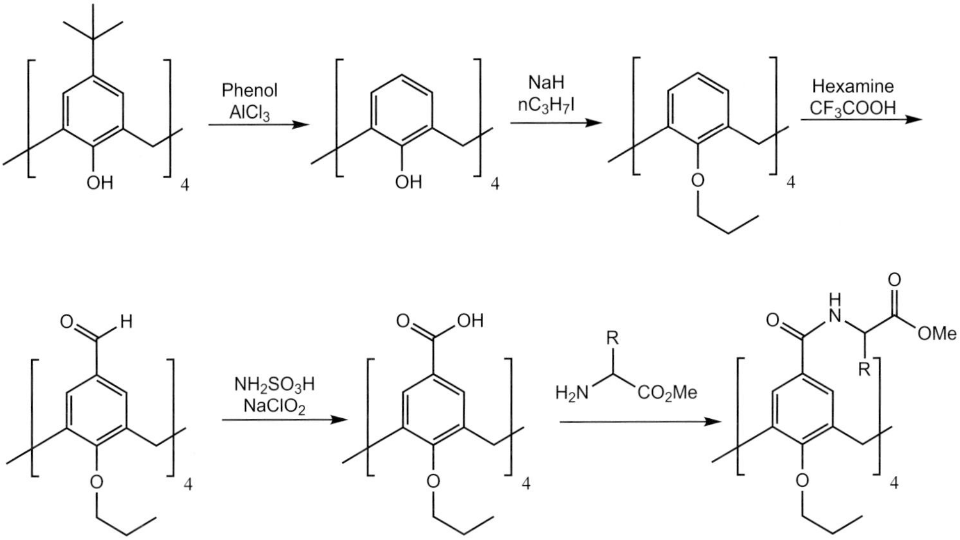

Scheme 7.2

were also evaluated as organic-soluble chiral NMR solvating agents [49]. The calix[4] arene derivative containing the *tert*-butyl ester of L-alanine was the best of those tested, but only produced a small degree of enantiomeric discrimination in the NMR spectra of 1,1′-binaphthalene-2,2′-dihydroxy, and for the N-(3,5-dinitrobenzoyl)

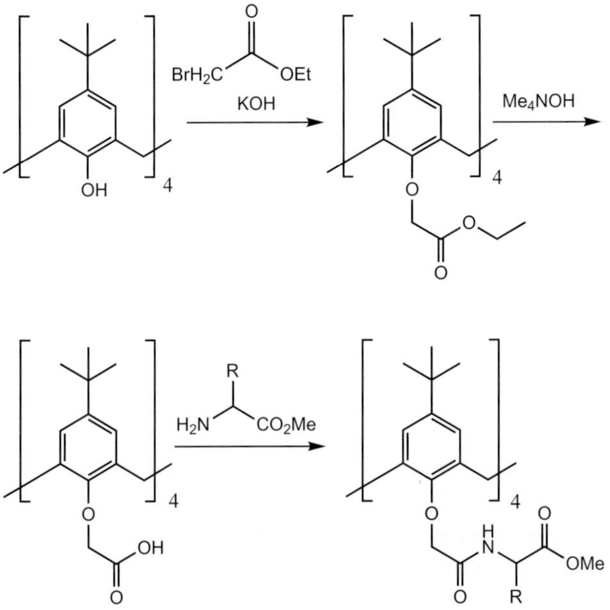

Scheme 7.3

derivatives of 1-phenylethylamine and 1-(1-naphthyl)ethylamine. Presumably the calix[4]arene and substrates were solvated too well by the organic solvent, thereby minimizing host–guest association needed for enantiomeric discrimination.

A synthetic procedure similar to Scheme 7.3 has been used to attach four *(S)*-di-2-naphthylprolinol groups to the lower rim of *p-tert*-butylcalix[4]arene (**22**). Enantioselective recognition of substrates such as 1-phenylethylamine, norephedrine, or phenylglycinol was observed in methanol. The separation of phenylglycinol by capillary electrophoresis was also facilitated by addition of **22** to the mobile phase [119]. The propranolol amide-derivatized p-*tert*-butylcalix[4]arene with an allyl group on the upper rim (**23**) was prepared by a similar scheme. Enantioselective fluorescent quenching was observed in the presence of phenylalaninol. The enantioselectivity of **23** toward phenylalanine was enhanced considerably in the presence of sodium or potassium ions [120].

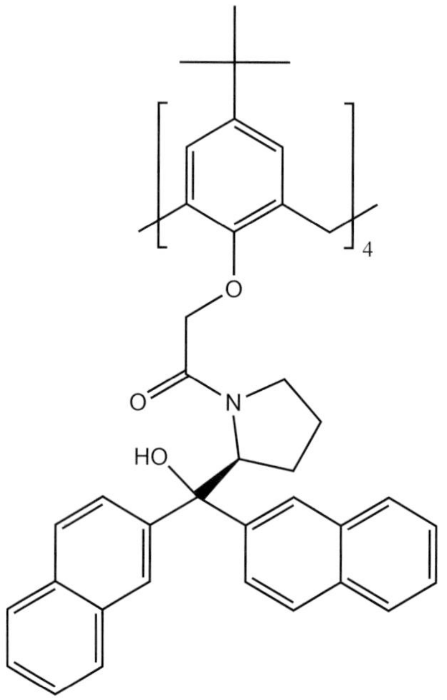

22

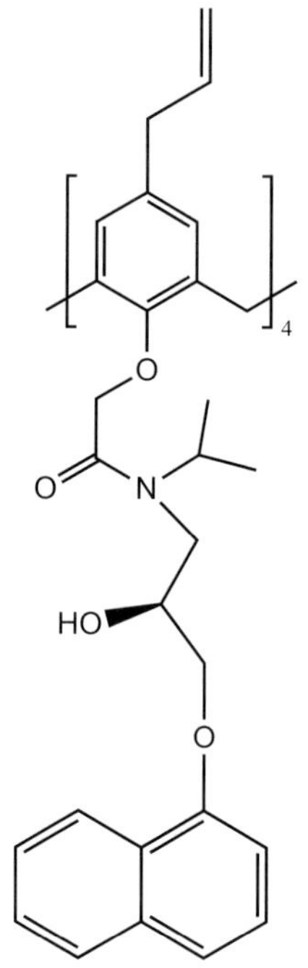

23

Other investigators have prepared calixarene derivatives with diamide groups at the lower rim and examined their chiral recognition properties. One example is the coupling of two dicyclopeptides such as cyclo(pro-ser) (**24**), cyclo(leu-ser), and cyclo(ala-ser) to the lower rim. Discrimination of the enantiomers of methyl lactate was observed using these calixarenes as a gas sensing device on a quartz crystal microbalance [121].

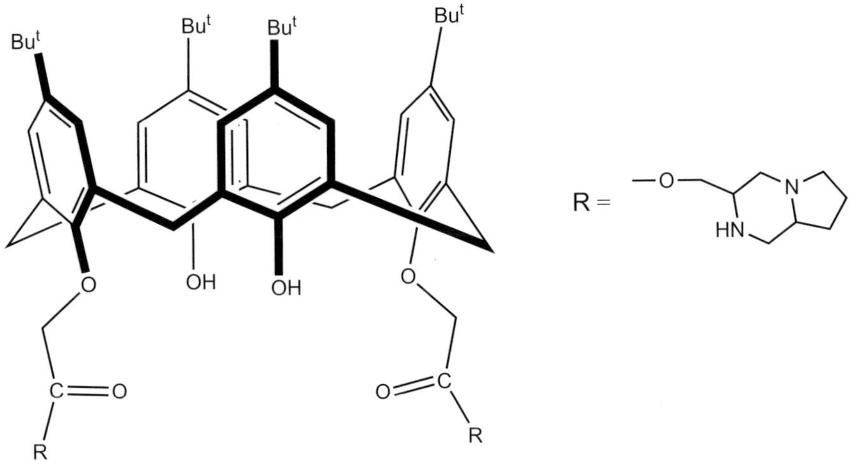

24

The coupling of 1-phenylethylamine and (1S,2S)-2-amino-1-(4-nitrophenyl)-1,3-propanediol to *p-tert*-butylcalix[4]arene (**25**) provided host compounds that enantioselectively interacted with amines. Enantiomeric discrimination was observed in the UV-vis spectra with 1-phenylethylamine and 1-(cyclohexyl)ethylamine and the ^{1}H NMR spectrum of 1-phenylethylamine in chloroform-*d* [122].

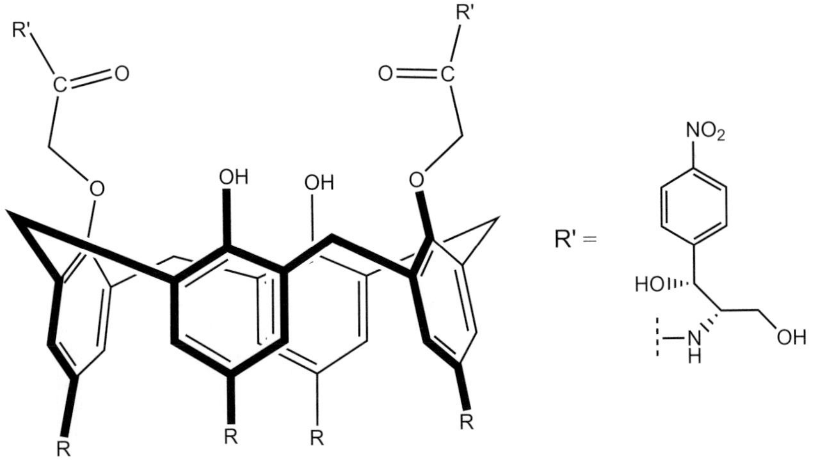

25

Inherently chiral calix[4]arene **26** has been synthesized and resolved into its two enantiomers by a selective crystallization with mandelic acid [123]. Using either of the pure enantiomers of **26**, enantiomeric discrimination was observed in the ^{1}H NMR spectra of mandelic acid. The D- or L-phenylalaninol derivative of a *p-tert*-butylcalix[6]arene (**27**) exhibited modest selectivity in extraction studies of amino acid methyl esters [124].

A calix[4]arene with two 1-cyclohexylethylamino groups attached through a Schiff base linkage (**28**) has been prepared. Binding constants measured by an ultraviolet titration showed discrimination of the enantiomers of 1-phenylethylamine [125]. A calix[4]arene with a chiral L-alanine attached through a Schiff base linkage (**29a**) associated enantioselectively with L- and D-threonine as observed by UV-vis spectroscopy. However, the corresponding L-alanine amide derivative prepared through a carboxylic acid intermediate (**29b**) did not cause any enantiomeric recognition of threonine [126].

Calix[5]arene **30** has been synthesized and evaluated for evidence of chiral recognition of ethyltrimethylammonium derivatives using ultraviolet visible and NMR spectroscopy. The macrocycle in **30** provided an additional constraint that produced a chiral space for binding of guest molecules. Large upfield shifts were observed for the resonances of substrates in chloroform-*d*, which indicates that the substrate is inserted deeply into the calixarene cavity and is shielded by the aromatic rings. Cation–π interactions were likely important in explaining the association [127].

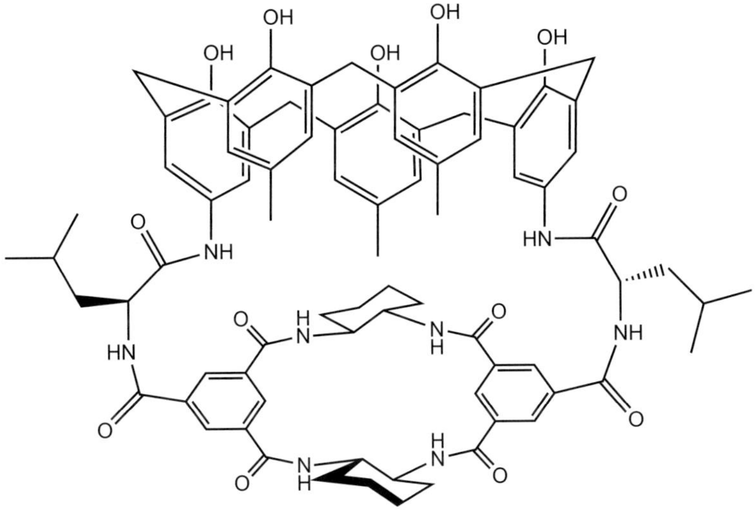

30

An optically pure tris(aminoethyl)amine ligand was bound to alternate phenol sites in a *p-tert*-butylcalix[6]arene to form **31**. The tris(aminoethyl)amino group locked the calix[6]arene into a cone conformation. Enantiomeric discrimination in the ^{1}H NMR spectrum of 4-methylimidazolidin-2-one and propane-1,2-diol was observed with **31** in chloroform-*d* [128].

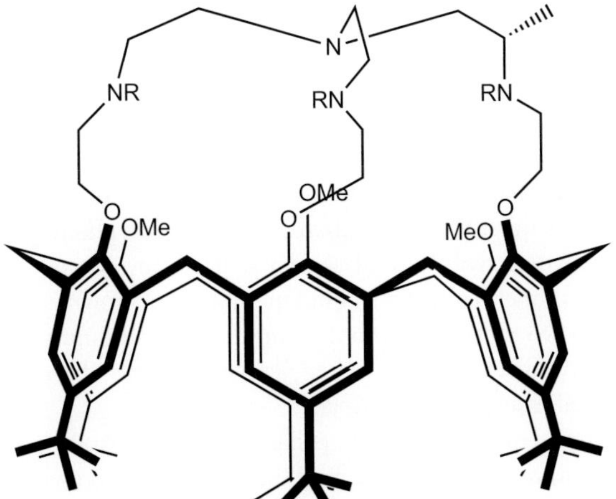

31

The reaction of *p-tert*-butylphenol with elemental sulfur and sodium hydroxide in tetraethylene glycol dimethyl ether leads to the formation of thiacalixarenes (**32**) with sulfur bridges between four, five, or six phenolic rings [129]. Using reactions comparable to those shown in Scheme 7.3 for regular calixarenes, 1-phenylethylamine or 1-(1-naphthyl)ethylamine was coupled to the bottom rim through amide linkages [130]. The calixarenes were used as gas chromatographic stationary phases to separate volatile derivatives of enantiomeric alcohols, amines, and amino acids. Whereas the thiacalixarenes exhibited enantioselective interactions with certain substrates, the corresponding calixarenes with methylene bridges between the phenol rings were ineffective at causing enantiomeric discrimination.

32

A series of homoazacalixarenes that incorporate amino acids as part of the macrocyclic framework have been prepared and examined for their enantioselective properties. The compounds are prepared using a bis(chloromethyl)phenol-formaldehyde tetramer or trimer that is then cyclized with an amino acid methyl ester of valine or cysteine. These systems demonstrated enantiomeric discrimination toward α-methylbenzyltrimethylammonium iodide. Association involved an interaction of the cationic nitrogen group with the π-basic aromatic rings in the calixarene cavity [131].

A rigid cone-shaped calixarene has been prepared from *syn*-dihydroxy[2.5]metacyclophane (Scheme 7.4) [132]. Chiral substituent groups such as the methyl esters of phenylalanine or phenylglycine were added at the N-terminal end through the phenol oxygen atoms using a procedure similar to that in Scheme 7.3 [133]. The phenylglycine derivative exhibited enantioselectivity toward amino acids, as evidenced by extraction studies and transport properties across a liquid membrane. A crown ether moiety has also been coupled to two alternate hydroxyl groups of the rigid calixarene framework [134]. Enantioselective recognition of 1-phenylethylamine was observed in extraction studies with the calixarene-crown couple.

Several other mixed calixarene–crown ether systems have been synthesized and evaluated for their chiral recognition properties. Each of these employs a crown or aza-crown unit that bridges two of the phenol hydroxy groups of the calixarene. The binaphthyl-based crown of calix[4]arene **33** has two indophenol-type units and functioned as a chromogenic solid phase receptor for phenylglycinols that adsorbed

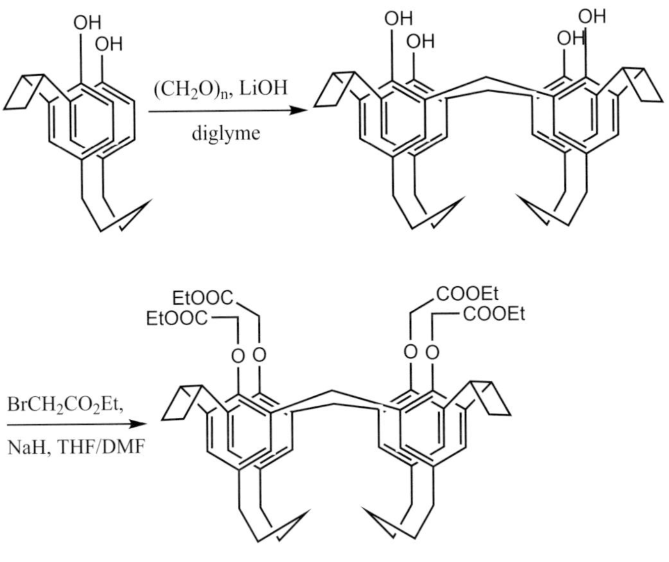

7.4 Calixarenes | 217

Scheme 7.4

from solution [135]. Compound **34**, which has a crown ether and naphthoic acid functionality attached through the phenol hydroxy groups, is fluorescent and showed enantioselective recognition toward leucinol [136]. The calixarene–crown couple with a 1,2-diphenyl-1,2-oxyamino residue (**35**) produced enantiomeric discrimination in the ^{1}H NMR spectrum of mandelic acid [137]. Calix[4](aza)crown couple **36** contains valine units and produced enantiomeric recognition in the dibenzoate derivatives of tartaric and amygdalic acid, as evidenced by ^{1}H and ^{13}C NMR and ultraviolet spectroscopy in chloroform-*d* [138].

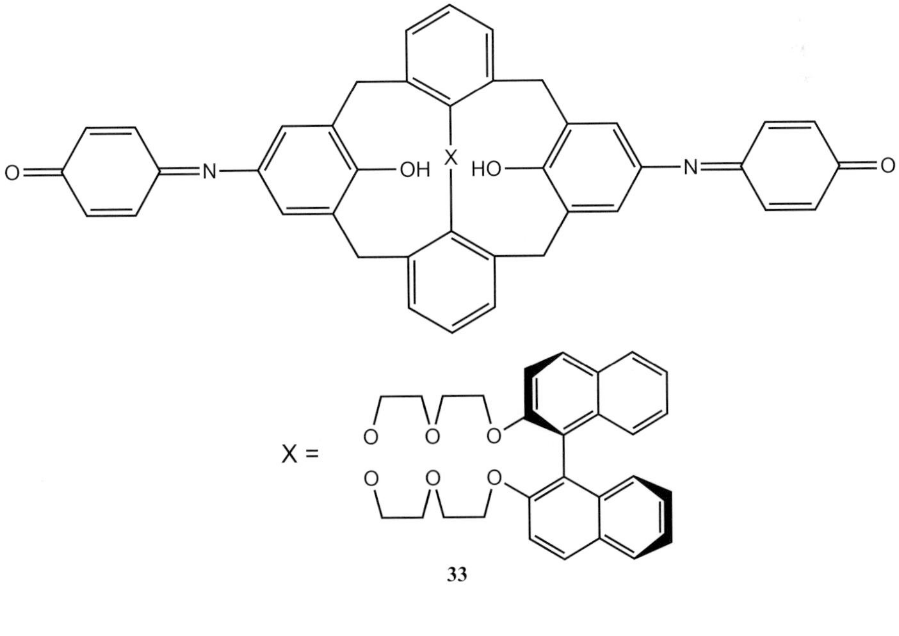

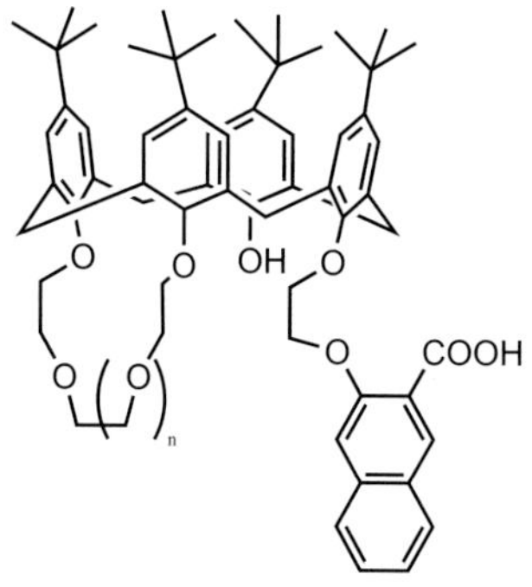

34 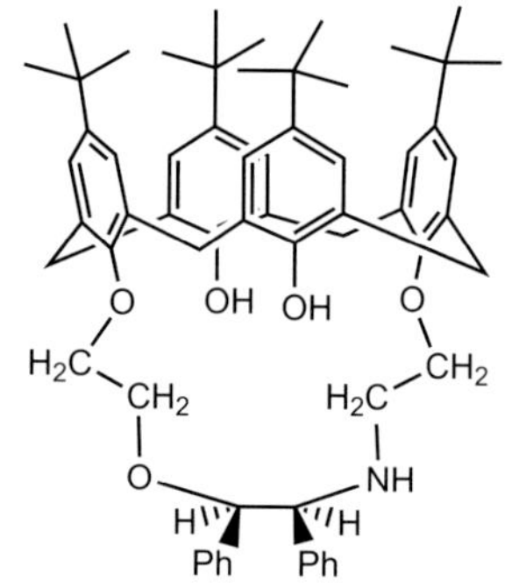 **35** 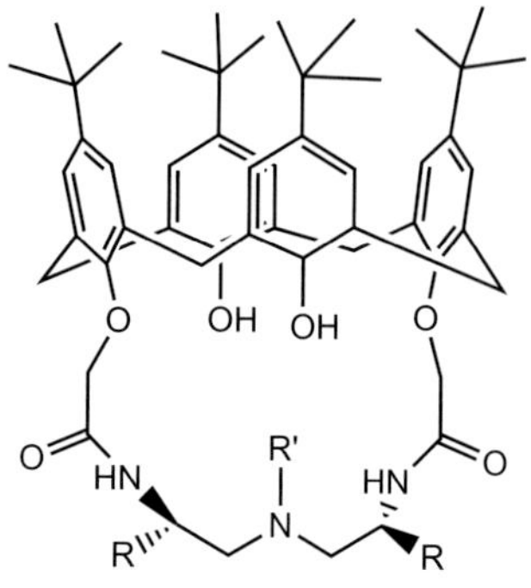 **36**

Certain calixarene compounds such as **37a** and **37b** exhibit the ability to form dimers in solution. A mixture of achiral **37a** and chiral **37b** has the possibility of forming homo- and heterodimers. The heterodimer of **37a** and **37b** did produce a small degree of enantiomeric discrimination in the ^{1}H NMR spectrum of racemic nopinone. Similarly, the heterodimer formed from the corresponding L-isoleucine and p-methylphenyl-substituted calixarenes produced enantiomeric splitting of the methyl signal in the NMR spectrum of 3-methylcyclopentanone [139].

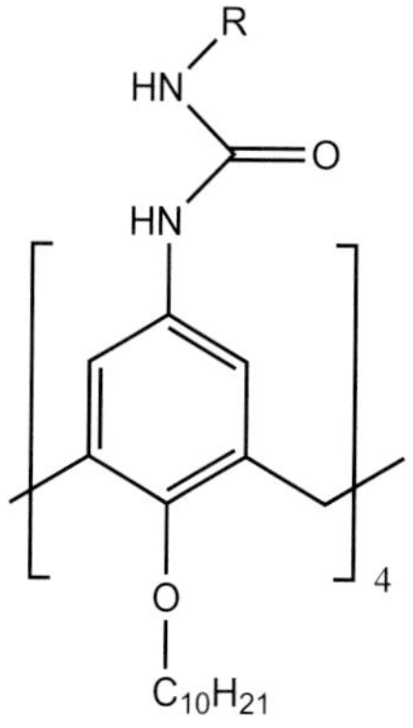

	R
a	-C$_6$H$_4$-p-C$_7$H$_{15}$
b	-SO$_2$C$_6$H$_4$-p-CH$_3$
c	-C(C$_4$H$_9$)C(O)OCH$_3$

37

7.5
Calix[4]resorcinarenes

Calix[4]resorcinarenes are prepared by the reaction of resorcinol and an aldehyde, leading exclusively to a tetrameric calix[4]resorcinarene (**38**). Unlike calixarenes – which are restricted to the use of formaldehyde, which results in a methylene bridge between the phenol rings – essentially any aldehyde can be used in the preparation of calix[4]resorcinarenes. Altering the nature of the bridge between the resorcinol rings readily facilitates the preparation of calix[4]resorcinarenes with a wide range of solubilities.

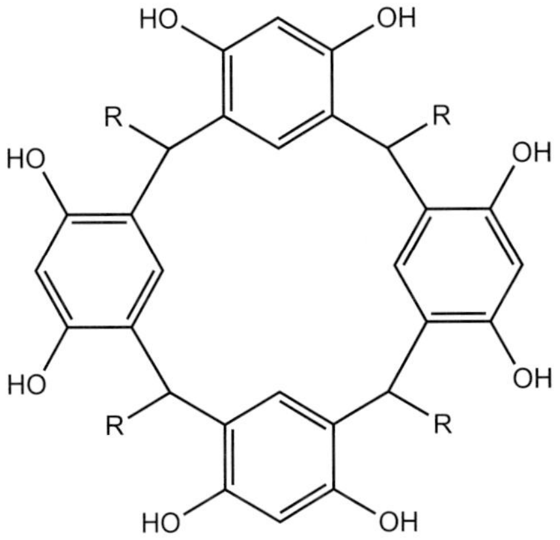

38

Chirality in calix[4]resorcinarenes is achieved either through inherently chiral systems or by attachment of optically pure chiral substituent groups. While many examples of inherently chiral calix[4]resorcinarenes have been prepared and studied [140], they have rarely been used for the purpose of chiral recognition. An exception involves the inherently chiral calix[4]resorcinarenes **39–41** [141]. These are chiral because of the alternate substitution pattern at the hydroxy groups. Enantiomeric discrimination of various N-trimethylammonium salts was observed by electrospray ionization mass spectrometry.

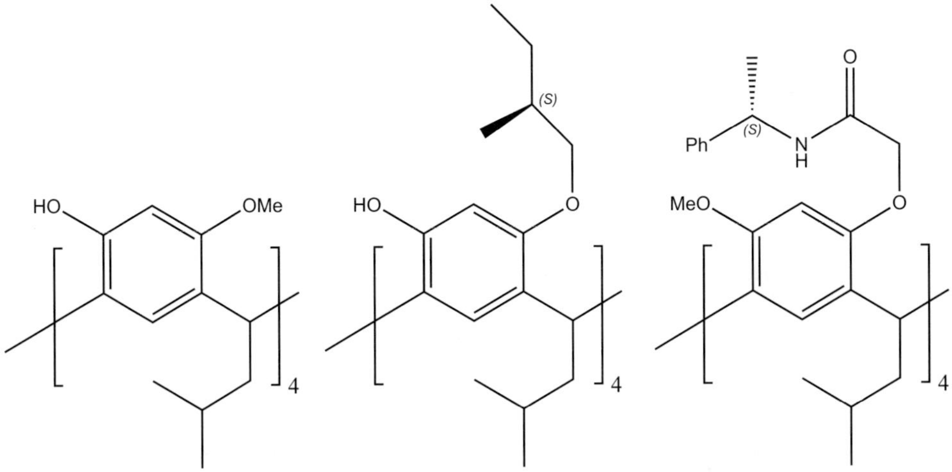

39   **40**   **41**

A common strategy for obtaining chiral calix[4]resorcinarenes is to attach an optically pure secondary amine between the two resorcinol hydroxyl groups using

Mannich conditions. Attachment of groups such as N-α-dimethylbenzylamine, N-methyl-1-(1-naphthyl)ethylamine, N-methyl-α-methylbenzylamine, and N-benzyl-α-methylbenzylamine to calix[4]resorcinarenes prepared using either acetaldehyde or octylaldehyde provided a set of organic-soluble compounds that were examined as chiral NMR solvating agents. No evidence for association between these calixresorcinarenes and a wide range of substrates compounds was observed in chloroform-d [49].

Attachment of a primary amine such as 1-phenylethylamine in the presence of two equivalents of formaldehyde leads to a cyclic product with oxazine rings (42) [142, 143]. Studies have, generally, found that the formation of one oxazine ring directs the formation of the remaining three oxazine rings in the same orientation so that only one geometric isomer forms. The 1-phenylethylamine derivative was shown by UV-vis spectroscopy [143] or surface tension isotherms in Langmuir monolayers on a water surface [142] to enantioselectively discriminate 1-phenylethylamine, 1-(cyclohexyl)ethylamine [143], or amino acids [142]. Calix[4]resorcinarene derivatives with norephedrine (43) [142, 143] and 1-phenylethylamine hydrochloride (44) [143] were also examined for their chiral recognition properties. In every case, only modest degrees of enantiomeric discrimination were observed. Presumably, this is because the organic solvent effectively solvates the substrate and calix[4]resorcinarene and reduces the extent of the association.

Y = PhCHCH$_3$
R = C$_{11}$H$_{23}$

42   43   44

In contrast, quite substantial host–guest complexation is observed with water-soluble calix[4]resorcinarenes containing a sulfonated bridge between the resorcinol rings and prolinylmethyl chiral moieties (45a–e) [144–149]. These calix[4]resorcinarenes have a hydrophobic cavity and the hydrophobic portion of water-soluble organic salts favorably associate by insertion into the cavity.

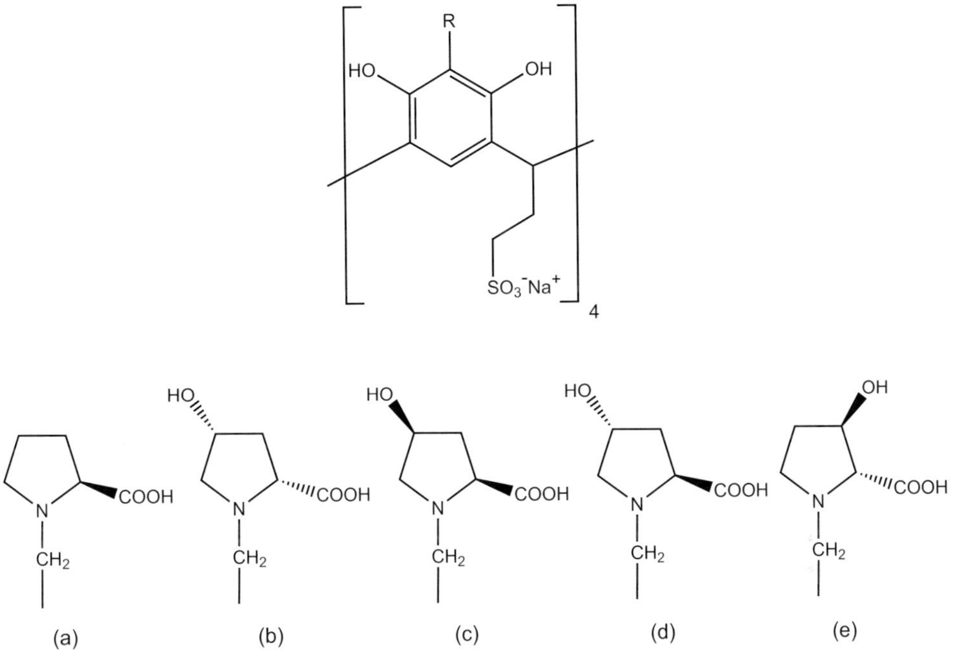

(a)   (b)   (c)   (d)   (e)

45 R = λ

The L-prolinylmethyl derivative (**45a**) is an effective water-soluble chiral NMR solvating agent. In initial work, **45a** was shown to be more effective for water-soluble substrates containing a phenyl ring than those with only aliphatic groups. Large shielding of the phenyl ring of the substrate indicated that association involved insertion of the hydrophobic aromatic ring of the substrate into the calix[4]resorcinarene cavity with the para-proton deepest in the cavity. The aliphatic substituent group of the substrate was then positioned to interact with the chiral prolinylmethyl moiety. The resonances of the ortho-hydrogen in the ^{1}H NMR spectrum generally showed larger enantiomeric discrimination than those of the meta- or para-hydrogen atoms, which was consistent with their relative proximities to the prolinylmethyl moieties [144].

Subsequent studies showed that even greater association, larger changes in chemical shifts, and often greater enantiomeric discrimination occurred in the ^{1}H NMR spectra of bicyclic substrates such as 1-(1-naphthyl)ethylamine, propranolol hydrochloride (**4**), and tryptophan methyl ester hydrochloride with **45a** [145, 146]. For 1-(1-naphthyl)ethylamine, propranolol, and tryptophan, the relative changes in chemical shifts for hydrogen atoms of the substrates indicated geometries as shown in Figure 7.3a–c, respectively. It was postulated that **45a** could accommodate an aromatic ring as shown in Figure 7.3b if it adopted a flattened cone conformation. Substrates such as carbinoxamine (**2**), doxylamine (**3**), pheniramine (**1**), chlorpheniramine, and brompheniramine also associate with **45a**. Steric hindrance from the para-substituted halide atom on **2**, chlorpheniramine, and brompheniramine blocks

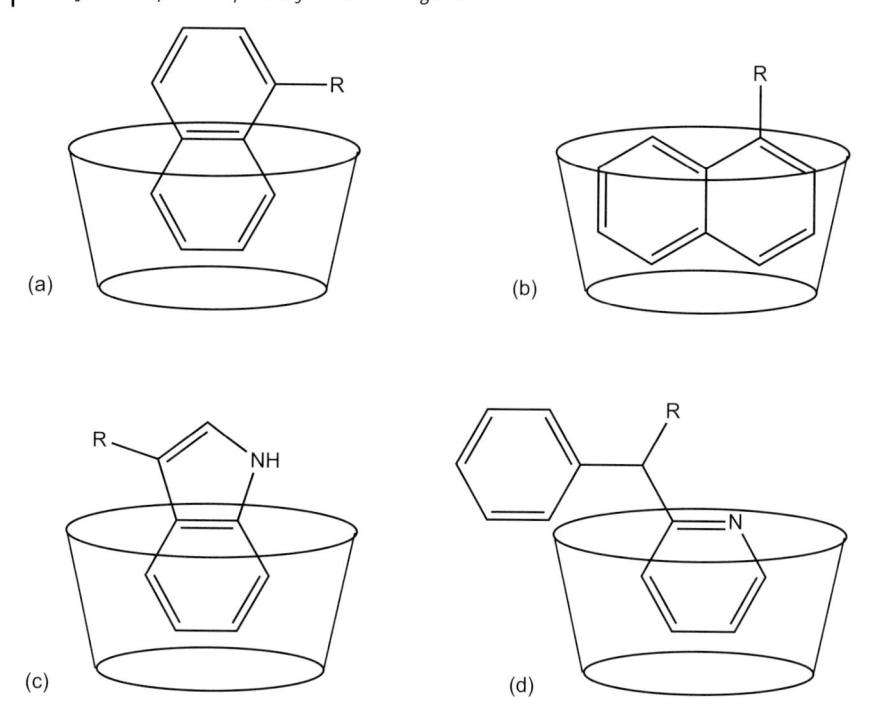

Figure 7.3 Model geometries of association of (a) 1-(1-naphthyl)ethylamine, (b) propranolol, (c) tryptophan and (d) chlorpheniramine with **45a**.

insertion of the phenyl ring into the resorcinarene cavity. The magnitude of the perturbations in chemical shifts indicated that the pyridyl ring inserts into the cavity as shown in Figure 7.3d [146]. For pheniramine and doxylamine, the magnitude of the changes in chemical shifts indicated that both rings associated to some degree within the cavity.

The association of substrates with **45a** requires an aromatic ring relatively free of steric hindrances. Studies have shown that mono- and ortho-substituted phenyl rings, and naphthyl rings that are mono-substituted, 1,2-, 1,8-, or 2,3-disubstituted, readily form host–guest complexes with **45a** by insertion of the aromatic ring into the cavity [147]. Compounds with indole, dihydroindole, and indane ring systems in which there are no other substituents on the six-membered ring also form host–guest complexes with **45a**.

More recent work has shown that calix[4]resorcinarenes with 3- and 4-hydroxyproline substituent groups (**45b–d**) are often more effective water-soluble chiral NMR solvating agents than **45a** [147–149]. Presumably the hydroxy groups on the proline moiety are involved in dipole–dipole interactions with the substrates that enhance the enantiomeric discrimination.

Another strategy for preparing chiral calix[4]resorcinarenes is to attach optically pure moieties through the hydroxy groups of the resorcinol rings. For example, L-valine-*tert*-butylamide substituent groups have been attached through the hydroxyl groups of a calix[4]resorcinarene prepared using 1-undecanal. The resulting calix[4]resorcinarene (**46**) was used as either a coated or bonded capillary gas chromatographic stationary phase to separate N(O,S)-trifluoroacetylmethyl esters of amino acids [150]. Octamide derivatives of calix[4]resorcinarenes (**47**) have been prepared by heating the corresponding octaethoxy methylated calix[4]resorcinarene in the liquid amine. Amberlite XAD resins were impregnated with the octamide calix[4]resorcinarenes and modest enantiomeric discrimination of sodium salts of tryptophan and phenylglycine was observed [151].

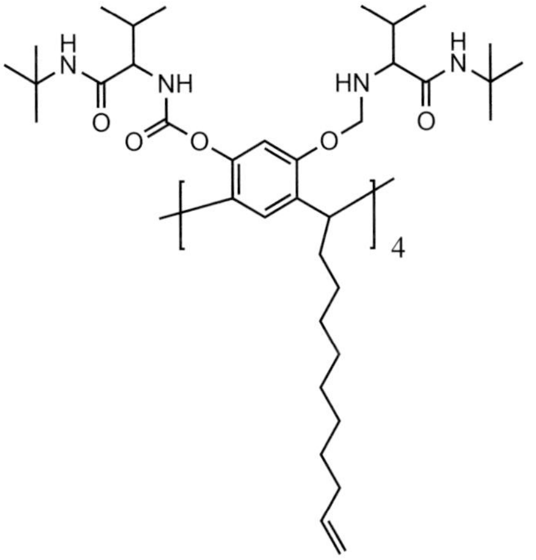

46

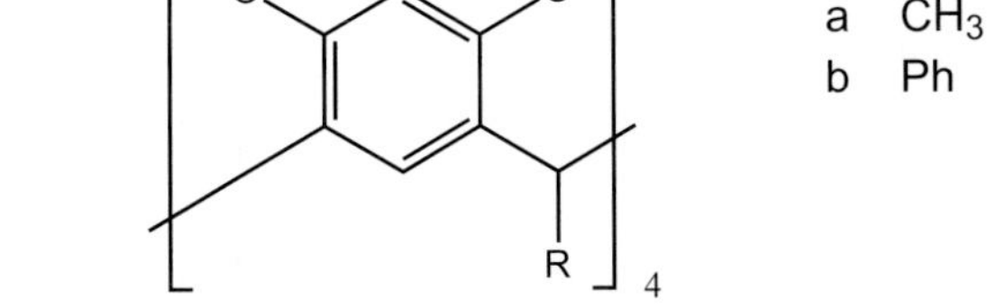

	R
a	CH$_3$
b	Ph

47

The hydroxy groups on adjacent resorcinol rings can be bridged with suitable reagents. One example is the bridged phosphamide derivatives **48**. These form Langmuir monolayer films on water and a pH-dependent enantiomeric discrimination was observed either through surface potential or surface pressure isotherms [152].

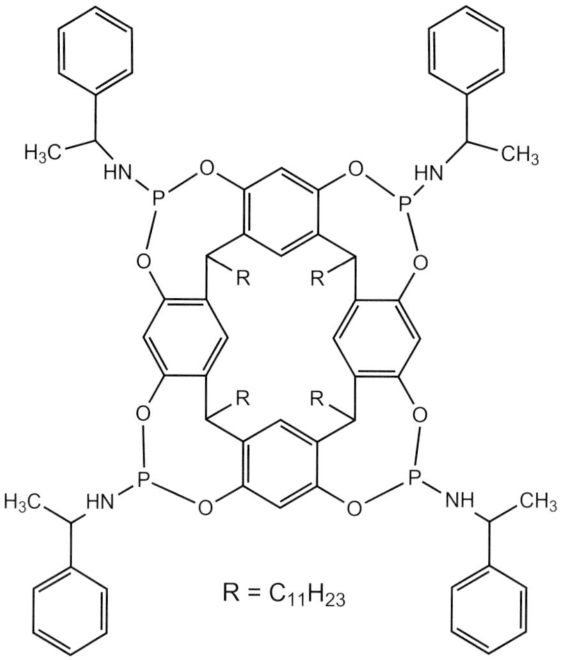

48

Two hydroxy bridged calix[4]resorcinarenes have been coupled into an elaborate hemicarciplex structure with binaphthyl [153–155] and (S,S)-1,4-di-O-tosyl-2,3-isopropylidene-L-threitol linkages [154]. The hemicarciplexes exhibited stereoselective encapsulation properties toward 1-bromo-2-methylbutane, 1,3-dibromobutane [153], and 2-butanol [154], and produced enantiomeric discrimination in the ^{1}H NMR spectra of aryl alkyl alcohols, sulfoxides, alkyl halides, and alkyl alcohols [155].

An alternative strategy for preparing calix[4]resorcinarenes involves the tetramerization of 2,4-dimethoxycinnamic acid amides with boron trifluoride etherate (Scheme 7.5) [156]. The valine amide derivative exhibited gas phase enantiomeric discrimination toward amino acids in studies using electrospray ionization mass spectrometry. Optically pure 2-aminobutane was used to displace the amino acid from the calix[4]resorcinarene to examine the extent of enantioselectivity. Both thermodynamic and kinetic factors were important in the displacement process [157].

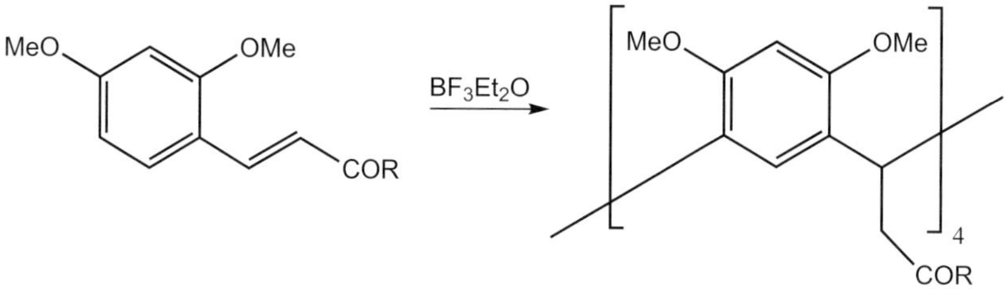

Scheme 7.5

7.6
Miscellaneous Receptor Compounds

A wide variety of specialized receptor compounds have been developed. Oftentimes these are designed for a single class of compounds or to accommodate a specific molecule. Ionic groups such as the guanidinium moiety or polar functionalities such as amides, ureas, thioureas, and carbamates are frequently incorporated into clip-shaped, cleft-shaped, or macrocyclic cavity compounds. The following section exemplifies the wide range of chiral receptor compounds that have been reported. A comprehensive review of the use of receptor compounds for the chiral recognition of anions has been published [158].

The cationic guanidinium group has been used in designing receptor molecules for carboxylate species. Compound **49**, which contains naphthyl substituent groups, produced a small degree of enantiomeric discrimination in the ^{1}H NMR spectra of mandelate, naproxenate, and N-acetyl tryptophan [159]. Compounds with lasalocid A or crown ether appendages (**50**), have been employed as zwitterionic amino acid carriers in membranes [160] or in extraction or NMR discrimination studies [161]. The ammonium ion associated with the crown moiety of **50** while the carboxylate ion associated with the guanidinium group.

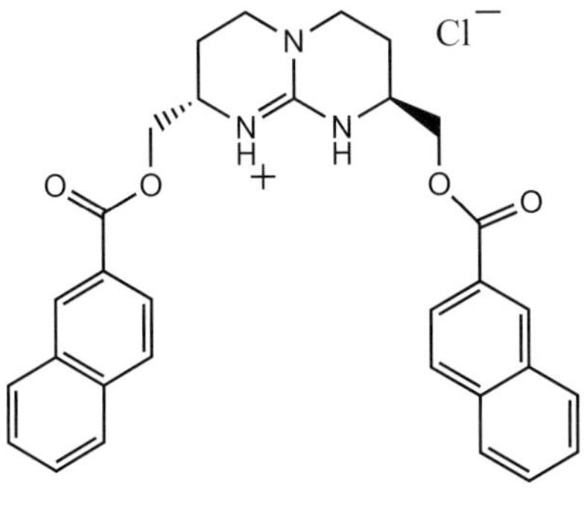

49

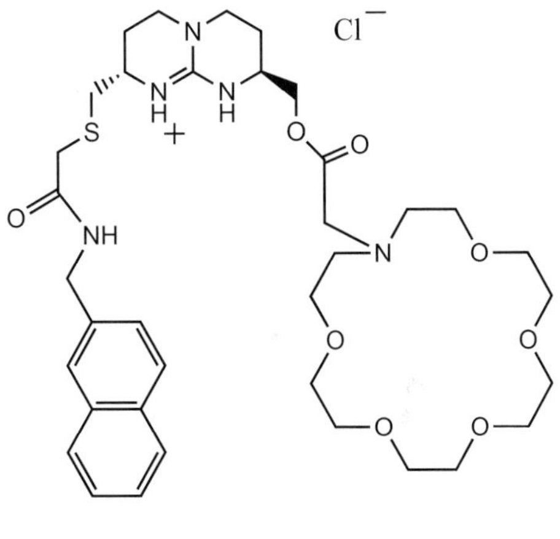

50

The axial chiral π-electron deficient tetracationic receptor **51** contains a 1,1'-binaphthyl unit and two cyclophane units. It was effective at enantiodifferentiating aromatic amino acids such as phenylalanine, tyrosine, and tryptophan in water and their N-acetylated derivatives in organic solvents [162]. The bisbinaphthyl macrocyclic species **52** exhibited enantioselective recognition of mandelic acid that could be monitored by fluorescence spectroscopy [163]. Macrocyclic receptor **53** with one binaphthyl and two xanthone units permitted the enantioselective extraction of amino acids in water in the presence of an 18-crown-6 ether. The ammonium group of the amino acid associated with the crown ether while the carboxylate ion associated with the macrocycle [164].

51   52

53

Macrocycle **54** contains an *N,N'*-bis(6-acylamino-2-pyridinyl)isophthalamide unit and was found to function as an effective chiral NMR solvating agent for carboxylic acids, oxazolidinones, lactones, alcohols, sulfoxides, sulfoximines, isocyanates, and epoxides [165]. The presence of the nitro group strengthened the hydrogen bonding properties of the amide groups of the macrocycle. In a subsequent report, **54** was bound to silica gel and used as a liquid chromatographic stationary phase to separate various compounds [166].

54

Compound **55** is an example of an acyclic thiourea receptor that exhibited chiral recognition toward amino acids [167]. The compound has a cleft with four hydrogen bond donors suitable for chiral recognition of carboxylate ions. Compounds **56** [168] and **57** [169] are examples of macrocyclic compounds that show selective binding toward certain amino acid enantiomers such as N-protected glutamate and aspartate. Binding of amino acids to **56** and several similar analogues was stronger in relatively polar solvents such as acetonitrile and methyl sulfoxide, whereas no binding occurred in chloroform [168]. This is opposite of the common behavior in which polar solvents effectively solvate the polar groups of the substrates and macrocycle to reduce binding. The unusual behavior was ascribed to a solvent-induced change in conformation that inhibited binding in chloroform. In contrast, **57** showed strong binding of monocarboxylates in chloroform but not methyl sulfoxide [169].

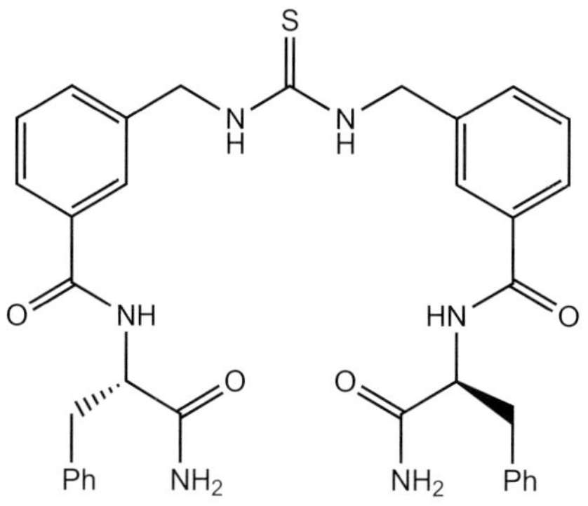

55

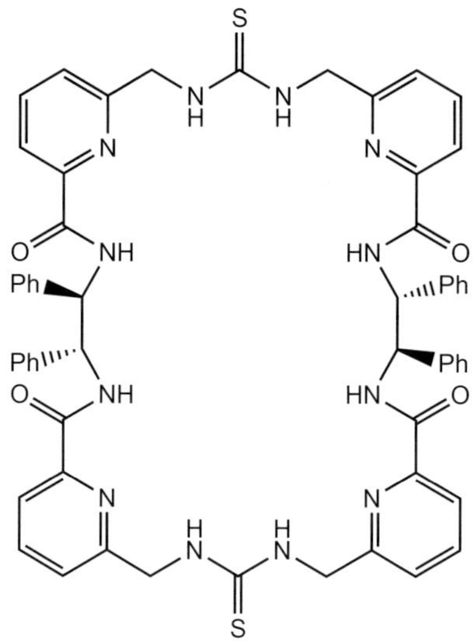

56

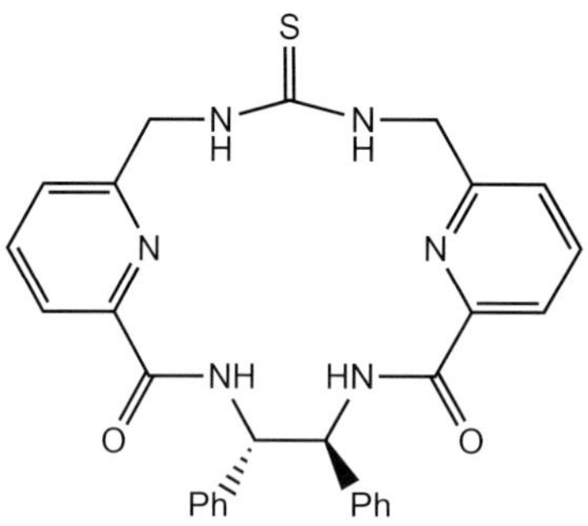

57

Compound **58** is a bischromenylurea macrocyclic compound with diphenyl-*p*-xylylenediamine spacers that has a cavity of suitable geometry to distinguish the enantiomers of naproxen [170]. Diphenylglycoluril-based receptors with appended amino acid groups, one of which is compound **59**, have been shown to exhibit enantioselective binding of catechol amines and amino acids. The compounds are water-soluble and enantiorecognition was monitored using UV-vis spectroscopy [171].

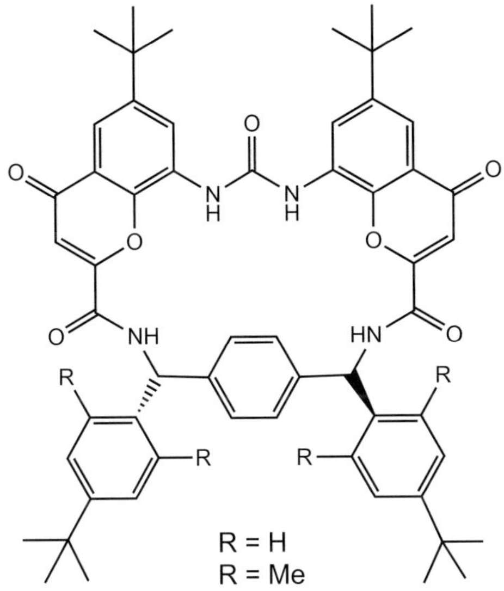

7.6 Miscellaneous Receptor Compounds | 229

58

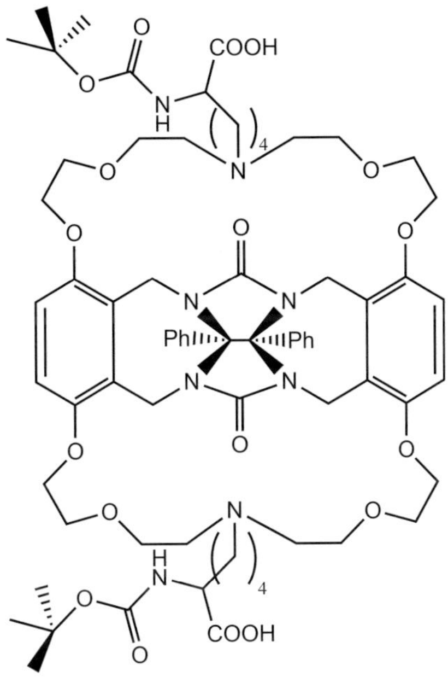

59

The cleft-shaped molecule N-(3,5-dinitrobenzoyl)-4-amino-3-methyl-1,2,3,4-tetrahydrophenanthrene (**60**) has been bound to silica gel and used as a highly versatile liquid chromatographic stationary phase [172]. The cleft geometry, amide functionality, and dinitrobenzoyl ring provide sites that facilitate association and enantiomeric recognition of substrates. The organic-soluble analog of **60** is effective as a chiral NMR solvating agent for various substrates, including epoxides, amides, lactones, lactams, alcohols, sulfoxides, and primary amines [173].

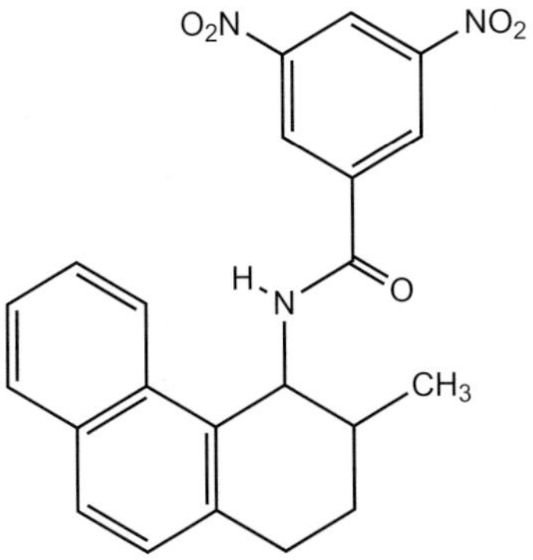

60

Compound **61** is a C_3-symmetric receptor that was attached to silica gel and used for the liquid chromatographic separation of neutral compounds having diverse structural features, including N-boc and N-3,5-dinitrobenzoyl protected amino acids [174]. A macrocycle assembled from chiral 1,2-diphenyl-1,2-diamine and 5-allyloxyisophthalic acid groups (**62**) was immobilized onto silica gel and found to preferentially bind the L-enantiomers of amino acids [175].

Compound **63** and similar analogs are patterned after chiral liquid chromatographic stationary phases. The ^{1}H NMR spectra of dinitrobenzoyl derivatives of aryl amides and carboxylic acid esters in chloroform-d exhibited enantiomeric discrimination in the presence of **63** [176]. Receptor **64** in which the R group is phenyl, naphthyl, or cyclohexyl forms three hydrogen bonds with carboxylic acids. Naproxen and other chiral carboxylic acids exhibited enantiomeric discrimination in the ^{1}H NMR spectrum in the presence of **64**. The naphthyl derivative was the most effective of the three receptor compounds [177].

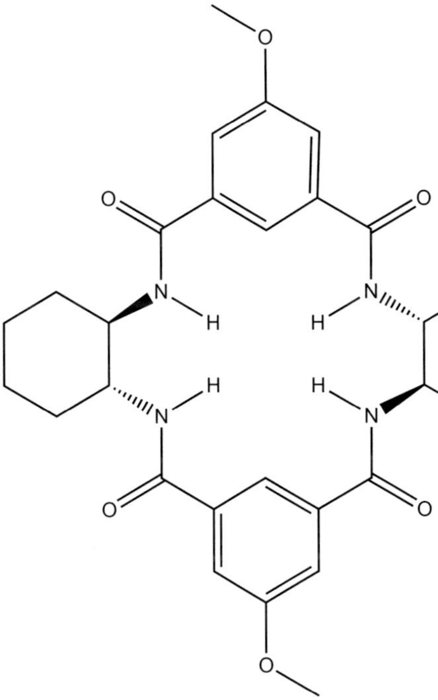

63

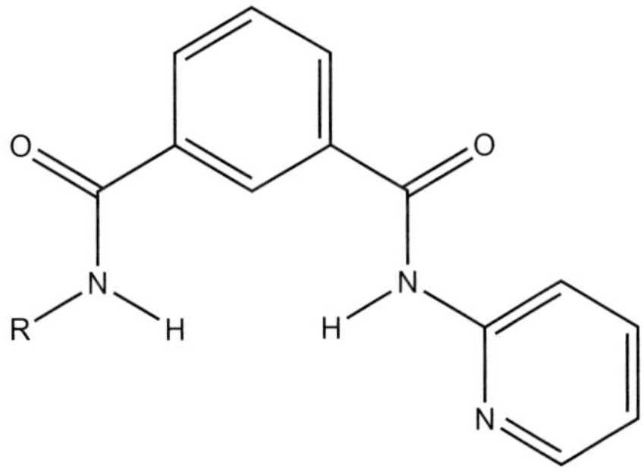

64

The acyclic receptor **65**, which is based on chiral deoxycholic acid, forms a molecular tweezer that is capable of enantiomerically discriminating amino acid methyl esters. The association was monitored using UV-vis spectroscopy [178]. Similarly, acyclic anthryl-containing receptors, one example of which is **66**, exhibited enantiomeric discrimination toward tetrabutylammonium mandelate. Enantiodifferentiation was observed using fluorescence and NMR spectroscopy [179]. Macrocyclic dioxopolyamines derived from one or two L-proline units showed chiral discrimination in the ^{1}H NMR spectra of mandelic acid and some of its derivatives as well as naproxen. The compound with two proline moieties (**67**) was more effective at causing enantiorecognition [180].

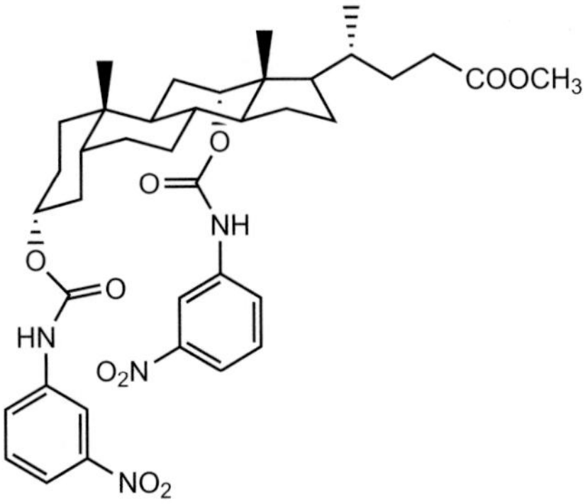

65

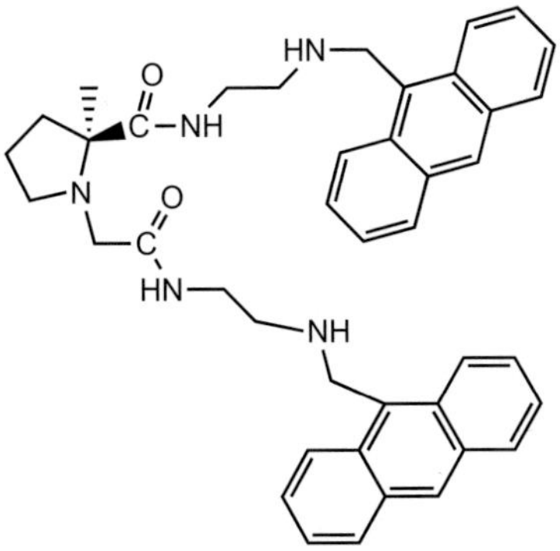

66

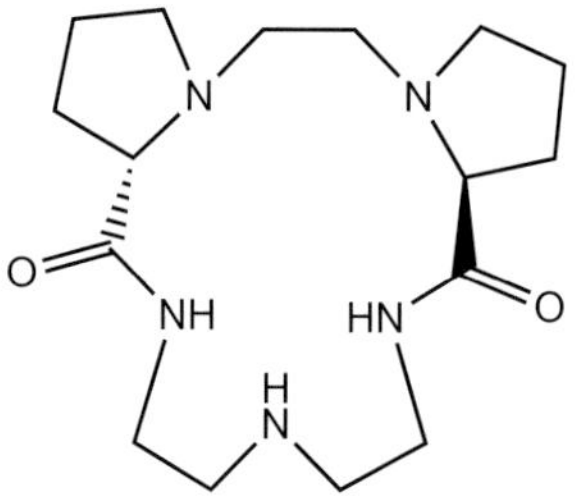

67

Carbamate derivatives of the cinchona alkaloids quinine, quinidine, cinchonidine, cinchonine, and the C9 epimers have been exploited for chiral recognition in liquid chromatography, capillary electrophoresis, extractions, and membranes. In some cases, the enantiomeric discriminating ability is unparalleled in magnitude relative to other enantioselective reagents. Geometric considerations show that these compounds have a distinct, pre-organized binding cleft into which the substrate molecules fits, which explains the remarkable selectivity for certain compounds [181].

The tripodal oxazoline **68** was effective in extraction experiments at enantiodiscriminating β-chiral aliphatic and aromatic primary ammonium ions. The three oxazoline rings tip up to hydrogen bond and secure ammonium ions in the cavity [182]. Compound **69**, which has two oxazoline rings attached to a pyridyl ring, binds effectively to secondary amines and is an effective chiral NMR solvating agent. The amines are analyzed as protonated cations and form hydrogen bonds to the nitrogen atoms of **69** [183].

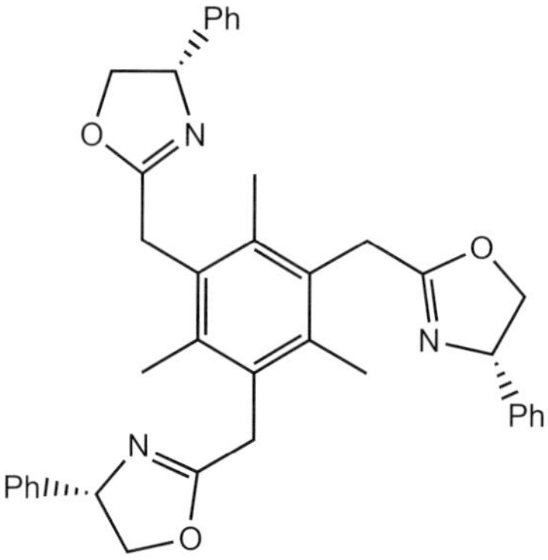

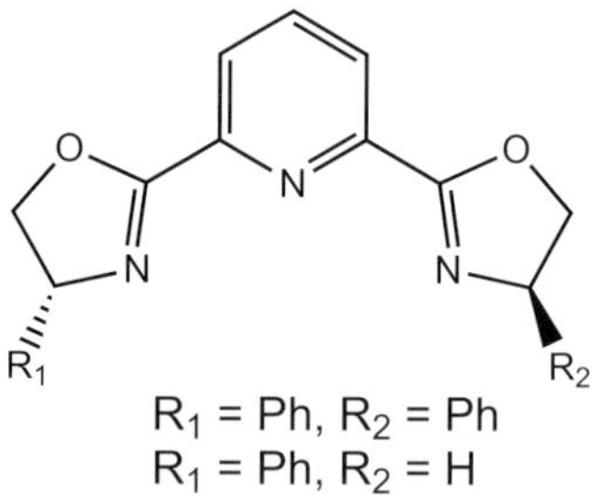

68   **69**

$R_1 = Ph, R_2 = Ph$
$R_1 = Ph, R_2 = H$

The phosphorylated macrocycle **70** caused enantiomeric discrimination of the diammonium ions of lysine and arginine. The host molecule is effective because it has both anionic groups and a hydrophobic binding site. Lysine and arginine ions have sufficiently long spacing between the two ammonium ions to associate simultaneously at both anionic phosphate groups. Mono cations or compounds with shorter distances between the two ammonium ions do not show enantiodifferentiation with **70** [184].

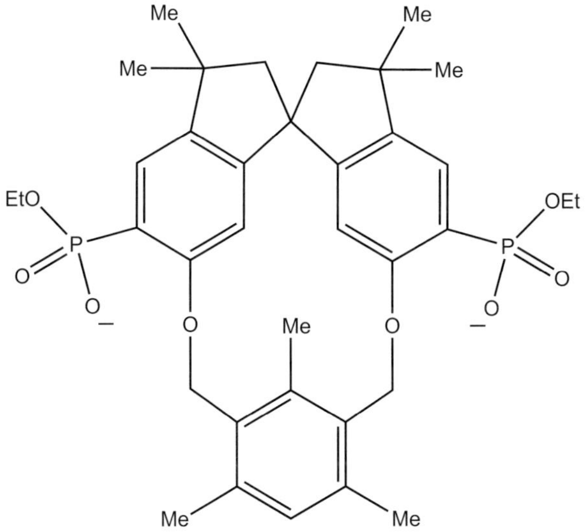

70

The aza macrocycle **71** is an effective chiral discriminating agent for dicarboxylates such as malate, tartrate, aspartate, and glutamate. Potentiometric titrations and electrospray ionization-mass spectrometry were used to examine the presence of enantiomeric differentiation [185].

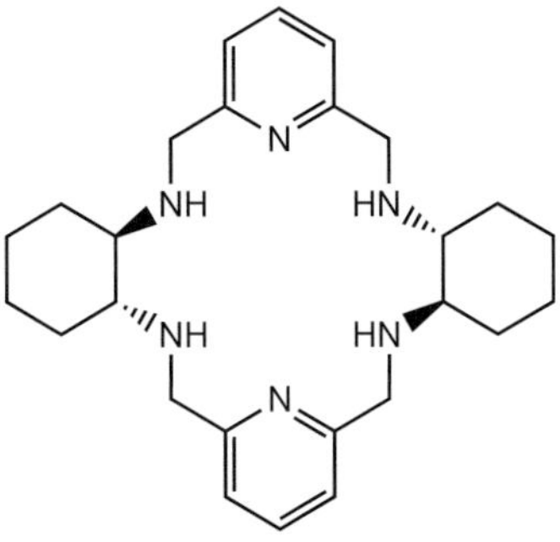

71

The *anti*-configuration of **72** is a clip-shaped molecule. The indole ring of tryptophan has a similar size and shape to the cavity of the compound. Circular dichroism titration data showed that tryptophan methyl ester hydrochloride exhibited enantioselective insertion into the clip [186].

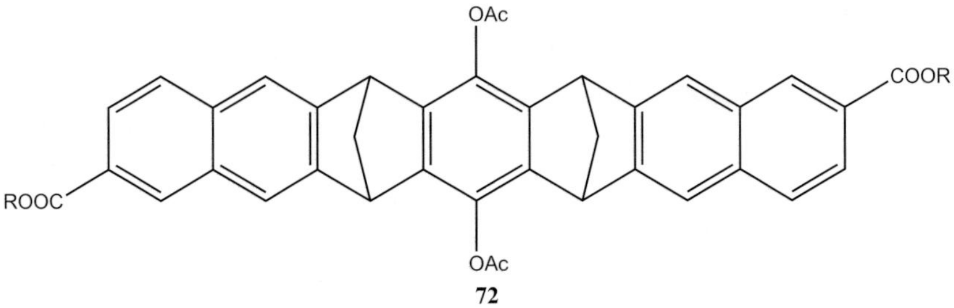

72

7 Artificial Receptor Compounds for Chiral Recognition

The water-soluble cyclophane **73**, which is derived from tartaric acid, causes enantiodifferentiation of carboxylic acids, 1-arylethanols, and aryl amines. Enantiomeric discrimination was observed in the ^{1}H NMR spectra and the aromatic rings of the cyclophane caused shielding of the guest substrate [187]. Compound **74** is a water-soluble cyclophane with a hydrophobic cavity. Enantiorecognition of substrates such as menthol and citronellol was observed in the ^{1}H NMR spectrum. The hydrophobic portion of the substrate associated in the cyclophane cavity [188]. Cryptophane **75** was specifically designed to accommodate and enantiomerically discriminate bromochlorofluoromethane. The ^{1}H resonance exhibited enantiodifferentiation in chloroform-d [189].

7.7
Metal-Containing Receptor Compounds

Metal complexes have been used extensively for chiral recognition, although few are designed in such a way that the metal site can be considered a true receptor cavity. Examples of metal complexes suitable for the analysis of chiral anions have been reviewed [158].

One interesting set of compounds are so-called zinc tweezer porphyrins, **76** being one example [190, 191]. Diamine substrates of suitable dimensions bind simultaneously at the two zinc ions and characteristic circular dichroism exciton spectra can be used for configurational assignments. Alternatively, mono amines can be derivatized with 4-N-Boc-aminomethyl-2-pyridine-carboxylic acid to create a diamine functionality of appropriate dimensions to bridge the zinc ions [190]. A zinc tweezer with biphenyl or naphthyl spacers showed excellent enantioselectivity toward lysine [192].

76

Zinc tweezer compounds have also been used as chiral NMR shift reagents. These are effective for compounds such as 1,2-diaminocyclohexane, aziridine, and isoxazoline, all of which can bind simultaneously to both zinc ions. The resonances of the substrates shift to exceptionally low frequencies (e.g., chemical shifts at -7 ppm) because of the large shielding of the porphyrin ring [191].

Compound **77** is a porphyrin dimeric ligand that when complexed with cobalt exists in a tweezer configuration. Differences in the circular dichroism spectra of limonene and *trans*-1,2-diaminocyclohexane were observed in the presence of **77**. The enantiodiscrimination of limonene was especially noteworthy. The porphyrin complex was also coated onto a quartz crystal microbalance and used for enantioselective gas-phase sensing of limonene [193].

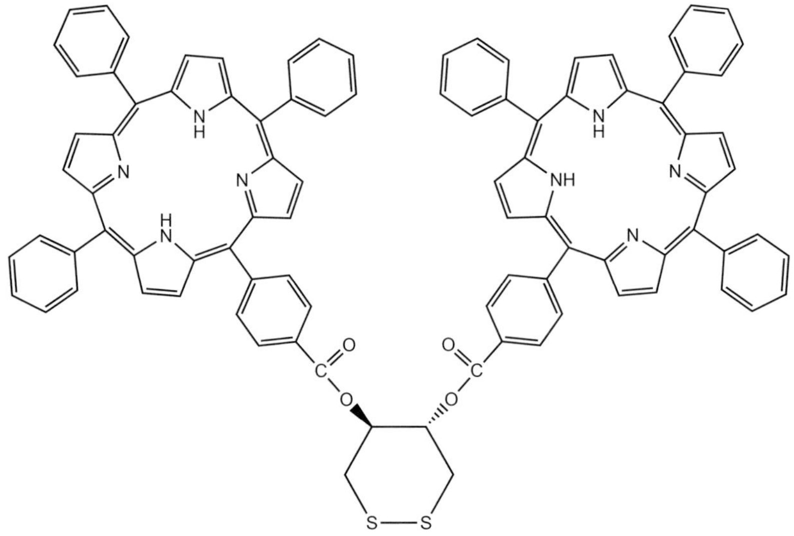

77

The bimetallic barium–zinc crown-salen complex (**78**) forms a folded geometry. The compound binds rigid bidentate guests such as amino acid esters in a sandwich-like conformation. Enantiodifferentiation was noted using UV-vis and circular dichroism spectra [194].

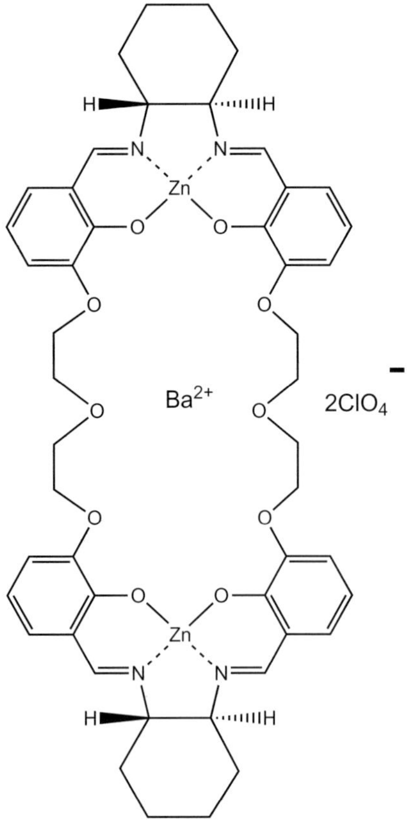

78

The rhodium complex [Cp*Rh(2'-deoxyadenosine)]$_3$(OTf)$_3$ (Cp* = η^5-C$_5$Me$_5$; OTf = CF$_3$SO$_3^-$) has a triangular dome-like cavity and can associate with suitable substrates through noncovalent π–π and hydrophobic interactions. The compound was a useful chiral NMR shift reagent for aromatic carboxylic acids, cyclohexyl acetic acids, and dipeptides and tripeptides [195].

Ruthenium receptor **79** contains a bipyridine-diamide moiety with two attached macrocyclic groups. Based on phosphorescence and ^{1}H NMR titrations, the complex shows selective binding of DD-dipeptides over the corresponding LD-, DL-, and LL-isomers. The system has also been used to screen 3375 different tripeptides in a 15 × 15 × 15 library. The tripeptides were bound on polystyrene beads and those with strong binding to the receptor could be identified colorimetrically. Stereoselective recognition occurred primarily for tripeptides with alanine residues [196].

79

References

1 Mitchell, C.R. and Armstrong, D.W. (2004) Cyclodextrin-based chiral stationary phases for liquid chromatography: a twenty-year overview in *Methods in Molecular Biology*, vol. 243 (eds G. Gubitz and M.G. Schmid), Humana Press, Totowa, NJ, pp. 61–112.

2 Choi, H.J. and Hyun, M.H. (2007) Liquid chromatographic chiral separations by crown ether-based chiral stationary phases. *J. Liq. Chromatogr. Relat. Technol.*, **30**, 853–875.

3 Wenzel, T.J. (2007) *Discrimination of Chiral Compounds Using NMR Spectroscopy*, John Wiley & Sons, Inc., Hoboken, NJ.

4 Schug, K.A. (2007) Solution phase enantioselective recognition and discrimination by electrospray ionization-mass spectrometry: state-of-the-art, methods, and an eye towards increased throughput measurements. *Comb. Chem. High Throughput Screening*, **10**, 301–316.

5 Ciucanu, I. and Kerek, F. (1984) A simple and rapid method for the permethylation

of carbohydrates. *Carbohydr. Res.*, **131**, 209–217.

6 Nowotny, H.-P., Schmalzing, D., Wistuba, D., and Schurig, V. (1989) Extending the scope of enantiomer separation on diluted methylated β-cyclodextrin derivatives by high-resolution gas chromatography. *J. High Resolut. Chromatogr.*, **12**, 383–393.

7 Kano, K. and Hasegawa, H. (2001) Chiral recognition of helical metal complexes by modified cyclodextrins. *J. Am. Chem. Soc.*, **123**, 10616–10627.

8 Uccello-Barretta, G., Balzano, F., Caporusso, A.M., Iodice, A., and Salvadori, P. (1995) Permethylated β-cyclodextrin as chiral solvating agent for the NMR assignment of the absolute configuration of chiral trisubstituted allenes. *J. Org. Chem.*, **60**, 2227–2231.

9 Wenz, G., Mischnick, P., Krebber, R., Richters, M., and Konig, W.A. (1990) Preparation and characterization of per-O-pentylated cyclodextrins. *J. High Resolut. Chromatogr.*, **13**, 724–728.

10 Konig, W.A., Lutz, S., Mischnick-Lubbecke, P., Brassat, B., von der Bey, E., and Wenz, G. (1988) Modified cyclodextrins – a new generation of chiral stationary phase for capillary gas chromatography. *Starch/Staerke*, **40**, 472–476.

11 Konig, W.A., Lutz, S., and Wenz, G. (1988) Modified cyclodextrins – novel highly enantioselective stationary phases for gas chromatography. *Angew. Chem. Int. Ed. Engl.*, **27**, 979–980.

12 Berthod, A., Li, W.Y., and Armstrong, D.W. (1990) Chiral recognition of racemic sugars by polar and nonpolar cyclodextrin-derivative gas chromatography. *Carbohydr. Res.*, **201**, 175–184.

13 Bicchi, C., Artuffo, G., D'Amato, A., Manzin, V., Galli, A., and Galli, M. (1992) Cyclodextrin derivatives in the GC separation of racemic mixtures of volatile compounds. Part V: heptakis 2,6-dimethyl-3-pentyl-β-cyclodextrins. *J. High Resolut. Chromatogr.*, **15**, 710–714.

14 Rong, D. and D'Souza, V.T. (1990) A convenient method for functionalization of the 2-position of cyclodextrins. *Tetrahedron Lett.*, **31**, 4275–4278.

15 Konig, W.A., Icheln, D., Runge, T., Pforr, I., and Krebs, A. (1990) Cyclodextrins as chiral stationary phases in capillary gas chromatography. Part VII: cyclodextrins with an inverse substitution pattern – synthesis and enantioselectivity. *J. High Resolut. Chromatogr.*, **13**, 702–707.

16 Konig, W.A., Icheln, D., Runge, T., Pfaffenberger, B., Ludwig, P., and Huhnerfuss, H. (1991) Gas chromatographic enantiomer separation of agrochemicals using modified cyclodextrins. *J. High Resolut. Chromatogr.*, **14**, 530–536.

17 Cousin, H., Trapp, O., Peulon-Agasse, V., Pannecoucke, X., Banspach, L., Trapp, G., Jiang, Z., Combret, J.C., and Schurig, V. (2003) Synthesis, NMR spectroscopic characterization and polysiloxane-based immobilization of the three regioisomeric monooctenylpermethyl-β-cyclodextrins and their application in enantioselective GC. *Eur. J. Org. Chem.*, 3273–3287.

18 Cousin, H., Cardinael, P., Oulyadi, H., Pannecoucke, X., and Combret, J.C. (2001) Synthesis of three isomeric mono-2-, 3-, or 6-hydroxy permethylated β-cyclodextrins and unambiguous high field NMR characterization. *Tetrahedron: Asymmetry*, **12**, 81–88.

19 Poon, Y.-F., Muderawan, I.W., and Ng, S.-C. (2006) Synthesis and application of mono-2^A-azido-2^A-deoxyperphenylcarbamoylated β-cyclodextrin and mono-2^A-azido-2^A-deoxyperacetylated β-cyclodextrin as chiral stationary phases for high-performance liquid chromatography. *J. Chromatogr. A*, **1101**, 185–197.

20 Cramer, F., Mackensen, G., and Sensse, K. (1969) Clathrate compounds. XX. Optical rotary dispersion spectra and conformation of glucose units in cyclodextrins. *Chem. Ber.*, **102**, 494–508.

21 Li, W.-Y., Jin, H.L., and Armstrong, D.W. (1990) 2,6-Di-O-pentyl-3-O-trifluoroacetyl cyclodextrin liquid stationary phases for capillary gas chromatographic separation of enantiomers. *J. Chromatogr.*, **509**, 303–324.

22 Kieser, B., Fietzek, C., Schmidt, R., Belge, G., Weimar, U., Schurig, V., and Gauglitz, G. (2002) Use of a modified cyclodextrin host for the enantioselective detection of a halogenated diether as chiral guest via optical and electrical transducers. *Anal. Chem.*, **74**, 3005–3012.

23 Hierlemann, A., Ricco, A.J., Bodenhofer, K., and Gopel, W. (1999) Effective use of molecular recognition in gas sensing: results from acoustic wave and in situ FT-IR measurements. *Anal. Chem.*, **71**, 3022–3035.

24 Kurzawski, P., Bogdanski, A., Schurig, V., Wimmer, R., and Hierlemann, A. (2008) Opposite signs of capacitive microsensor signals upon exposure to the enantiomers of methyl propionate compounds. *Angew. Chem. Int. Ed.*, **47**, 913–916.

25 Armstrong, D.W., Stalcup, A.M., Hilton, M.L., Duncan, J.D., Faulkner, J.R. Jr., and Chang, S.-C. (1990) Derivatized cyclodextrins for normal-phase liquid chromatographic separation of enantiomers. *Anal. Chem.*, **62**, 1610–1615.

26 Aburatani, R., Okamoto, Y., and Hatada, K. (1990) Optical resolving ability of 3,5-dimethylphenylcarbamates of oligosaccharides and cyclodextrins. *Bull. Chem. Soc. Jpn.*, **63**, 3606–3610.

27 Hargitai, T. and Okamoto, Y. (1993) Evaluation of 3,5-dimethylphenyl carbamoylated α-, β-, and γ-cyclodextrins as chiral stationary phases for HPLC. *J. Liq. Chromatogr.*, **16**, 843–858.

28 Takeichi, T., Toriyama, H., Shimura, S., Takayama, Y., and Morikawa, M. (1995) Cyclodextrin carbamates as novel chiral stationary phases for capillary gas chromatography. *J. High Resolut. Chromatogr.*, **18**, 179–189.

29 Uccello-Barretta, G., Ferri, L., Balzano, F., and Salvadori, P. (2003) Partially versus exhaustively carbamoylated cyclodextrins: NMR investigation on enantiodiscriminating capabilities in solution. *Eur. J. Org. Chem.*, 1741–1748.

30 Stalcup, A.M., Chang, S.-C., Armstrong, D.W., and Pitha, J. (1990) (S)-2-Hydroxypropyl-β-cyclodextrin, a new chiral stationary phase for reversed-phase liquid chromatography. *J. Chromatogr.*, **513**, 181–194.

31 Takeo, K., Uemura, K., and Mitoh, H. (1988) Derivatives of α-cyclodextrin and the synthesis of 6-O-α-D-glucopyranosyl-α-cyclodextrin. *J. Carbohydr. Chem.*, **7**, 293–308.

32 Takeo, K., Mitoh, H., and Uemura, K. (1989) Selective chemical modification of cyclomalto-oligosaccharides via *tert*-butyldimethylsilylation. *Carbohydr. Res.*, **187**, 203–221.

33 Dietrich, A., Maas, B., Karl, V., Kreis, P., Lehmann, D., Weber, B., and Mosandl, A. (1992) Stereoisomeric flavor compounds. Part LV: stereodifferentiation of some chiral volatiles on heptakis(2,3-di-O-acetyl-6-O-tert-butyldimethylsilyl)-β-cyclodextrin. *J. High Resolut. Chromatogr.*, **15**, 176–179.

34 Dietrich, A., Maas, B., Messer, W., Bruche, G., Karl, V., Kaunzinger, A., and Mosandl, A. (1992) Stereoisomeric flavor compounds. Part LVIII: the use of heptakis(2,3-di-O-methyl-6-O-tert-butyldimethylsilyl)-β-cyclodextrin as a chiral stationary phase in flavor analysis. *J. High Resolut. Chromatogr.*, **15**, 590–593.

35 Miranda, E., Sanchez, F., Sanz, J., Jimenez, M.I., and Martinez-Castro, I. (1998) 2,3-Di-O-pentyl-6-O-tert-butyldimethylsilyl-β-cyclodextrin as a chiral stationary phase in capillary gas chromatography. *J. High Resolut. Chromatogr.*, **21**, 225–233.

36 Chankvetadze, B., Endresz, G., and Blaschke, G. (1996) Charged cyclodextrin derivatives as chiral selectors in capillary electrophoresis. *Chem. Soc. Rev.*, **25**, 141–153.

37 Stella, V. and Rajewski, R. (1992) Derivatives of cyclodextrins exhibiting enhanced aqueous solubility and the use thereof, U.S. Patent 5,134,127.

38 Owens, P.K., Fell, A.F., Coleman, M.W., and Berridge, J.C. (1996) Method development in liquid chromatography with a charged cyclodextrin additive for chiral resolution of *rac*-amlodipine utilizing a central composite design. *Chirality*, **8**, 466–476.

39 Talt, R.J., Thompson, D.O., Stella, V.J., and Stobaugh, J.F. (1994) Sulfobutyl ether β-cyclodextrin as a chiral discriminator for use with capillary electrophoresis. *Anal. Chem.*, **66**, 4013–4018.

40 Tanaka, Y., Yanagawa, M., and Terabe, S. (1996) Separation of neutral and basic enantiomers by cyclodextrin electrokinetic chromatography using anionic cyclodextrin derivatives as chiral pseudo-stationary phases. *J. High Resolut. Chromatogr.*, **19**, 421–433.

41 Bernstein, S., Joseph, J.P., and Nair, V. (1977) Cyclodextrin sulfate salts as complement inhibitors, U.S. Patent 4,020,160.

42 Stalcup, A.M. and Gahm, K.H. (1996) A sulfated cyclodextrin chiral stationary phase for high-performance liquid chromatography. *Anal. Chem.*, **68**, 1369–1374.

43 Vincent, J.B., Sokolowski, A.D., Nguyen, T.V., and Vigh, G. (1997) A family of single-isomer chiral resolving agents for capillary electrophoresis. 1. Heptakis(2,3-diacetyl-6-sulfato)-β-cyclodextrin. *Anal. Chem.*, **69**, 4226–4233.

44 Zhu, W. and Vigh, G. (1997) A family of single-isomer, sulfated γ-cyclodextrin chiral resolving agents for capillary electrophoresis. 1. Octakis(2,3-diacetyl-6-sulfato)-γ-cyclodextrin. *Anal. Chem.*, **69**, 310–317.

45 Wenzel, T.J., Amonoo, E.P., Shariff, S.S., and Aniagyei, S.E. (2003) Sulfated and carboxymethylated cyclodextrins and their lanthanide complexes as chiral NMR discriminating agents. *Tetrahedron: Asymmetry*, **14**, 3099–3104.

46 Reuben, J., Rao, C.T., and Pitha, J. (1994) Distribution of substituents in carboxymethyl ethers of cyclomaltoheptaose. *Carbohydr. Res.*, **258**, 281–285.

47 Dignam, C.F., Randall, L.A., Blacken, R.D., Cunningham, P.R., Lester, S.-K.G., Brown, M.J., French, S.C., Aniagyei, S.E., and Wenzel, T.J. (2006) Carboxymethylated cyclodextrin derivatives as chiral NMR discriminating agents. *Tetrahedron: Asymmetry*, **17**, 1199–1208.

48 Provencher, K.A. and Wenzel, T.J. (2008) Carboxymethylated cyclodextrins and their paramagnetic lanthanide complexes as water-soluble chiral NMR solvating agents. *Tetrahedron: Asymmetry*, **19**, 1797–1803.

49 Smith, K.J., Wilcox, J.D., Mirick, G.E., Wacker, L.S., Ryan, N.S., Vensel, D.A., Readling, R., Domush, H.L., Amonoo, E.P., Shariff, S.S., and Wenzel, T.J. (2003) Calix[4]arene, calix[4]resorcarene, and cyclodextrin derivatives and their lanthanide complexes as chiral NMR shift reagents. *Chirality*, **15**, S150–S158.

50 Kano, K. and Hasegawa, H. (2000) Chiral recognition of Ru(phen)$_3^{2+}$ by anionic cyclodextrins. *Chem. Lett.*, 698–699.

51 Parmerter, S.M., Allen, E.E. Jr., and Hull, G.A. (1969) Cyclodextrin with cationic properties, U.S. Patent 3,453,257.

52 Schulte, G., Chankvetadze, B., and Blaschke, G. (1997) Enantioseparation in capillary electrophoresis using 2-hydroxypropyltrimethylammonium salt of β-cyclodextrin as a chiral selector. *J. Chromatogr. A*, **771**, 259–266.

53 Bunke, A. and Jira, Th. (1996) Chiral capillary electrophoresis using a cationic cyclodextrin. *Pharmazie*, **51**, 672–673.

54 Roussel, C. and Favrou, A. (1995) Cationic β-cyclodextrin: a new versatile chiral additive for separation of drug enantiomers by high-performance liquid chromatography. *J. Chromatogr. A*, **704**, 67–74.

55 O'Keeffe, F., Shamsi, S.A., Darcy, R., Schwinte, P., and Warner, I.M. (1997) A persubstituted cationic β-cyclodextrin for chiral separations. *Anal. Chem.*, **69**, 4773–4782.

56 Haynes, J.L. III, Shamsi, S.A., O'Keefe, F., Darcey, R., and Warner, I.M. (1998) Cationic β-cyclodexin derivative for chiral separations. *J. Chromatogr. A*, **803**, 261–271.

57 Tang, W., Muderawan, I.W., Ong, T.T., and Ng, S.-C. (2005) A family of single-isomer positively charged cyclodextrins as chiral selectors for capillary electrophoresis: mono-6^A-butylammonium-6^A-deoxy-β-cyclodextrin tosylate. *Electrophoresis*, **26**, 3125–3133.

58 Boger, J., Corcoran, R.J., and Lehn, J.-M. (1978) Cyclodextrin chemistry. Selective modification of all primary hydroxyl groups of α- and β-cyclodextrins. *Helv. Chim. Acta*, **61**, 2190–2218.

59 Brown, S.E., Coates, J.H., Duckworth, P.A., Lincoln, S.F., Easton, C.J., and May, B.L. (1993) Substituent effects and chiral discrimination in the complexation of benzoic, 4-methylbenzoic and (RS)-2-

phenylpropanoic acids and their conjugate bases by β-cyclodextrin and 6^A-amino-6^A-deoxy-β-cyclodextrin in aqueous solution: potentiometric titration and ^{1}H nuclear magnetic resonance spectroscopic study. *J. Chem. Soc. Faraday Trans.*, **89**, 1035–1040.

60 Kano, K., Kitae, T., and Takashima, H. (1996) Use of electrostatic interaction for chiral recognition. Enantioselective complexation of anionic binaphthyls with protonated amino-β-cyclodextrin. *J. Inclusion Phenom. Mol. Recognit. Chem.*, **25**, 243–248.

61 Rekharsky, M.V. and Inoue, Y. (2002) Complexation and chiral recognition thermodynamics of 6-amino-6-deoxy-β-cyclodextrin with anionic, cationic, and neutral chiral guests: counterbalance between van der Waals and coulombic interactions. *J. Am. Chem. Soc.*, **124**, 813–826.

62 Kitae, T., Nakayama, T., and Kano, K. (1998) Chiral recognition of α-amino acids by charged cyclodextrins through cooperative effects of Coulomb interaction and inclusion. *J. Chem. Soc., Perkin Trans. 2*, 207–212.

63 Wenzel, T.J., Bogyo, M.S., and Lebeau, E.L. (1994) Lanthanide-cyclodextrin complexes as probes for elucidating optical purity by NMR spectroscopy. *J. Am. Chem. Soc.*, **116**, 4858–4865.

64 Wenzel, T.J., Miles, R.D., Zomlefer, K., Frederique, D.E., Roan, M.A., Troughton, J.S., Pond, B.V., and Colby, A.L. (2007) Dysprosium(III)-diethylenetriaminepentaacetate complexes of aminocyclodextrins as chiral NMR shift reagents. *Chirality*, **12**, 30–37.

65 Gong, Y., Xiang, Y., Yue, B., Xue, G., Bradshaw, J.S., Lee, H.K., and Lee, M.L. (2003) Application of diaza-18-crown-6-capped β-cyclodextrin bonded silica particles as chiral stationary phases for ultrahigh pressure capillary liquid chromatography. *J. Chromatogr. A*, **1002**, 63–70.

66 Lovely, A.E. and Wenzel, T.J. (2006) Chiral NMR discrimination of secondary amines using (18-crown-6)-2,3,11,12-tetracarboxylic acid. *Org. Lett.*, **8**, 2823–2826.

67 Lovely, A.E. and Wenzel, T.J. (2006) Chiral NMR discrimination of pyrrolidines using (18-crown-6)-2,3,11,12-tetracarboxylic acid. *Tetrahedron: Asymmetry*, **17**, 2642–2648.

68 Lovely, A.E. and Wenzel, T.J. (2006) Chiral NMR discrimination of piperidines and piperazines using (18-crown-6)-2,3,11,12-tetracarboxylic acid. *J. Org. Chem.*, **71**, 9178–9182.

69 Lovely, A.E. and Wenzel, T.J. (2008) Chiral NMR discrimination of amines: analysis of secondary, tertiary and prochiral amines using (18-crown-6)-2,3,11,12-tetracarboxylic acid. *Chirality*, **20**, 370–378.

70 Zhang, X.X., Bradshaw, J.S., and Izatt, R.M. (1997) Enantiomeric recognition of amine compounds by chiral macrocyclic receptors. *Chem. Rev.*, **97**, 3313–3361.

71 Kyba, E.B., Koga, K., Sousa, L.R., Siegel, M.G., and Cram, D.J. (1973) Chiral recognition in molecular complexing. *J. Am. Chem. Soc.*, **95**, 2692–2693.

72 Sousa, L.R., Hoffman, D.H., Kaplan, L., and Cram, D.J. (1974) Total optical resolution of amino esters by designed host-guest relationships in molecular complexation. *J. Am. Chem. Soc.*, **96**, 7100–7101.

73 Dotsevi, G., Sogah, Y., and Cram, D.J. (1975) Chromatographic optical resolution through chiral complexation of amino ester salts by a host covalently bound to silica gel. *J. Am. Chem. Soc.*, **97**, 1259–1261.

74 Lingenfelter, D.S., Helgeson, R.C., and Cram, D.J. (1981) Host-guest complexation. 23. High chiral recognition of amino acid and ester guests by hosts containing one chiral element. *J. Org. Chem.*, **46**, 393–406.

75 Cram, D.J. (1988) The design of molecular hosts, guests, and their complexes. *Science*, **240**, 760–767.

76 Hyun, M.H., Han, S.C., Lipshutz, B.H., Shin, Y.-J., and Welch, C.J. (2001) New chiral crown ether stationary phase for the liquid chromatographic resolution of α-amino acid enantiomers. *J. Chromatogr. A*, **910**, 359–365.

77 Stoddart, J.F. (1979) From carbohydrates to enzyme analogues. *Chem. Soc. Rev.*, **8**, 85–142.

78 Joly, J.-P. and Moll, N. (1990) Resolution of free aromatic amino acid enantiomers by host-guest complexation using reversed-phase liquid chromatography. *J. Chromatogr.*, **521**, 134–140.

79 Weinstein, S.E., Vining, M.S., and Wenzel, T.J. (1997) Lanthanide-crown ether mixtures as chiral NMR shift reagents for amino acid esters, amines and amino alcohols. *Magn. Reson. Chem.*, **35**, 273–280.

80 Joly, J., Nazhaoui, M., and Dumont, B. (1994) Synthesis and complexation behavior of some crown ethers derived from D-hexopyranosides and D-mannitol towards racemic phenylglycine salts. *Bull. Soc. Chim. Fr.*, **131**, 369–380.

81 Wenzel, T.J., Thurston, J.E., Sek, D.C., and Joly, J.-P. (2001) Utility of crown ethers derived from methyl β-D-galactopyranoside and their lanthanide couples as chiral NMR discriminating agents. *Tetrahedron: Asymmetry*, **12**, 1125–1130.

82 Girodeau, J.-M., Lehn, J.-M., and Sauvage, J.-P. (1975) A polyfunctional chiral macrocyclic polyether derived from L-(+)-tartaric acid. *Angew. Chem. Intl. Ed. Engl.*, **14**, 764.

83 Behr, J.-P., Girodeau, J.-M., Hayward, R.C., Lehn, J.-M., and Sauvage, J.-P. (1980) Molecular receptors. Functionalized and chiral macrocyclic polyethers derived from tartaric acid. *Helv. Chim. Acta*, **63**, 2096–2111.

84 Anantanarayan, A., Carmicheal, V.A., Dutton, P.J., Fyles, T.M., and Pitre, M.J. (1986) Synthesis of crown ethers from (+)-tartaric acid: improved procedures. *Synth. Commun.*, **16**, 1771–1776.

85 Behr, J.-P., Lehn, J.-M., Moras, D., and Thierry, J.C. (1981) Chiral and functionalized face-discriminated and side-discriminated macrocyclic polyethers. Syntheses and crystal structures. *J. Am. Chem. Soc.*, **103**, 701–703.

86 Behr, J.-P., Lehn, J.-M., and Vierling, P. (1982) Molecular receptors. Structural effects and substrate recognition in binding of organic and biogenic ammonium ions by chiral polyfunctional macrocyclic polyethers bearing amino-acid and other side-chains. *Helv. Chim. Acta*, **65**, 1853–1867.

87 Gehin, D., Kollman, P.A., and Wipff, G. (1989) Anchoring of ammonium cations to an 18-crown-6 binding site: molecular mechanics and dynamics study. *J. Am. Chem. Soc.*, **111**, 3011–3023.

88 Yasaka, Y., Yamamoto, T., Kimura, K., and Shono, T. (1980) Simple evaluation of enantiomer-selectivity of crown ether using membrane electrode. *Chem. Lett.*, 769–772.

89 Machida, Y., Nishi, H., Nakamura, K., Nakai, H., and Sato, T. (1998) Enantiomer separation of amino compounds by a novel chiral stationary phase derived from crown ether. *J. Chromatogr. A*, **805**, 85–92.

90 Hyun, M.H., Jin, J.S., and Lee, W. (1998) Liquid chromatographic resolution of racemic amino acids and their derivatives on a new chiral stationary phase based on crown ether. *J. Chromatogr. A*, **822**, 155–161.

91 Hyun, M.H. (2003) Characterization of liquid chromatographic chiral separation on chiral crown ether stationary phase. *J. Sep. Sci.*, **26**, 242–250.

92 Steffeck, R.J., Zelechonok, U., and Gahm, K.H. (2002) Enantioselective separation of racemic secondary amines on a chiral crown ether-based liquid chromatography stationary phase. *J. Chromatogr. A*, **947**, 301–305.

93 Hyun, M.H. (2006) Preparation and application of HPLC chiral stationary phases based on (+)-(18-crown-6)-2,3,11,12-tetracarboxylic acid. *J. Sep. Sci.*, **29**, 750–761.

94 Wenzel, T.J. and Thurston, J.E. (2000) (+)-(18-Crown-6)-2,3,11,12-tetracarboxylic acid and its ytterbium(III) complex as chiral NMR discriminating agents. *J. Org. Chem.*, **65**, 1243–1248.

95 Wenzel, T.J. and Thurston, J.E. (2000) Enantiomeric discrimination in the NMR spectra of underivatized amino acids and α-methyl amino acids using (+)-(18-crown-6)-2,3,11,12-tetracarboxylic acid. *Tetrahedron Lett.*, **41**, 3769–3772.

96 Wenzel, T.J., Freeman, B.E., Sek, D.C., Zopf, J.J., Nakamura, T., Yongzhu, J., Hirose, K., and Tobe, Y. (2004) Chiral recognition in NMR spectroscopy using crown ethers and their ytterbium(III)

complexes. *Anal. Bioanal. Chem.*, **378**, 1536–1547.

97 Machida, Y., Kagawa, M., and Nishi, H. (2003) Nuclear magnetic resonance studies for the chiral recognition of (+)-(R)-18-crown-6-tetracarboxylic acid to amino compounds. *J. Pharm. Biomed. Anal.*, **30**, 1929–1942.

98 Kaneda, T., Ishizaki, Y., Misumi, S., Kai, Y., Hirao, G., and Kasai, N. (1988) Synthesis, coloration, and crystal structure of the "dibasic" chromoacerand-piperazine 1:1 salt complex. *J. Am. Chem. Soc.*, **110**, 2970–2972.

99 Sawada, M., Okumura, Y., Shizuma, M., Takai, Y., Hidaka, Y., Yamada, H., Tanaka, T., Kaneda, T., Hirose, K., Misumi, S., and Takahashi, S. (1993) Enantioselective complexation of carbohydrate or crown ether hosts with organic ammonium ion guests detected by FAB mass spectrometry. *J. Am. Chem. Soc.*, **115**, 7381–7388.

100 Naemura, K., Asada, M., Hirose, K., and Tobe, Y. (1995) Preparation of enantiomeric recognition of chiral azophenolic crown ethers having three chiral barriers on each of the homotopic faces. *Tetrahedron: Asymmetry*, **6**, 1873–1876.

101 Naemura, K., Matsunaga, K., Fuji, J., Ogasahara, K., Nishikawa, Y., Hirose, K., and Tobe, Y. (1998) Temperature dependence of enantioselectivity in complexations of optically active phenolic crown ethers with chiral amines in solution. *Anal. Sci.*, **14**, 175–182.

102 Naemura, K., Nishioka, K., Ogasahara, K., Nishikawa, Y., Hirose, K., and Tobe, Y. (1998) Preparation and temperature-dependent enantioselectivities of homochiral phenolic crown ethers having aryl chiral barriers: thermodynamic parameters for enantioselective complexation with chiral amines. *Tetrahedron: Asymmetry*, **9**, 563–574.

103 Hirose, K., Goshima, Y., Wakebe, T., Tobe, Y., and Naemura, K. (2007) Supramolecular method for the determination of absolute configuration of chiral compounds: theoretical derivatization and a demonstration for a phenolic crown ether–2-amino-1-ethanol system. *Anal. Chem.*, **79**, 6295–6302.

104 Sawada, M., Takai, Y., Yamada, H., Hirayama, S., Kaneda, T., Tanaka, T., Kamada, K., Mizooku, T., Takeuchi, S., Ueno, K., Hirose, K., Tobe, Y., and Naemura, K. (1995) Chiral recognition in host-guest complexation determined by the enantiomer-labeled guest method using fast atom bombardment mass spectrometry. *J. Am. Chem. Soc.*, **117**, 7726–7736.

105 Sawada, M., Takai, Y., Yamada, H., Nishida, J., Kaneda, T., Arakawa, R., Okamoto, M., Hirose, K., Tanaka, T., and Naemura, K. (1998) Chiral amino acid recognition detected by electrospray ionization (ESI) and fast atom bombardment (FAB) mass spectrometry (MS) coupled with the enantiomer-labelled (EL) guest method. *J. Chem. Soc., Perkin Trans. 2*, 701–710.

106 Sawada, M., Takai, Y., Imamura, H., Yamada, H., Takahashi, S., Yamaoka, H., Hirose, K., Tobe, Y., and Tanaka, J. (2001) Chiral recognizable host-guest interactions detected by fast-atom bombardment mass spectrometry: application to the enantiomeric excess determination of primary amines. *Eur. J. Mass Spectrom.*, **7**, 447–459.

107 Hirose, K., Fujiwara, A., Matsunaga, K., Aoki, N., and Tobe, Y. (2002) Preparation of phenolic chiral crown ethers and podands and their enantiomer recognition ability toward secondary amines. *Tetrahedron: Asymmetry*, **14**, 555–566.

108 Tsubaki, K., Tanima, D., Nuruzzaman, M., Kusumoto, T., Fuji, K., and Kawabata, T. (2005) Visual enantiomeric recognition of amino acid derivatives in protic solvents. *J. Org. Chem.*, **70**, 4609–4616.

109 Bradshaw, J.S., Jolley, S.T., and Izatt, R.M. (1982) Preparation of chiral diphenyl-substituted polyether-diester compounds. *J. Org. Chem.*, **47**, 1229–1232.

110 Davidson, R.B., Bradshaw, J.S., Jones, B.A., Dalley, N.K., Christensen, J.J., Izatt, R.M., Morin, F.G., and Grant, D.M. (1984) Enantiomeric recognition of organic ammonium salts by chiral crown ethers based on the pyridine-18-crown-6 structure. *J. Org. Chem.*, **49**, 353–357.

111 Habata, Y., Bradshaw, J.S., Young, J.J., Castle, S.L., Huszthy, P., Pyo, T., Lee, M.L.,

and Izatt, R.M. (1996) New pyridino-18-crown-6 ligands containing two methyl, two *tert*-butyl, or two allyl substituents on chiral positions next to the pyridine ring. *J. Org. Chem.*, **61**, 8391–8396.
112 Pocsfalvi, G., Liptak, M., Huszthy, P., Bradshaw, J.S., Izatt, R.M., and Vekey, K. (1996) Characterization of chiral host-guest complexation in fast atom bombardment mass spectrometry. *Anal. Chem.*, **68**, 792–795.
113 Kontos, Z., Huszthy, P., Bradshaw, J.S., and Izatt, R.M. (1999) Enantioseparation of racemic organic ammonium perchlorates by a silica gel bound optically active di-*tert*-butylpyridino-18-crown-6-ligand. *Tetrahedron: Asymmetry*, **10**, 2087–2099.
114 Lakatos, S., Fetter, J., Bertha, F., Huszthy, P., Toth, T., Farkas, V., Orosz, G., and Hollosi, M. (2008) Preparation of a new chiral acridino-18-crown-6 ether-based stationary phase for enantioseparation of racemic protonated primary aralkyl amines. *Tetrahedron*, **64**, 1012–1022.
115 Gutsche, C.D. (1989) *Calixarenes*, Royal Society of Chemistry, Cambridge, UK.
116 Vysotsky, M., Schmidt, C., and Bohmer, V. (2000) Chirality in calixarenes and calixarene assemblies. *Adv. Supramol. Chem.*, **7**, 139–233.
117 Brewster, R.E., Caran, K.L., Sasine, J.S., and Shuker, S.B. (2004) Peptidocalixarenes in molecular recognition. *Curr. Org. Chem.*, **8**, 867–881.
118 Pena, M.S., Zhang, Y., and Warner, I.M. (1997) Enantiomeric separations by use of calixarene electrokinetic chromatography. *Anal. Chem.*, **69**, 3239–3242.
119 Jennings, K. and Diamond, D. (2001) Enantioselective molecular sensing of aromatic amines using tetra-(*S*)-di-2-naphthylprolinol calix[4]arene. *Analyst*, **126**, 1063–1067.
120 Lynam, C., Jennings, K., Nolan, K., Kane, P., McKervey, M.A., and Diamond, D. (2002) Tuning and enhancing enantioselective quenching of calixarene hosts by chiral guest amines. *Anal. Chem.*, **74**, 59–66.
121 Guo, W., Wang, J., Wang, C., He, J.-Q., He, X., and Cheng, J.-P. (2002) Design, synthesis, and enantiomeric recognition of dicyclopeotide-bearing calix[4]arenes: a promising family for chiral gas sensor coatings. *Tetrahedron Lett.*, **43**, 5665–5667.
122 Kocabas, E., Durmaz, M., Alpaydin, S., Sirit, A., and Yilmaz, M. (2008) Chiral mono and diamide derivatives of calix[4]arenes for enantiomeric recognition of chiral amines. *Chirality*, **20**, 26–34.
123 Shirakawa, S., Moriyama, A., and Shimizu, S. (2007) Design of a novel inherently chiral calix[4]arene for chiral molecular recognition. *Org. Lett.*, **9**, 3117–3119.
124 Erdemir, S., Tabakci, M., and Yilmaz, M. (2006) Synthesis and chiral recognition abilities of new calix[6]arenes bearing amino alcohol moieties. *Tetrahedron: Asymmetry*, **17**, 1258–1263.
125 Durmaz, M., Alpaydin, S., Sirit, A., and Yilmaz, M. (2006) Chiral Schiff base derivatives of calix[4]arene: synthesis and complexation studies with chiral and achiral amines. *Tetrahedron: Asymmetry*, **17**, 2322–2327.
126 Gu, J., He, W., Shi, X., and Ji, L. (2008) New chiral calixarene derivatives: synthesis and their chiral recognition toward amino acids by UV-Vis spectroscopy. *Chem. Res. Chin. Univ.*, **24**, 106–109.
127 Haino, T., Fukuoka, H., Iwamoto, H., and Fukazawa, Y. (2008) Synthesis and enantioselective recognition of a calix[5]arene-based chiral receptor. *Supramol. Chem.*, **20**, 51–57.
128 Garrier, E., Le Gac, S., and Jabin, I. (2005) First enantiopure calix[6]aza-cryptand: synthesis and chiral recognition properties towards neutral molecules. *Tetrahedron: Asymmetry*, **16**, 3767–3771.
129 Kumagai, H., Hasegawa, M., Miyanari, S., Sugawa, Y., Sato, Y., Hori, T., Ueda, S., Kamiyama, H., and Miyano, S. (1997) Facile synthesis of a *p-tert*-butylthiacalix[4]arene by the reaction of *p-tert*-butylphenol with elemental sulfur in the presence of a base. *Tetrahedron Lett.*, **38**, 3971–3972.
130 Narumi, F., Iki, N., Suzuki, T., Onodera, T., and Miyano, S. (2000) Syntheses of chirally modified thiacalix[4]arenes with enantiomeric amines and their application to chiral stationary phases for gas chromatography. *Enantiomer*, **5**, 83–93.

131 Ito, K., Noike, M., Kida, A., and Ohba, Y. (2002) Syntheses of chiral homoazacalix[4]arenes incorporating amino acid residues: molecular recognition for racemic quaternary ammonium ions. *J. Org. Chem.*, **67**, 7519–7522.

132 Okada, Y., Ishii, F., Kasai, Y., and Nishimura, J. (1992) Synthesis and characterization of a calixarene analog locked in the cone conformation. *Chem. Lett.*, 755–758.

133 Okada, Y., Kasai, Y., and Nishimura, J. (1995) The selective extraction and transport of amino acids by calix[4]arene-derived esters. *Tetrahedron Lett.*, **36**, 555–558.

134 Okada, Y., Mizutani, M., Ishii, F., and Nishimura, J. (1997) A new class of chiral calix[4]arenes as receptors with planar chirality. *Tetrahedron Lett.*, **38**, 9013–9016.

135 Kubo, Y., Hirota, N., Maeda, S., and Tokita, S. (1998) Naked-eye detectable chiral recognition using a chromogenic receptor. *Anal. Sci.*, **14**, 183–189.

136 Luo, J., Zheng, Q.-Y., Chen, C.-F., and Huang, Z.-T. (2005) Facile synthesis and optical resolution of inherently chiral fluorescent calix[4]crowns: enantioselective recognition towards chiral leucinol. *Tetrahedron*, **61**, 8517–8528.

137 Liu, X.-X. and Zheng, Y.-S. (2006) Chiral nitrogen-containing calix[4]crown – an excellent receptor for chiral recognition of mandelic acid. *Tetrahedron Lett.*, **47**, 6357–6360.

138 He, Y., Xiao, Y., Meng, L., Zeng, Z., Wu, X., and Wu, C.-T. (2002) New type chiral calix[4]crowns: synthesis and chiral recognition. *Tetrahedron Lett.*, **43**, 6249–6253.

139 Rebek, J. Jr. (2000) Host-guest chemistry of calixarene capsules. *Chem. Commun.*, 637–643.

140 Bohmer, V., Kraft, D., and Tabatabai, M. (1994) Inherently chiral calixarenes. *J. Inclusion Phenom. Mol. Recognit. Chem.*, **19**, 17–39.

141 Mehdizadeh, A., Letzel, M.C., Klaes, M., Agena, C., and Mattay, J. (2004) Chiral discrimination on the host-guest complexation of resorc[4]arenes with quarternary amines. *Eur. J. Mass Spectrom.*, **10**, 649–655.

142 Pietraszkiewicz, M., Prus, P., and Fabionowski, W. (1998) Chiral recognition studies of amino acids by chiral calix[4]resorcinarenes in Langmuir films. *Pol. J. Chem.*, **72**, 1068–1075.

143 Iwanek, W. and Urganiak, M. (1999) Chiral calixarenes derived from resorcinol. Part 6. Chiral discrimination by resorcarenes. *Pol. J. Chem.*, **73**, 2067–2072.

144 Yanigihara, R., Tominaga, M., and Aoyama, Y. (1994) Chiral host-guest interaction. A water-soluble calix[4]resorcarene having L-proline moieties as a non-lanthanide chiral NMR shift reagent for chiral aromatic guests in water. *J. Org. Chem.*, **59**, 6865–6867.

145 Dignam, C.F., Richards, C.J., Zopf, J.J., Wacker, L.S., and Wenzel, T.J. (2005) An enantioselective NMR shift reagent for cationic aromatics. *Org. Lett.*, **7**, 1773–1776.

146 Dignam, C.F., Zopf, J.J., Richards, C.J., and Wenzel, T.J. (2005) Water-soluble calix[4]resorcarenes as enantioselective NMR shift reagents for aromatic compounds. *J. Org. Chem.*, **70**, 8071–8078.

147 O'Farrell, C.M., Chudomel, J.M., Collins, J.M., Dignam, C.F., and Wenzel, T.J. (2008) Water-soluble calix[4]resorcinarenes with hydroxyproline groups as chiral NMR solvating agents. *J. Org. Chem.*, **73**, 2843–2851.

148 O'Farrell, C.M. and Wenzel, T.J. (2008) Water-soluble calix[4]resorcinarenes as chiral NMR solvating agents for phenyl-containing compounds. *Tetrahedron: Asymmetry*, **19**, 1790–1796.

149 O'Farrell, C.M., Hagan, K.A., and Wenzel, T.J. (2009) Water-soluble calix[4]resorcinarenes as chiral NMR solvating agents for bicyclic aromatic compounds. *Chirality*, **21**, 911–921.

150 Ruderisch, A., Pfeiffer, J., and Schurig, V. (2001) Synthesis of an enantiomerically pure resorcinarene with pendant L-valine residues and its attachment to a polysiloxane (Chirasil-Calix). *Tetrahedron: Asymmetry*, **12**, 2025–2030.

151 Seyhan, S., Ozbayrak, O., Demirel, N., Merdivan, M., and Pirinccioglu, N. (2005) Chiral separation of amino acids by chiral octamide derivatives of calixarenes derived from resorcinol

by impregnation on a polymeric support. *Tetrahedron: Asymmetry*, **16**, 3735–3738.
152. Prus, P., Pietraszkiewicz, M., and Bilewicz, R. (2001) Calix[4]resorcinarene: molecular recognition in Langmuir films. *Mater. Sci. Eng., C*, **18**, 157–159.
153. Judice, J.K. and Cram, D.J. (1991) Stereoselectivity in guest release from constrictive binding in a hemicarceplex. *J. Am. Chem. Soc.*, **113**, 2790–2791.
154. Park, B.S., Knobler, C.B., Eid, C.N. Jr., Warmuth, R., and Cram, D.J. (1998) Chiral and somewhat hydrophilic hemicarceplexes. *Chem. Commun.*, 55–56.
155. Yoon, J. and Cram, D.J. (1997) Chiral recognition properties in complexation of two asymmetric hemicarcerands. *J. Am. Chem. Soc.*, **119**, 11796–11806.
156. Botta, B., Monache, G.D., Salvatore, P., Gasparrini, F., Villani, C., Botta, M., Corelli, F., Tafi, A., Gacs-Baitz, E., Santini, A., Carvalho, C.F., and Misiti, D. (1997) Synthesis of C-alkylcalix[4]arenes. 4. Design, synthesis, and computational studies of novel chiral amido[4]resorcinarenes. *J. Org. Chem.*, **62**, 932–938.
157. Botta, B., Caporuscio, F., D'Acquarica, I., Monache, G.D., Subissati, D., Tafi, A., Botta, M., Filippi, A., and Speranza, M. (2006) Gas-phase enantioselectivity of chiral amido[4]resorcinarene receptors. *Chem. Eur. J.*, **12**, 8096–8105.
158. Stilbor, I. and Zlatuskova, P. (2005) Chiral recognition of anions. *Top. Curr. Chem.*, **255**, 31–63.
159. Echavarren, A., Galan, A., Lehn, J.-M., and de Mendoza, J. (1989) Chiral recognition of aromatic carboxylate anions by an optically active abiotic receptor containing a rigid guanidinium binding subunit. *J. Am. Chem. Soc.*, **111**, 4994–4995.
160. Breccia, P., Gool, M.V., Perez-Fernandez, R., Martin-Santamaria, S., Gago, F., Prados, P., and de Mendoza, J. (2003) Guanidinium receptors as enantioselective amino acid membrane carriers. *J. Am. Chem. Soc.*, **125**, 8270–8284.
161. Galán, A., Andreu, D., Echavarren, A.M., Prados, P., and de Mendoza, J. (1992) A receptor for the enantioselective recognition of phenylalanine and tryptophan under neutral conditions. *J. Am. Chem. Soc.*, **114**, 1511–1512.
162. Asakawa, M., Brown, C.L., Pasini, D., Stoddart, J.F., and Wyatt, P.G. (1996) Enantioselective recognition of amino acids by axially-chiral π-electron-deficient receptors. *J. Org. Chem.*, **61**, 7234–7235.
163. Xu, M.-H., Lin, J., Hu, Q.-S., and Pu, L. (2002) Fluorescent sensors for the enantioselective recognition of mandelic acid: signal amplification by dendritic branching. *J. Am. Chem. Soc.*, **124**, 14239–14246.
164. Hernandez, J.V., Oliva, A.I., Simon, L., Muniz, F.M., Grande, M., and Moran, J.R. (2004) Ternary enantioselective complexes from α-amino acids, 18-crown-6 ether and a macrocyclic xanthone-based receptor. *Tetrahedron Lett.*, **45**, 4831–4833.
165. Ema, T., Tanida, D., and Sakai, T. (2006) Versatile and practical chiral shift reagent with hydrogen-bond donor/acceptor sites in a macrocyclic cavity. *Org. Lett.*, **8**, 3773–3775.
166. Ema, T., Tankida, D., Sugita, K., Sakai, T., Miyazawa, K., and Ohnishi, A. (2008) Chiral selector with multiple hydrogen-bonding sites in a macrocyclic cavity. *Org. Lett.*, **10**, 2365–2368.
167. Kyne, G.M., Light, M.E., Hursthouse, M.B., de Mendoza, J., and Kilburn, J.D. (2001) Enantioselective amino acid recognition using acyclic thiourea receptors. *J. Chem. Soc., Perkin Trans. 1*, 1258–1263.
168. Rossi, S., Kyne, G.M., Turner, D.L., Wells, N.J., and Kilburn, J.D. (2002) A highly enantioselective receptor for N-protected glutamate and anomalous solvent-dependent binding properties. *Angew. Chem. Int. Ed.*, **41**, 4233–4235.
169. Ragusa, A., Hayes, J.M., Light, M.E., and Kilburn, J.D. (2007) A combined computational and experimental approach for the analysis of the enantioselective potential of a new macrocyclic receptor for N-protected α-amino acids. *Chem. Eur. J.*, **13**, 2717–2728.
170. Gonzalez, S., Palaez, R., Sanz, F., Jimenez, M.B., Moran, J.R., and Caballero, M.C. (2006) Macrocyclic chiral receptors toward enantioselective

recognition of naproxen. *Org. Lett.*, **8**, 4679–4682.

171 Escuder, B., Brown, A.E., Feiters, M.C., and Nolte, R.J.M. (2004) Enantioselective binding of amino acids and amino alcohols by self-assembled chiral basket-shaped receptors. *Tetrahedron*, **60**, 291–300.

172 Pirkle, W.H. and Welch, C.J. (1994) Chromatographic and ^{1}H NMR support for a proposed chiral recognition model. *J. Chromatogr. A*, **683**, 347–353.

173 Koscho, M.E. and Pirkle, W.H. (2005) Investigation of a broadly applicable chiral selector used in enantioselective chromatography (Whelk-O 1) as a chiral solvating agent for NMR determination of enantiomeric composition. *Tetrahedron: Asymmetry*, **16**, 3345–3351.

174 Gasparrini, F., Misiti, D., Villani, C., Borchardt, A., Burger, M.T., and Still, W.C. (1995) Enantioselective recognition by a new chiral stationary phase at receptorial level. *J. Org. Chem.*, **60**, 4314–4315.

175 Gasparrini, F., Misiti, K., Pierini, M., and Villani, C. (2002) A chiral A_2B_2 macrocyclic minireceptor with extreme enantioselectivity. *Org. Lett.*, **4**, 3993–3996.

176 Uccello-Barretta, G., Balzano, F., Martinelli, J., Berni, M., Villani, C., and Gasparrini, F. (2005) NMR enantiodiscrimination by cyclic tetraamidic chiral solvating agents. *Tetrahedron: Asymmetry*, **16**, 3746–3751.

177 Bilz, A., Stork, T., and Helmchen, G. (1997) New chiral solvating agents for carboxylic acids: discrimination of enantiotopic nuclei and binding properties. *Tetrahedron: Asymmetry*, **8**, 3999–4002.

178 Hue, C.H., Mu, Q.M., and Chen, S.H. (2001) Design and synthesis of chiral molecular tweezers based on deoxycholic acid. *Chin. Chem. Lett.*, **12**, 413–416.

179 Xu, K.-X., Wu, X.-J., He, Y.-B., Liu, S.-Y., Qing, G.-Y., and Meng, L.-Z. (2005) Synthesis and chiral recognition of novel chiral fluorescence receptors bearing 9-anthryl moieties. *Tetrahedron: Asymmetry*, **16**, 833–839.

180 Yang, X., Wu, X., Fang, M., Yuan, Q., and Fu, E. (2004) Novel rigid chiral macrocyclic dioxopolyamines derived from L-proline as chiral solvating agents for carboxylic acids. *Tetrahedron: Asymmetry*, **15**, 2491–2497.

181 Lammerhofer, M. and Lindner, W. (2008) Liquid chromatographic enantiomer separation and chiral recognition by cinchona alkaloid-derived enantioselective separation materials. *Adv. Chromatogr.*, **46**, 1–107.

182 Kim, S.-G., Kim, K.-H., Kim, Y.K., Shin, S.K., and Ahn, K.H. (2003) Crucial role of three-center hydrogen bonding in a challenging chiral molecular recognition. *J. Am. Chem. Soc.*, **125**, 13819–13824.

183 Sada, K., Tateishi, Y., and Shinkai, S. (2004) A chiral pybox ligand as a new chiral shift reagent for secondary dialkylammonium cations. *Chem. Lett.*, 582–583.

184 Finocchiaro, P., Failla, S., and Consiglio, G. (2005) Phosphorylated macrocycles: structures, complexing properties, and molecular recognition. *Russ. Chem. Bull., Int. Ed.*, **54**, 1355–1372.

185 Gonzalez-Alvarez, A., Alfonso, I., Diaz, P., Garcia-Espana, E., Gotor-Fernandez, V., and Gotor, V. (2008) A simple helical macrocyclic polyazapyridinophane as a stereoselective receptor of biologically important dicarboxylates under physiological conditions. *J. Org. Chem*, **73**, 374–382.

186 Fukuhara, G., Madenci, S., Polkowska, J., Bastkowski, F., Klarner, F.-G., Origane, Y., Kaneda, M., Mori, T., Wada, T., and Inoue, Y. (2007) Inherently chiral molecular clips: synthesis, chiroptical properties, and application to chiral discrimination. *Chem. Eur. J.*, **13**, 2473–2479.

187 Takahashi, I., Odashima, K., Koga, K., and Kitajima, H. (1992) Utility of an optically active paracyclophane as a new type of shift reagent for nuclear magnetic resonance (NMR) spectra based on the diastereomeric host-guest complex formation. *Chem. Express*, **7**, 653–656.

188 Wilcox, C.S., Webb, T.H., Zawacki, F.J., Glagovich, N., and Suh, H. (1993) Selectivity in molecular recognition of steroids, alkanes and alicyclic substrates in aqueous media. *Supramol. Chem.*, **1**, 129–137.

189 Canceill, J., Lacombe, L., and Collet, A. (1985) Analytical optical resolution of

bromochlorofluoromethane by enantioselective inclusion into a tailor-made "cryptophane" and determination of its maximum rotation. *J. Am. Chem. Soc.*, **107**, 6993–6996.

190 Huang, X., Nakanishi, K., and Berova, N. (2000) Porphyrins and metalloporphyrins: versatile circular dichroic reporter groups for structural studies. *Chirality,* **12**, 237–255.

191 Ema, T., Ouchi, N., Doi, T., Korenaga, T., and Sakai, T. (2005) Highly sensitive chiral shift reagent bearing two zinc porphyrins. *Org. Lett.*, **7**, 3985–3988.

192 Hayashi, T., Aya, T., Nonoguchi, M., Mizutani, T., Hisaeda, Y., Kitagawa, S., and Ogoshi, H. (2002) Chiral recognition and chiral sensing using zinc porphyrin dimers. *Tetrahedron*, **58**, 2803–2811.

193 Paolesse, R., Monti, D., La Monica, L., Venanzi, M., Froiio, A., Nardis, S., Di Natale, C., Martinelli, E., and D'Amico, A. (2002) Preparation and self-assembly of chiral porphyrin diads on the gold electrodes of quartz crystal microbalances: a novel potential approach to the development of enantioselective chemical sensors. *Chem. Eur. J.*, **8**, 2476–2483.

194 Gao, F., Ruan, W.-J., Chen, J.-M., Zhang, Y.-H., and Zhu, Z-.A. (2005) Spectroscopy, NMR and DFT studies on molecular recognition of crown ether bridged chiral heterotrinuclear salen Zn(II) complex. *Spectrochim. Acta A*, **62**, 886–895.

195 Ogo, S., Nakamura, S., Chen, H., Isobe, K., Watanabe, Y., and Fish, R.H. (1998) A new, aqueous ^{1}H NMR shift reagent based on host-guest molecular recognition principles for organic compound structural analysis: non-covalent π-π and hydrophobic interactions using a supramolecular host, [Cp*Rh(2′-deoxyadenosine)]$_3$(OTf)$_3$. *J. Org. Chem.*, **63**, 7151–7156.

196 Chang, K.-H., Liao, J.-H., Chen, C.-T., Mehta, B.K., Chou, P.-T., and Fang, J.-M. (2005) Stereoselective recognition of tripeptides guided by encoded library screening: construction of chiral macrocyclic tetraamide ruthenium receptor for peptide sensing. *J. Org. Chem.*, **70**, 2026–2032.

8
Fullerene Receptors Based on Calixarene Derivatives
Pavel Lhoták and Ondřej Kundrát

8.1
Introduction

In 1985, the discovery of fullerenes [1] (Figure 8.1) as the third allotrope (besides diamond and graphite) of one of the most important elements on Earth – carbon – opened up a completely unknown and unexpected world of fascinating molecules. These closed cage-like molecules containing only hexagonal and pentagonal rings [2] of carbon atoms represent one of the most striking and exciting discoveries in chemistry of twentieth century [3]. As seen in Figure 8.2, up to 1990 fullerenes remained more or less a chemical curiosity without any bigger scientific impact because of their general inaccessibility. The experiments of Huffman and Krätschmer [4] with resistive heating of graphite in early 1990s gave the scientific community the first macroscopic samples of fullerenes and, consequently, induced an explosive increase in the number of scientific papers dealing with this topic.

Fullerenes were quickly recognized as highly useful compounds possessing many fascinating chemical and physical properties (electrochemical behavior, spectroscopic properties, topology, chemical transformations, superconductivity, free-radical scavenging, etc.). Their importance was underlined by the awarding of the 1996 Nobel Prize in Chemistry to Robert F. Curl, Richard E. Smalley, and Harold W. Kroto for their pioneering work in fullerene discovery. Currently, fullerenes are still among the most intensively studied artificial molecules. A noticeable development of fullerene chemistry was marked by the stage where a great number of potential applications and practical uses in everyday life were proposed.

Albeit fullerenes exhibit almost unlimited possibilities they still suffer from some drawbacks, the most significant of them being the price. Fullerenes can be produced by several different methods starting from graphite: resistive heating, arc vaporization, and/or inductive heating. More recently, other methods, based on combustion of benzene in sooting flames or pyrolysis of suitable hydrocarbons, have also been described [5]. Depending on the method used, a highly complex reaction mixture in the form of soot and slag is produced, in which fullerenes C_{60} and C_{70}, higher fullerenes, and carbon nanotubes usually constitute only several percents (by weight)

Artificial Receptors for Chemical Sensors. Edited by V.M. Mirsky and A.K. Yatsimirsky
Copyright © 2011 WILEY-VCH Verlag GmbH & Co. KGaA, Weinheim
ISBN: 978-3-527-32357-9

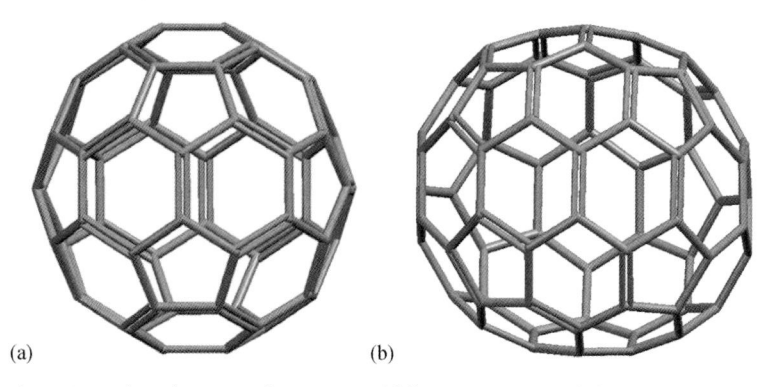

Figure 8.1 Three-dimensional structures of fullerenes: (a) C_{60} and (b) C_{70}.

while the most part is amorphous carbon. To obtain pure C_{60} and C_{70}, highly complicated isolation and purification processes have to be carried out, mostly based on sequential combination of extractions and chromatographic steps.

Despite incredible progress in fullerene chemistry, purification remains a challenge to scientists and to a large extent determines fullerene prices. It is estimated that about 80% of fullerene operational cost is constituted by time/energy-consuming isolation and purification procedures. Hence, it is very important and highly desirable to develop novel isolation and purification methods based on a supramolecular approach, that is, to use selective complexation agents capable of noncovalent interactions with fullerenes [6].

Recently, many host molecules known for trapping fullerenes both in solution and in solid state have appeared in the literature. Some of these host–guest complexes are based mainly on π–π interactions (porphyrins) [7], while others exploit the shape complementarity (concave/convex) with fullerene spheres enjoying van der Waals interactions and/or solvophobic effects (cyclodextrins) [8]. Very promising results

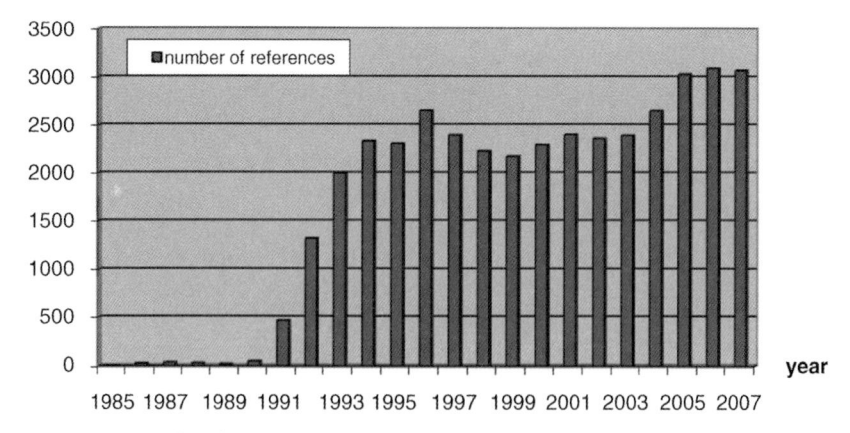

Figure 8.2 Number of papers published, searching for the keyword "fullerene." Source: SciFinder Scholar™ 2007.

have been achieved with a family of macrocyclic compounds possessing aromatic cavities preorganized for capturing fullerenes via π–π interactions – especially if the host–guest phenomenon can be simultaneously strengthened by additional interactions arising from mutual complemental curvatures of interacting species. In this context, electron-rich aromatic oligomers, such as calixarenes [9], homooxacalixarenes [10], resorcinarenes [11], and cyclotriveratrylenes [12] are of great importance because of their potential applications in complexation, extraction, purification, or chemical modification of fullerene molecules. This chapter demonstrates some basic applications of calix[n]arenes in the host–guest chemistry of fullerenes.

8.2
Calixarenes

Calix[n]arenes [13] are macrocyclic compounds that can be formed very easily by condensation of para-substituted phenols with formaldehyde or sulfur. Depending on the reaction conditions, the commonly used starting compound, p-tert-butylphenol, can be easily transformed into macrocycles with various cavity sizes (4–8 phenolic rings, Figure 8.3). These molecules became extremely useful in host–guest chemistry due to their ability to complex ions/neutral molecules insides their cavities.

Calixarenes have many interesting and very attractive features compared to other families of macrocyclic compounds: (i) one-pot high-yield preparation on multi-gram scale, (ii) easy chemical transformation (derivatization) of the basic skeleton, (iii) various size of cavity, (iv) precisely defined three-dimensional shape of the cavity (in case of calix[4]arene), and (v) excellent complexation properties. All these features make calixarenes ideal candidates for the design of more sophisticated supramolecular systems, self-assemblies, receptors, and so on. Because of their unique tunable 3D shapes leading to four different isolable conformers (atropoisomers) – cone, partial cone, 1,2-alternate, and 1,3-alternate (Figure 8.4) – calix[4]arenes are frequently used as building blocks and/or molecular scaffolds for the construction of more elaborate host molecules, sensors, and molecular systems with well-preorganized structures.

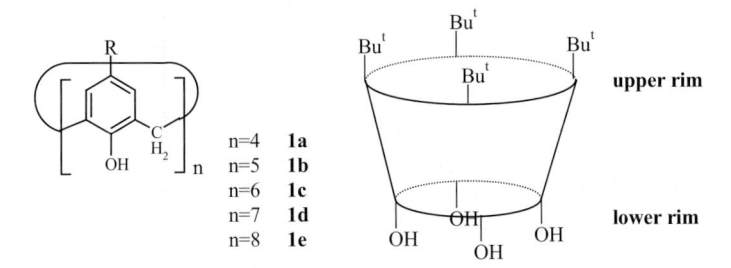

Figure 8.3 Calix[n]arenes.

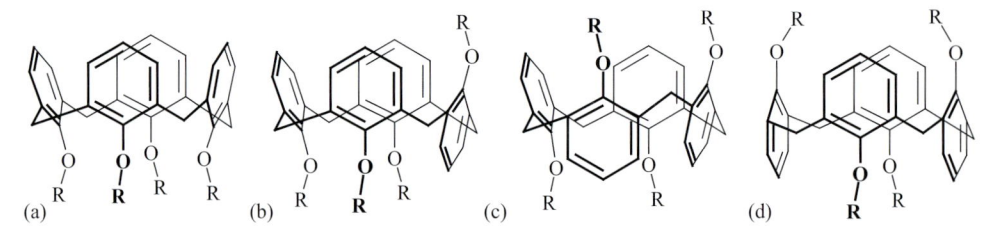

Figure 8.4 Four basic conformations (atropoisomers) of calix[4]arene – (a) cone, (b) partial cone, (c) 1,2-alternate and (d) 1,3-alternate.

8.3
Solid State Complexation by Calixarenes

Obviously, matching the curvature of fullerenes with bowl-shaped calixarenes could lead to attractive interactions between both systems. Depending on the specific conditions (cavity size, nature of substituents, substitution pattern, conformation of calixarene) these interactions could be described as van der Waals (dispersion) forces and/or concave–convex π–π interactions [14]. The potential of calixarenes in the host–guest chemistry of fullerenes was recognized as early as in 1992 when the first paper describing fullerene complexation appeared [15]. Calix[8]arene bearing eight sulfonic acids on the lower rim was used to solubilize C_{60} in aqueous solutions. In this case, hydrophobic effects rather than π–π interactions are responsible for the inclusion phenomenon [16].

Probably, the most spectacular example of fullerene complexation was demonstrated by the groups of Atwood [17] and Shinkai [18], who independently discovered application of calix[8]arene **1e** (R = Bu^t) for fullerene purification. The whole process is very simple and leads to efficient isolation of C_{60} from crude fullerite soot. As shown in Figure 8.5, a toluene solution of crude carbon soot was mixed with a toluene solution of calix[8]arene **1e** (R = Bu^t) to give a 1:1 complex with fullerene. As this complex is sparingly soluble, it can be recovered as precipitate by

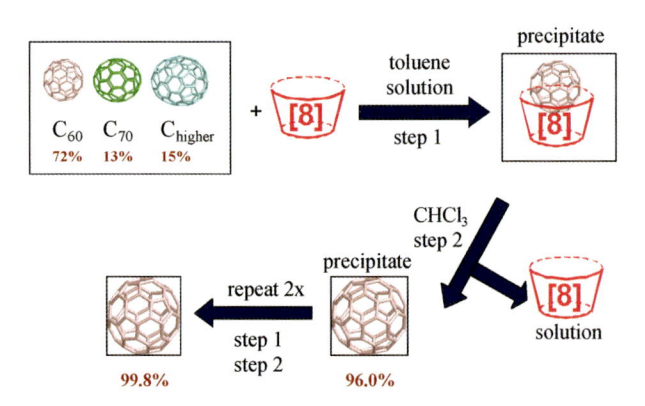

Figure 8.5 Pufirication of C_{60} from crude arc soot (schematically).

simple filtration. The complexation is surprisingly selective for C_{60}, and as a result the purity of C_{60} can be enhanced up to 96% by such a simple precipitation/filtration protocol. Moreover, the complex itself is not stable in chloroform. Hence, simple washing of the resulting complex directly on filter paper leads to its destruction and, finally, to dissolution of **1e** ($R = Bu^t$), leaving pure fullerene C_{60} on the filter (C_{60} is almost insoluble in $CHCl_3$). The same operation (precipitation/destruction) can be repeated once again using the precipitate from the first purification. A second cycle gives C_{60} with 99.0% purity, while the next round of purification gives even 99.8% purity of fullerene C_{60}.[1)]

The choice of toluene as a solvent is critical for the whole process. Xylenes and mesitylene rapidly decompose the complex at ambient temperature, yielding free fullerenes and **1e** ($R = Bu^t$) after evaporation. It was found that stability of the complex is limited only to the solid state and its formation could not be studied directly in solution where dissociation into individual components occurs.

As the structure of complex was not fully understood, Shinkai et al. carried out a broad study of factors influencing the complexation ability of the calix[n]arene skeleton. Many derivatives of **1e** bearing other substituents on the upper rim (H, Me, Et, Pr, i-Pr, n-Bu, s-Bu, n-pentyl, t-pentyl, t-octyl, Ph) were synthesized and studied for their inclusion ability [19]. It was shown that only **1e** ($R = Pr^i$) and **1e** ($R = Bu^t$) can precipitate C_{60} from toluene, whereas C_{70} did not form any similar precipitate. On the other hand, using benzene as a solvent, the precipitate of complex **1e** ($R = Bu^t$) with C_{70} having 1:2 stoichiometry was isolated, and was characterized by elemental analysis and CP-MAS ^{13}C NMR spectroscopy.

While the 1:1 stoichiometry could be explained by a simple "ball inside the cavity" concept, the existence of 1:2 stoichiometry was very surprising as the calix[8]arene cavity is not big enough for two such balls. In this context, the fact that even much smaller calix[6]arenes **1c** ($R = H$) and **1c** ($R = Bu^t$) also formed complexes with fullerenes C_{60} or C_{70} possessing 1:2 (calix:fullerene) stoichiometry seemed very confusing [19]. The obvious discrepancies between a simple "ball and socket" approximation and the above contradictory results were finally resolved by Atwood [20]. Controlled slow evaporation of a toluene–CH_2Cl_2 solution of **1c** ($R = H$) and C_{60} or C_{70} gave black crystals suitable for X-ray diffraction analysis. It was revealed that both complexes **1c** ($R = H$)·$(C_{60})_2$ and **1c** ($R = H$)·$(C_{70})_2$ are almost identical – they are isostructural – despite the anisotropic shape of C_{70}. Calixarene **1c**

---

1) Isolation of C_{60} from fullerite soot (according Reference [18]): carbon soot (200 mg) containing 72% of C_{60}, 13% of C_{70} and 15% of other compounds (higher fullerenes, nanotubes) was dissolved in 60 ml of toluene and mixed with a toluene solution (200 ml) of **1e** (390 mg). The resulting dark solution was left overnight (15 h) and the precipitate was collected by filtration. HPLC analysis revealed that the precipitate contains C_{60} in 96% purity. The precipitate was dissolved in 200 ml of hot toluene and left at room temperature overnight. The purity of C_{60} in the precipitate thus obtained increased to 99%. The same operation can be repeated once again, yielding a precipitate of complex with 99.8% purity of C_{60}. The precipitate from the last step was stirred in 30 ml of $CHCl_3$ for 1 h to yield 102 mg of C_{60} precipitate while **1e** remained in solution.

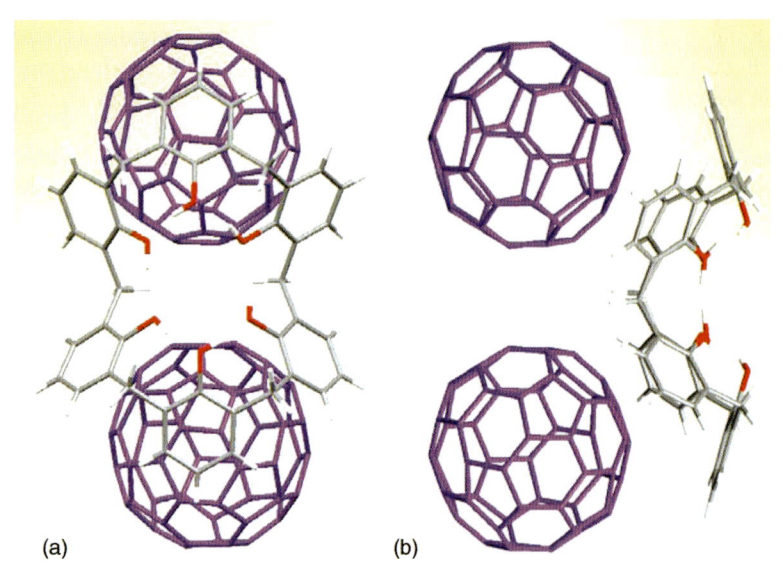

Figure 8.6 (a) X-ray structure of **1c** (R = H)·(C$_{60}$)$_2$ complex; (b) side view of the same complex.

(R = H) is in a rather uncommon double-cone conformation, forming two shallow cavities, each of which is occupied by fullerene guests (Figure 8.6).

As we can see, the C$_{60}$ molecules are in close proximity, forming a dimer; at the same time, all the hydroxyl groups of calix[6]arene are oriented to one side of the macrocycle, thus enabling strong hydrogen bonding in the solid state. These structural features seem to be general for this kind of complex. Dimerization or even oligomerization of fullerenes is favored over a single fullerene molecule sitting in a single cavity as this complexation mode would lead to disruption of hydrogen bonds on the lower rim of calixarene. As emphasized recently by Raston [21], who examined a **1** (R = H)·(C$_{70}$)$_2$ complex using synchrotron source X-ray diffraction, fullerenes are in fact included in an *endo* cavity, and the striking feature of the assembly is the extensive CH···π interactions involving methyl moieties of the Bu^t groups with fullerenes.

These findings also shed the light on the structure of Shinkai's/Atwood's original complex. IR measurements confirmed the presence of hydrogen bonds, albeit partly disrupted, as indicated by shift of OH stretching from 3400 to 3200 cm^{-1} [22]. It was evident from solid state NMR (^{13}C CP-MAS) that the complex contained two different *tert*-butyl groups in approximately 2 : 6 ratio. The combination of these facts together with X-ray powder diffraction data and molecular mechanics led to a reconsideration of the structure of the **1e** (R = Bu^t):C$_{60}$ complex [23]. The complex itself has a micelle-like structure with a fullerene trimer being surrounded by three molecules of **1e** (R = Bu^t), again accepting double cone conformation (Figure 8.7). The whole assembly with a unique 3 : 3 stoichiometry [**1e** (R = Bu^t)]$_3$·(C$_{60}$)$_3$ is held together by the peculiar net of fullerene–fullerene interactions, van der Waals contacts, and

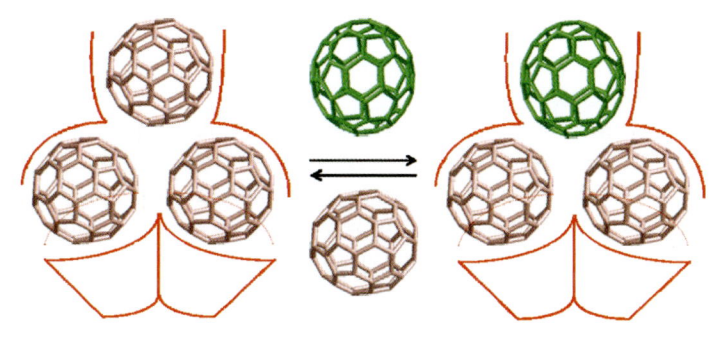

Figure 8.7 Proposed structure of [**1e** (R = Bu^t)]$_3$·(C$_{60}$)$_3$ complex.

hydrogen bonds. In any event, definitive solid-state structure confirmation of the proposed assembly remains elusive. As stated above, pure C$_{70}$ does not create a similar precipitate (complex) with **1e** (R = Bu^t) in toluene; on the other hand, some complexation of C$_{70}$ occurs in the presence of excess C$_{60}$. This can be explained by the replacement of one C$_{60}$ molecule in the trimer complex. A maximum incorporation of C$_{70}$ (approximately 15%) was proven by HPLC. Incorporation of more C$_{70}$ probably leads to destruction of the trimeric structure.

Based on the X-ray crystallographic studies, calix[5]arene possesses the best fit between the size and shape of the cavity and the size and curvature of fullerenes C$_{60}$ or C$_{70}$. Both systems can enjoy mutual concave/convex complementarity, and, consequently, depending on the substitution, the 1:1 or 1:2 complexes can be formed. Thus, basic calix[5]arene **1b** (R = H) forms a 1:1:1 complex with C$_{60}$ and toluene in which fullerene is situated directly in the cavity of calixarene (Figure 8.8a) [24]. The macrocyclic skeleton is held in the *cone* conformation via a circular hydrogen bond array on the lower rim. Individual complexes are packed within the crystal lattice in such a way that fullerenes are ordered into zigzag arrays with short fullerene–fullerene contacts. A similar 1:1 ball-and-socket complex

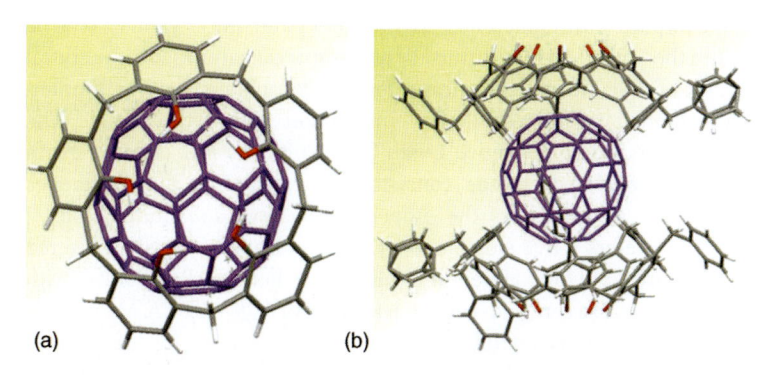

Figure 8.8 (a) Structure of **1b** (R = H)·C$_{60}$·toluene complex (toluene molecule omitted for clarity); (b) X-ray structure of **1b** (R = Bn)2·C$_{60}$ complex with 2:1 stoichiometry.

(without solvent molecule) was obtained also for C_{70}. In this case, fullerene–fullerene interactions lead to the formation of zigzag sheets well-separated by layers of **1b** (R = H) molecules [25].

Substitution of the upper rim of calix[5]arene by benzyl groups [compound **1b** (R = Bn)] afforded a 2 : 1 complex with C_{60} (from toluene solution). Here, fullerene is surrounded by two molecules of **1b** (R = Bn) stabilized by hydrogen bonding in the *cone* conformation to maximize the contacts between concave/convex surfaces (Figure 8.8b) [26]. The same type of 2: 1 encapsulation was found also for both benzyl substituted homooxacalix[3]arene **2** [26] and upper rim iodo-substituted calix[5]arene **3a** [27]. Interestingly, even a small change in calixarene substitution can completely reverse the inclusion mode, as documented by derivatives **3b** and **3c**, both of which form only 1: 1 complexes with C_{60} in the solid state [27]. The C_{60} inclusion by **1b** (R = Bn) was also studied directly in toluene solution, using high-precision densitometry to measure the change in partial molar volume on complexation [28]. The results indicated that a complex with 1: 1 stoichiometry is formed by the displacement of two toluene molecules from the cavity of **1b** (R = Bn).

3a: X= I
3b: X= Me
3c: X= H

The cavity of calix[4]arene derivatives is obviously too small to create the ball-and-socket type of complexes with fullerenes. On the other hand, the C_{60} and C_{70} can be included within the "endo" cavities formed by packing of calixarenes into the crystal lattice. The lack of suitable shape (concave/convex) complementarity is counterbalanced by extended fullerene–fullerene interaction within the crystal. The crystallization of tetrabromo derivative **4** with C_{60} results in a $4 \cdot C_{60}$ complex with alternating one-dimensional strands consisting of calixarenes and fullerenes (Figure 8.9a) [29]. Alternatively, the upper-rim benzyl substituted calix[4]arene **1a** (R = Bn) prefers a 2 : 1 stoichiometry, **1a** (R = Bn)·$(C_{60})_2$, with a layered arrangement (Figure 8.9b) [30]. Here, each fullerene is here in close contact with five other fullerenes, with a centroid–centroid distance of 9.79–10.02 Å. Many other structural arrangements based on fullerene–fullerene interactions, such as dimers, columns, zigzag chains, sheets, honeycomb, and so on, can be found in recent reviews [6, 31].

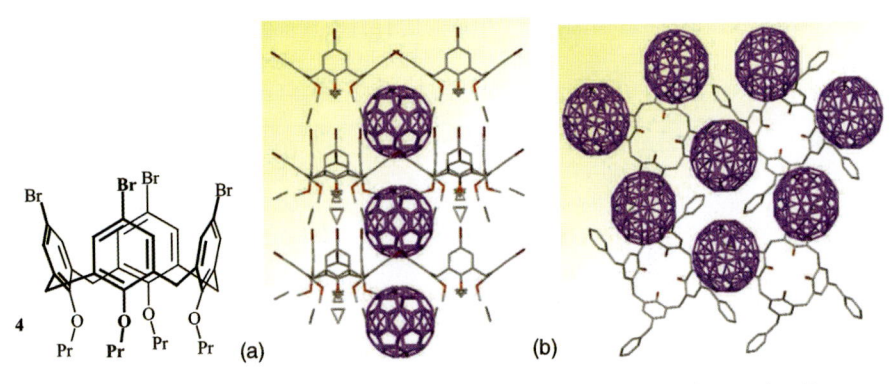

Figure 8.9 (a) Columnar structural motifs in crystal packing of **4**·C_{60} complex; (b) crystal packing of **1a** (R = Bn)·$(C_{60})_2$ with layered arrangement.

8.4
Complexation in Solution

While the previous section focused on selected examples of solid-state inclusion complexes with calix[n]arenes, the "real" host–guest chemistry of fullerenes with high impact on potential applications should occur in solution. As we have demonstrated, bigger calixarenes such as **1e** or **1c** can form interesting complexes in the solid state. Of course, this ability has also been studied in solution, but the corresponding complexation constants were not impressive [32]. Calixarene **1c** was transformed into aniline substituted derivatives **5** and **6** and their capacity to form inclusion complexes was studied by UV-vis spectroscopy. Absorption changes of the C_{60}–calixarene system at 539 nm showed only small association constants, $K_{60} = 7.9$ M^{-1} for **5** and $K_{60} = 110$ M^{-1} for **6** (in toluene) [33]. Generally, the application of bigger calixarenes **1c–e** for fullerene recognition in solution suffers from several drawbacks: (i) low solubility, (ii) the absence of selective derivatization methods, and (iii) conformational freedom that cannot be removed by simple alkylation of the lower rim. For all these reasons calix[5]arenes seem to be ideal candidates for fullerene complexation, as their cavities match the concave shape of C_{60}/C_{70}, and their chemistry is much more advanced compared to their bigger relatives [13].

Simple calix[5]arene derivatives **3a–c**, used for solid state inclusion of fullerenes [27], also behave like fullerene receptors in the solution [34]. A color change

Table 8.1 Association constants K (M^{-1}) measured by UV-vis titration at room temperature.

		Toluene	Benzene or tetralin[a]	CS$_2$	o-Dichlorobenzene
3a	C$_{60}$	2120 ± 110	1840 ± 130	660 ± 30	308 ± 41
	C$_{70}$	520 ± 60	210 ± 20	260 ± 60	240 ± 40
3b	C$_{60}$	1673 ± 70	1507 ± 84	600 ± 3	277 ± 14
	C$_{70}$	380 ± 30	90 ± 10	280 ± 30	230 ± 20
3c	C$_{60}$	588 ± 70	459 ± 74	284 ± 70	207 ± 11
	C$_{70}$	310 ± 30	80 ± 20	160 ± 10	190 ± 20

a) Benzene for C$_{60}$, tetralin for C$_{70}$.

from purple to pale yellow was observed in CS$_2$ solution of C$_{60}$ upon addition of receptors. The isosbestic point at 478 nm (in the case of **3a**) as well as a Job plot provided evidence for a 1:1 stoichiometry in solution. The corresponding association constants are collected in Table 8.1; obviously, they are solvent dependant. The solubility of the guest increases along the series benzene < toluene < CS$_2$ < o-dichlorobenzene while the constants decrease in the same order. This indicates that the more weakly solvated guest (fullerene) is more strongly bound by the host molecule (calix). A thermodynamic study on supramolecular complex formation revealed that desolvation of the guest plays a crucial role in toluene [35]. The plot of free energy of the C$_{60}$ complexation (ΔG) versus the fullerene solubility gave good linear relationships (correlation coefficient $r^2 > 0.95$). This is a consequence of competition between complex formation and desolvation of guest in these nonpolar solvents [36].

The highest complexation constants ($K_{60} = 2120$ M^{-1}, $K_{70} = 520$ M^{-1} in toluene) were achieved using receptor **3a**, bearing two iodine atoms with high polarizability. This substitution of the upper rim enhances the interactions with fullerenes if compared with methyl groups (**3b**) or H-atoms (**3c**), suggesting that van der Waals (dispersion) forces play a crucial role in the complexation process. Interestingly, in all cases C$_{60}$ is bound preferentially over C$_{70}$, with maximum selectivity being around 4. While the complexation stoichiometry of **3a–c** in solution is always 1:1, an X-ray study of the solid-state complex of **3a** revealed an entirely different situation. A C$_{60}$ molecule is encapsulated within the cavity composed of two calixarene molecules, with their upper rims being oriented towards each other, thus leading to the formation of the 2:1 complex [27].

The formation of 2:1 complexes in the solid state led to the design of more sophisticated receptors for solution recognition based on bis-calix[5]arenes. Gutsche [37] has synthesized derivative **7** with direct connection of calix[5]arene's upper rims. Unfortunately, very low complexation constants ($K_{60} = 43$ M^{-1}, $K_{70} = 233$ M^{-1} in CS$_2$) were obtained, presumably as a result of competition between the fullerene and solvent complexation into the cavity. In contrast, a very nice solid-state complex with encapsulated C$_{60}$ was obtained by crystallization from toluene–CS$_2$ mixture. Much better results were achieved using double calixarenes **8–12**, where two calix[5]arene molecules are interconnected via spacer to provide a sizable cavity.

7

8a: X= I, Y=
8b: X= Me, Y= —C≡C—(Me-aryl)—C≡C—

9a: X= I, Y=
9b: X= Me, Y= —C≡C—

10a: X= I, Y=
10b: X= Me, Y= —C≡C—C≡C—

11: X= Me, Y= —(biphenyl)—

12: X= Me, Y= ≡—(phenyl)—≡

While simple calixarene-based hosts possess higher binding affinity towards C_{60}, all double-calixarene receptors show remarkable preference for C_{70}. As shown in Table 8.2, the highest association constants were obtained in toluene for **8a** [36, 38]. Interestingly, the selectivity factor C_{70}/C_{60} of **9a** is again solvent dependant, with the peak value around 10 in toluene, and gradually decreasing in the order toluene > benzene > CS_2 solution. This trend seems to be general as it was observed also for other bis-calixarenes receptors. In all cases studied, formation of 1 : 1 stoichiometry was observed.

These receptors were also studied for their ability to complex the dumb-bell-shaped dimer of C_{60}–C_{120} fullerene (**13**) [39]. Standard UV-vis titrations in o-dichlorobenzene solution documented the formation of 1 : 1 complexes. The best results were achieved

Table 8.2 Association constants K (M^{-1}) measured by UV-vis titration at room temperature.

Solvent	Fullerene	8a	9a	10a
Toluene	C_{60}	76 000 ± 5000	8 300 ± 200	2700 ± 100
	C_{70}	163 000 ± 16 000	85 000 ± 13 000	5500 ± 200
Benzene	C_{60}	47 000 ± 2000	5 700 ± 200	3000 ± 100
	C_{70}	72 000 ± 7000	49 000 ± 1000	4100 ± 500
CS_2	C_{60}	5 400 ± 800	1 500 ± 100	670 ± 10
	C_{70}	9 600 ± 300	6 600 ± 200	1030 ± 40
o-Dichlorobenzene	C_{60}	3 000 ± 200	1 200 ± 20	440 ± 10
	C_{70}	4 100 ± 100	1 900 ± 20	490 ± 20

with receptors **10b** ($K_{120} = 12\,300\,\text{M}^{-1}$) and **12** ($K_{120} = 10\,700\,\text{M}^{-1}$) with cavities extended by linear spacers. Interestingly, **8b** ($K_{120} = 3900\,\text{M}^{-1}$, meta-substituted spacer) and **11** ($K_{120} = 2700\,\text{M}^{-1}$, 4,4′-disubstituted biphenyl spacer) are much less effective in the complexation of C_{120}. The application of similar receptors could overcome problems with the poor solubility of C_{120} fullerene.

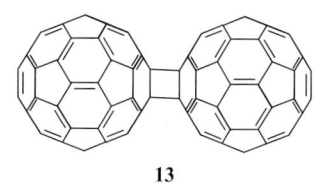

13

Recently, the binding abilities of double calixarene receptors towards higher fullerenes (C_{76}, C_{78}, and C_{84}) were investigated in organic solutions [40]. As these fullerenes are still available only on the mg scale due to their difficult purification, the application of host–guest chemistry could offer an alternative approach to their separation.

Table 8.3 shows complexation constants towards higher fullerenes measured in two different solvents [41]. The strongest binding was achieved with receptors **8b** and **9b**, and, not surprisingly, the complexation phenomenon is strongly solvent dependant. The fullerene molecules are much better solvated (and hence soluble) in CS_2 than in toluene. As a consequence, the desolvation energy that has to be paid during the complexation is much lower in the case of toluene, leading finally to comparatively better complexation. As follows from Table 8.3, the comparable constants in toluene are about one order of magnitude higher than those in CS_2 (compare $K_{78} = 3 \times 10^5\,\text{M}^{-1}$ in toluene versus $K_{78} = 3.9 \times 10^4\,\text{M}^{-1}$ in CS_2 for **8b**).

As we have already mentioned, the cavity of calix[4]arene is too small to effectively complex fullerenes. On the other hand, assembly of multiple calix[4]arenes can lead to the formation of an extended cavity with some complexation ability towards C_{60} or C_{70}. Among these receptors, the "head-to-head" bis-calixarene **14** was found to form 2:1 complexes with C_{60} or C_{70}. Two molecules of receptor are orthogonally oriented in a "tennis-ball" fashion to create a dimer with intertwining *tert*-butyl groups while fullerene is encapsulated within this cavity [42]. Recently, the inclusion complex of water-soluble tetrasodium salt of sulfonated thiacalix[4]arene (**15**) with C_{60} fullerene

Table 8.3 Association constants K (M^{-1}) with higher fullerenes measured by UV-vis titration at room temperature.

	Solvent	C_{60}	C_{70}	C_{76}	C_{78}	C_{84}
8b	Toluene	68 000 ± 5000	118 000 ± 12 000	200 000 ± 14 000	300 000 ± 50 000	53 000 ± 2000
	CS_2	2 530 ± 60	4 600 ± 100	10 500 ± 400	39 000 ± 6000	5 700 ± 100
9b	Toluene	9 000 ± 200	91 000 ± 9000	220 000 ± 30 000	230 000 ± 40 000	42 000 ± 2000
	CS_2	1 380 ± 60	3 500 ± 300	16 000 ± 800	42 000 ± 3000	9 300 ± 400
10b	Toluene	4 100 ± 100	7 300 ± 800	41 700 ± 2000	28 000 ± 3000	36 000 ± 4000
	CS_2	720 ± 80	1 100 ± 100	900 ± 100	42 ± 7	106 ± 4

was studied by photoluminescence and quantum-chemical methods (AM1, HF/6-31G*). The stoichiometry of the complex was found to be 2 : 1 with C_{60} residing in the dimeric cavity of the calixarene [43].

Yet another strategy to build calixarene-based cavities suitable for fullerene complexation in solution is represented by the "true" supramolecular approach. Based on this strategy, the fullerene molecule is trapped by a supramolecular capsule formed by self-assembly of suitable starting building blocks. Thus, two molecules of calix[5]arene **16** substituted on the upper rim with a 2,2′-bipyridine unit can be held together via complexation of Ag^+ cation. The tetrahedral coordination sphere of silver cation helps to preorganize calixarene cavities in close proximity, and the cavity thus formed is large enough to accommodate C_{60} or C_{70} fullerenes [44]. ESI-MS measurements of the complexes revealed a preference for formation of a 1 : 1 C_{60} inclusion complex. Figure 8.10 shows a very similar approach based on the metal-induced preorganization of a suitable cavity [45]. Pure compound **18** interacts weakly with fullerenes in 1,2-dichloroethane ($K_{60} = 98\,\text{M}^{-1}$ and $K_{70} = 250\,\text{M}^{-1}$). The addition of one equivalent of $[Cu(MeCN)_4]PF_6$ leads to the intramolecular complex formation and a forced change in the conformation of receptor. As a consequence, this better preorganization gives much better association constants towards fullerenes, with preference for C_{60} ($K_{60} = 3800\,\text{M}^{-1}$ and $K_{70} = 950\,\text{M}^{-1}$).

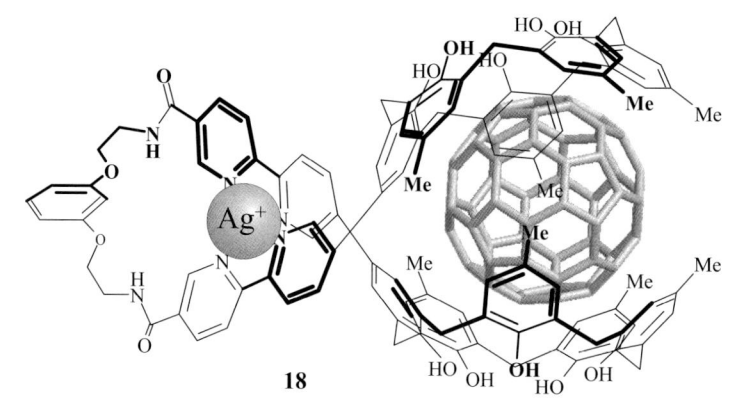

262 | *8 Fullerene Receptors Based on Calixarene Derivatives*

Figure 8.10 Formation of metal-induced cavity for fullerene complexation.

A supramolecular strategy based on a hydrogen bonding self-assembly of ureido-derivative **17** has been reported by the same group [46]. The self-association of **17** was studied using ^{1}H NMR titrations in a CDCl$_3$–CS$_2$ mixture. The formation constant of self-assembled dimer **17·17** was determined to be 32 M^{-1}, while in the presence of a small amount of C$_{60}$ it increased up to 104 M^{-1}, indicating the formation of ternary complex **17·C$_{60}$·17**. As the whole self-assembly is held via hydrogen bonds, the complex can be destroyed easily via the protonation of urea groups by addition of trifluoroacetic acid.

The development of chemical sensors for fullerene recognition has led to the design of calix[5]arene-based receptors with a luminophore moiety covalently appended near the fullerene binding site. As it is known that fullerenes can act as acceptors in an energy transfer processes, they might act as luminescence quenchers. Sensors **19** and **20**, possessing Re(bipy) complexes on the upper rim of a calix[5]arene cavity, are based on this strategy [47]. Thus, free compound **19** exhibits strong orange fluorescence (365 nm) in toluene solution, which is immediately extinguished upon the addition of C_{60} or C_{70}. The luminescence titrations experiments with **20** and Re(bipy)(CO)$_3$Cl indicated that quenching could be attributed to two independent processes associated with energy transfer from the luminophore to the fullerene: (i) intermolecular quenching arising from simple collision between the free fullerene and the luminophore and (ii) intramolecular quenching between the complexed fullerene and the receptor. As expected, this kind of luminescence receptor shows very high sensitivity towards C_{60}/C_{70}, even at concentrations of less than 10^{-5} M.

Another promising approach towards fullerene receptors is based on immobilization of suitable calixarenes on a gold surface [48]. These SAMs (self-assembled monolayers) were formed on gold electrodes and studied by cyclic voltammetry as redox probes [49]. The best results were achieved using calixarene derivatives bearing thioctic acid residues on the lower rim. It was shown that calix[4,6,8]arenes formed regular monolayers on a gold surface, being anchored via S…Au interactions, while their cavities are exposed to the bulk solution. On the other hand, only the calix[8]arene derivative, possessing a suitable cavity, was effective in trapping fullerene C_{60}, leading to responses in electrochemical behavior.

8.5
Calixarenes as Molecular Scaffolds

We have already mentioned that calix[4]arenes possess a cavity that is too small to include fullerenes effectively. On the other hand, the chemistry of these macrocycles is currently well established and numerous regio/stereo-selective substitution reactions can be used for their selective derivatization. A very attractive and useful feature of calix[4]arenes is their tunable three-dimensional shapes (Figure 8.4), which can be easily immobilized by simple chemical transformations on the lower rim of calixarene skeleton (alkylation, acylation) [13]. As a result, calix[4]arenes are frequently used in the role of molecular scaffold, enabling the construction of more elaborated supramolecular systems, including various receptors. Among them, derivatives, which could be called molecular tweezers for fullerene picking, have been designed and studied.

The design of these molecules is based on the finding that the curved π surfaces of fullerenes can interact with planar porphyrin or metalloporphyrin moieties by attractive π–π interactions [50]. While these interactions were originally recognized in the solid state, suitably designed bis- or multi-porphyrin systems can interact with fullerenes also in solution [51]. The first calixarene derivative of this type – compound **21** – showed selective complexation of C_{70} in C_6D_6 while C_{60} did not induce any chemical shift changes [52]. A similar design of molecular tweezers with highly preorganized and rigid structure has been applied for derivative **22** [53]. This calix[4]arene, bearing on the upper rim two porphyrin units directly connected via the *meso* positions, exhibited good complexation ability towards C_{70}.

¹H NMR complexation studies showed a much higher complexation constant in C_6D_6 ($K_{70} = 4500 \pm 600\ M^{-1}$) than in toluene-$d_8$ ($K_{70} = 1000 \pm 200\ M^{-1}$), again in agreement with previously observed solvent influence. Interestingly, the chemical induced shifts upon addition of C_{60} were too small to allow quantification of the interactions.

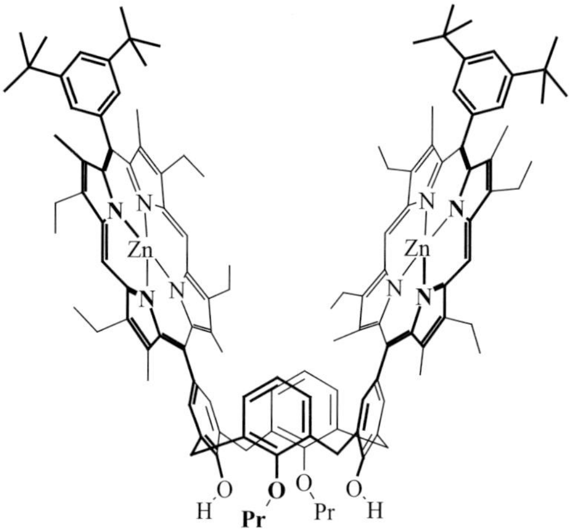

21

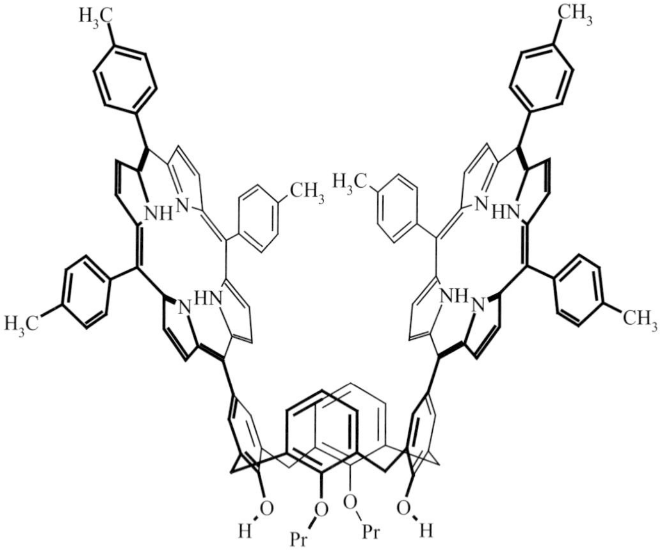

22

Another approach towards molecular tweezers capable of fullerene recognition is represented by compounds **23–25** [54]. The calix[4]arene or thiacalix[4]arene [55] skeleton was used as a scaffold to hold two or four tetraphenylporphyrin units on the lower rim.

23a, X = CH$_2$, R = H, M = 2H
23b, X = S, R = H, M = 2H
24a, X = CH$_2$, R = -CH$_2$CO-porf, M = 2H
24b, X = S, R = -CH$_2$CO-porf, M = 2H
25a, X = CH$_2$, R = H, M = Zn^{2+}
25b, X = S, R = H, M = Zn^{2+}

^{1}H NMR spectroscopy revealed that the considerable complexation induced chemical shifts were observed for porphyrin protons, while the calixarene signals remained unchanged. These results indicate that there is a direct contact/interaction between the porphyrins and fullerene. Moreover, a Job plot analysis confirmed the formation of 1 : 1 complexes of the receptors with fullerenes. In contrast, the model compound **26** bearing only one porphyrin unit showed no measurable interactions with C$_{60}$ or C$_{70}$. This suggests that the preorganization of porphyrin moieties on the lower rim of the calixarene is an essential requirement for the effective complexation of fullerenes, and the whole system behaves like a pair of tweezers (Figure 8.11), where the calixarene itself does not interact with fullerenes but, rather, holds porphyrin units at a suitable mutual distance corresponding roughly to the diameter of C$_{60}$/C$_{70}$. Table 8.4 shows the complexation constants for C$_{60}$ and C$_{70}$ in toluene-d_8 measured by ^{1}H NMR spectroscopy. Interestingly, in nearly all cases C$_{70}$ was preferred over C$_{60}$ with the highest selectivity factor 14 for compound **25b**. In this context, it is somewhat unexpected that almost identical receptors **25a** and **25b**, which differ only in the central scaffold unit (calix[4]arene versus thiacalix[4]arene), exhibit substantially different selectivity in fullerene recognition [54].

8 Fullerene Receptors Based on Calixarene Derivatives

27, Ar = (3,5-di-tert-butylphenyl)

28, Ar = (pentafluorophenyl)

The complexation phenomenon can also be simply followed using UV-vis spectroscopy. After addition of C_{60} or C_{70} the original split Soret band undergoes a significant hypochromic shift, and well-defined isosbestic points appeared. The complexation was also evidenced by fluorescence spectroscopy. While the lifetime of **23a** did not show any changes upon fullerene addition, steady state fluorescence was strongly quenched as a consequence of photoinduced electron transfer between porphyrin moiety and fullerene (Figure 8.12).

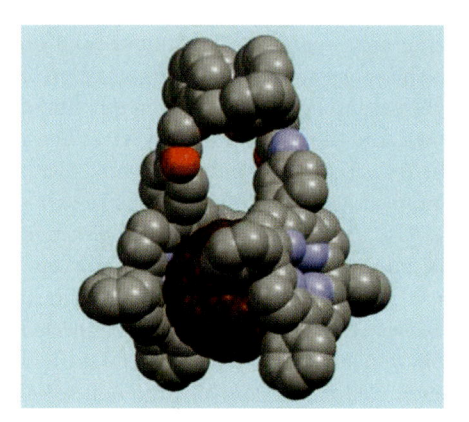

Figure 8.11 Computer generated figure of a molecular tweezer-based fullerene receptor.

Table 8.4 Association constants K (M^{-1}) with fullerenes measured by ^{1}H NMR titration in toluene-d_8 at 298 K.

	C_{60}	C_{70}
23a	4290	21 100
23b	2340	15 600
24a	3510	3330
24b	3420	6350
25a	8600	27 950
25b	2710	37 400

A similar design based on calix[4]arene-bisporphyrin conjugates has been described recently [56]. Two porphyrin groups were distally appended onto the lower rim of calix[4]arene and the complexation of fullerenes was optimized by varying the spacer linkage and the nature of arylporphyrins. The highest complexation constants in toluene were obtained with derivative **27**, which has 3,5-bis-*tert*-butylphenyl substituents on a tetraarylporphyrin moiety ($K_{60} = 26\,000 \pm 4000$ M^{-1}, $K_{70} = 234\,000 \pm 20\,000$ M^{-1}), with a selectivity factor C_{70}/C_{60} equal to 9. The corresponding Zn(II) or Cu(II) metalloporphyrin derivatives showed lower complexation constants ($K_{70}[Zn] = 207\,000 \pm 25\,000$ M^{-1}, $K_{70}[Cu] = 120\,000 \pm 9000$ M^{-1}) with slightly higher selectivity for C_{70} (<11). Interestingly, using receptor **28**, which possesses electron-withdrawing substituents (pentafluorophenyl groups), led not only to much lower constants but also to worse selectivity towards C_{70} (selectivity factor <2). The thermodynamics of fullerene complexation have been determined in terms of ΔH

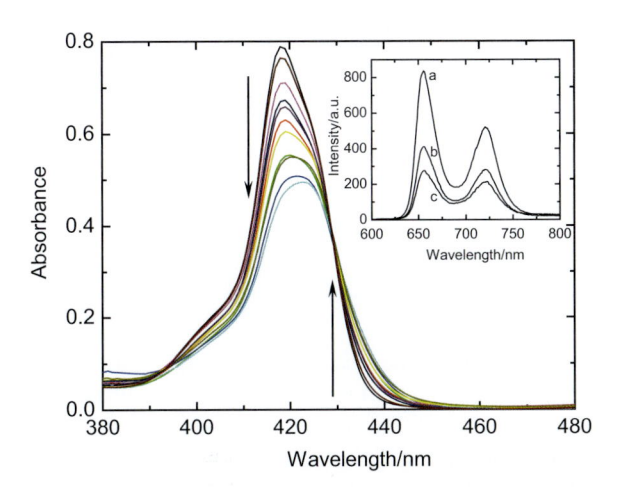

Figure 8.12 Difference UV-vis spectra showing the Soret band of **23a** after addition of C_{70}. Arrows follow changes due to increasing concentrations of C_{70}. The isosbestic point is at 429 nm. Inset: example of steady-state fluorescence spectra of **23a** (a) in the presence of various concentrations of C_{70} (b) and (c); all spectra in toluene.

and ΔS; the results are consistent with previously mentioned enthalpy-driven, solvation-dependent process.

Recently, molecular tweezers **29** based on calix[4]arene and thiacalix[4]arene derivatives immobilized in the *1,3-alternate* conformation were reported [57]. In principle, these compounds could bind two fullerenes at both sides of the molecule. In any event, contrary to expectation, the UV-vis and ^{1}H NMR studies proved the formation of only 1:1 complexes. In all cases, the receptors have substantially stronger affinity towards fullerene C_{70} than C_{60}. The resulting selectivity factor is again solvent-dependant, with highest values in benzene (>10).

29, X = -S- or -CH$_2$-
Y = -CH=N- or -CH$_2$-NH-

8.6
Outlook

The examples described above demonstrate the amazing progress made during the last two decades in the joint supramolecular chemistry of calixarenes and fullerenes. On the other hand, despite some remarkable and obvious successes, the authors are confident that we are still only at the beginning of a long road, the end of which cannot be simply foreseen. Starting from simple complexation studies scientists are heading toward the design of more sophisticated fullerene–calixarene assemblies with useful functions on a macroscopic level. By learning more and more about the mutual interactions of both systems, their shape and interaction complementarity, it should be possible to design novel supramolecular systems with precisely defined three-dimensional architecture, giving birth to novel materials with tailor-made properties and many practical applications. Hence, we dare say that continuing progress in the chemistry of these compounds will bring many fascinating discoveries in the near future.

References

1. Kroto, H.W., Heath, J.R., O'Brien, S.C., Curl, R.F., and Smalley, R.E. (1985) C60: buckminsterfullerene. *Nature*, **318**, 162–163.
2. Fowler, P.W. and Manolopoulos, D.E. (2006) *An Atlas of Fullerenes*, Dover Publication, Mineola.
3. (a) For books on fullerenes see, for example: Hirsch, A. (1994) *The Chemistry of Fullerenes*, Thieme, New York;(b) Hirsch, A. and Brettreich, M. (2005) *Fullerenes-Chemistry and Reactions*, Willey-VCH Verlag GmbH, Weinheim; (c) Langa, F. and Nierengarten, J.F. (eds) (2007) *Fullerenes: Principles and Applications*, RCS Publishing, Cambridge.
4. Krätschmer, W., Lamb, L.D., Fostiropoulos, K., and Huffmann, D.R. (1990) Solid C_{60}: a new form of carbon. *Nature*, **347**, 354–358.
5. Taylor, R. and Burley, G.A. (2007) Production, isolation and purification of fullerenes, in *Fullerenes: Principles and Applications* (eds F. Langa and J.F. Nierengarten) RCS Publishing, Cambridge, pp. 1–14.
6. Makha, M., Purich, A., Raston, C.L., and Sobolev, A.N. (2006) Structural diversity of host-guest and intercalation complexes of fullerene C_{60}. *Eur. J. Inorg. Chem.*, 507–517.
7. (a) Tashiro, K. and Aida, T. (2007) Metalloporphyrin hosts for supramolecular chemistry of fullerenes. *Chem. Soc. Rev.*, **36**, 189–197; (b) Wang, Y.-B. and Lin, Z. (2003) Supramolecular interactions between fullerenes and porphyrins. *J. Am. Chem. Soc.*, **125**, 6072–6073.
8. Andersson, T., Nilsson, K., Sundahl, M., Westman, G., and Wennerström, O. (1992) C_{60} embedded in cyclodextrin: a water-soluble fullerene. *J. Chem. Soc., Chem. Commun.*, 604–606.
9. Delgado, J.L. and Nierengarten, J.F., Fullerenes and calixarenes, in *Calixarenes in the Nanoworld* (eds J. Vicens, J. Harrowfield, and L. Baklouti) Springer, Dordrecht, ch 9.
10. (a) Ikeda, A., Yoshimura, M., Udzu, H., Fukuhara, C., and Shinkai, S. (1999) Inclusion of [60]fullerene in a homooxacalix[3]arene-based dimeric capsule cross-linked by a Pd^{II}-pyridine interaction. *J. Am. Chem. Soc.*, **121**, 4296–4297; (b) Tsubaki, K., Tanaka, K., Kinoshita, T., and Fuji, K. (1998) Complexation of C_{60} with hexahomooxacalix[3]arenes and supramolecular structures of complexes in the solid state. *Chem. Commun.*, 895–896.
11. (a) Fox, O.D., Cookson, J., Wilkinson, E.J.S., Drew, M.G.B., McLean, E.J., Teat, S.J., and Beer, P.D. (2006) Nanosized polymetallic resorcinarene-based host assemblies that strongly bind fullerenes. *J. Am. Chem. Soc.*, **128**, 6990–7002; (b) Rose, K.N., Barbour, L.J., Orr, G.W., and Atwood, J.L. (1998) Self-assembly of carcerand-like dimers of calix[4]resorcinarene facilitated by hydrogen bonded solvent bridges. *Chem. Commun.*, 407–408.
12. (a) Huerta, E., Cequier, E., and de Mendoza, J. (2007) Preferential separation of fullerene[84] from fullerene mixtures by encapsulation. *Chem. Commun.*, 5016–5018; (b) Rio, Y. and Nierengarten, J.-F. (2002) Water soluble supramolecular cyclotriveratrylene-[60]fullerene complexes with potential for biological applications. *Tetrahedron Lett.*, **43**, 4321–4324.

13 (a) For books on calixarenes see: Vicens, J., Harrowfield, J., and Baklouti, L. (eds) (2007) *Calixarenes in the Nanoworld*, Springer, Dordrecht; (b) Asfari, Z., Böhmer, V., Harrowfield, J., and Vicens, J. (eds) (2001) *Calixarenes 2001*, Kluwer Academic Publishers, Dordrecht; (c) Mandolini, L. and Ungaro, R. (2000) *Calixarenes in Action*, Imperial College Press, London; (d) Gutsche, C.D. (1998) *Calixarenes Revisited, Monographs in Supramolecular Chemistry*, vol. 6 (ed. J.F. Stoddart), The Royal Society of Chemistry, Cambridge.

14 Kawase, T. and Kurata, H. (2006) Ball-, bowl-, and belt-shaped conjugated systems and their complexing abilities: exploration of the concave-convex π-π interaction. *Chem. Rev.*, **106**, 5250–5273.

15 Williams, R.M. and Verhoeven, J.W. (1992) Supramolecular encapsulation of fullerene C_{60} in a water-soluble calixarene: a core-shell charge-transfer complex. *Recl. Trav. Chim. Pays-Bas*, **111**, 531–532.

16 Barcza, A.B., Rohonczy, J., Rozlosnik, N., Gilanyi, T., Szabo, B., Lovas, G., Braun, T., Samu, J., and Bacza, L. (2001) Aqueous solubilization of [60]fullerene *via* inclusion complex formation and the hydration of C_{60}. *J. Chem. Soc., Perkin Trans. 2*, 191–196.

17 Atwood, J.L., Koutsantonis, G.A., and Raston, C.L. (1994) Purification of C_{60} and C_{70} by selective complexation with calixarenes. *Nature*, **368**, 229–231.

18 Suzuki, T., Nakashima, K., and Shinkai, S. (1994) Very convenient and efficient purification method for fullerene (C_{60}) with 5,11,17,23,29,35,41,47-octa-*tert*-butylcalix[8]arene-49,50,51,52,53,54,55,56-octol. *Chem. Lett.*, 699–702.

19 Suzuki, T., Nakashima, K., and Shinkai, S. (1995) Influence of *para*-substituents and solvents on selective precipitation of fullerenes by inclusion in calix[8]arenes. *Tetrahedron Lett.*, **36**, 249–252.

20 Atwood, J.L., Barbour, L.J., Raston, C.L., and Sudria, I.B.N. (1998) C_{60} and C_{70} compounds in the pincerlike jaws of calix[6]arene. *Angew. Chem. Int. Ed.*, **37**, 981–983.

21 Makha, M., Raston, C.L., Sobolev, A.N., and Turner, P. (2006) Exclusive endo-cavity interplay of *t*-Bu-calix[6]arene with C_{70}. *Cryst. Growth Design*, **6**, 224–228.

22 Williams, R.M., Zwier, J.M., and Verhoeven, J.W. (1994) Interactions of fullerenes and calixarenes in the solid state studied with ^{13}C CP-MAS NMR. *J. Am. Chem. Soc.*, **116**, 6965–6966.

23 Raston, C.L., Atwood, J.L., Nichols, P.J., and Sudria, I.B.N. (1996) Supramolecular encapsulation of aggregates of C_{60}. *Chem. Commun.*, 2615–2616.

24 Atwood, J.L., Barbour, L.J., Heaven, M.W., and Raston, C.L. (2003) Controlling van der Waals contacts in complexes of fullerene C_{60}. *Angew. Chem. Int. Ed.*, **42**, 3254–3257.

25 Atwood, J.L., Barbour, L.J., Heaven, M.W., and Raston, C.L. (2003) Association and orientation of C_{70} on complexation with calix[5]arene. *Chem. Commun.*, 2270–2271.

26 Atwood, J.L., Barbour, L.J., Nichols, P.J., Raston, C.L., and Sandoval, C.A. (1999) Symmetry-aligned supramolecular encapsulation of C_{60}:[$C_{60(L)_2}$], L=*p*-benzylcalix[5]arene or *p*-benzylhexahomooxacalix[3]arene. *Chem. Eur. J.*, **5**, 990–996.

27 Haino, T., Yanase, M., and Fukazawa, Y. (1997) Crystalline supramolecular complex of C60 with calix[5]arenes. *Tetrahedron Lett.*, **38**, 3739–3742.

28 Isaacs, N.S., Nichols, P.J., Raston, C.L., Sandoval, C.A., and Young, D.J. (1997) Solution volume studies of a deep cavity inclusion complex of C_{60}: [*p*-benzylcalix[5]arene $_{C_{60}}$]. *Chem. Commun.*, 1839–1840.

29 Barbour, L.J., Orr, G.W., and Atwood, J.L. (1998) Supramolecular assembly of well-separated, linear columns of closely-spaced C_{60} molecules facilitated by dipole induction. *Chem. Commun.*, 1901–1902.

30 Makha, M., Raston, C.L., Sobolev, A.N., Barbour, L.J., and Turner, P. (2006) Endo- versus exo-cavity interplay of *p*-benzylcalix[4]arene with spheroidal molecules. *Cryst. Eng. Commun.*, **8**, 306–308.

31 Hardie, M.J. and Raston, C.L. (1999) Confinement and recognition of icosahedral main group cage molecules: fullerene C_{60} and *o*-, *m*-, *p*-dicarbadodecaborane(12). *Chem. Commun.*, 1153–1163.

32 Bhattacharya, S., Chattopadhyay, S., Nayak, S.K., and Banerjee, M. (2005) Solution NMR studies of supramolecular complexes of [60]- and [70]fullerenes with mono O-substituted calix[6]arene. *Spectrochim. Acta A*, **62**, 729–735.

33 Araki, K., Akao, K., Ikeda, A., Suzuki, T., and Shinkai, S. (1996) Molecular design of calixarene-based host molecules for inclusion of C_{60} in solution. *Tetrahedron Lett.*, **37**, 73–76.

34 Haino, T., Yanase, M., and Fukazawa, Y. (1997) New supramolecular complex of C_{60} based on calix[5]arene – its structure in the crystal and in solution. *Angew. Chem. Int. Ed. Engl.*, **36**, 259–260.

35 Yanase, M., Matsuoka, M., Tatsumi, Y., Suzuki, M., Iwamoto, H., Haino, T., and Fukazawa, Y. (2000) Thermodynamic study on supramolecular complex formation of fullerene with calix[5]arenes in organic solvents. *Tetrahedron Lett.*, **41**, 493–497.

36 Haino, T., Yanase, M., Fukunaga, C., and Fukazawa, Y. (2006) Fullerene encapsulation with calix[5]arenes. *Tetrahedron*, **62**, 2025–2035.

37 Wang, J., Bodige, S.G., Watson, W.H., and Gutsche, C.D. (2000) Complexation of fullerenes with 5,5′-biscalix[5]arene. *J. Org. Chem.*, **65**, 8260–8263.

38 Haino, T., Yanase, M., and Fukazawa, Y. (1998) Fullerenes enclosed in bridged calix[5]arenes. *Angew. Chem. Int. Ed.*, **37**, 997–998.

39 Haino, T., Seyama, J., Fukunaga, C., Murata, Y., Komatsu, K., and Fukazawa, Y. (2005) Calix[5]arene-based receptor for dumb-bell-shaped C_{120}. *Bull. Chem. Soc. Jpn.*, **78**, 768–770.

40 Haino, T., Fukunaga, C., and Fukazawa, Y. (2006) A new calix[5]arene-based container: selective extraction of higher fullerenes. *Org. Lett.*, **8**, 3545–3548.

41 Haino, T., Fukunaga, C., and Fukazawa, Y. (2007) Complexation of higher fullerenes by calix[5]arene-based host molecules. *J. Nanosci. Nanotechnol.*, **7**, 1386–1388.

42 Iglesias-Sánchez, J.C., Fragoso, A., de Mendosa, J., and Prados, P. (2006) Aryl-aryl linked bi-5,5′-*p-tert*-butylcalix[4]arene tweezer for fullerene complexation. *Org. Lett.*, **8**, 2571–2574.

43 Kunsági-Máté, S., Szabó, K., Bitter, I., Nagy, G., and Kollár, L. (2004) Complex formation between water-soluble sulfonated calixarenes and C_{60} fullerene. *Tetrahedron Lett.*, **45**, 1387–1390.

44 Haino, T., Araki, H., Yamanaka, Y., and Fukazawa, Y. (2001) Fullerene receptor based on calix[5]arene through metal-assisted self-assembly. *Tetrahedron Lett.*, **42**, 3203–3206.

45 Haino, T., Yamanaka, Y., Araki, H., and Fukazawa, Y. (2002) Metal-induced regulation of fullerene complexation with double-calix[5]arene. *Chem. Commun.*, 402–403.

46 Yanase, M., Haino, T., and Fukazawa, Y. (1999) A self-assembling container for fullerenes. *Tetrahedron Lett.*, **40**, 2781–2784.

47 Haino, T., Araki, H., Fujiwara, Y., Tanimoto, Y., and Fukazawa, Y. (2002) Fullerene sensors based on calix[5]arene. *Chem. Commun.*, 2148–2149.

48 Pan, G.B., Liu, J.M., Zhang, H.M., Wan, L.J., Zheng, Q.Y., and Bai, C.L. (2003) Conformation of calix[8]arene and a C_{60}/calix[8]arene complex on a Au(111) surface. *Angew. Chem. Int. Ed.*, **42**, 2747–2751.

49 Zhang, S. and Echegoyen, L. (2005) Supramolecular incorporation of fullerenes on gold surfaces: comparison of C_{60} incorporation by self-assembled monolayers of different calix[n]arene (n=4,6,8) derivatives. *J. Org. Chem.*, **70**, 9874–9881.

50 Boyd, P.D.W. and Reed, C.A. (2005) Fullerene-porphyrin constructs. *Acc. Chem. Res.*, **38**, 235–242.

51 (a) Shoji, Y., Tashiro, K., and Aida, T. (2006) Sensing of chiral fullerenes by a cyclic host with an asymmetrically distorted p-electronic component. *J. Am. Chem. Soc.*, **128**, 10690–10691; (b) Ouchi, A., Tashiro, K., Yamaguchi, K., Tsuchiya, T., Akasaka, T., and Aida, T. (2006) A self-regulatory host in an oscillatory guest motion: complexation of fullerenes with a short-spaced cyclix dimer of an organorhodium porphyrin. *Angew. Chem. Int. Ed.*, **45**, 3542–3546; (c) Wu, Z.Q., Shao, X.B., Li, C., Hou, J.L., Wang, K., Jiang, X.K., and Li, Z.T. (2005) Hydrogen-bonding-driven preorganized zinc porphyrin receptors for

efficient complexation of C_{60}, C_{70} and C_{60} derivatives. *J. Am. Chem. Soc.*, **127**, 17460–468; (d) Iwamoto, H., Yamaguchi, M., Hiura, S., and Fukazawa, Y. (2004) Synthesis of molecular tweezers bearing two porphyrins and its complexation toward electron acceptors. *Heterocycles*, **63**, 2005–2011;(e)Yamaguchi, T., Ishii, N., Tashiro, K., and Aida, T. (2003) Supramolecular peapods composed of metalloporphyrin nanotube and fullerenes. *J. Am. Chem. Soc.*, **125**, 13934–13935.

52 Arimura, T., Nishioka, T., Suga, Y., Murata, S., and Tachiya, M. (2002) Inclusion properties of a new metallo-porphyrin dimer derived from a calix[4]arene: tweezers for C_{70}. *Mol. Cryst. Liq. Cryst.*, **379**, 413–418.

53 Kas, M., Lang, K., Stibor, I., and Lhotak, P. (2007) Novel fullerene receptors based on calix-porphyrin conjugates. *Tetrahedron Lett.*, **48**, 477–481.

54 Dudic, M., Lhotak, P., Stibor, I., Petrickova, H., and Lang, K. (2004) (Thia)calix[4]arene-porphyrin conjugates: novel receptors for fullerene complexation with C_{70} over C_{60} selectivity. *New J. Chem.*, **28**, 85–90.

55 (a) For recent reviews on thiacalixarenes see: Lhoták, P. (2004) Chemistry of thiacalixarenes. *Eur. J. Org. Chem.*, 1675–1692; (b) Morohashi, N., Narumi, F., Iki, N., Hattori, T., and Miyano, S. (2006) Thiacalixarenes. *Chem. Rev,* **106**, 5291–5316.

56 Hosseini, A., Taylor, S., Accorsi, G., Armaroli, N., Reed, C.A., and Boyd, P.D.W. (2006) Calix[4]arene-linked bisporphyrin host for fullerenes: binding strength, solvation effects, and porphyrin-fullerene charge transfer bands. *J. Am. Chem. Soc.*, **126**, 15903–15913.

57 Kundrat, O., Kas, M., Tkadlecova, M., Lang, K., Cvacka, J., Stibor, I., and Lhoták, P. (2007) Thiacalix[4]arene-porphyrin conjugates with high selectivity towards fullerene C_{70}. *Tetrahedron Lett.*, **48**, 6620–6623.

9
Guanidinium Based Anion Receptors

Carsten Schmuck and Hannes Yacu Kuchelmeister

9.1
Introduction

Anion recognition plays a decisive role in a multitude of essential biological processes like enzyme activity, protein folding, or DNA regulation [1]. The vast majority of enzyme substrates that have been characterized are negatively charged species [2]. Therefore the importance of anion recognition for biological systems can hardly be neglected.

Nevertheless, the design of artificial anion receptors is a demanding task. In comparison to cationic species their relatively large size demands for larger receptors. Even simple inorganic anions occur in a manifold of geometries and forms: there are spherical (halides), tetrahedral (PO_4^{3-}, SO_4^{2-}), planar (NO_3^-), linear (SCN^-, N_3^-), and more complex examples (oligophosphates). In regard to cations of comparable size anions have a higher free energy of solution in water (e.g., $\Delta G_F^- = -434.3\,\text{kJ}\,\text{mol}^{-1}$, $\Delta G_K^+ = -337.2\,\text{kJ}\,\text{mol}^{-1}$) [3]. Therefore anion receptors have to compete more effectively with the surrounding medium. Finally, organic and also most inorganic anions often only exist in a small pH window, which may lead to problems with proton transfer due to the large basicity of the anion, especially in less polar solvents.

It is remarkable that in Nature the amino acid arginine is particularly often involved in the binding of anionic substrates by proteins [4]. Site-directed mutagenesis experiments in active sites of certain enzymes showed that – in the microenvironment of a protein – the energetic stabilization of carboxylates by the guanidinium moiety in the arginine side chain outmatches the analog interaction with the primary ε-ammonium group of lysine by $21\,\text{kJ}\,\text{mol}^{-1}$ [5]. As has been shown by X-ray analysis, this enormous stabilization is due to the Y-shaped, planar orientation of the guanidinium group, which is perfectly preorganized for a binding pattern with two strong, parallel hydrogen bonds that cooperate with the ion pairing (Figure 9.1) [6].

Furthermore, the extraordinarily high basicity of free guanidine (pK_a 13.5) or of the guanidine group in the arginine side chain (pK_a 12.5) [7], which may even be

---

Artificial Receptors for Chemical Sensors. Edited by V.M. Mirsky and A.K. Yatsimirsky
Copyright © 2011 WILEY-VCH Verlag GmbH & Co. KGaA, Weinheim
ISBN: 978-3-527-32357-9

Figure 9.1 The positively charged guanidinium moiety **2** can form two hydrogen bonds to anions like carboxylate (**1**) or phosphate (**3**).

increased by substitution, assures constant protonation throughout a large pH interval including physiological conditions. This is due to delocalization of the positive charge over three nitrogen atoms. The conservation of positive charge and hydrogen bond donor capacity over a broad pH range make guanidinium a versatile recognition motif.

A prominent example for molecular recognition in biological systems is the "arginine fork" (**4**) of the HIV-1 Tat protein depicted in Figure 9.2: a guanidinium

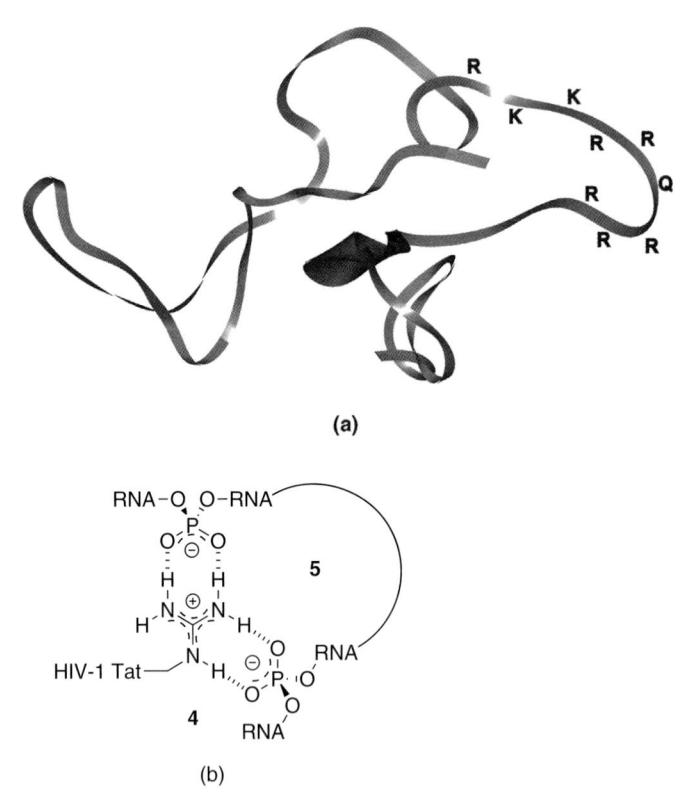

Figure 9.2 (a) Tertiary structure of the HIV-1 Tat protein as determined by NMR analysis. For better visualization only the binding region RKKRRQRRR (residues 49–57) is labeled; (b) the "arginine-fork" binding motif **4** recognizes RNA bulges and loops (**5**) that have two phosphate groups in close proximity.

group acts as binding motif for two phosphate groups that are in close proximity to each other and is thus able to recognize RNA bulges and loops [8].

Nature's success has inspired supramolecular chemists over the last 30 years to utilize the guanidinium group as a "work horse" for anion recognition. Its structure is easily modifiable and the unique binding pattern allows for predictable host–guest alignment. The following sections will give some examples of guanidinium based receptors, starting with an overview of the last 30 years paired with instructive background information before focusing on recent advances in the recognition of various types of anions. We do not intend to give a comprehensive overview of all work that has been done in this field, but we will, rather, concentrate on instructive and prominent examples in combination with newer findings of guanidinium based anion recognition.

9.2
Instructive Historical Examples

Artificial receptors based on simple alkylguanidinium cations designed over the last three decades have been comprehensively summarized [9]. Therefore, this chapter will focus on a few selected, older examples.

In the late 1970s Lehn and coworkers were the first to study the complexation behavior of macrocycles **6–8** that contain two or three guanidinium moieties (Figure 9.3) [10]. The association constants towards PO_4^{3-} in water are rather small (50–250 M^{-1}; pH titration) due to the competitive solvation by water. Ultimately, the binding energy results from the difference between the energy released upon complexation minus the energy that is necessary to remove the solvate shell from host and guest. The latter is very large for guanidinium derived ions in water.

With its positive charge and six potential hydrogen bond donor sites guanidinium is one of the most hydrophilic functional groups known [11]. A neutron diffraction

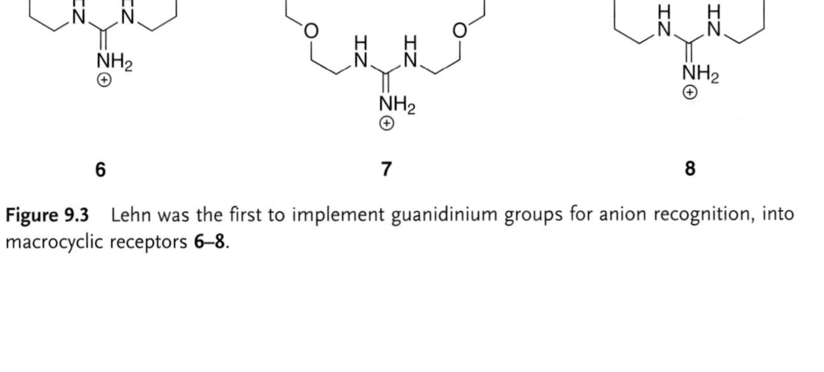

Figure 9.3 Lehn was the first to implement guanidinium groups for anion recognition, into macrocyclic receptors **6–8**.

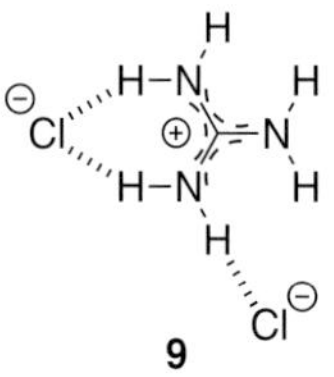

Figure 9.4 Dominant form of guanidinium chloride ion pairing in solution as determined by simulations.

experiment combined with molecular dynamics simulations demonstrated that the hydration shell is distributed anisotropically around the guanidinium core [12]. There is a strong tendency for the water molecules to form linear hydrogen bonds with the guanidinium NH. The second proton of each NH_2 group is additionally able to interact with the same water oxygen at an angle of approximately 50°. Owing to the directionality of the hydrogen bonds the water molecules are constrained to stay in the plane of the guanidinium core. By direct averaging there are approximately 4.5 water molecules noncovalently bound to a guanidinium cation. As indicated in Figure 9.4, anions like chloride compete exactly for these binding sites and therefore before binding can take place these water molecules need to be removed.

The solvation by water is so efficient that the stability of the ion pair lactate–guanidinium has a binding constant of merely $K \leq 6\,M^{-1}$ in water based on spectropolarimetry measurements [13]. A larger statistical analysis of various data of organic and inorganic ions assigns an average energetic stability of $\leq 5\,kJ\,mol^{-1}$ to a single salt bridge in water, which corresponds to an association constant of $K \leq 7\,M^{-1}$ [14]. Moreover these numbers are extrapolated for infinite dilute solutions (ionic strength $= 0$). In a more realistic scenario with millimolar salt concentration the strength of Coulomb based interactions is further decreased.

Proteins circumvent the problem: the interaction between receptor and substrate takes place in a hydrophobic pocket or in areas of low dielectric constant inside the enzyme. The complexation of substrates by artificial receptors, on the other hand, takes place in an environment that is normally fully exposed to the solvent. The lack of hydrophobic shielding is a general problem of supramolecular chemistry because noncovalent interactions – in contrast to covalent bonds – are highly dependent on external parameters such as solvent composition, polarity, or temperature [15].

In the early 1980s Schmidtchen and also de Mendoza and their coworkers utilized the charged, bicyclic guanidinium core **10** for the complexation of anions (Figure 9.5). Implementation of the guanidinium moiety into a decalin framework replaces hydrogen donor sites with organic moieties, thus improving the solubility in nonpolar solvents. The binding studies were carried out in chloroform as a less competing solvent. Schmidtchen's receptor (R = propenyl) binds to *p*-nitrobenzoate with $K = 7 \times 10^6\,M^{-1}$ (UV titration) [16]. De Mendoza's binding studies demonstrate the dependence of the association constant on the receptor's counterion: poorly coordinating species such as hexafluoro phosphate or tetraphenyl borate compete less with the substrate, which results in significantly stronger binding [17].

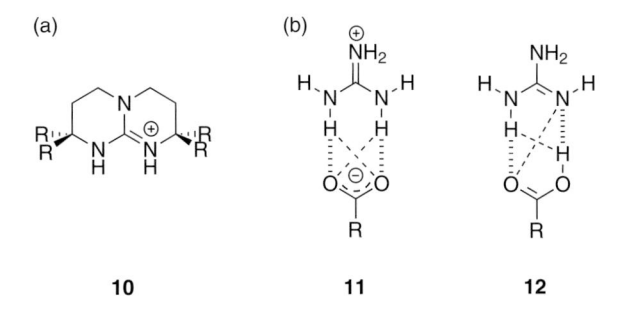

Figure 9.5 (a) Schmidtchen and De Mendoza implemented the guanidinium group in a decalin framework: the NH groups are preorganized in the favorable *syn* conformation; (b) the guanidinium-carboxylate complex adopts a DD-AA orientation (**11**) with favorable secondary interactions and not the corresponding DA-AD orientation in an alternative neutral complex (**12**).

The conformational freedom of the receptors is restricted: The guanidinium moiety is preorganized in a *syn* conformation in which the NH protons are optimally arranged for binding to two *syn* lone pairs of oxoanions. This leads to a stable ionic hydrogen-bonded DD-AA (donor–donor–acceptor–acceptor) complex (**11**) with attractive secondary interactions between donor and adjacent acceptor atoms. The large pK_a difference between guanidinium cation and carboxylic acids (≈9 units) prevents proton transfer that would otherwise lead to the loss of ion pairing and a less stable DA-AD complexation (**12**) with repulsive secondary interactions [18].

This bicyclic guanidinium served as a framework for manifold receptors for various anions in subsequent years; some of them are presented later in this chapter. However, this approach is mainly limited to nonpolar organic solvents.

Hamilton and coworkers designed the tweezer receptor **13** (Figure 9.6) that incorporates two guanidinium arms fixed to an aromatic template via amide bonds. UV and NMR binding studies revealed that **13** is able to bind diphenyl phosphate (**14**) in the more polar organic solvent acetonitrile with $K = 4.6 \times 10^4 \, M^{-1}$ [19]. The carbonyl group preorganizes the guanidinium cation by means of intramolecular hydrogen bonds. Additionally, it increases the acidity of the NH protons, thereby enhancing their hydrogen donor ability. The pK_a of the guanidinium moiety may vary

Figure 9.6 Hamilton's tweezer receptor **13** can catalyze phosphodiester cleavage supposedly by stabilizing the pentagonal transition state (**14**) of the nucleophilic (Nu) substitution.

depending on the adjacent group: in general the pK_a will increase in the order acyl ≥ phenyl ≥ alkyl substitution. The implementation of the second guanidinium group augments the number of potential binding sites for attractive interactions between substrate and receptor. At neutral pH phosphate (as HPO_4^{2-}) and sulfate present a tetrahedral binding motif with two negative charges. Carboxylate and nitrate, on the other hand, represent a trigonal planar geometry with just one charge. Therefore, two guanidinium moieties in the right geometrical orientation are ideal for binding to phosphate and sulfate whereas one guanidinium is sufficient for carboxylate and nitrate binding.

In the complex between **13** and **14** the combination of several interactions results in high affinity. In general, the best approach to achieve substrate binding in polar solvents by artificial hosts is to use not only one but several noncovalent interactions at the same time. Though one interaction alone might not be strong enough by itself the combination of many weak interactions may lead to high binding constants ("Gulliver effect") [20].

Another goal of supramolecular chemistry is the mimicry of enzymes, that is, receptors that are not only able to recognize a given substrate but also to catalyze a reaction. To achieve catalytic activity the receptor needs to bind more strongly to the transition state than to the substrate. Receptor **13** accelerates the cleavage of phosphodiesters in RNA by a factor of 700 in comparison to the uncatalyzed scenario by stabilizing the trigonal-bipyramidal intermediate **15** (as a model for the transition state) via four hydrogen bonds and two salt bridges [21].

Hamilton's later studies revealed some thermodynamic aspects of the interaction between guanidinium based receptors (**16** and **17** and similar urea and thiourea based systems; Figure 9.7) and dicarboxylates in polar solvents (from DMSO to water) by means of isothermal titration calorimetry (ITC) [22]. As expected, stronger binding occurs with increasing hydrogen acidity whereas binding constants decrease in more polar solvents. More interestingly, in this system the association between guanidinium and carboxylate is enthalpy driven in DMSO; however, in more polar solvents like methanol or water entropy becomes the driving force due to the liberation of

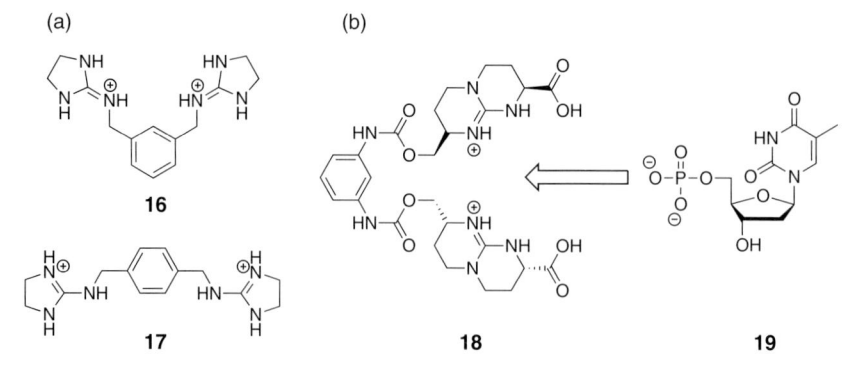

Figure 9.7 (a) Thermodynamic studies (ITC) carried out by Hamilton and coworkers revealed that – depending on the solvent – **16** and **17** bind to dicarboxylates driven by entropy; (b) Schmidtchen's receptor **18** binds to thymidine-5′-phosphate (**19**) in water with excellent affinity.

solvent molecules during the binding event. These results once more stress the crucial influence of the medium on noncovalent interactions.

Schmidtchen and coworkers designed the tweezer receptor **18** that employs the bicyclic guanidinium scaffold **10** into each of the two arms. This work provides the first example of specific binding of a mononucleotide in water (thymidine-5′-phosphate **19**, $K = 10^6\,\text{M}^{-1}$, ^{1}H NMR titration) [23]. The enormously high binding affinity can be explained by the ideal geometrical alignment of the bicyclic guanidines perpendicular to each other and perfectly preorganized for the complexation of tetrahedral shaped anions by two hydrogen bonded salt bridges.

Based on the bicyclic guanidinium receptors Schmidtchen also studied extensively the thermodynamic driving forces of molecular recognition events using ITC measurements [24]. In his work he emphasizes the importance of entropy and solvent effects on supramolecular binding events, which are often neglected in the classical lock-and-key approach to receptor design that strives for enthalpically favorable complementary between host and guest [25]. The change of entropy is determined by two opposing effects: on the one hand receptor and substrate lose translational and rotational degrees of freedom upon complexation (negative ΔS^0, entropically disadvantageous) while, on the other hand, solvent molecules and counterions are liberated from host and guest, which is favorable for entropy (positive ΔS^0).

The measurements show that receptors with a bulky organic periphery that are less tightly solvated show a favorable enthalpy change on complexation while entropy is diminished because only little solvent is liberated. More strongly solvated receptors showed smaller binding enthalpy but a more favorable entropy. Furthermore, it becomes clear once more that the counterion exerts a strong influence on complexation, which results in differences in the binding constant of up to a power of ten. As for the solvent molecules, the stronger the counterion coordinates to the receptor the smaller is the gain in enthalpy.

These observations are an example of enthalpy–entropy compensation for weak interactions in polar solvents, that is, the gain in enthalpy caused by groups that have a high affinity for the substrate is decreased by an opposite influence of the entropy and vice versa [26]. Notably, ΔS should be taken into consideration in the design of new artificial receptors because it might either become an obstacle or the driving force for molecular recognition.

Anslyn and coworkers have prepared the tripodal receptor **20** that binds to citrate (**21**) through three H-bond assisted salt bridges between carboxylate and guanidinium (Figure 9.8). Sterics determine the preorganization of the three arms on one side of the aromatic template, thus yielding a cavity neatly formed for the complexation of small molecules. This receptor is a beautiful example of how the accumulation of weak noncovalent interactions may overcome the competitive solvation. Compound **20** not only shows a high affinity for its substrate in water ($6.9 \times 10^3\,\text{M}^{-1}$, ^{1}H NMR titration) but is also very selective [27]. It can be used to successfully determine the citrate content of crude orange juices in the presence of other carboxylates by means of fluorescence and absorption analysis of an indicator displacement assay [28].

Figure 9.8 Anslyn's tripodal receptor **20** is preorganized for the formation of a 1 : 1 complex with citrate (**21**). High binding constants are achieved, even in water.

However, in buffered solution the affinity decreases by more than two orders of magnitude ($K < 10^2$ M^{-1}) compared to pure water due to the increased ionic strength of the solvent that accompanies the higher salt concentration. It may be concluded that a 1 : 1 ion pairing of alkylguanidinium cations with carboxylates is normally not strong enough for the formation of satisfactory stable complexes in highly competitive solvents like buffered water. For possible future applications artificial receptors have to be able to compete with *in vivo* conditions. Therefore, receptors should be able to effectively bind the desired substrate in the presence of a 120 mM aqueous sodium chloride solution. This is where receptors based on simple alkylguanidinium cations reach their limits. A more sophisticated design with additional interactions between substrate and receptor is needed.

Following this idea Schmuck and coworkers introduced guanidiniocarbonyl-pyrroles of type **22** as a new binding motif for the recognition of carboxylates in aqueous solvent (Figure 9.9) [29]. The structurally rather simple receptor **24** has a binding constant of $K = 2.8 \times 10^3$ M^{-1} (^{1}H NMR titration) for the complexation of acetate in 40% water in DMSO whereas simple alkylguanidinium cations show no detectable affinity at all under these conditions.

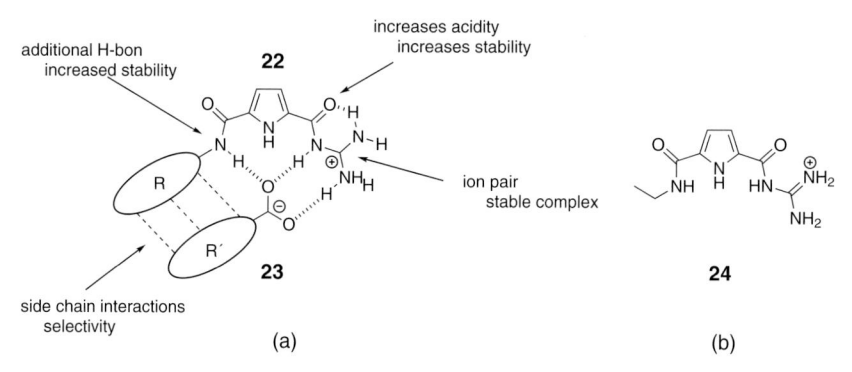

Figure 9.9 (a) Guanidiniocarbonyl pyrrole cations (**22**) efficiently bind carboxylates (**23**) even in aqueous solvents due to a combination of ion pair formation and additional H-bonds; (b) prototype receptor **24** binds to acetate with good affinity in aqueous solvent.

The carboxylate binding site (CBS) was designed based on theoretical calculations. It incorporates additional H-bond donors that further enhance the stability of the complex between receptor and substrate. As mentioned before, the increased acidity of acyl guanidiniums (p$K_a \approx$ 6–7) favors the formation of hydrogen bonded ion pairs. The binding motif is planar and rather rigid and therefore ideally preorganized for the binding of planar anions. Additional secondary interactions between receptor side chain and carboxylate can easily be introduced to increase affinity and achieve selectivity.

To better understand the complexation behavior a semi-quantitative estimate for the various energetic contributions of the individual binding interactions has been derived from a systematic variation of the receptor structure [30]. Of course, it is impossible to exactly determine the binding energy of an individual bond or type of interaction within an array of several noncovalent interactions as the replacement of one interaction affects the remaining ones. Nevertheless, if the change in complex stability that accompanies the switching on and off, respectively, of a specific interaction is significantly large, the importance of this interaction for the overall binding can at least be estimated.

While the parent guanidinium chloride **25** (Figure 9.10) does not bind at all to N-acetyl alanine in aqueous DMSO, the increased acidity of acylguanidinium (**26**) increases the binding affinity (50 M^{-1}). An additional hydrogen bridge from the pyrrole NH (**27**) augments the association constant significantly (130 M^{-1}). Further stabilization is the result of secondary interactions and therefore dependent on the side chain substitution. An amide group at position 5 of the pyrrole ring (as in **24**) increases the binding constant to 770 M^{-1}. While the glycine derivative **28** has more or less the same association constant as **24** (680 M^{-1}) valine substitution (**29**) increases the binding strength by a factor of two (1610 M^{-1}). Further investigations showed that the size and electronic structure of the aromatic ring is also of importance [31]: pyrrole systems are superior to analogous benzene derivatives,

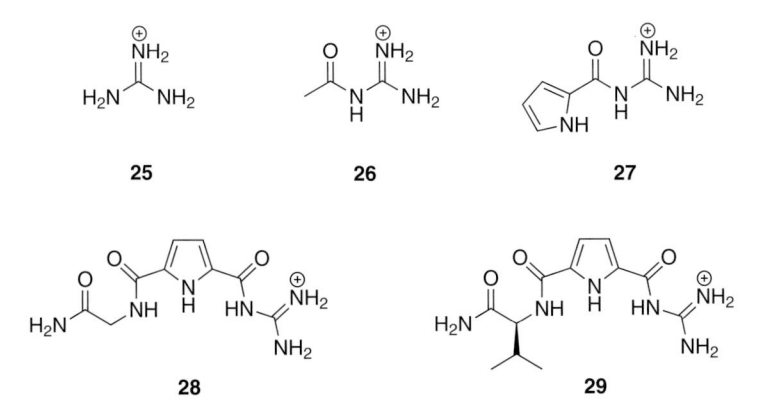

Figure 9.10 Various guanidinium compounds have been studied for their ability to bind N-acetyl alanyl carboxylate. Besides the ion pairing, mainly the amide NH at position 5 of the pyrrole (in **28** or **29**) increases the complex stability significantly.

which in turn show higher binding affinities than pyridine or furan derivatives. The lone pair on the heteroatom in the ring of the latter two provides additional repulsive interactions to the carboxylate [32].

Based on these data the energetic contributions of the diverse interactions were assigned semi-quantitatively. The hydrogen bonds are not equally important for the binding but differ significantly in their energetic contribution. Thus the acylguanidinium group with its hydrogen bonded salt bridge contributes $10\,\mathrm{kJ\,mol^{-1}}$ to the overall ΔG, the pyrrole NH another $2\,\mathrm{kJ\,mol^{-1}}$, and the amide NH $4\,\mathrm{kJ\,mol^{-1}}$. Owing to the high flexibility of the terminal carbamoyl group its substitution determines whether the binding energy is further increased (by up to $2\,\mathrm{kJ\,mol^{-1}}$). These numbers are of course only valid for this receptor type under the assumption that no pronounced cooperativity or anticooperativity takes place upon substrate complexation.

Recent state-of-the-art theoretical calculations are in good agreement with these experimental findings [33]. DFT calculations show that hydrogen bond 1 between **30** and **31** is the strongest interaction due to the close proximity between O and N atoms, well within the sum of their van der Waals radii (Figure 9.11a). The distance between the atoms involved in hydrogen bond 4 on the other hand is longer than the sum of their van der Waals radii. Hydrogen bonds 2 and 3 are bifurcated interactions. The CO of an additional amide group like in **24** accepts a hydrogen bond from the amide NH of the substrate. As depicted in Figure 9.11, the guanidiniocarbonyl-pyrrole moiety can exit in three conformations arising from rotation around the pyrrole carbonyl amide bonds. Calculations suggest that the out-out conformation **32** is the most stable because of intramolecular dipole–dipole interactions. The second most stable conformer is the out-in one (**33**). The in-in conformation **34** is the most unfavorable. The calculations show that substrate binding occurs from the out-in conformer **33**. The more stable out-out conformer **32** is not suitable for substrate

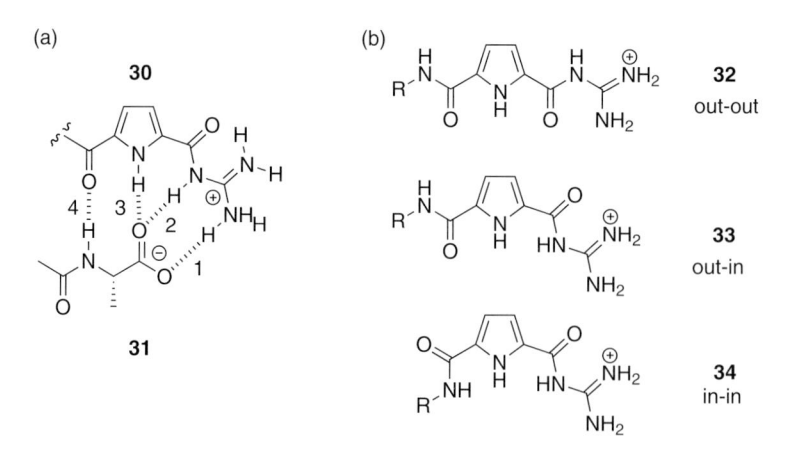

Figure 9.11 (a) Numbering of hydrogen bonds involved in complexation between Schmuck's CBS and N-acety alanyl carboxylate; (b) conformations of guanidiniocarbonyl-pyrrole hosts that arise from rotation around the pyrrole carbonyl amide bonds.

binding. Additionally, upon binding amino acid derivatives the CO at position 5 of the pyrrole ring in **33** accepts a hydrogen bond from the N-acetyl amide group of the substrate.

Energetic considerations based on comparison between DFT and MD calculations seem to indicate that most of the binding energy can be linked to a strong binding enthalpy (approx. $8\,kJ\,mol^{-1}$) that emerges from the formation of three hydrogen bonds between the guanidiniocarbonyl-pyrrole cation and the negatively charged carboxylate of the substrate. The second amide group at position 5 can form an additional hydrogen bond to the substrate that further stabilizes the complex (approx. $2\,kJ\,mol^{-1}$).

In summary, the guanidiniocarbonyl-pyrrole moiety has proved to be a very potent binding motif for oxoanions. Therefore, this binding motif served as a scaffold for many artificial receptors for various substrates that will in part be presented later.

9.3
Recent Advances in Inorganic Anion Recognition

The following paragraphs focus mainly on recent developments of anion recognition by artificial guanidinium-containing receptors, that is, the period from 2003 to today. Initially, for anion recognition the guanidinium moiety was used mainly to bind to tetrahedral anions like sulfate and phosphate. Around the turn of the millennium carboxylate recognition became increasingly important, a trend that continues to this day. This section begins with examples of receptors for inorganic anions like phosphate, nitrate and sulfate that have been published in the last five years. Carboxylate recognition is described in later sections.

Despite the highly symmetric trigonal planar binding motif it is difficult to design nitrate receptors. Nitrate competes quite poorly for hydrogen donor sites with other anions because it is a rather weak base that shows only a little tendency for the formation of robust hydrogen bonded frameworks in solution. De Mendoza and coworkers were the first to design a guanidinium-based receptor for the recognition of nitrate [34]. Macrocycles **35–37** contain one guanidinium and two urea moieties, thus affording six potential hydrogen donor sites while introducing just one positive charge. The H-bond donors are complementary in number and orientation to the lone pair orbitals of nitrate while the implementation of just one positive charge into the receptor guarantees charge complementary [35]. The receptor **37** binds nitrate with $K = 7.4 \times 10^4\,M^{-1}$ (acetonitrile, ITC) and moderate selectivity ($NO_3^-/Cl^- = 1.3$). Both enthalpy and entropy are favorable; the latter is the main driving force. The solid state structure given in Figure 9.12 visualizes the embracing of nitrate in the predicted orientation.

Following Beer's work [36] Tárraga, Molina, and coworkers designed the anion receptor **38** that incorporates ferrocene and neutral guanidine moieties (Figure 9.13) [37]. The guanidine groups serve as linker and recognition element while the ferrocene unit introduces redox functionality allowing for electrochemical sensing of the guest [38]. The authors report the ability of this receptor to perform

284 | *9 Guanidinium Based Anion Receptors*

35 n = 4
36 n = 5
37 n = 6

(a)

(b)

Figure 9.12 (a) Macrocycles **35–37** offer six hydrogen donor sites that point into the cavity for the complexation of nitrate; (b) X-ray structure of the nitrate complex with **37** [34]. Hydrogen bonds are indicated by dotted lines.

redox-ratiometric measurements: in contrast to most ferrocene-based redox sensors that only rely on changes in the oxidation potential of the ferrocene moiety, in this case the appearance of a second oxidation potential leads to an individual perturbation of two redox potentials upon complexation. The receptor is able to recognize and discriminate various anions (F^-, AcO^-, HSO_4^-, and $H_2PO_3^-$) that elicit different electrochemical responses in differential pulse voltammetry measurements [39] in a solution of 10% water in DMSO. ITC revealed a binding constant

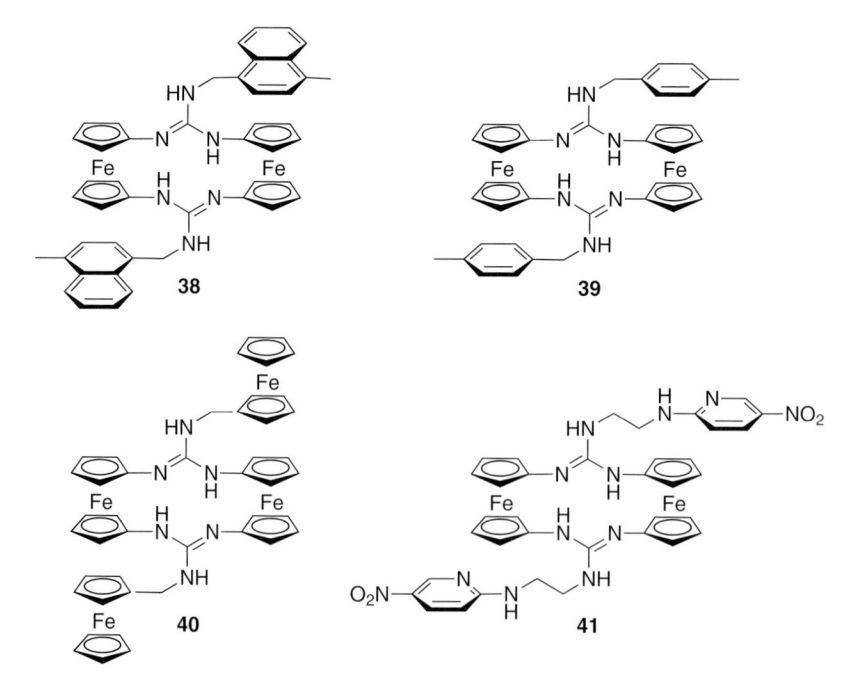

Figure 9.13 Tárraga, Molina, and their coworkers' guanidine-based receptors are able to sense simple anions electrochemically in organic solvents.

for $H_2PO_3^-$ in DMSO of $K = 2.8 \times 10^4 \, M^{-1}$. Whereas host **38** with the neutral guanidine does not react at all when adding Cl^- and NO_3^-, the protonated receptor is able to differentiate between these anions in dichloromethane.

Based on this first receptor Tárraga, Molina, and coworkers developed receptors **39–41** that behave similarly upon the addition of simple anions and are even able to distinguish between different amino acids (Glu, Trp, Leu, Phe) [40]. UV-vis data in dichloromethane are consistent with most ferrocenyl chromophores: two charge-transfer bands in the visible area that are perturbed upon complexation. For receptor **41** the change in color is visible with the naked eye when adding F^-, AcO^-, $H_2PO_4^-$, or $HP_2O_7^{3-}$ and quantifiable by means of UV titration. Binding constants in dichloromethane are in the range of $K \approx 10^5 \, M^{-1}$. Receptor **38** can selectively sense NO_3^- through fluorescence emission quenching.

Sulfate recognition by guanidinium based receptors has been much less studied than carboxylate or phosphate complexation. Sulfate is less basic than phosphate and therefore the affinity for hydrogen donors is less pronounced. The tetrahedral geometry with its two negative charges, on the other hand, provides a wider range of possible interactions. Accordingly, several minima were localized for hydrogen bridged sulfate-guanidinium dimers and trimers in a theoretical study [41]. Just like for phosphate binding, two guanidinium moieties are necessary to compensate the charges and hydrogen acceptor sites.

Kobiro, Inoue, and coworkers introduced sulfate receptor **42** based on a bicyclic guanidinium scaffold and a DMAP moiety as UV chromophore for spectral analysis [42]. Various studies (UV, CD, fluorescence, and NMR) led to the proposed binding mode depicted in Figure 9.14. Upon the addition of sulfate the 2:1 receptor-substrate complex **43** is formed first. When the amount of substrate surmounts 0.5 equivalents a 1:1 complex (**44**) arises. The two binding constants for the stepwise formation of the 1:1 and the 2:1 complex were measured in acetonitrile as $K_1 = 1.5 \times 10^6 \, M^{-1}$ and $K_2 = 4.8 \times 10^4 \, M^{-1}$ (^1H NMR titration), respectively. Still, the recognition of sulfate is a challenging problem, let alone the complexation in more demanding solvents such as alcohols or water.

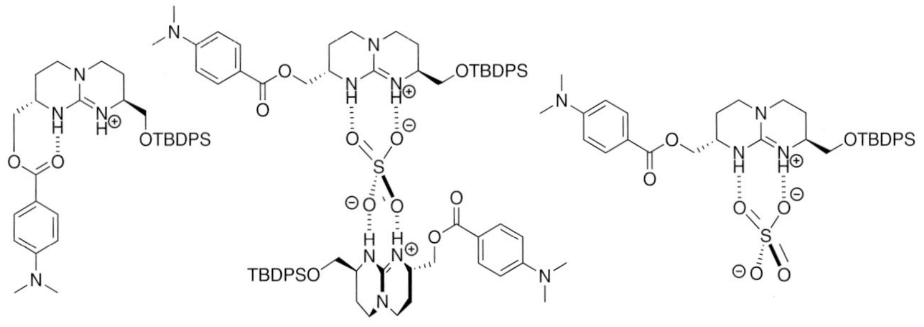

Figure 9.14 Guanidinium-based sulfate receptor **42** binds to sulfate in acetonitrile with a stoichiometry that depends on the amount of sulfate present.

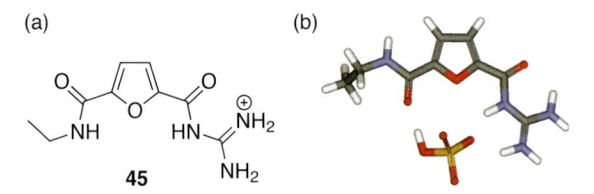

Figure 9.15 (a) Schmuck's furan based receptor **45** binds to hydrogen sulfate only in acidic solution; (b) energy-minimized structure for the complex between **45** and hydrogen sulfate.

Another approach for sulfate recognition by abiotic receptors has been conducted by Schmuck's group [43]. The guanidiniocarbonyl furan receptor **45** binds to hydrogen sulfate at pH 4.6 in a buffered 1:1 solution of water–DMSO with an association constant of $K = 6 \times 10^2 \, M^{-1}$ (UV titration) (Figure 9.15). Compared to a pyrrole the furan moiety reduces the electron density of the acylguanidinium, enhancing its acidity. With a pK_a of 5.5 the guanidine is protonated only at a pH below 5. Therefore complexation can only occur with weakly basic anions that are still deprotonated under acidic conditions like hydrogen sulfate but not dihydrogen phosphate or carboxylate.

The calculated structure shows that the oxoanion is oriented such that only one oxygen atom interacts with the guanidinium cation, in contrast to the normal bidentate binding mode, in order to avoid the repulsive interaction between sulfate and furan oxygen lone pairs. Additionally to the hydrogen bonded salt bridge between sulfate and guanidinium the OH group of the hydrogen sulfate forms a hydrogen bond with the CO of the amide group at position 5 of the receptor. The limiting factor of such furan based receptors, however, will be their low stability under the inevitable acidic conditions.

Anslyn and coworkers continued their work on the design of artificial receptors for the construction of chemosensors in supramolecular analytical chemistry [44]. The metallo-receptor **46** (Figure 9.16) was reported as a selective binder for monoprotonated phosphate and arsenate [45]. In a buffered solution of 2% methanol in water it binds to HPO_4^{2-} and $HAsO_4^{2-}$ with $K = 1.5 \times 10^4$ and $1.7 \times 10^4 \, M^{-1}$ (UV-vis

Figure 9.16 Metallo-receptor **46** selectively binds to HPO_4^{2-} and $HAsO_4^{2-}$ with high binding constant in water. The main binding energy comes from the metal center while the guanidinium arms ameliorate binding strength and selectivity.

titration), respectively. Other anions (AcO$^-$, NO$_3^-$, HCO$_3^-$, Cl$^-$, or SO$_4^{2-}$) are bound two orders of magnitude weaker. ITC studies show that the main part of the free energy that is released upon binding results from the complexation of the anion with the Cu(II) center, while the cationic arms tune the interactions in regard to stability and selectivity [46]. The complexation is driven by a favorable change of entropy and enthalpy. A host that incorporates ammonium instead of guanidinium groups binds more strongly but less selectively. Its binding is driven by entropy only. This disparity is due to the increased rigidity and preorganization and the lower grade of solvation of the guanidinium moiety in **46**, which leads to a complexation that is mainly driven by enthalpy. The higher solvated ammonium side chain, on the other hand, leads to a binding event that is driven by entropy.

Anslyn's receptor **46** could be used as a chemosensor to determine the phosphate concentration in protein-free samples of horse serum and saliva, using a colorimetric indicator-displacement assay, in agreement with clinically approved methods and literature data [47].

9.4
Organic and Biological Phosphates

This section covers receptor types that bind to more complex substrates like organic phosphates and biologically relevant ones, for example, nucleotides. Schmidtchen and coworkers continued their detailed thermodynamic studies of receptors for anionic guests. The complexation behavior of tetrasubstituted bicyclic guanidinium-based receptors **47** and **48** with phosphates of varying size (**49–52**) was measured via ITC (Figure 9.17) [48]. The implementation of carboxamide groups in **48** should introduce additional hydrogen bond donor sites in order to enhance the enthalpic binding strength.

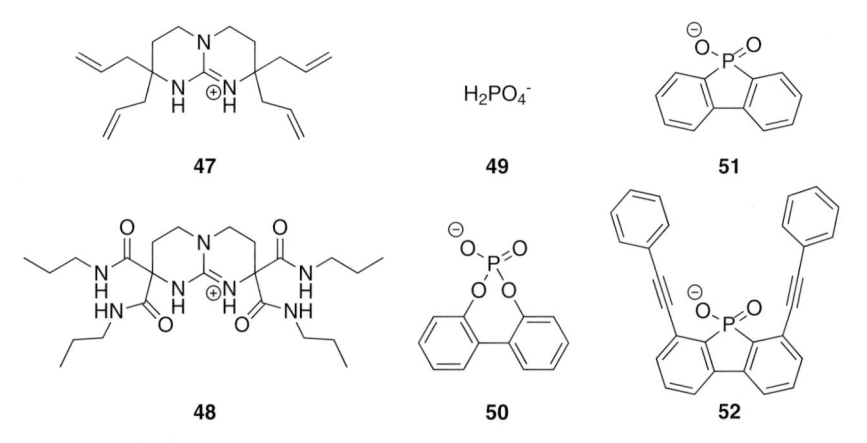

Figure 9.17 Schmidtchen could show that receptor **48** binds more strongly to phosphates **49–52** in acetonitrile than **47** due to a more favorable entropy contribution.

A comparative study of both receptors shows, as expected, that **48** binds more strongly to all substrates. At first glance this seems to verify the proposed concept of stronger enthalpic contribution. However, inspection of the underlying thermodynamic state function demonstrates that enthalpy is even worse than for the tetraallyl-substituted receptor **47**. The enhanced affinity is due to a much more favorable change of entropy during complexation. This is, however, not due to the release of the solvation shell of the polar amido functions because then the change in entropy should correlate to the size of the phosphate guests. It had been found previously that the more extended the interface between the binding partners the greater is the release of solvent molecules on the event of complexation [49]. The biggest difference in ΔS in this binding study, however, was found for the smallest substrates. A possible explanation, rather than desolvation, is an increase in configurational entropy of the binding partners as a consequence of low structural definition in the host–guest complex.

Although entropic contributions are nowadays attended to in supramolecular binding events, entropy is mainly associated with the liberation of solvent molecules, neglecting the change of entropy due to the diversity of the binding mode and the accompanying configurational entropic contributions that constitute a substantial share of the overall entropy output. A qualitative estimate can be deduced by testing closely related host–guest pairs under strictly identical conditions. In a recent study **48** was varied by attaching different moieties to the carboxamido group to tune affinity and selectivity and to get more insight into entropic factors [50].

In general the testing of **48** and two additional receptors (**53** and **54**; Figure 9.18) against **49** and **50** delivers similar results: binding energy mainly arises from positive association entropy and less from enthalpic interactions (hydrogen bonding).

The exception is the anilide host **54** with dihydrogen phosphate as guest. It is the only anchor group in this series that provides a negative entropy contribution, which in turn is compensated by the dramatically enhanced attractive binding enthalpy. Overall, a lower affinity results compared to all other host–guest pairs for the benefit of a better structured complex. Although dihydrogen phosphate shows the largest enthalpic contribution upon complexation of all tested oxoanions (probably due to the greater potential for hydrogen bonding) the structure-giving role of the phosphate guest has to be supported by synergistic mutual interactions of the phenyl rings in order to explain these results.

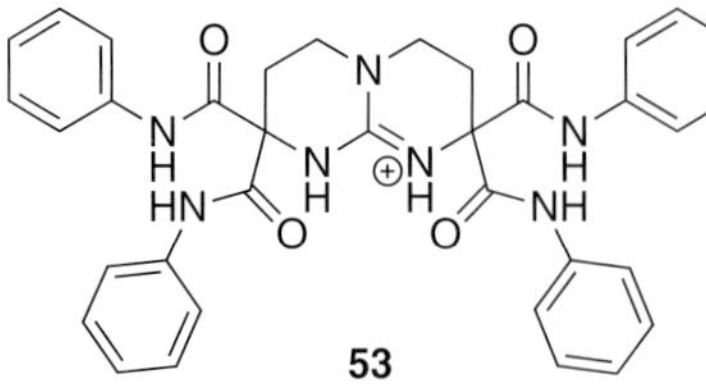

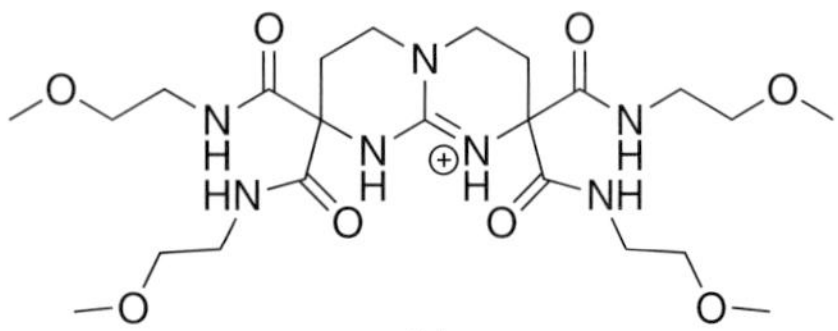

Figure 9.18 Schmidtchen's second generation of receptors bind to **50** and dihydrogen phosphate driven by entropy, too, just like the parent compound **48**. The exception is the host–guest pair **54**-$H_2PO_4^-$, for which complexation is driven by enthalpy.

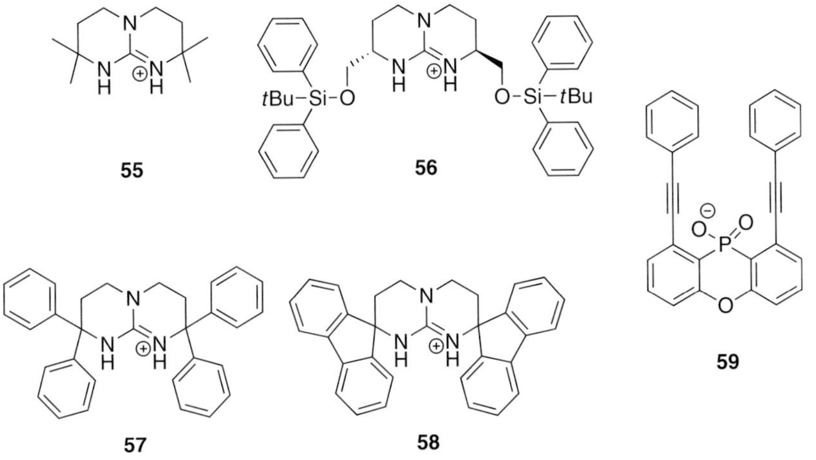

Figure 9.19 In a further study Schmidtchen prepared another four receptors (**55–58**) that he then tested for their binding behavior towards phosphates of increasing steric hindrance.

The affinity for the guests in acetonitrile is high with $K = 10^5$ to 10^6 M^{-1} (ITC). In the more polar solvent methanol the affinity decreases to $K = 10^4$ M^{-1} (ITC). In contrast to acetonitrile complexation in methanol is endothermic while entropy is greatly raised. Schmidtchen concluded that the complexation follows a general ion pairing process that takes place under substantial desolvation of both ionic partners without generating a singular host–guest complex structure. Unlike the good affinity, selectivity is poor, probably due to the easy accessibility and high conformational flexibility of the binding site.

Following these results Schmidtchen provided a trend analysis to study experimentally the role of the binding mode diversity (structural "fuzziness") on molecular recognition of the guanidinium-oxoanion pair [51]. Five different but structurally closely related receptors (**47, 55–58**) served as hosts for three rigid phosphinate substrates (**51, 52,** and **59**) with decreasing accessibility of their binding site (Figure 9.19).

All associations are strongly driven by enthalpy. The change of complexation entropy can be related to the tightness of the mutual fit of the host–guest partners that approaches a minimum limit and can be interpreted as a unique lock-and-key binding mode. The positive entropy changes correlate inversely with the binding interface area. Therefore, desolvation effects can be ruled out as main producers of favorable entropy. Schmidtchen suggests a broad configurational variety of the structure of the host–guest complex resulting from thermal population of various distinct binding modes (potential minima) rather than a single binding mode.

Since complexation is mainly driven by enthalpy a molecular scenario becomes plausible in which guanidinium and phosphinate assemble with shielding of their binding sites. Classic ionic hydrogen bonding then dominates the total interaction. Binding constants of all host–guest combinations are in a comparable range; the receptors are not selective. The highest binding constant is found between receptor **56**

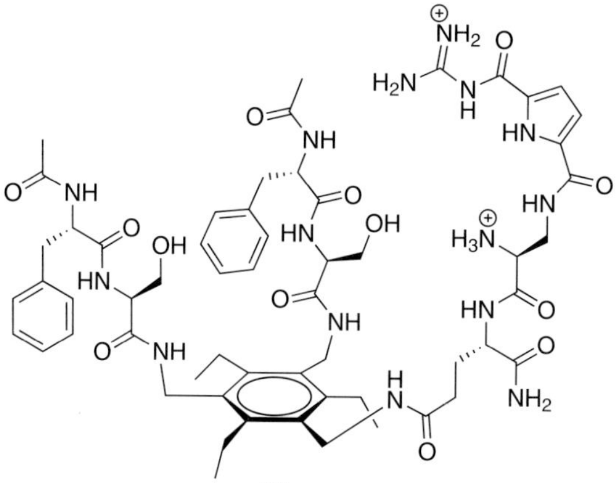

Figure 9.20 Biomimetic receptor **60** can bind carbohydrate phosphates like galactose-1-phosphate (**61**) in water with moderate binding constant.

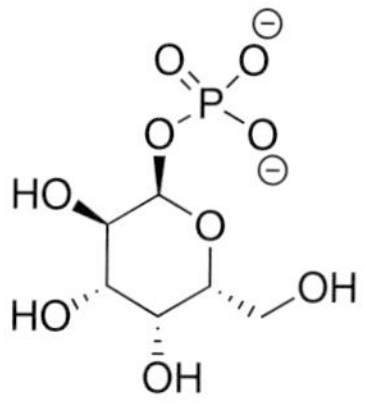

and substrate **59** (6.3×10^5 M^{-1}, acetonitrile, ITC). This is the complex between the sterically most demanding substrate and a receptor with large, but to some extent flexible, side chains. The receptor can deeply penetrate the guest but at the same time shield the recognition site. The pronounced van der Waals interaction results in the strongest exothermicity. A worse geometric fit in turn allows for more motional flexibility between host and guest. The lower degree of structural definition is then reflected by a more positive association entropy.

Schmuck and coworkers have designed the biomimetic receptor **60** for anionic carbohydrates (Figure 9.20) [52]. Nature uses a combination of polar and nonpolar interactions for the recognition of sugars: whereas nonpolar aromatic amino acids like phenyl alanine or tryptophan are often found to interact with the CH framework, more polar residues like asparagine, serine, lysine, and especially arginine form hydrogen bonds to the carbohydrate's OH groups [53]. Based on this principle the artificial receptor **60** was designed: the implementation of phenylalanine could provide nonpolar interactions whereas serine and Schmuck's guanidiniocarbonyl-pyrrole binding motif might lead to polar interaction.

Binding studies showed that **60** can bind to carbohydrate phosphates, uronic acids, or more complex substances like AMP or cAMP with association constants in the range of 10^3 M^{-1}. Comparison with simple anions like methyl phosphate and acetate or various epimeric sugars and regioisomeric substrates does not lead to a general trend in selectivity besides the fact that phosphates are bound more strongly than carboxylates. Thus the highest binding constants in 20% buffered water in DMSO are found for galactose-1-phosphate (**61**) and methyl phosphate with $K = 3.4 \times 10^3$ and 3.8×10^3 M^{-1} (UV titration), respectively.

Following a combinatorial approach Anslyn and coworkers prepared the receptor library **62** (4913 members) for the selective recognition of ATP based on a 2,4,6-triethylbenzene core (Figure 9.21) [54]. Two guanidinium moieties were incorporated

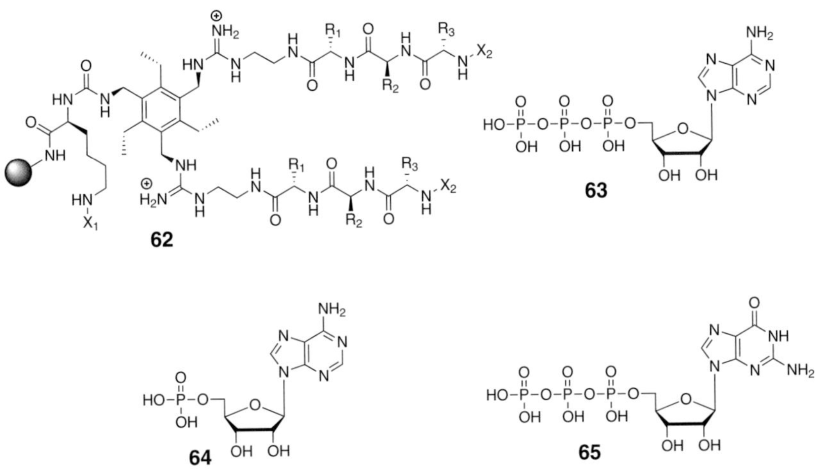

Figure 9.21 Receptors in the combinatorial library **62** incorporate two peptidic arms with three amino acids each. Fluorophores can be attached to X_1 and X_2. Receptors with good affinity to ATP (**63**) in buffered water and selectivity over AMP (**64**) and GTP (**65**) were found.

as anchor groups for phosphate recognition, whereas the combinatorial tripeptidic arms (AA^1–AA^3) are supposed to generate selectivity for the binding of the nucleotide base adenine. Furthermore, fluorophores were attached as binding indicators. Immobilization of the receptors on a solid support – in this case PEG resin – allows for the incorporation of the sensors in arrays [55].

It was found that the receptors with the peptide sequence Ser-Tyr-Ser bind to ATP (**63**) in buffered water with $K = 3.4 \times 10^3 \, M^{-1}$ (pH 7.1, fluorescence). More importantly the receptor only binds very weakly to the competing analytes AMP (**64**) and GTP (**65**). The lack of binding affinity to AMP suggests the necessity of a triphosphate to strongly bind to the guanidinium moieties. The lack of response to GTP indicates a specificity of the peptide arms for adenine over guanosine as nucleotide base. Furthermore, the presence of serine and tyrosine in the most effective receptor suggests π-stacking between the phenol of tyrosine and adenine and hydrogen bonding between the serine OH and/or the ribose or adenine.

In later studies the receptor library was used in an array for the detection of ATP [56]: 30 randomly chosen beads were placed in a micromachined chip-based array platform [57]. Not a specific interaction between one receptor with the substrate but the pattern of the responses on contact with the substrate of the unscreened library in the array serves as discriminating factor. An indicator-displacement assay was used as signaling protocol upon delivery of the substrate to the sensors. It could be demonstrated that the sensor array is able to distinguish between AMP, GTP, and ATP. A closer look at the receptors revealed that the highest response factors for ATP were again found for hosts containing serine and aromatic amino acids. Anslyn's approach is a nice example for the successful combination of supramolecular chemistry principles with pattern based recognition.

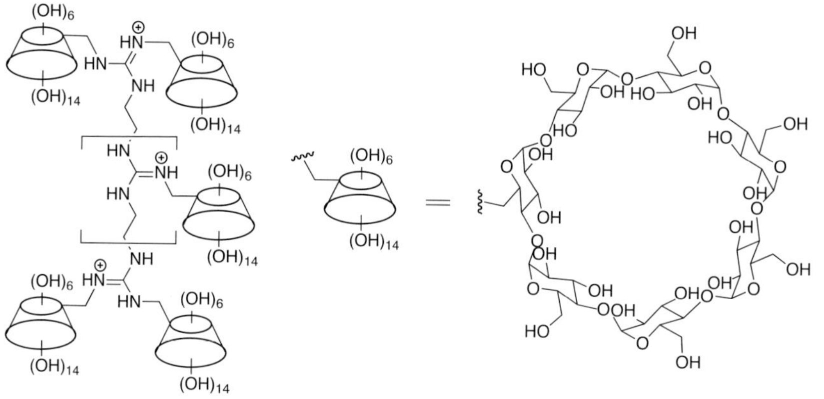

Figure 9.22 Marsura's receptors are composed of guanidinium and cyclodextrin moieties. The first allows for hydrogen bonded ion pairing while the latter adds hydrophobic interactions.

Marsura and coworkers have developed multivalent hosts of type **66** with a defined number of guanidinium centers and cyclodextrins based on a bis(guanidinium) tetrakis(β-cyclodextrin) tetrapod by means of non-polymeric cyclodextrin oligomer "bottom-up" synthesis (Figure 9.22) [58]. This receptor is able to form electrostatic hydrogen bonds between guanidinium and phosphate as well as hydrophobic inclusion complexes within the cyclodextrins.

The receptor **66** with $n = 0$ forms 1 : 2 host–guest complexes with ATP, ADP, and AMP with binding constants around $K \approx 2 \times 10^6 \, \text{M}^{-2}$ (^{1}H NMR titration) in aqueous solution. Binding strengths are very high, but there is no selectivity between the different substrates. The two substrate molecules in the 1 : 2 complex are bound with similar binding constants, indicating a non-cooperative or statistical binding with two independent complexation steps. Closer analysis of NMR data revealed that both ribose and nucleobase residues of the guest are inside the cyclodextrin cavity. Electrostatic interactions between guanidinium and phosphate were verified by ^{31}P NMR.

9.5
Polycarboxylate Binding

While early investigations have demonstrated that the guanidinium moiety is an excellent binding motif for simple carboxylates, more recent studies have aimed at more complex substrate structures. This section provides examples of successful attempts in polycarboxylate binding.

Schmidtchen and coworkers have designed the chiral macrocyclic host **67** that contains two bicyclic guanidinium anchor groups connected to each other by four urea units that are supposed to assist the complexation of carboxylate anions

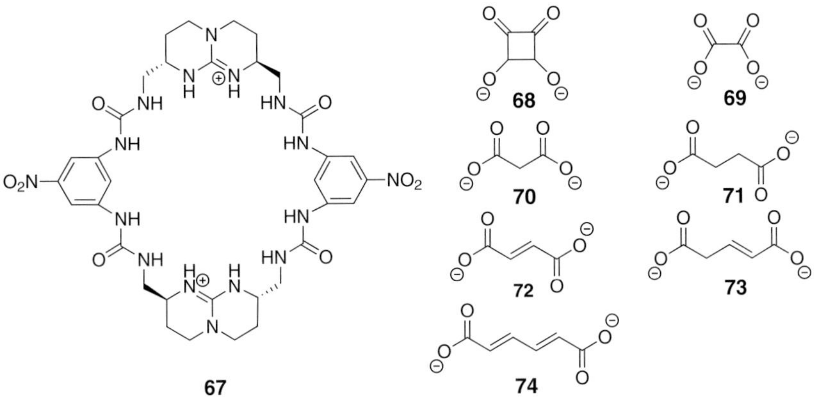

Figure 9.23 Macrocyclic receptor **67** has been tested by Schmidtchen for its binding behavior towards squarate (**68**), oxalate (**69**), malonate (**70**), succinate (**71**), fumarate (**72**), glutaconate (**73**) and *trans,trans*-mucanoate (**74**).

(Figure 9.23) [59]. ITC measurements with simple dicarboxylates **68–73** of varying sizes in acetonitrile suggest higher order complexes besides the regular 1 : 1 binding for all substrates, indicating that receptor and substrate are not complementary. A host–guest stoichiometry of 1 : 2 was often observed. Although no size-dependent correlation could be found, a general energy pattern is valid for all substrates: complexation is driven by both moderate enthalpic and rather high entropic contributions, which is typical for unspecific ion pairing. Different contributions of ΔH and ΔS in the complexes lead to association constants in the range of $K \approx 10^4$–10^6 M^{-1}. For the rigid olefinic substrates fumarate (**72**) and glutaconate (**73**) a second energy pattern and consequently a second binding mode was detected: the enthalpic contribution on binding is dramatically enhanced, especially for fumarate. Although this effect is counteracted to some extent by negative entropy contributions it results in the highest binding constant for fumarate with $K = 1.7 \times 10^7$ M^{-1} (acetonitrile). Both substrates seem to bind in a way that strongly restricts internal mobility in the complexes. For glutaconate complexation, however, in the complex even more degrees of freedom are reduced and binding is accompanied by a lower enthalpic gain. This leads to a much lower affinity for glutaconate. Since these two substrates differ only by one methylene group, solvation effects can be ruled out as explanation. Recently, Schmidtchen and coworkers have expanded these studies to a total of 18 different dianionic substrates of varying size and rigidity, all but one of them carboxylates [60]. The results described above could be validated.

Independent from these measurements Schmidtchen showed that the chiral receptor **67** can differentiate between L- and D-tartrate by means of entropic differentiation of the diastereomeric complexes [59]. This is reflected in the binding constants of $K = 3.1 \times 10^6$ M^{-1} for the L- and $K = 8.8 \times 10^5$ M^{-1} for the D-form. The same entropic discrimination exists for L- and D-aspartate; however, enthalpy–entropy compensation totally consumes the effect in this case: the two enantiomers have a very similar binding constant.

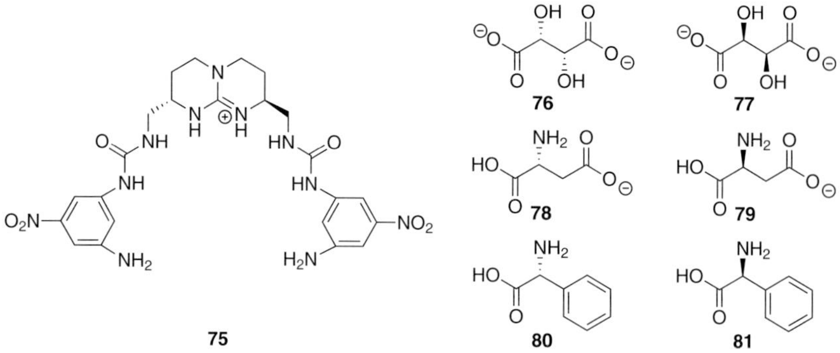

Figure 9.24 Enantiodifferentiation of tartrate (**76**, **77**), aspartate (**78**, **79**) and phenylglycine (**80**, **81**) by Schmidtchen's macrocycle **67** and the open-chain analog **75** has been compared by Schmidtchen and coworkers.

To learn about the consequences of structural relaxation the open-chain analog **75** has been prepared (Figure 9.24) [60]. ITC measurements were carried out in regard to the enantiodifferentiation of tartrate, aspartate, and phenylglycine. Despite similar binding motifs in **67** and **75** and comparable complex affinities of $K \approx 10^4-10^7 \, M^{-1}$ (ITC, acetonitrile), the calorimetric analysis reveals no significant differentiation in enantiomer binding. The data shows in this case that this effect is not a result of enthalpy–entropy compensation. The stereogenic centers in host and guest seem to be too remote to influence each other. These results are in agreement with the general guideline for enantio-recognition of carboxylates that proposes a well-defined geometry of the complex.

Schmuck and coworkers have synthesized the artificial receptor **82** based on Anslyn's tripodal chemosensor **20** [27] by implementing three guanidiniocarbonyl-pyrrole moieties as carboxylate binding sites instead of ammonium or simple guanidinium groups [61]. Receptor **82** binds to the aromatic tricarboxylate **83** in buffered water with a remarkably high binding constant of $3.4 \times 10^5 \, M^{-1}$ (pH 6.3, UV titration). The complex structure was determined by NOESY measurements and molecular modeling. As depicted in Figure 9.25b the substrate (yellow) lies atop of the benzene ring within van der Waals distance, probably allowing for attractive π-stacking. Each carboxylate group is bound to a side chain via a hydrogen bonded salt bridge. Furthermore the three arms provide an extensive hydrophobic shielding for the substrate once it is bound within the inner cavity. The microenvironment is thus rendered much more hydrophobic than the bulk solvent.

For the more flexible and less symmetric citrate (**21**) as substrate an only slightly smaller binding constant of $K = 1.6 \times 10^5 \, M^{-1}$ (UV titration) is found. Compared to Anslyn's receptor, which features a binding constant of $K = 7 \times 10^3 \, M^{-1}$ in pure water but which drops two orders of magnitude in the presence of buffer, it is evident just how potent Schmuck's binding motif for carboxylates is. Even in a buffered aqueous solution of sodium chloride (1000-fold excess of chloride ions) the receptor

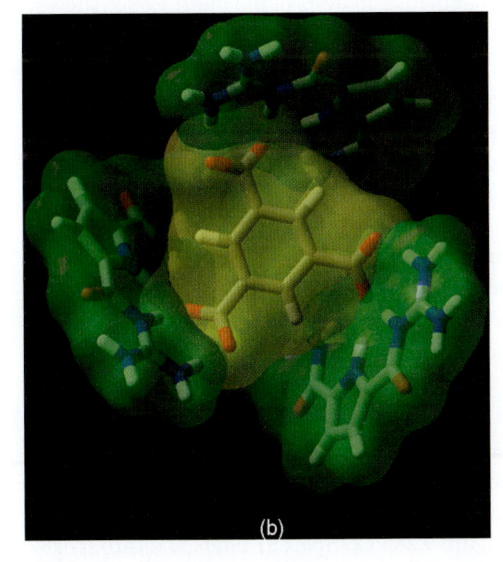

Figure 9.25 (a) Receptor **82** binds to the aromatic tricarboxylate **83** and citrate in buffered water with excellent binding affinity; (b) solvent-accessible surfaces of receptor (green) and substrate **83** (yellow) within the complex. The substrate's carboxylates are completely shielded from the solvent, allowing for a strong interaction with the guanidiniocarbonyl-pyrrole moieties even in water.

still binds to citrate with excellent $K = 8.4 \times 10^4 \, M^{-1}$ relative to all other anions present in the buffer mixture.

Compound **82** was used in a naked eye indicator displacement assay for the detection of citrate [62]. The indicator, carboxyfluorescein (**84**), has just slightly less affinity to the receptor than citrate, allowing for optimal selectivity. Neither malate (**85**) nor tartrate (**86**), which are both closely related in terms of biological occurrence as well as their recognition elements (carboxylates and OH groups), can displace the carboxyfluorescein from its complex with **82** (Figure 9.26). Other monoanions like

Figure 9.26 Carboxyfluorescein (**84**) for the naked eye detection of citrate, but not malate (**85**) or tartrate (**86**), by **82**.

acetate have no influence either. A quantitative analysis of the fluorescence shows a selectivity of citrate over malate and tartrate of 4 : 1 and 9 : 1, respectively, even after the addition of a large excess of 14 equivalents of each substrate.

Anslyn and coworkers have reported the Cu(II) and guanidinium containing receptor **87** as binding motif for polycarboxylates (Figure 9.27) [63]. The tetracationic host should be able to accommodate a small guest molecule with up to four carboxylate groups. Two of them can chelate the Cu(II) center and one each bind to the two guanidinium moieties. Binding studies in buffered water with a series of mono- to tetracarboxylates show that tetracarboxylate **88** is bound strongest with $K = 1.9 \times 10^4 \, M^{-1}$ (pH 7.4, ITC and UV titration). The tricarboxylate tricarballyate (**89**) and the dicarboxylates glutarate (**90**) and succinate (**71**) are bound worse by one order of magnitude. Acetate as a monocarboxylate is bound worse by two orders of magnitude.

To assign the corresponding energetic attributions of the recognition event to the distinct binding motifs, host **87** was cleaved to give the copper center in the form of the bis(aminomethyl)pyridine Cu(II) ligand–metal complex **91** and the bis-guanidinium residue **92** where the portion equipped with the copper ligand is capped with an acyl amide group (Figure 9.28). Complex **91** binds to dicarboxylates with $K \approx 10^2 \, M^{-1}$ and to acetate even more weakly, by about one order of magnitude; **92** binds to these three substrates with $K \leq 50 \, M^{-1}$.

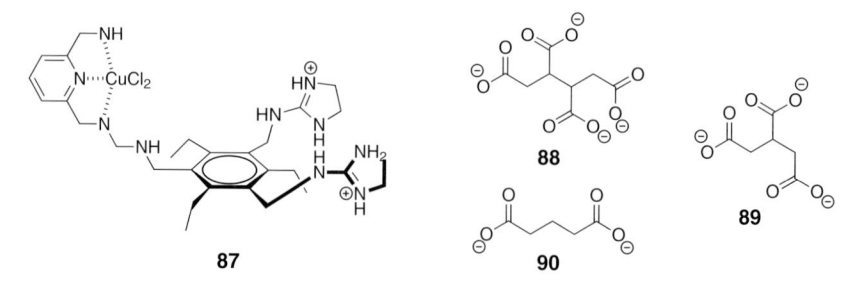

Figure 9.27 Anslyn's receptor **87** with a copper and a guanidinium based recognition site. Various substrates were tested (**88–90**).

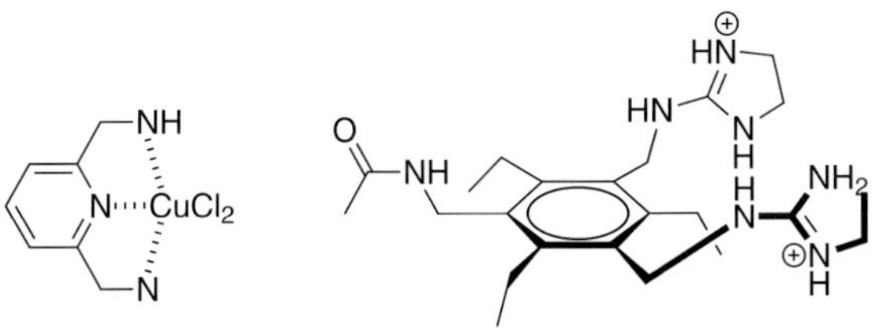

Figure 9.28 Separated binding motifs of receptor **87**: copper (**91**) and guanidinium (**92**) residues.

Thermodynamic data obtained by ITC measurements show that the release of Gibbs free energy upon binding of succinate or glutarate by **87** is greater than the sum of free energy for the isolated functional groups (**91** and **92**). The individual weak noncovalent interactions are exploited optimally and indeed surmount the binding strength of their sum. This is an effect known as positive cooperativity [64]. Whereas negative cooperativity is rather common, the opposite effect has only rarely been observed for synthetic receptors in water.

Cooperativity can be observed when a system of covalently linked binding sites has a fundamental binding quality that is not present in the isolated binding motifs. ITC measurements suggest that cooperativity in **87** stems from enthalpy. Anslyn proposes that the close proximity of positive charges in the unbound host leads to an additional destabilization that does not occur in the isolated recognition sites in **91** and **92**. Upon binding this electrostatic strain is relieved to some extent. From the opposite point of view, the lack of negative enthalpic cooperativity shows that the distance between the binding motifs in the host seems to be ideally complementary to the functional groups of the guest. Otherwise, an enthalpic penalty would be the consequence. The good match between receptor and substrate leads to a strong and tight complexation. The defined complex structure in turn reduces rotational and vibrational degrees of freedom and a negative entropic cooperativity is the result. This effect, however, is overruled by the enthalpic cooperativity in Anslyn's receptor.

9.6
Amino Acid Recognition

Another important area of interest is the recognition of amino acids and peptides for the obvious reason of their biological abundance and functions. In recent years the vast majority of publications dealing with anion recognition by artificial guanidinium based receptors have concentrated on peptidic substrates. The following sections discuss these systems, beginning here with amino acid recognition.

Schmuck and coworkers have designed N'-substituted guanidiniocarbonyl-pyrroles of type **93** (Figure 9.29) [65]. In additional to the CBS, further interactions with the backside of the substrate should be induced by attaching a side chain to N'. This can lead to an increase of complex stability and selectivity.

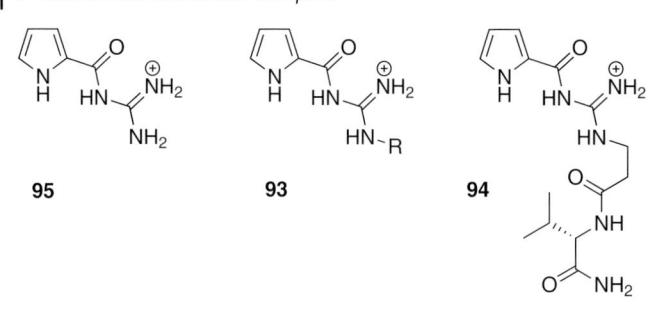

Figure 9.29 Substitution of Schmuck's guanidiniocarbonyl-pyrrole binding motif at N′ (R) gives rise to receptors of type **93** that bind to amino acids with good affinity.

Compound **94** binds to N-acetylated amino acid carboxylates in buffered water with $K \geq 10^3 \, M^{-1}$ (pH 6.0). The stability depends on the side chain of the amino acid: valine is bound almost twice as well as alanine ($K = 1.8 \times 10^3 \, M^{-1}$ versus $1.0 \times 10^3 \, M^{-1}$). The unsubstituted receptor **95** cannot discriminate between those two substrates, indicating that the side chain is responsible for the selectivity indeed. According to molecular mechanic calculations the stronger association could be due to favorable hydrophobic interactions between the isopropyl side chains in water. Yet, this receptor is not able to recognize more polar substrates such as charged, unprotected amino acids.

Schmuck and coworkers have also reported amino acid receptors **96** and **97**, similar to their carboxylate binding motif **22**, in which the direction of the amide group in position 5 of the pyrrole is reversed in order to study the effect of the exchange of hydrogen bond donor and acceptor (Figure 9.30) [66]. Binding studies with N-Ac-Ala and O-Ac-Lac in a solution of 40% water in DMSO revealed rather weak association constants ranging from 220 to 460 M^{-1} (^{1}H NMR titration). Contrary the original recognition motif these new receptors bind more strongly to alanine than to lactate [30].

Molecular modeling suggests a binding mode in which the CO of the reversed amide is now able to form an additional hydrogen bond with the substrate NH. This is in agreement with the observed induced shift changes of the NMR experiments. This new interaction, however, brings the two acetyl groups of host and guest in close proximity, which leads to destabilizing steric interactions and consequently to a decrease in binding affinity.

In another attempt to recognize amino acids Schmuck and coworkers have synthesized the tris-cationic receptor **98** [67]. ^{1}H NMR titrations revealed an association constant for N-acetyl-alanine of $K = 2.1 \times 10^3 \, M^{-1}$ in 10% DMSO in water. The more polar dianionic N-acetyl-aspartate is bound as a 1:1 complex with only 480 M^{-1}. On the other hand, N-acetyl-glutamate, which differs structurally only by one methylene group in the side chain, shows an entirely different binding behavior. Receptor **98** forms a 2:1 complex that reveals positive cooperativity. While K_1 (460 M^{-1}) is comparable to aspartate the second binding constant K_2 (3300 M^{-1}) is increased by a factor of 7. A possible explanation for the binding process is depicted

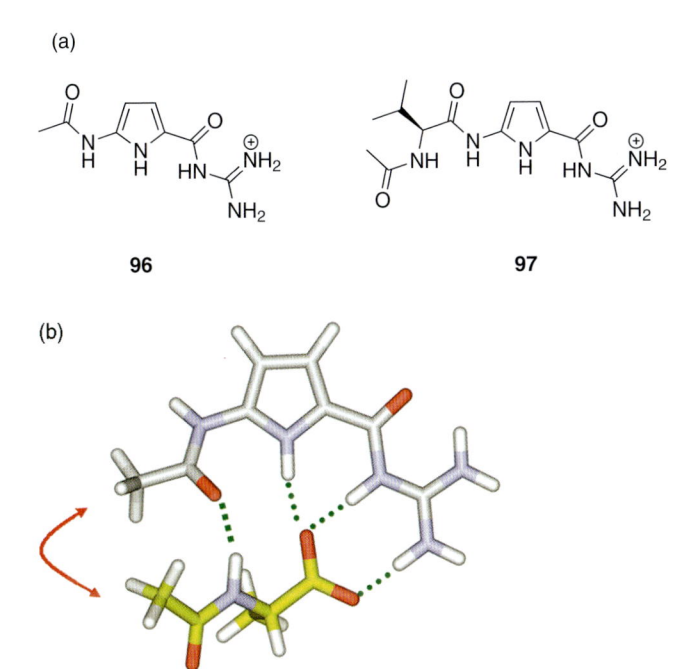

Figure 9.30 (a) Hydrogen bond donor and acceptor of Schmuck's CBS (**22**) are reversed at the amide group in position 5 in receptors **96** and **97**; (b) calculated energy minimized structure for the complex between N-Ac-Ala-O⁻ (yellow) and **96** (gray). Hydrogen bonds are shown in green, unfavorable steric interactions in red.

in Figure 9.31: the small distance between two carboxylates in aspartate compared to glutamate prevents the formation of a 2:1 complex due to unfavorable steric/electrostatic interactions. The tris-cation **98** is hence capable of differentiating between glutamate and aspartate, which is remarkable regarding their structural similarity and flexibility.

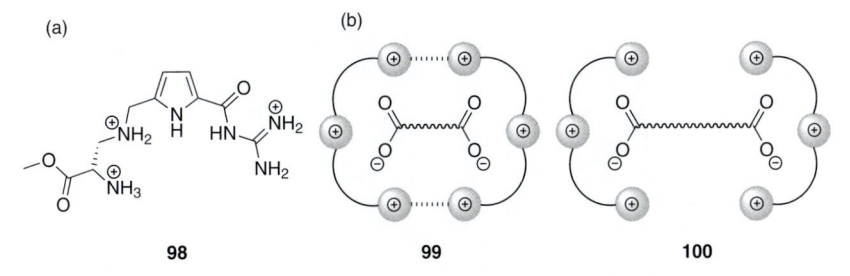

Figure 9.31 (a) Tris-cationic receptor **98** can differentiate between aspartate and glutamate in water; (b) steric and/or electrostatic interactions (dotted line) prevent the formation of a 2:1 complex for aspartate (**99**) but not glutamate (**100**).

Figure 9.32 Bis cationic receptors **101–105** with varying spacer length for the recognition of N-acetylated amino acids.

Another study on amino acid recognition was conducted by Schmuck's group [68]: a series of bis-cationic receptors (**101–105**) was prepared by attaching simple primary ammonium groups via flexible linkers of varying length to the guanidiniocarbonyl-pyrrole core (Figure 9.32).

Figure 9.33 depicts the results of UV binding studies with various N-acetylated amino acids (Val, Asp, Glu, Ala, Phe) in 10% DMSO in water. Complex stability depends significantly on both the anion and the lengths of the linker in the host. With increasing length the complex stability increases for all substrates until it reaches its optimum for the C4 linker before it decreases again. The distance

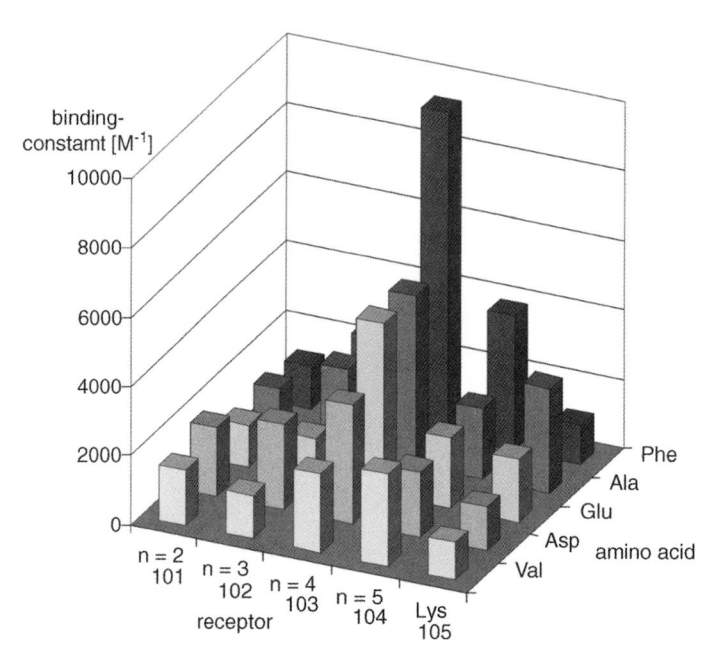

Figure 9.33 Binding constants for bis-cationic receptors **101–105** and various N-acetylated amino acid carboxylates calculated from UV-titration experiments in buffer water (K in M^{-1}).

between ammonium and guanidiniocarbonyl-pyrrole is probably too short for strain-free interaction with the substrate in the smallest hosts **101** and **102**, a view that is supported by molecular mechanic calculations. With four carbon atoms in **103** the linker is just long enough to allow for optimal interaction. For **104** and the lysine derivative **105** with five C atoms each, an additional entropic cost is required to orient the longer and now more flexible linker. This probably compensates to some extent the attractive interaction of the ammonium group. An entropic penalty was to be expected for more flexible receptor design and is, therefore, reflected in the progression **103** to **105**.

Of all substrates N-acetyl-phenylalanine binds most tightly to **103** with a high binding constant of $1.1 \times 10^4 \, M^{-1}$ (UV titration). NOE measurements and molecular modeling calculations indicate a well-defined complex structure with an additional cation–π interaction between the aromatic ring in the amino acid side chain and the guanidiniocarbonyl-pyrrole cation.

9.7
Dipeptides as Substrate

To advance from single amino acid to dipeptide recognition further binding sites that provide additional interactions with the substrate are necessary for strong and selective binding. Schmuck and coworkers have solved this problem by further derivatizing the side chain of their binding motif to the needs of the peptidic substrate [69]. Based on theoretical calculations an imidazole moiety was implemented in receptor **106** via an aromatic linker to provide an additional hydrogen bond from the imidazole NH to the dipeptide backbone (Figure 9.34).

Nonpolar N-acetylated dipeptidic substrates (Gly-Gly, Ala-Ala, Val-Ala, and Val-Val) were studied as substrates in 10% DMSO in water by means of UV titration. The highest binding constant was observed for N-Ac-Val-Val-OH with $K = 5.4 \times 10^4 \, M^{-1}$. Notably, nonpolar N-acetylated amino acids (Ala, Gly) bind ten-times less strongly to the host. However, **106** is not able to differentiate between the dipeptides because the

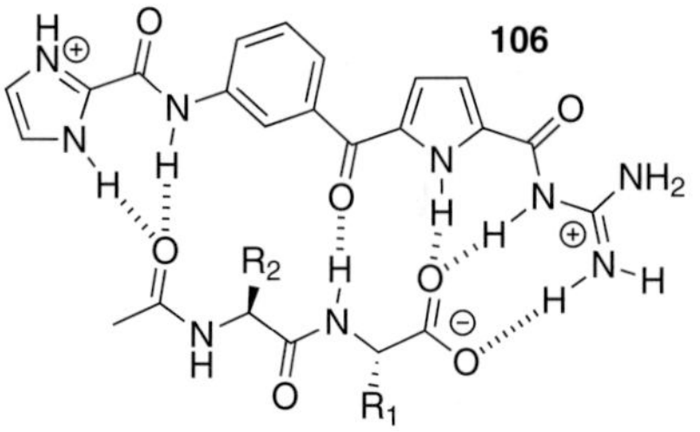

Figure 9.34 Receptor **106** binds unselectively to nonpolar N-acetylated amino acids (**107**) with good affinity in aqueous solution.

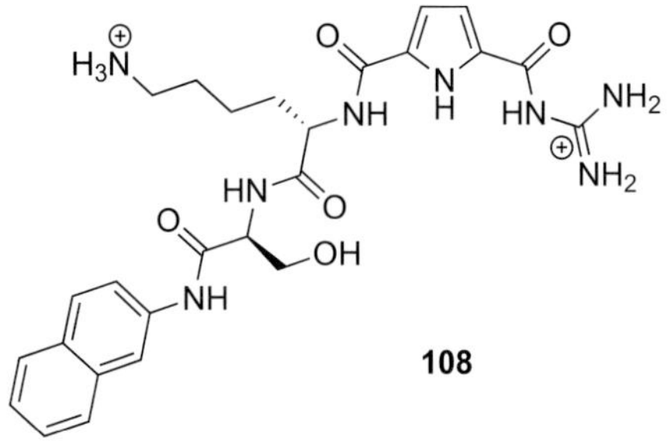

Figure 9.35 Receptor **108** binds to N-acetylated dipeptides unselectively but with good affinity in aqueous solution.

interaction between substrate and receptor is mainly limited to the amide backbone and not to the amino acid side chains. A modest preference for large and bulky amino acids can be observed, for example, Val-Val is bound strongest followed by Ala-Ala, and finally Gly-Gly. This trends most probably only reflects the overall hydrophobic character of the corresponding dipeptides.

In another approach for dipeptide recognition Schmuck and coworkers studied receptor **108**, which contains a lysine attached to the guanidiniocarbonyl-pyrrole to provide additional noncovalent binding sites, a serine to enhance solubility in aqueous solution, and a naphthyl group for potential additional hydrophobic interactions (Figure 9.35) [70].

UV titration studies were carried out in a buffered solution of 20% DMSO in water with a series of nonpolar N-acetylated amino acids, dipeptides, and tripeptides as substrate. Single amino acids are bound rather weakly by the receptor, whereas dipeptides and tripeptides bind stronger due to additional interactions with the host compound ($K > 10^3 \, M^{-1}$). Some preference can be observed for alanine at the C-terminus of the peptide sequence over more bulky amino acids (Phe, Val, Glu). The highest binding constant was found for N-Ac-D-Phe-Ala-OH with $K = 5 \times 10^4 \, M^{-1}$ (pH 6.0). Molecular mechanics calculations suggest a binding mode where the naphthyl ring is π-stacking with the guanidiniocarbonyl-pyrrole cation, thus pointing away from the substrate. The close proximity of the alanine methyl group to the aromatic moiety might explain its preference over larger amino acids at this position. The lysine moiety interacts with the N-terminal acetyl group of the substrate. Although **108** shows good affinity to the desired substrates it lacks pronounced selectivity. Furthermore, the naphthyl group does not impart the planned decisive role in the binding mode, demonstrating just how difficult it is to rationally design receptors for complex anionic targets.

Nevertheless, Schmuck's group continued with the rational approach to receptor design: the artificial receptor **109** was synthesized for the selective recognition of alanine-containing dipeptides [71]. Molecular modeling was utilized to test various building blocks *in silico* for their complementary to the peptide sequence Ala-Ala. The most promising candidate **109** consisting of the CBS and a cyclotribenzylene unit,

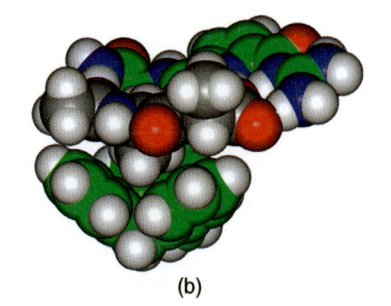

Figure 9.36 (a) Design of receptor **109** with a preference for alanine-containing dipeptide carboxylates in buffered aqueous solution; (b) energy-minimized structure for the complex formed between **109** (green) and the dipeptide D-Ala-D-Ala (gray) from the side, showing the interaction between the methyl group and the cyclotribenzylene cavity.

was synthesized. The aromatic moiety is supposed to induce selectivity: it forms a hydrophobic cavity just large enough for a methyl group but not for larger alkyl chains [72].

UV binding studies in a buffered solution of 10% DMSO in water indeed demonstrate a strong complexation of N-Ac-Ala-Ala, with a binding constant of $K = 3.3 \times 10^4 \, M^{-1}$ (pH 6.1), verifying the predicted affinity. Even more importantly, **109** is very selective over other peptides like N-Ac-Val-Val ($K < 10^3 \, M^{-1}$). The predicted complex structure is depicted in Figure 9.36b. This work nicely describes the advantages of modern theoretical methods that are able to decrease tedious trial and error synthesis of artificial receptors, at least for small substrates. Unfortunately, the larger and more complex the substrate the more time consuming, more difficult, and more expensive it is to obtain such predictions without a high error-proneness. Notably, molecular modeling is an enthalpic model only that neglects all entropic contributions of the binding event. More sophisticated state-of-the-art methodology has to be used to include entropic factors into the calculations. Nevertheless, this rather facile approach – receptor design based on molecular mechanics – remains a helpful tool for the prediction of the geometric fit between substrate and receptor.

9.8
Polypeptide Recognition

In Nature the recognition of C-terminal oligopeptides plays a decisive role in various processes like the mode of action of the antibiotic vancomycin [73] or in Ras-protein induced oncogenesis [74]. Thus the development of artificial receptors for the specific complexation of biologically relevant oligopeptides under physiological conditions is of utmost importance, for example, for the targeting of cellular

Figure 9.37 Chiral bicyclic tetrakis-guanidinium receptor **110** recognizes helical peptides with good binding constant in aqueous methanol solution.

processes [75] or the finding of new therapeutics [76]. However, due to the high flexibility of the peptidic substrate it is very challenging to obtain selective receptors that form a stable host–guest complex in the competing environment of an aqueous solution.

In a collaboration between Hamilton, Giralt, De Mendoza, and their coworkers tetraguanidinium **110** was devised as binder for a protein surface (Figure 9.37) [77]. Proteins often function as part of networks regulated by protein–protein interactions [78]. The design of specific inhibitors of protein aggregation should therefore lead to new approaches for the treatment of a few serious diseases [75]. However, the design of such binders is difficult due to the physicochemical properties of protein–protein interfaces that are very large, flat, and rich in well-solvated hydrophilic residues like Asp, Glu, and Arg [79].

The shape and charge of **110** is complementary to helical peptides that feature an anionic residue in every fourth position. It is able to bind to the peptide sequence **111** in a solution of 10% water in methanol with $K = 1.6 \times 10^5 \, M^{-1}$ (CD titration) (Figure 9.38) [80].

The tumor suppressor protein P53, a key therapeutic target for cancer treatment [81], only operates when aggregated into a tetramer. The tetramerization domain has two overlapping helical tetra-anionic sequences on its surface formed by four glutamate moieties on the one side and three glutamate plus one aspartate

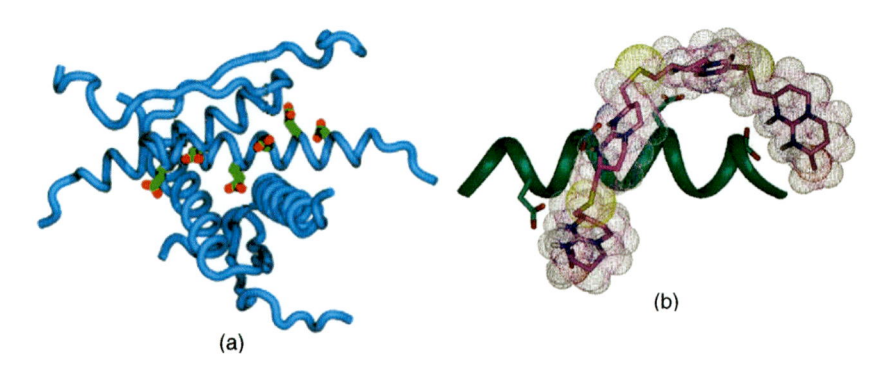

Figure 9.38 (a) Ribbon representation of the tetramerization domain of a monomer of protein P53 [77]; (b) energy-minimized structure of an α-helical peptide backbone with four aspartate residues (D) at each fourth position (N-Ac-AAADQLDALDAQDAAY-CONH$_2$, **111**) and receptor **110** [9b].

residues on the other. Hence, it was reasoned that **110** might be able to bind to the protein domain. By means of [^{1}H,^{15}N]HSQC titration a binding constant of $K = 2 \times 10^4$ M^{-1} was found. Saturation transfer difference spectroscopy [82] confirmed that **110** is indeed bound to the surface of the protein. The same technique revealed that the binding mode in the protein differs slightly from the preferred interaction to every fourth residue in the free peptide **111**. Sterics and the presence of the longer glutamate residues are supposed to be the reason for the elongated binding mode. These results demonstrate that the ideal binding mode can be perturbed by minute changes in the layout of the binding site when suitable alternatives of low-energy modes are available. Hence, this work represents another case of binding mode diversity and is at the same time an excellent example of successful structure-based receptor design.

Driving the rational approach to molecular recognition one step further Schmuck, Schrader, and coworkers invented ditopic receptors (**112–115**) for the RGD (Arg-Gly-Asp) peptide sequence (Figure 9.39) [83]. The RGD loop plays a key role within several cell–cell and cell–matrix adhesion processes [84]. Some severe diseases can be associated with this binding motif and malfunctions thereof [85]. Therefore a model system for studying the supramolecular aspects of RGD recognition has been prepared to obtain knowledge about critical factors governing this biological recognition event. The design consists of Schrader's *m*-xylylene bisphosphonate bis-anion for arginine recognition [86] and Schmuck's CBS. Flexible peptidic linkers (**112–114**) and a rigid aromatic spacer (**115**) have both been used to connect the two binding motifs.

UV titrations in buffered water showed no affinity of receptors **112–114** to both tested substrates, the free H-RGD-OH and the N- and C-protected tripeptide

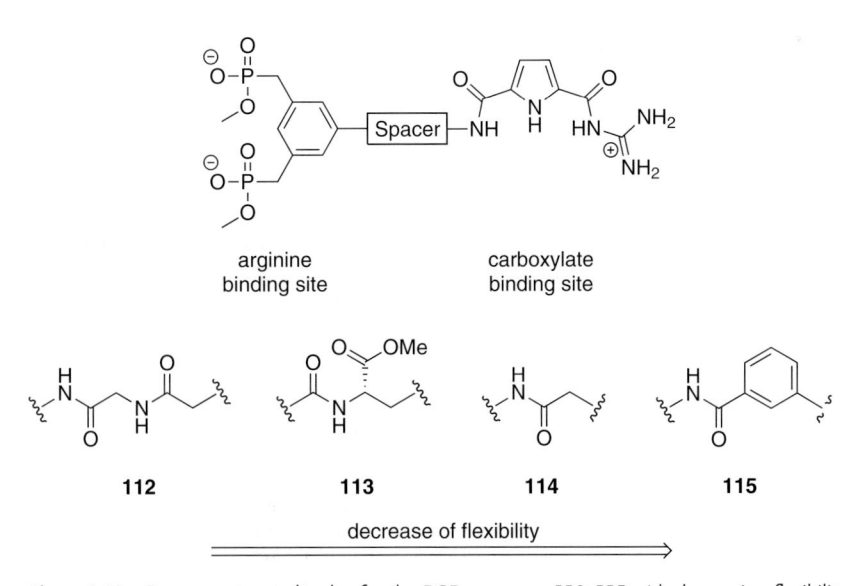

Figure 9.39 Four receptor molecules for the RGD sequence **112–115** with decreasing flexibility.

Ac-RGD-NH$_2$ that resembles the internal RGD loop found in proteins. Both **112** and **113** most likely suffer from efficient intramolecular self-association while **114** probably forms dimers of itself. Molecular mechanics calculations suggest that the glycine spacer is too short so that distances between the complementary recognition motifs in the receptor and the guest do not match.

The rigid receptor **115** has a weak affinity ($K < 1000$ M^{-1}) to the free tripeptide. For the protected analog, on the other hand, a binding constant of $K = 2.7 \times 10^3$ M^{-1} is reported (pH 6.0). This is in agreement with biological assay studies that state a lower affinity of the free peptide in comparison with derivatives with an internal RGD sequence or a protected N-terminus [87]. No complexation can be observed with Ac-RGG-NH$_2$, Ac-GGD-NH$_2$, and Ac-GGG-NH$_2$, thus verifying the selectivity of the host.

Molecular modeling calculations suggest an almost perfect match in distance between binding sites and the formation of a hydrogen bond network with ten inter- and two intramolecular hydrogen bonds. As expected the guanidiniocarbonyl-pyrrole cation binds the carboxylate moiety of the aspartate side chain while the bisphosphonate unit interacts with the arginine side chain.

To obtain information about the role of the bisphosphonate motif additional binding studies were carried out with the methyl ester precursor **116** (Figure 9.40). Surprisingly the association constant was almost doubled ($K = 4.7 \times 10^3$ M^{-1}). This effect is probably due to facilitated desolvation of the precursor with its two charges less than **115**. Additional hydrophobic interactions might also influence the binding behavior.

These studies demonstrate again that it is possible to create tailor-made receptors even for larger peptides. However, the problem of inter- and intramolecular self-association is exemplary of the problems that arise from the ever increasing complexity of the molecular recognition event when the substrate becomes larger and larger, thus rendering such a rational approach extremely difficult. Theoretical methods such as force field calculations are not yet reliable enough to completely design an artificial receptor for a large substrate rationally.

Combinatorial chemistry presents an alternative approach for the preparation of host molecules for complex targets. Preparation of combinatorial libraries and subsequent screening and selection has been shown to be able to identify suitable host molecules for various substrate types. Application of this random trial and error methodology on supramolecular problems was pioneered by Still and coworkers [88] and has been utilized for recognition of such different compounds as proteins [89], synthetic oligomers [90], small molecules [91], or oligosaccharides [92].

Kilburn and coworkers have prepared a library containing 2197 tripeptide receptors of type **117** (Figure 9.41) [93]. Despite its inherent flexibility the tweezer-like design has proven to be highly selective for peptide sequences in nonpolar [94] as well as in aqueous environment [95]. A bis(aminoalkyl)guanidinium was incorporated as head group for the specific recognition of the C-terminus of the peptide while the two arms consist of three amino acids each for additional interactions with the substrate.

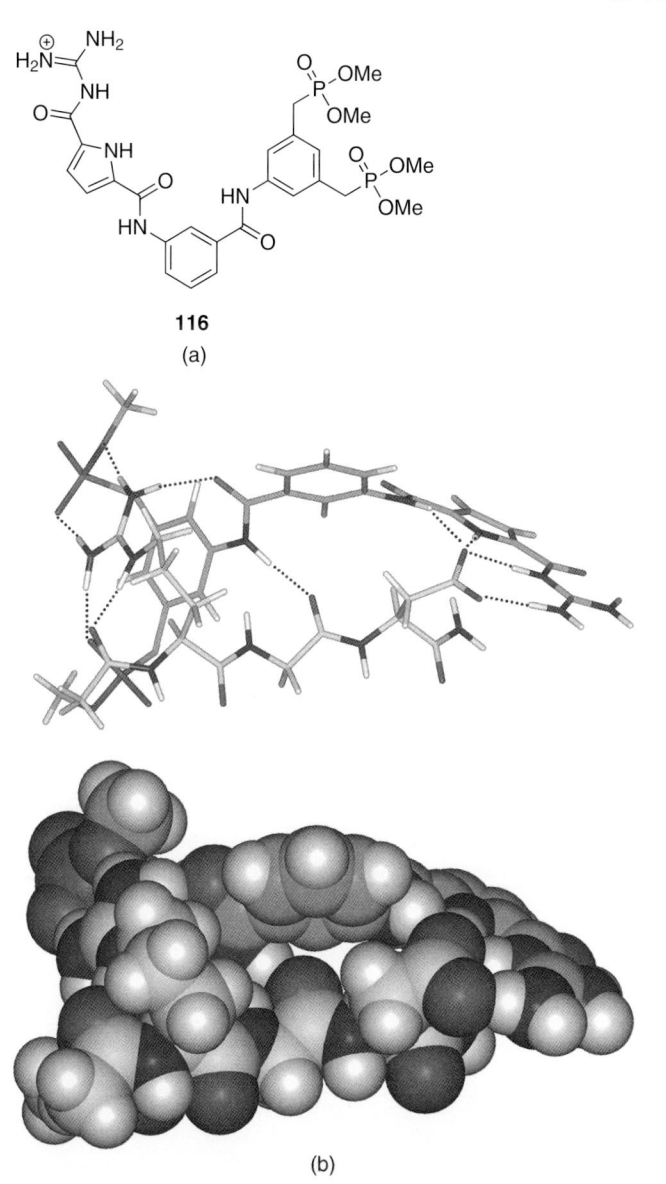

Figure 9.40 (a) The less polar bisphosphonate precursor **116** shows enhanced binding affinity to N-Ac-RGD-CONH$_2$; (b) calculated energy-minimized structure of the complex between receptor **115** and N-Ac-RGD-CONH$_2$.

On-bead screening against the N-dye-labeled tripeptide Glu(OtBu)-Ser(OtBu)-Val-OH (**118**) and subsequent measurements with the resynthesized receptor in free solution provided high affinity for the receptor with the sequence AA^1–AA^3 = Pro-Leu-Met with an association constant in a buffered solution of 15% DMSO in water of

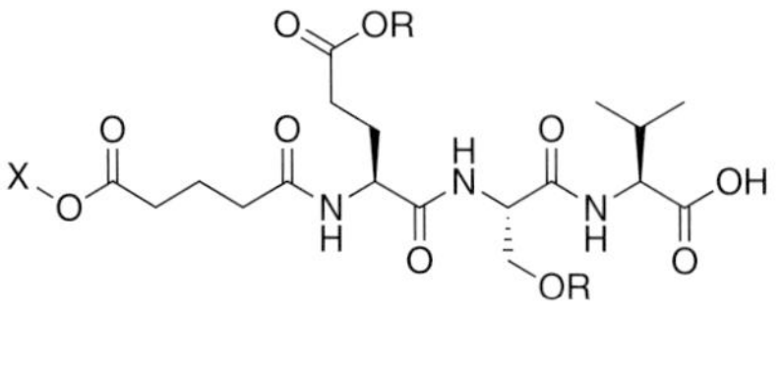

Figure 9.41 Combinatorial library of tweezer receptors (**117**) derived from a guanidinium scaffold to identify a selective receptor for dye-labeled (X) N-Ac-Glu(OtBu)-Ser(OtBu)-Val-OH (**118**).

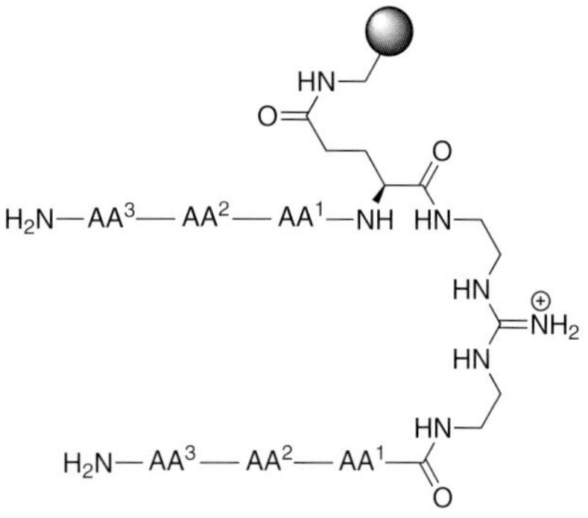

$K = 8.2 \times 10^4 \, M^{-1}$ (pH 8.8, UV titration) [96]. However, no complexation was observed when testing the unprotected much more polar derivative **119** under these conditions, in good agreement with the importance of hydrophobic effects in polar solvents. A certain selectivity can be stated when considering the binding studies with dye-labeled D-Ala-D-Ala that did not yield any results.

To increase structural diversity Kilburn's next step was to synthesize a large combinatorial library (15 625 members) of unsymmetrical tweezer receptors (**120**), that is, the two peptidic arms of the hosts were differently substituted (Figure 9.42) [97]. The library was aimed at the recognition of the tripeptide Lys-D-Ala-D-Ala-OH. This peptide sequence is interesting in terms of its relevance to cell wall maturation upon treatment with vancomycin, leading to bacteria

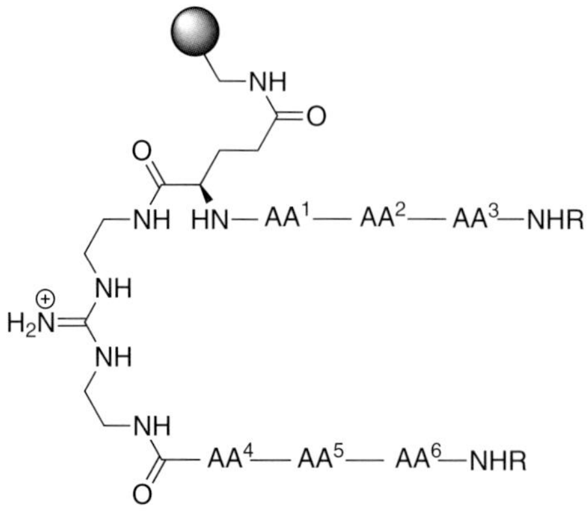

Figure 9.42 Two-armed tweezer receptor library **120** for the binding of dye labeled (X) N-Ac-Lys-D-Ala-D-Ala-OH (**121**) in aqueous solution. The arms are synthesized sequentially to give a structurally more diverse library of unsymmetrical receptors.

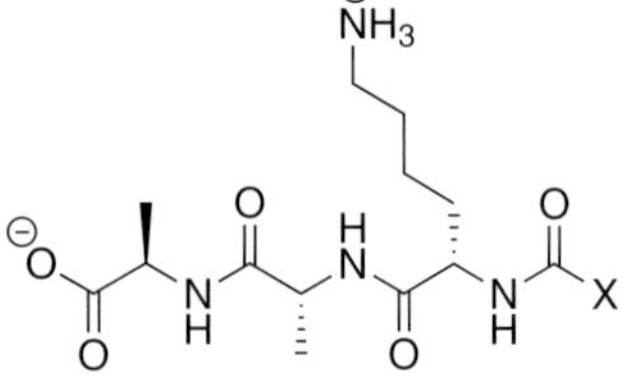

death [73]. During the synthesis of the bacterial cell wall, linear peptidoglycans are crosslinked via a transamidation reaction involving the tetrapeptide sequence D-Glu-Lys-D-Ala-D-Ala-OH, which is also the point of attack of the glycopeptide antibiotic vancomycin.

On-bead screening with dye-labeled N-Ac-Lys-D-Ala-D-Ala-OH (**121**) and sequencing of the hits leads to the desired receptors. Subsequent resynthesis followed by on-bead UV measurements in a buffered solution of less than 10% DMSO in water (pH 8.5) revealed a binding constant of $K = 1.4 \times 10^3$ M^{-1} for the receptor with the peptide sequence $AA^1-AA^3 = $ Gly-Val-Val and $AA^4 - AA^6 = $ Met-His-Ser. Unfortunately resynthesis of the receptor and UV-titrations in free solution could not reproduce these results – no binding data could be obtained. It has been shown before that the solid support may have a macroscopic influence on the binding event [98]. Furthermore, the limits of such large libraries become clear: only positive hits can be selected qualitatively and have to be resynthesized for additional analysis.

The best possibility of finding an artificial peptide receptor is most probably the combination of rational design with the power of combinatorial chemistry. Schmuck and coworkers have prepared the medium-sized combinatorial library **122** of one-armed cationic peptide receptors (512 members) (Figure 9.43) [99]. Next to a guanidiniocarbonyl-pyrrole group that acts as CBS a variable tripeptide chain is attached for the formation of a β-sheet with the backbone of tetrapeptide substrates. Furthermore, the additional electrostatic and steric interactions between the amino acid side chains both in substrate and receptor should further enhance the binding stability and substrate selectivity.

The first target was the hydrophobic tetrapeptide N-Ac-Val-Val-Ile-Ala-OH (**123**), which represents the C-terminal sequence of the amyloid-β-peptide (Aβ) that is responsible for the formation of the protein plaque in the brain of patients that suffer from Alzheimer's disease [100]. This peptide sequence promotes the formation of self-aggregated β-sheets of Aβ through a combination of hydrogen bonds and

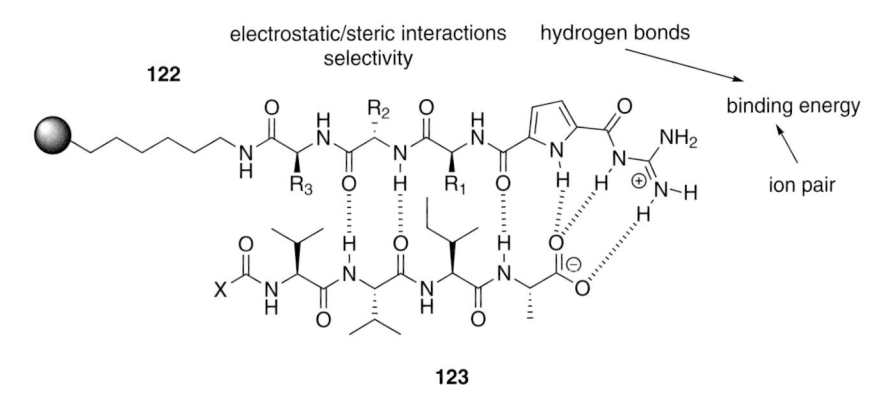

Figure 9.43 Schematic representation of complex formation between the receptor library **122** and the fluorescent labeled (X) tetrapeptide substrate **123**.

hydrophobic interactions [101]. Owing to the nonpolar nature of the substrate eight residues were carefully chosen among the proteinogenic amino acids for positions AA^1–AA^3 of the receptor to provide a representative range of varying polar, charged, and hydrophobic moieties: Lys(Boc), Tyr(tBu), Ser(tBu), Glu(OBzl), Phe, Val, Leu, and Trp.

Since the single members of the library were synthesized spatially separated by means of the IRORI radiofrequency tagging technology [102], in combination with the manageable number of receptors, the entire library can be measured on-bead not only qualitatively but quantitatively as well. The side chains of the amino acids in **122** were left protected to allow for better hydrophobic interactions to the rather nonpolar substrate. Binding constants for the fluorescence-labeled substrate **123** in buffered water vary from $K = 20\,M^{-1}$ to $4.2 \times 10^3\,M^{-1}$ (pH 6.0). The highest binding constant was obtained for the sequence AA^1–AA^3 = Lys(Boc)-Ser(OtBu)-Phe. Binding constants were confirmed in solution by UV titrations. When testing the substrate with a methyl ester protected C-terminus only weak and rather unspecific interactions to the receptor library can be observed. Side chain interactions alone are obviously not strong enough for the formation of stable complexes. The unprotected negatively charged **123**, on the other hand, is bound selectively only by some but not all members of the library although the salt bridge is the same for all of them. Hence it can be reasoned that the CBS is responsible for strong complexation and the peptidic arm for selectivity. A closer look at the data allows for a correlation between complex stability and structure: hydrophobic interactions with the first amino acid of the substrate (Val) are most important. This is in excellent agreement with previous studies [103]. Molecular modeling suggests β-sheet formation between the tetrapeptide and the receptors. Additional hydrophobic interactions occur between the side chains and a salt bridge between carboxylate and guanidinium, just as predicted. Additional NMR data is in good agreement with these results.

The best receptors of this screening inhibit *in vitro* the formation of amyloid plaques of Aβ(1–42) but not of Aβ(1–40), which has the wrong C-terminal sequence. The receptor both retarded the formation of amyloid plaque as well as significantly reduced the amount of fibrils formed [104].

Next, Schmuck and coworkers applied the combinatorial receptor library **122** for recognition of the non-hydrophobic tetrapeptide N-Ac-D-Glu-Lys-D-Ala-D-Ala-OH (**124**, EKAA; Figure 9.44) [105]. This substrate is interesting in terms of its relevance to bacterial treatment with vancomycin (vide supra).

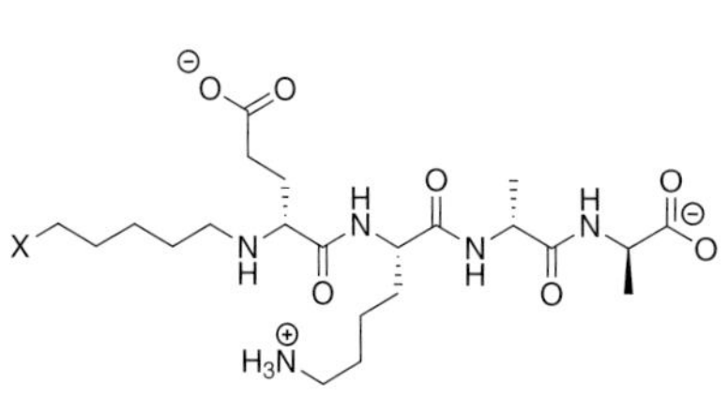

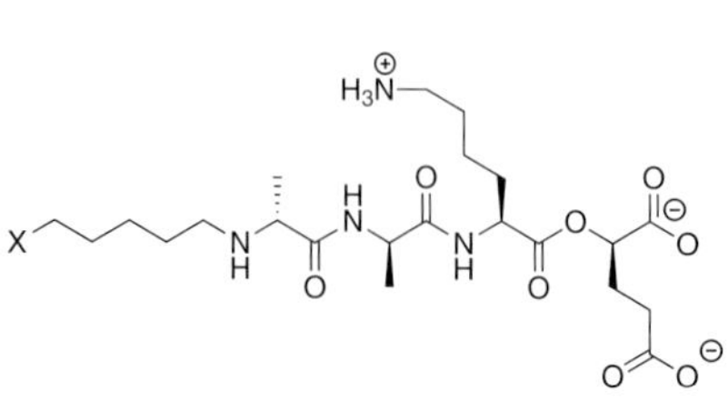

Figure 9.44 Dye labeled (X) substrates EKAA (**124**) and AAKE (**125**).

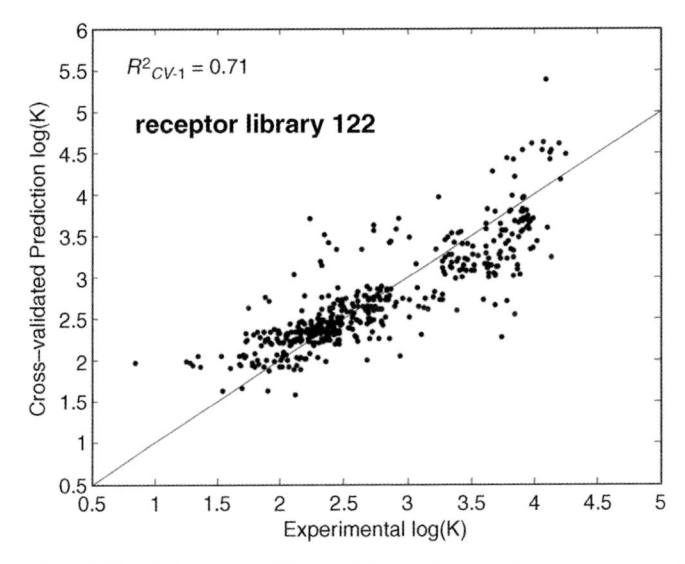

Figure 9.45 Mathematical QSAR model correlating experimental versus predicted values of log K.

In contrast to the former binding study all receptors were now fully deprotected to allow for better interaction with the polar substrate. Binding studies in buffered water were performed as described above and could identify efficient receptors that bind **124** with up to $K = 1.7 \times 10^4 \, M^{-1}$ (pH 6.0, CBS-Lys-Lys-Phe). The diversity of binding strength of the library members was large, again. The reverse substrate sequence AAKE (**125**) was tested, too; it is less efficiently bound than the original substrate with a maximum binding constant of $K = 6 \times 10^3 \, M^{-1}$ [106]. Hence, it could be demonstrated that these flexible one-armed cationic receptors are able to bind selectively to polar tetrapeptides. Binding studies on-bead were verified by ^1H NMR and/or UV titrations in solution.

Based on the quantitative analysis data from the on-bead screening a statistical quantitative structure–activity relationship (QSAR) could be established. Using 49 physicochemical parameters per amino acid position a suitable mathematical QSAR model was set up as depicted in Figure 9.45. Analysis of the model showed that the binding is solely determined by electrostatic interactions. Furthermore, a virtual library with 8000 members was analyzed based on the statistical QSAR model but did not lead to receptors with significantly better binding properties than the best one identified from the experimental screening. Hence, not the size but the correctly chosen diversity of the library determines whether efficient receptors can be found. These results undermine the usefulness of small and focused libraries that give much more information than large random libraries with their limited hit or non-hit outputs.

Finally, the best receptor found experimentally for the EKAA sequence (CBS-KKF) was then tested against a combinatorial library containing 320 members of closely related tetrapeptides (Figure 9.46) [107]. The substrates presented only three

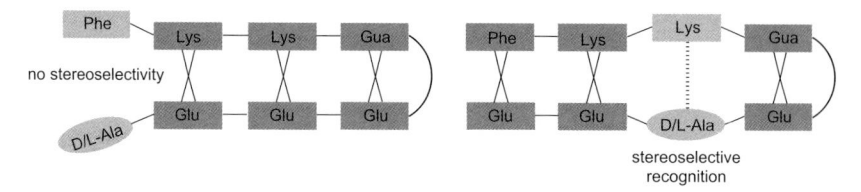

Figure 9.46 Screening of a small focused library revealed stereoselectivity when the position where the D-Ala/L-Ala exchange takes place is fixed at both sides by strong electrostatic interactions between receptor and substrate.

different side chains (Ala, Lys, and Glu) and differed only in the absolute configuration of one building block (D/L-Ala). Binding constants in buffered water ranged from $K < 50\,\mathrm{M}^{-1}$ to $2.7 \times 10^4\,\mathrm{M}^{-1}$ (pH 6.1, fluorescence). Furthermore a sequence-dependent stereoselectivity of the receptor could be observed. The receptor differentiates between D- and L-alanine but only when the position where the exchange takes place is fixed at both sides by strong charge interactions between host and guest.

These results once more stress the potential of fully analyzable small libraries. Of course, the limited structural diversity requires careful design to provide the correct diversity needed to answer a certain question.

In most recent work, Schmuck, Schlücker, and their coworkers could monitor the complex formation between the receptor with the sequence CBS-Lys-Lys-Phe-NH$_2$ and N-Ac-Glu-Glu-Glu-Glu-OH as substrate in aqueous solution by means of UV resonance Raman spectroscopy [108]. A Raman titration experiment could qualitatively assign the spectral changes upon complexation to a change of conformer structure of the receptor and to electronic changes caused by hydrogen enforced ion pair formation. In combination with theoretical calculation Raman spectroscopy presents a promising new tool for the study of supramolecular complex formation.

9.9
Conclusion

The guanidinium cation has been shown to be a versatile functional group with unique properties and has therefore been extensively used for the recognition of various anions over the past few decades. It was implemented in many different molecular scaffolds that were used as hosts in nonpolar as well as polar environments. Even under the most demanding physiological conditions impressive improvements have been made in recent years. Nowadays there are receptors that bind selectively and strongly to complex polar substrates even in water. However, guanidinium–oxoanion interaction remains a large driving force even when additional recognition sites are implemented into the receptor design.

Significant progress has been made in the understanding of the physicochemical principles with regard to recognition events with guanidinium based receptors.

Concepts like preorganization, desolvation, or thermodynamics of binding, to name a few, have been presented in this chapter. An ever more impressive set of tools for both synthesis and analysis of artificial receptors is continuously developing: intelligent combinatorial approaches, for example, help improve synthetic and analytic aspects while theoretical methods like molecular modeling assist in advancing our understanding of the interactions at the molecular level.

Despite the great achievements in these respective fields there are still huge gaps in our understanding of the complex processes in artificial and natural systems that rely on noncovalent interactions. Achieving Nature's efficiency in strong and selective binding to a given substrate and making use of this complexation like enzymes do remains a goal that has yet to be fully realized. Moreover, although there now exist quite a few examples of molecular recognition under physiological conditions, this topic remains a demanding and active area of research. The field of anion recognition requires further development and the guanidinium cation has proved to be, and will continue to remain, a work horse of supramolecular chemistry.

Nevertheless, further tuning of the guanidinium moiety for stronger and more selective anion recognition by derivatization is prone to reach its limit one day. Compared to the large improvements already made, the design of novel elaborate small receptor systems will probably result in small improvements only. We are convinced that in the future the microenvironment of supramolecular receptors will come more and more into focus. It has already been shown that the behavior of hydrogen bonded systems can be influenced by their microenvironment [109]. It is therefore easily imaginable that an efficient binding motif like the guanidinium cation can be implemented into a water-soluble macrosystem – such as, for example, dendrimers or polycyclodextrin compounds – that hydrophobically shields the "active site" of the receptor against the competitive influence of the aqueous environment. We will most likely continue to see some interesting new developments in the field of anion recognition by cationic guanidium hosts in the years to come.

References

1. (a) Kim, D.H. and Park, J. (1996) *Bioorg. Med. Chem. Lett.*, 2967; (b) Luo, R., David, L., Hung, H., Devaney, J., and Gilson, M.K. (1999) *J. Phys. Chem. B*, **103**, 727; (c) García-Pérez, M., Pinto, M., and Subirana, J.A. (2003) *Biopolymers*, **69**, 432.
2. (a) Lange, L.G., Riordan, J.F., and Vallèe, B.L. (1974) *Biochem.*, **13**, 4361; (b) Schmidtchen, F.P. (1988) *Nachr. Chem. Tech. Lab.*, **36**, 8.
3. Goldman, S. and Bates, R.G. (1972) *J. Am. Chem. Soc.*, **94**, 1476.
4. (a) Riordan, J.F. (1979) *Mol. Cell. Biochem.*, **26**, 71; (b) Shimoni, L. and Glusker, J.P. (1995) *Protein Sci.*, **4**, 65; (c) Anslyn, E.V. and Hannon, C.L. (1993) The guanidinium group: its biological role and synthetic analogs, in *Bioorganic Chemistry Frontiers*, vol. 3, Springer Publishers, Berlin.
5. (a) Clarke, A.R., Atkinson, T., and Holbrook, J.J. (1989) *Trends Biochem. Sci.*, **14**, 101; (b) Inoue, Y., Kuramitsu, S., Inoue, K., Kagamiyama, H., Hiromi, K., Tanase, S., and Morino, Y. (1989) *J. Biol. Chem.*, **264**, 9673; (c) White, P.W. and Kirsch, J.F. (1992) *J. Am. Chem. Soc.*, **114**, 3567.
6. (a) Tsikaris, V., Cung, M.T., Panou-Pomonis, E., and Sakarellos-Daitsiotis,

M. (1993) *J. Chem. Soc. Perkin Trans. 2*, 1345; (b) Yokomori, Y. and Hodgson, D.J. (1988) *Int. J. Peptide Protein Res.*, **31**, 289; (c) Adams, J.M. and Small, R.W.H. (1974) *Acta Crystallogr., Sect. B*, **30**, 2191.
7. Cox, K.A., Gaskell, S.J., Morris, M., and Whiting, A. (1996) *J. Am. Soc. Mass Spectrom.*, **7**, 552.
8. Calnan, B.J., Tidor, B., Biancalana, S., Hudson, D., and Frankel, A.D. (1991) *Science*, **252**, 1167.
9. (a) Gale, P.A., García-Garrido, S.E., and Garric, J. (2008) *Chem. Soc. Rev.*, **37**, 151; (b) Blondeau, P., Segura, M., Pérez-Fernández, R., and De Mendoza, J. (2007) *Chem. Soc. Rev.*, **36**, 198; (c) Gale, P.A. and Quesada, R. (2006) *Coord. Chem. Rev.*, **250**, 3219; (d) Schmuck, C. (2006) *Coord. Chem. Rev.*, **250**, 3053; (e) Schmidtchen, F.P. (2006) *Coord. Chem. Rev.*, **250**, 2918; (f) Houk, R.J.T., Tobey, S.L., and Anslyn, E.V. (2005) *Top. Curr. Chem.*, **255**, 199; (g) Schmidtchen, F.P. (2005) *Top. Curr. Chem.*, **255**, 1; (h) Schug, K.A. and Lindner, W. (2005) *Chem. Rev.*, **105**, 67; (i) Best, M.D., Tobey, S.L., and Anslyn, E.V. (2003) *Coord. Chem. Rev.*, **240**, 3; (j) Gale, P.A. (2003) *Coord. Chem. Rev.*, **240**, 191; (k) Gale, P.A. (2000) *Coord. Chem. Rev.*, **199**, 181; (l) Fitzmaurice, R.J., Kyne, G.M., Douheret, D., and Kilburn, J.D. (2002) *J. Chem. Soc., Perkin Trans. 1*, 841; (m) Snowden, T.S. and Anslyn, E.V. (1999) *Curr. Opin. Chem. Biol.*, **3**, 740; (n) Beer, P.D. and Schmitt, P. (1997) *Curr. Opin. Chem. Biol.*, **1**, 475; (o) Bianchi, A., Bowman-James, K., and Garcia-España, E. (1997) *Supramolecular Chemistry of Anions*, Wiley-VCH Verlag GmbH, New York; (p) Schmidtchen, F.P. and Berger, M. (1997) *Chem. Rev.*, **97**, 1609; (q) Seel, C., Galán, A., and De Mendoza, J. (1995) *Top. Curr. Chem.*, **175**, 101.
10. Dietrich, B., Fyles, T.M., Lehn, J.M., Please, L.G., and Fyles, D.L. (1978) *J. Chem. Soc., Chem. Commun.*, 934.
11. Wolfenden Andersson, R.L., Cullis, P.M., and Southgate, C.C.B. (1981) *Biochemistry*, **20**, 849.
12. Mason, P.E., Neilson, G.W., Enderby, J.E., Saboungi, M.-L., Dempsey, C.E., MacKerell, A.D. Jr. and Brady, J.W. (2004) *J. Am. Chem. Soc.*, **126**, 11462.
13. Horvath, P., Gergely, A., and Noszal, B. (1996) *J. Chem. Soc., Perkin Trans. 2*, 1419.
14. Schneider, H.-J. (1994) *Chem. Soc. Rev.*, **22**, 227.
15. (a) Jeffrey, G.A. (1997) *An Introduction to Hydrogen Bonding*, Oxford University Press, New York; (b) Israelachvili, J. (1992) *Intermolecular & Surface Forces*, 2nd edn, Academic Press, London; (c) Widom, B., Bhimalapuram, P., and Koga, K. (2003) *Phys. Chem. Chem. Phys.*, **5**, 3085.
16. Müller, G., Riede, J., and Schmidtchen, F.P. (1988) *Angew. Chem. Int. Ed. Engl.*, **27**, 1516.
17. Echavarren, A., Galán, A., Lehn, J.-M., and De Mendoza, J. (1989) *J. Am. Chem. Soc.*, **111**, 4994.
18. Jorgensen, W.L. and Pranata, J. (1990) *J. Am. Chem. Soc.*, **112**, 2008.
19. Dixon, R.P., Geib, S.J., and Hamilton, A.D. (1992) *J. Am. Chem. Soc.*, **114**, 365.
20. Prins, L.-J., Reinhoudt, D.N., and Timmerman, P. (2001) *Angew. Chem. Int. Ed.*, **40**, 2383.
21. Jubian, V., Dixon, R.P., and Hamilton, A.D. (1992) *J. Am. Chem. Soc.*, **114**, 1120.
22. Linton, B.R., Goodman, M.S., Fan, E., Van Arman, S.A., and Hamilton, A.D. (2001) *J. Org. Chem.*, **66**, 7313.
23. Schmidtchen, F.P. (1989) *Tetrahedron Lett.*, **30**, 4493.
24. (a) Schmidtchen, F.P. (1990) *Tetrahedron Lett.*, **31**, 2269; (b) Kurzmeier, H. and Schmidtchen, F.P. (1990) *J. Org. Chem.*, **55**, 3749; (c) Berger, M. and Schmidtchen, F.P. (1999) *J. Am. Chem. Soc.*, **121**, 9986.
25. Fischer, E. (1894) *Chem. Ber.*, **27**, 2985.
26. Haj-Zaroubi, M., Mitzel, N.W., and Schmidtchen, F.P. (2002) *Angew. Chem. Int. Ed.*, **41**, 104.
27. Metzger, A., Lynch, V.M., and Anslyn, E.V. (1997) *Angew. Chem. Int. Ed. Engl.*, **36**, 862.
28. Metzger, A. and Anslyn, E.V. (1998) *Angew. Chem. Int. Ed. Engl.*, **37**, 649.
29. Schmuck, C. (1999) *Chem. Commun.*, 843.
30. Schmuck, C. (2000) *Chem. Eur. J.*, **6**, 709.
31. Schmuck, C. and Machon, U. (2005) *Chem. Eur. J.*, **11**, 1109.
32. (a) For similar conclusions in different systems see: Kavallieratos, K., Bertao, C.M., and Crabtree, R.H. (1999) *J. Org. Chem.*, **64**, 1675; (b) Kyne, G.M., Light,

M.E., Hursthouse, M.B., De Mendoza, J., and Kilburn, J.D. (2001) *J. Chem. Soc., Perkin Trans. 1*, 1258; (c) Chang, S.-Y., Kim, H.S., Chang, K.-J., and Jeong, K.-S. (2004) *Org. Lett.*, **6**, 181.

33 Moiani, D., Cavallotti, C., Famulari, A., and Schmuck, C. (2008) *Chem. Eur. J.*, **14**, 5207.

34 Blondeau, P. and De Mendoza, J. (2007) *New J. Chem.*, **31**, 736.

35 (a) For theoretical studies see: Hay, B.P., Gutowski, M., Dixon, D.A., Garza, J., Vargas, R., and Moyer, B.A. (2004) *J. Am. Chem. Soc.*, **126**, 7925; (b) Hay, B.P., Firman, T.K., and Moyer, B.A. (2005) *J. Am. Chem. Soc.*, **127**, 1810.

36 Beer, P.D., Drew, M.G.B., and Smith, D.K. (1997) *J. Organomet. Chem.*, **543**, 259.

37 Otón, F., Tárraga, A., and Molina, P. (2006) *Org. Lett.*, **8**, 2107.

38 For a review on electrochemical molecular sensing see: Beer, P.D., Gale, P.A., and Chen, G.Z. (1999) *J. Chem. Soc., Dalton Trans.*, 1897.

39 Serr, B.R., Andersen, K.A., Elliot, C.M., and Anderson, O.P. (1988) *Inorg. Chem.*, **27**, 4499.

40 Otón, F., Espinosa, A., Tárraga, A., De Arellano, C.R., and Molina, P. (2007) *Chem. Eur. J.*, **13**, 5742.

41 Rozas, I. and Kruger, P.E. (2005) *J. Chem. Theory Comput.*, **1**, 1055.

42 Kobiro, K. and Inoue, Y. (2003) *J. Am. Chem. Soc.*, **125**, 421.

43 Schmuck, C. and Machon, U. (2006) *Eur. J. Org. Chem.*, 4385.

44 (a) Snowden, T.S. and Anslyn, E.V. (1999) *Curr. Opin. Chem. Biol.*, **3**, 740–746; (b) Anslyn, E.V. (2007) *J. Org. Chem.*, **72**, 687.

45 Tobey, S.L., Jones, B.D., and Anslyn, E.V. (2003) *J. Am. Chem. Soc.*, **125**, 4026.

46 Tobey, S.L. and Anslyn, E.V. (2003) *J. Am. Chem. Soc.*, **125**, 14807.

47 Tobey, S.L. and Anslyn, E.V. (2003) *Org. Lett.*, **5**, 2029.

48 Jadhav, V.D. and Schmidtchen, F.P. (2005) *Org. Lett.*, **7**, 3311.

49 Deanda, F., Smith, K.M., Liu, J., and Pearlman, R.S. (2004) *Mol. Pharm.*, **1**, 23.

50 Jadhav, V.D., Herdtweck, E., and Schmidtchen, F.P. (2008) *Chem. Eur. J.*, 6098.

51 Haj-Zaroubi, M. and Schmidtchen, F.P. (2005) *ChemPhysChem*, **6**, 1181.

52 Schmuck, C. and Heller, M. (2007) *Org. Biomol. Chem.*, **5**, 787.

53 (a) Watson, S.R., Imai, Y., Fennie, J., Geoffroy, J.S., and Rosen, S.D. (1990) *J. Cell. Biol.*, **110**, 2221; (b) Jin, L., Abrahams, J.P., Skinner, R., and Petitou, M. (1997) *Proc. Natl. Acad. Sci. USA*, **94**, 14683.

54 Schneider, S.E., O'Neil, S.N., and Anslyn, E.V. (2000) *J. Am. Chem. Soc.*, **122**, 542.

55 (a) Lavigne, J.J., Savoy, S., Clevenger, M.D., Ritchie, J.E., McDoniel, B., Yoo, S.-J., Anslyn, E.V., McDevitt, J.T., Shear, J.B., Niekirk, D.J., Yamada, H., and Toyoshima, K.J. (1998) *J. Pharm. Sci.*, **87**, 552; (b) Dickinson, T.A. and Walt, D.R. (1997) *Anal. Chem.*, **69**, 3413.

56 McCleskey, S.C., Griffin, M.J., Schneider, S.E., McDevitt, J.T., and Ansyn, E.V. (2003) *J. Am. Chem. Soc.*, **125**, 1114.

57 Goodey, A., Lavigne, J.J., Savoy, S., Rodriguez, M., Curey, T., Tsao, A., Simmons, G., Yoo, S., Sohn, Y., Anslyn, E.V., Shear, J.B., Niekirk, D.J., and McDevitt, J.T. (2001) *J. Am. Chem. Soc.*, **123**, 2559.

58 Menual, S., Duval, R.E., Cuc, D., Mutzenhardt, P., and Marsura, A. (2007) *New J. Chem.*, **31**, 995.

59 Jadhav, V.D. and Schmidtchen, F.P. (2006) *Org. Lett.*, **8**, 2329.

60 Jadhav, V.D. and Schmidtchen, F.P. (2008) *J. Org. Chem.*, **73**, 1077.

61 Schmuck, C. and Schwegmann, M. (2005) *J. Am. Chem. Soc.*, **127**, 3373.

62 Schmuck, C. and Schwegmann, M. (2006) *Org. Biomol. Chem.*, **4**, 836.

63 Hughes, A.D. and Anslyn, E.V. (2007) *Proc. Natl. Acad. Sci. USA*, **104**, 6538.

64 (a) Badjic, J.D., Neslon, A., Cantrill, S.J., Turnbull, W.B., and Stoddart, J.F. (2005) *Acc. Chem. Res.*, **38**, 723; (b) Tobey, S.L. and Anslyn, E.V. (2003) *J. Am. Chem. Soc.*, **125**, 10963; (c) Williams, D.H. and Westwell, M.S. (1998) *Chem. Soc. Rev.*, **27**, 57; (d) Jencks, W.P. (1981) *Proc. Natl. Acad. Sci. USA*, **78**, 4046.

65 Schmuck, C. and Bickert, V. (2003) *Org. Lett.*, **5**, 4579.

66 Schmuck, C. and Dudaczek, J. (2005) *Tetrahedron Lett.*, **46**, 7101.

67 Schmuck, C. and Geiger, L. (2005) *J. Am. Chem. Soc.*, **127**, 10486.
68 Schmuck, C. and Bickert, V. (2007) *J. Org. Chem.*, **72**, 6832.
69 Schmuck, C. and Geiger, L. (2004) *J. Am. Chem. Soc.*, **126**, 8898.
70 Schmuck, C. and Hernandez-Folgado, L. (2007) *Org. Biomol. Chem.*, **5**, 2390.
71 Schmuck, C., Rupprecht, D., and Wienand, W. (2006) *Chem. Eur. J.*, **12**, 9186.
72 (a) Collet, A., Dutata, J.P., Lozach, B., and Canceill, J. (1993) *Top. Curr. Chem.*, **165**, 103; (b) Collet, A. (1987) *Tetrahedron*, **24**, 5725.
73 (a) Süssmuth, R.D. (2002) *ChemBioChem*, **3**, 295; (b) Nicolaou, K.C., Boddy, C.N.C., Bräse, S., and Wissinger, N. (1999) *Angew. Chem. Int. Ed.*, **38**, 2096; (c) Williams, D.H. and Bardesley, B. (1999) *Angew. Chem. Int. Ed.*, **38**, 1172; (d) Walsh, C. (1999) *Science*, **284**, 442.
74 (a) Wittinghofer, A. and Waldmann, H. (2000) *Angew. Chem. Int. Ed.*, **39**, 4192; (b) Hinterding, K., Alonso-Díaz, D., and Waldmann, H. (1998) *Angew. Chem. Int. Ed.*, **37**, 688.
75 Peczuh, M.W. and Hamilton, A.D. (2000) *Chem. Rev.*, **100**, 2479.
76 (a) For review articles on artificial peptide receptors see: Schneider, H.-J. (2000) *Adv. Supramol. Chem.*, **6**, 185; (b) Schneider, H.-J. (1993) *Angew. Chem. Int. Ed. Engl.*, **105**, 890; (c) Webb, T.H. and Wilcox, C.S. (1993) *Chem. Soc. Rev.*, **22**, 383.
77 Alvatella, X., Martinell, M., Gairí, M., Mateu, M.G., Feliz, M., Hamilton, A.D., De Mendoza, J., and Giralt, E. (2004) *Angew. Chem. Int. Ed.*, **43**, 196.
78 Jeong, H., Tombor, B., Albert, R., Oltval, Z.N., and Barabási, A.-L. (2000) *Nature*, **407**, 651.
79 (a) Laskowski, R.A., Luscome, N.M., Swindells, M.B., and Thornton, J.M. (1996) *Protein Sci.*, **5**, 2438; (b) Sheinerman, F.B., Norel, R., and Hönig, B. (2000) *Curr. Opin. Struct. Biol.*, **10**, 153; (c) DeLano, W.L. (2002) *Curr. Opin. Struct. Biol.*, **12**, 14.
80 Peczuch, M.W., Hamilton, A.D., Sánchez-Quesada, J., De Mendoza, J., Haack, T., and Giralt, E. (1997) *J. Am. Chem. Soc.*, **119**, 9327.
81 Chène, P. (2001) *Oncogene*, **20**, 2611.
82 (a) Mayer, M. and Meyer, B. (2001) *J. Am. Chem. Soc.*, **123**, 6108; (b) Mayer, M. and Meyer, B. (1999) *Angew. Chem.*, **111**, 1902; (c) Klein, J., Meinecke, R., Mayer, M., and Meyer, B. (1999) *J. Am. Chem. Soc.*, **121**, 5336.
83 Schmuck, C., Rupprecht, D., Junkers, M., and Schrader, T. (2007) *Chem. Eur. J.*, **13**, 6864.
84 Hynes, R.O. (2002) *Cell*, **110**, 673.
85 (a) Kuphal, S., Bauer, R., and Bosserhoff, A.-K. (2005) *Cancer Metastasis Rev.*, **24**, 195; (b) Triantafilou, K., Takada, Y., and Triantafilou, M. (2001) *Crit. Rev. Immunol.*, **21**, 311.
86 Rensing, S. and Schrader, T. (2002) *Org. Lett.*, **4**, 2161.
87 (a) Susuki, Y., Hojo, K., Okazaki, I., Kamata, H., Sasaki, M., Maeda, M., Nomizu, M., Yamamoto, Y., Nakagawa, S., Mayumi, T., and Kawasaki, K. (2002) *Chem. Pharm. Bull.*, **50**, 1229; (b) Meissner, R.S., Perkins, J.J., Duong, L.T., Hartman, G.D., Hoffman, W.F., Huff, J.F., Ihle, N.C., Leu, C.-T., Nagy, R.M., Naylor-Olsen, A., Rodan, G.A., Rodan, S.B., Whitman, D.B., Wesolowski, G.A., and Duggan, M.E. (2002) *Bioorg. Med. Chem. Lett.*, **12**, 25.
88 (a) Boyce, R., Li, G., Nestler, H.P., Suenaga, T., and Still, W.C. (1994) *J. Am. Chem. Soc.*, **116**, 7955; (b) Gennari, C., Nestler, H.P., Salom, B., and Still, W.C. (1995) *Angew. Chem. Int. Ed. Engl.*, **34**, 1765.
89 (a) Shuttleworth, S.J., Connors, R.V., Fu, J., Liu, J., Lizarzaburu, M.E., Qiu, W., Sharma, R., Wanska, M., and Zhang, A.J. (2005) *Curr. Med. Chem.*, **12**, 1239; (b) Maly, D.J., Huang, L., and Ellman, J.A. (2002) *ChemBioChem*, **1**, 16.
90 Cho, C.Y., Moran, E.J., Cherry, S.R., Stephans, J.C., Fodor, S.P.A., Adams, C.L., Sunderam, A., Jacobs, J.W., and Schultz, P.G. (1993) *Science*, **261**, 1303.
91 (a) Uttamchandani, M., Walsh, D.P., Yao, S.Q., and Chang, Y.T. (2005) *Curr. Opin. Chem. Biol.*, **9**, 4; (b) Thompson, L.A. and Ellman, J.A. (1996) *Chem. Rev.*, **96**, 555.
92 Furka, A., Sebestyen, F., Asgedom, M., and Dibo, G. (1991) *Int. J. Pept. Protein Res.*, **37**, 487.

93 Jensen, K.B., Braxmeier, T.M., Demarcus, M., and Kilburn, J.D. (2002) *Chem. Eur. J.*, **8**, 1300.

94 (a) Lwik, D.W.P.M., Weingarten, M.D., Boekema, M., Brouwer, A.J., Still, W.C., and Liskamp, R.M.J. (1998) *Angew. Chem.*, **110**, 1947; (b) Lwik, D.W.P.M., Mulders, S.J.E., Cheng, Y., Shao, Y., and Liskamp, R.M.J. (1996) *Tetrahedron Lett.*, **37**, 8253; (c) Gennari, C., Nestler, H.P., Salom, B., and Still, W.C. (1995) *Angew. Chem.*, **107**, 1894; (d) Wennemers, H., Yoon, S.S., and Still, W.C. (1995) *J. Org. Chem.*, **60**, 1108.

95 (a) Monnee, M.C.F., Brouwer, A.J., and Lisamp, R.M.J. (2004) *QSAR Comb. Sci.*, **23**, 546; (b) Davies, M., Bonnat, M., Guillier, F., Kilburn, J.D., and Bradley, M. (1998) *J. Org. Chem.*, **63**, 8696; (c) Torneiro, M. and Still, W.C. (1997) *Tetrahedron*, **53**, 8739.

96 Srinivasan, N. and Kilburn, J.D. (2004) *Curr. Opin. Chem. Biol.*, **8**, 305.

97 Shepherd, J., Gale, T., Jensen, K.B., and Kilburn, J.D. (2006) *Chem. Eur. J.*, **12**, 713.

98 Conza, M. and Wennemers, H. (2003) *Chem. Commun.*, 866.

99 (a) Schmuck, C. and Heil, M. (2003) *ChemBioChem*, **4**, 1232; (b) Schmuck, C. and Heil, M. (2003) *Org. Biomol. Chem.*, **1**, 633.

100 (a) Zerovnik, E. (2002) *Eur. J. Biochem.*, **269**, 3362; (b) Hardy, J. and Selkoe, D.J. (2002) *Science*, **297**, 353; (c) Austen, A. and Manca, M. (2000) *Chem. Br.*, **36**, 28; (d) Gopinath, L. (1998) *Chem. Br.*, **34**, 38; (e) Lansbury, P.T. Jr. (1996) *Acc. Chem. Res.*, **29**, 317; (f) Yankner, B.A. (1996) *Neuron*, **16**, 921.

101 Jarret, J.T., Berger, E.P., and Lansbury, P.T. Jr. (1993) *Biochemistry*, **32**, 4693.

102 Czarnik, A.W. (1997) *Curr. Opin. Chem. Biol.*, **1**, 60.

103 Lansbury, P.T. Jr., Costa, P.R., Griffiths, J.M., Simon, E.J., Auger, M., Halverson, K.J., Kocisko, D.A., Hendsch, Z.S., Ashbury, T.T., Spencer, R.G.S., Tidor, B., and Griffin, R.G. (1995) *Nat. Struct. Biol.*, **2**, 990.

104 Schmuck, C., Frey, P., and Heil, M. (2005) *ChemBioChem*, **6**, 628.

105 Schmuck, C., Heil, M., Scheiber, J., and Baumann, K. (2005) *Angew. Chem. Int. Ed.*, **44**, 7208.

106 Schmuck, C. and Heil, M. (2006) *Chem. Eur. J.*, **12**, 1339.

107 Schmuck, C. and Wich, P. (2006) *Angew. Chem. Int. Ed. Engl.*, **45**, 4277.

108 Küstner, B., Schmuck, C., Wich, P., Jehn, C., Srivastava, S.K., and Schlücker, S. (2007) *Phys. Chem. Chem. Phys.*, **9**, 4598.

109 (a) De Greef, T.F.A., Nieuwenhuizen, M.M.L., Stals, P.J.M., Fitié, C.F.C., Palmans, A.R.A., Sijbsesma, R.P., and Meijer, E.W. (2008) *Chem. Commun.*, 4306; (b) Merschky, M. (2010) Dissertation, Essen.

10
Artificial Receptors Based on Spreader-Bar Systems
Thomas Hirsch

The basic idea for artificial receptors formed in monolayer systems was formulated in the work of Sagiv [1]. He describes a monolayer of *n*-octadecyltrichlorosilane (OTS) on glass, assembled together with a dye, characterized by a polar moiety at one end of the molecule and a nonpolar part. A mixed monolayer of OTS with incorporated dye molecules is formed. After treatment of the surface by chloroform the physisorbed dye was washed out and the silane remains on the substrate. By immersion of this silane covered glass substrate in a solution of the dye a readsorbtion of this molecule was found. In addition, not only the molecule present during the self-assembly of the silane was able to adsorb to this surface, other molecules with the same geometrical properties as the displaced molecule could be entrapped (Figure 10.1).

One drawback of this concept is that after desorption the stability of this structured monolayer is weak, because of lateral diffusion of the molecules forming the SAM (self-assembled monolayer). Another limitation is in the choice of molecules that can be used for this memory effect. They have to be from the same type as the monolayer-forming compound with a small polar and a large nonpolar moiety.

It is typical for many systems ordered on the nanometer scale that even small structural changes lead to a total loss of function of the whole system. The stabilization by crosslinking leads to other problems, such as complicated chemistry and/or poor compatibility of subsequent preparation steps, resulting in major limitations in the selection of molecules that can be used.

Known techniques for forming micro-structured ultrathin layers, which include photolithography [2, 3], electron beam lithography [4] or microcontact printing (μ-CP) [5, 6], and soft lithography [7, 8], are limited in their resolution and cannot reproducibly achieve stable patterns with dimensions at the nanometer scale. The μ-CP technique is inherently limited by the physical interaction of a macroscopic stamp with the surface, often leading to a less structured organic layer with significant defect density; moreover, very precise structures achieved with μ-CP have only been described so far by using of long-chain alkanethiols [9, 10]. Therefore, the *top down* approach breaks down when molecular precision is desired.

This challenge was a strong motivation for the development of *bottom-up* approaches based on subsequent assembly of complete structures molecule by molecule.

Artificial Receptors for Chemical Sensors. Edited by V.M. Mirsky and A.K. Yatsimirsky
Copyright © 2011 WILEY-VCH Verlag GmbH & Co. KGaA, Weinheim
ISBN: 978-3-527-32357-9

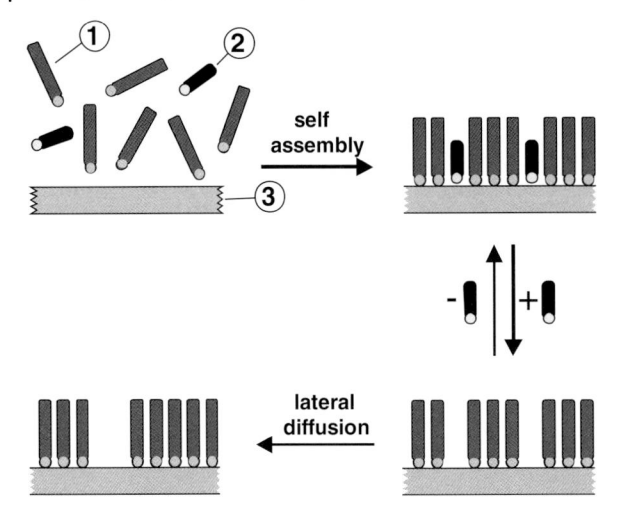

Figure 10.1 Adsorption of a silane (1) together with a polar molecule (2) onto a glass surface (3). A memory effect for the surface was found. The polar molecules could be washed out and rebound. This characteristic is lost with time because of lateral diffusion of the molecules within the monomolecular film.

Single-molecule manipulation has been demonstrated successfully using scanning probe microscopy, but this technique is extremely time consuming and therefore too expensive for any industrial, and many laboratorial, applications [10–15].

A combination of the speed and versatility of lithographic techniques with the resolution of single-molecule manipulation can be realized by introducing a technique using the way that biological systems explore: self-assembly. Moreover, according to the current state of technology, self-assembly is probably the only possible way to fabricate nanoscale assemblies simply and economically effective.

The natural phenomenon of self-assembly has been explored for producing supramolecular alignments and have been adapted to form even nanoscale patterns [16–20]. The best studied systems are SAMs formed spontaneously by chemisorption of thiol-terminated molecules onto a gold surface [21–23]. The high stability and low defect density of these molecular arrangements is a consequence of the attractive van der Waals forces between the methylene groups and the covalent bond between gold and sulfur. The chain length of the alkanethiol determines the insulating properties of the SAM [24].

Multi-component SAMs formed by co-deposition of two or more adsorbates from solution have been investigated for their patterning potential [17, 18, 25–28]; it has been shown that, depending on the molecules used, the resulting monolayer content is a homogeneous mixture or separated phases of these compounds [26, 29]. Mixed monolayers consist of electro-inactive insulating long-chain thiols, and conductive aromatic thiols have also been used to demonstrate a template directed growth of polymer nanostructures; a subsequent electropolymerization of aniline occurred at sites occupied by the latter sort of thiols only [30].

The spreader-bar technique enables the formation of mixed monolayers consisting of two types of molecules by immersion of a gold substrate into one solution containing both constituents. One type of molecule defines the receptive properties of the resulting monomolecular film. It has to have a more or less rigid structure, similar to the shape of the desired analyte and it is called the spreader-bar molecule. The second type of molecule, the so-called matrix molecule, forms a monomolecular film on the substrate and has the function to stabilize the position of the spreader-bar molecules on the surface, so that their movements are hindered. Both types of molecules are chemisorbed to the gold surface and result in a monolayer with defined cavities. In similar way to molecularly imprinted polymers this technique acts as a two-dimensional imprinting and the structures formed should be able to interact with analyte molecules in a solution; theses structures are stable against lateral diffusion because the template will remain in the surface (Figure 10.2).

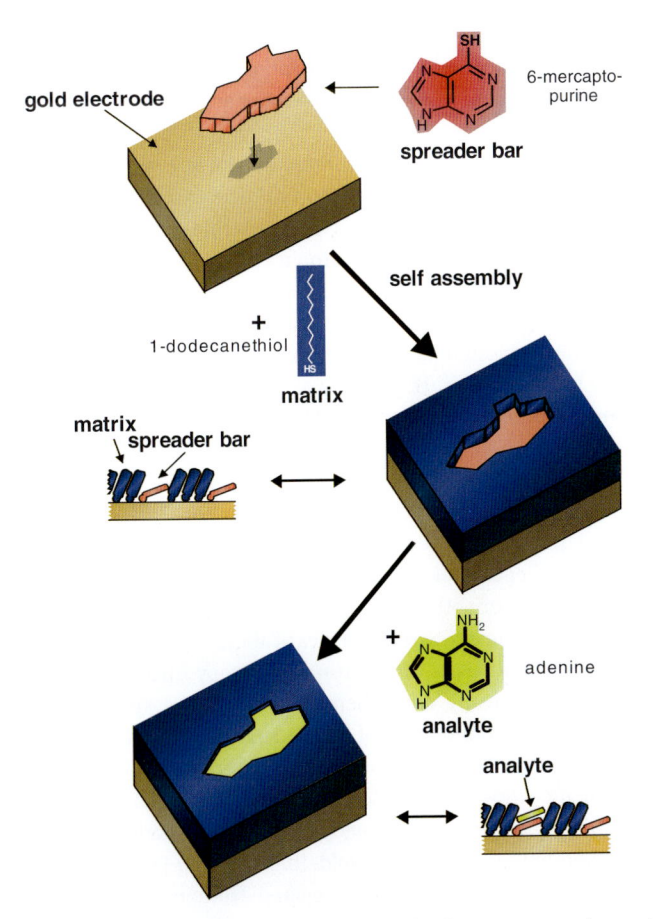

Figure 10.2 Principle of a spreader-bar stabilized, nanostructured monolayer [31]. Reproduced by permission of The Royal Society of Chemistry.

The molar ratio of two different thiols within the monolayer can be controlled by the molar ratio of the two molecules in the deposition solution, in that the surface composition is determined by thermodynamics or by kinetics [32–34]. The mixed monolayer from the spreader-bar type formed on a gold substrate by co-adsorption of 1-dodecanethiol (C12) together with a large planar rigid molecule with a developed π-electron system, a thiolated derivate of 5,10,15,20-tetrakis(4-sulfonatophenyl)porphyrin (TMPP), has been studied extensively by XPS (X-ray photoelectron spectroscopy) [35]. The relative fractions of C12 and TMPP in the solution were varied. The deposition time was 72 h, which is far beyond a kinetic limit for surface/solution exchange of thiolated compounds on gold surface [36]. Therefore, one can expect a quasi-equilibrium ratio of the template and matrix molecules in the film. According to the NEXAF spectra and the XPS data, the film formed from the 1:10 000 (1-dodecanethiol:TMPP) solution contains only TMPP moieties, the "1:1000" film is mostly TMPP with some percentage of 1-dodecanethiol, the "1:100" film contains a minor amount of TMPP, and the spectra for the "1:10" film exhibit only 1-dodecanethiol features. Thus, the amount of TMPP species in the mixed film can be adjusted precisely. According to NEXAFS (near edge X-ray absorption fine structure) difference curves, the TMPP molecules both in the one-component and mixed films have an in-plane (strongly inclined) geometry.

These spreader-bar systems have been used as templates for the generation of metallic nanoparticles by selective reduction of platinum into the nanopores formed by spreader-bar molecules of TMPP in a matrix of 1-dodecanethiol on gold electrodes [37]. Direct visualization of nanoparticles by electron microscopy was performed successfully (Figure 10.3). The nanoparticles were formed by reduction of copper or platinum. There should be no physical or chemical limitations for preparation of nanoparticles from many other noble and transition metals according to this method. The particles are relatively homogeneous. Depending on the deposition charge, the size of nanoparticles can be controlled in the range 20–1000 nm.

It should be possible to also form smaller nanoparticles (the size is probably limited only by the size of the spreader-bar used, that is, about 2 nm for TMPP), but the electron microscopy employed was not able to visualize nanoparticles smaller than 20 nm. The total area of the formed platinum nanoelectrodes, estimated from an investigation of underpotential deposition of copper, is about 1.5 times higher than the geometrical electrode area. For smooth nanoparticles, this corresponds to a ratio of nanoparticle radius to mean distance between particles of about 0.35; this is similar to the value obtained from digital analysis of electron microscopy images (0.22).

To prove that the thiol moiety really does chemically attach the spreader-bar molecules to the substrate, mixed monolayers of 1-hexadecanethiol and the large planar molecule aluminum phthalocyanine chloride (AlPC) were tested. The presence of 12 conjugated aromatic rings in the molecule and the suggestion of strong π-electron interaction with the metal led to the expectation that planar adsorption of these molecules occurs, even without exploiting gold–thiol binding [38]. XPS and NEXAFS of the pure AlPC films and the films prepared from mixed AlPC–1-hexadecanethiol solutions were studied [39]. In the spectra of pure AlPC films, characteristic emissions of all the elements comprising the AlPC molecule, including

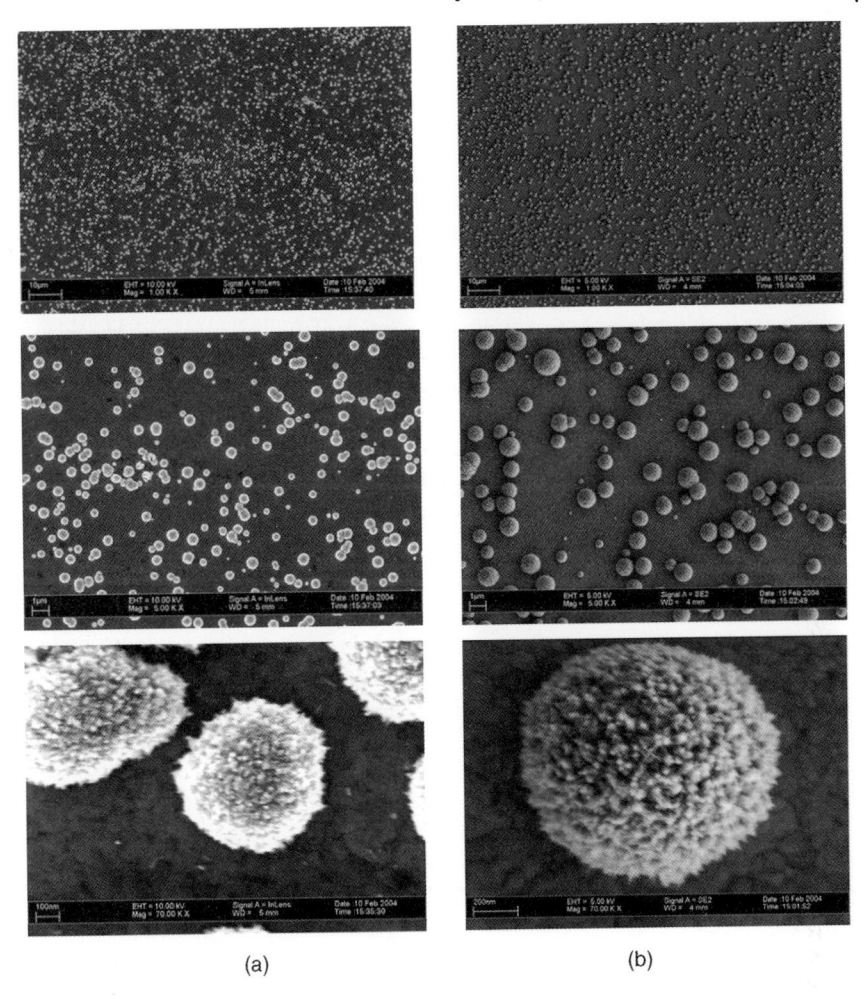

Figure 10.3 Scanning electron microscopy of gold electrodes coated with a nanostructured monolayer consisting of TMPP and 1-dodecanethiol after electrochemical platinum deposition. The deposition charge was 41 C m^{-2} in column (a) and 160 C m^{-2} in column (b). [37]. Reproduced with permission. Copyright Wiley-VCH Verlag GmbH & Co. KGaA.

those related to nitrogen, chlorine, and aluminum could be found. The situation changed crucially as soon as the Au substrates were immersed in the mixed AlPC–1-hexadecanethiol solutions. Despite the strong AlPC excess (by factors of 10 and 100), the respective XPS and NEXAFS spectra are characteristic of a one-component C16 SAM and do not exhibit any features related to the AlPC molecules. In particular, the C 1s XPS spectra exhibit a relatively sharp emission at about 285.0 eV, which is characteristic of the intact alkanethiolate SAMs [40], whereas no emissions were observed in the N 1s, O 1s, Cl 2p, and Al 2p ranges. The effective thickness of both films prepared from the mixed AlPC–1-hexadecanethiol solutions was estimated at about 18.9 Å, which is the expected value for the C16 SAM on Au [41]. The N K-edge NEXAFS spectra exhibited identical smooth and structure-less curves, without any features related to the excitation from the N1s core level to nitrogen-derived unoccupied molecular orbitals. Thus, it can be concluded that the obtained layers represent well-ordered and densely packed C16 SAMs, which do not contain any AlPC molecules, within the detection limit of XPS and NEXAFS spectroscopy (several % of the monolayer surface). This result implies that even for large planar molecules with developed π-electron systems special anchor groups (e.g., thiol) are important.

The first artificial interface according to the spreader-bar principle, with high affinity for barbiturate, was created by co-adsorption of thiobarbiturate (the template) and dodecanethiol (the matrix) onto a gold substrate [42]. This process leads to the formation of binding sites with a structure complementary to that of thiobarbituric acid. Binding of barbiturate and of other species to the respective surface was detected by capacitance measurements: an increase of the dielectric thickness decreases the electrode capacitance [43]. A high selectivity of this artificial chemoreceptor for barbiturate was observed (Figure 10.4); while there is a considerable response to the addition of barbiturate, diethylbarbiturate addition caused no effect. Moreover, the effects due to additions of barbiturate in the presence of a high concentration of diethylbarbiturate were exactly the same as in the absence of diethylbarbiturate. The capacitance response to pyridine was also much lower compared to barbiturate.

Thiol-modified purines and pyrimidines (spreader-bar) co-adsorbed with 1-dodecanethiols (matrix) onto a gold surface form self-assembled nanostructured

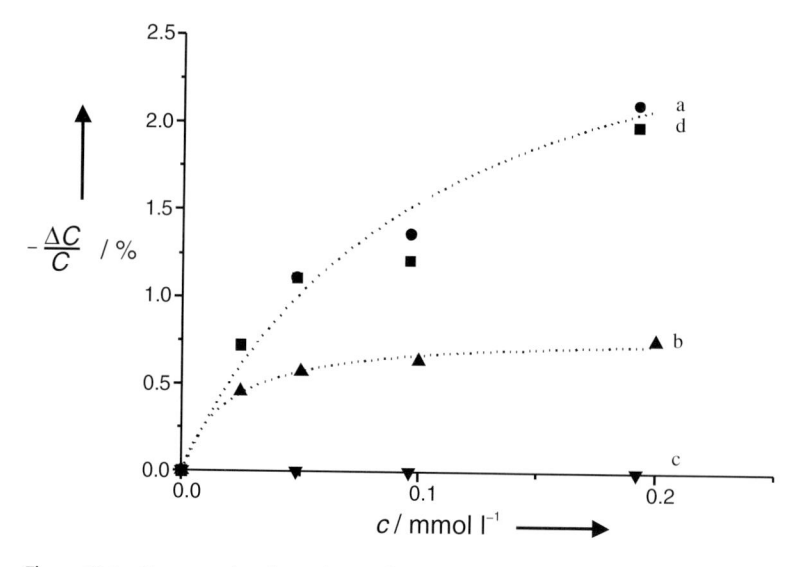

Figure 10.4 Concentration dependence of capacitance changes due to binding of barbiturate with the artificial receptors. Curves signify binding of (a) barbiturate, (b) pyridine, (c) diethylbarbiturate, and (d) barbiturate in the presence of 1 mmol l^{-1} diethylbarbiturate. Template: thiobarbiturate, matrix: dodecanethiol. Electrolyte: 5 mmol l^{-1} phosphate buffer, 100 mmol l^{-1} KCl, pH 5.5. [42]. Reproduced with permission. Copyright Wiley-VCH Verlag GmbH & Co. KGaA.

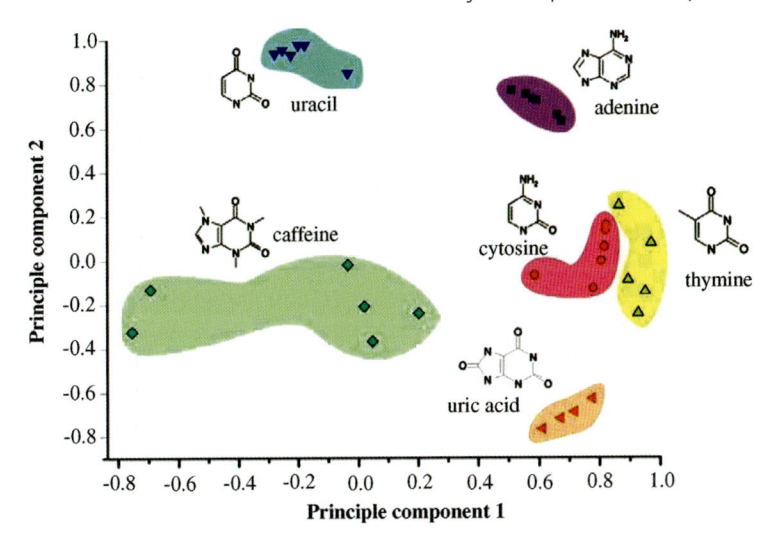

Figure 10.5 Patterns of different concentrations of caffeine, uracil, adenine, cytosine, thymine, and uric acid on an array of artificial receptors formed by thiolated derivatives of purines (ASH, GSH) and pyrimidines (CSH, TSH, USH) presented in a plot of principal components. Capacitive transducing was used [31]. Reproduced by permission of The Royal Society of Chemistry.

monolayers showing recognition properties towards different purines and pyrimidines, depending on the type of spreader-bar used. The structures were investigated by FTIR spectroscopy, contact angle measurements, ellipsometry, impedance spectroscopy, voltammetry, XPS, NEXAFS, and XAM. A monolayer assembled from a solution consisting of 6-mercaptopurine and 1-dodecanethiol (6-mercaptopurine: 1-dodecanethiol = 100 : 1); both species formed a mixed SAM on the gold surface. Analysis of the NEXAFS and XPS data suggest that about 74% of the 6-mercaptopurine molecules are in the mixed film (Figure 10.5).

Orientation of the 6-mercaptopurine molecules in the mixed film (23°) differed from that in the one-component SAM while the orientation of the 1-dodecanethiol species (32°) was identical with that in a one-component 1-dodecanethiol film. Interestingly, the effect of alkanethiol on the orientation of 6-mercaptopurine was observed even at very low alkanethiol concentration in the coating solution, where the concentration of alkanethiol in monolayers was less than the FTIR-detection limit. This confirms the initial suggestion and interpretation of FTIR spectra. XAM images taken with a resolution of 50 nm did not exhibit any domain structure. The results show that the coating conditions optimized for affinity properties correspond to a narrow range of the spreader-bar/matrix ratio in the coating solution leading to the presence of both types of molecule on the surface.

The spreader-bar approach provides a simple method for producing a huge number of receptors with different selectivities, so as a first application a sensor-array of five receptors was introduced, which offers the possibility of detecting different purines and pyrimidines by pattern recognition [31].

10 Artificial Receptors Based on Spreader-Bar Systems

For gold electrodes covered by monolayer of a single component, either of matrix or spreader-bar molecules, no recognition abilities were found. For example, the changes in capacitive current at 80 Hz due to adsorption of purines and pyrimidines from a solution of 300 µmol l^{-1} on 1-dodecanethiol coated electrodes were 0.7% for adenine and even less for every other substance.

By varying of the chain length of the matrix molecules it was found that the spreader-bar systems of 2-thiobarbituric acid together with longer alkanethiols show higher affinity towards barbituric acid than systems with shorter alkanethiol chain length. For systems with very short matrix molecules the affinity was lost completely (Figure 10.6). An explanation is that the cavities formed by the spreader-bar molecule and the surrounding matrix molecules are not deep enough for the analyte to be stabilized near the receptor. The same system was used to test the selectivity by measurement of the change in electrical capacitance for addition of molecules of different size. Molecules that are smaller than the spreader-bar molecule show an increase in electrical capacitance. The highest signal changes for the same concentration of analyte have been found for barbituric acid, which has a similar structure and size to the spreader-bar molecule 2-thiobarbituric acid. The system shows no affinity for bigger molecules, such as 2-diethylbarbituric acid. These results indicate that the model of the cavities within the monolayer is sufficient to describe the affinity properties of spreader-bar systems.

The influence of the chain length of the matrix molecules was also tested by electrochemical investigation of the desorption kinetics of quinone in two-

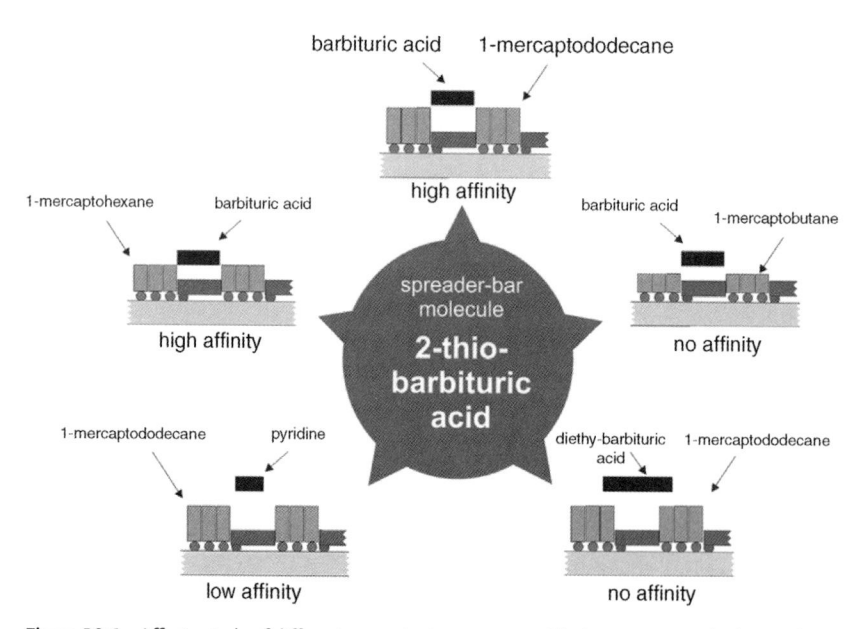

Figure 10.6 Affinity study of different spreader-bar systems with the same spreader-bar molecule on gold electrodes based on monitoring of changes in the electrical capacitance by addition of the same concentration of analyte.

dimensional imprinted monolayers with different alkanethiols [44]. The dissociation rate depends strongly on the surrounding of the template. With increasing chain length from C_{14} to C_{18} of the hydrophobic alkane thiols, desorption of 60% of the quinones will increase from about 30 min to about 60 min. The lower rate constant can be explained by capping of the noncovalently linked quinones by the long alkane chains. This effect can also be found for the improved binding capabilities for long-chain alkanethiols, which assist synergistically the binding of the quinones.

The behavior of mixed monolayers consisting of dodecanethiol and one of thiolated purines or pyrimidines was quite different: an adsorption of adenine, cytosine, thymine, uracil, caffeine, or uric acid resulted in changes in capacitive current of over 25% (Figure 10.7).

The pattern recognition technique, based on principal component analysis or neuronal nets, allows one to reach a very high selectivity of chemical analysis, but requires a pre-formation of an array of chemical sensors with essentially distinguished properties of single sensors. With an array of electrodes, modified with mixed monolayers consisting of five receptors formed by thiolated derivatives of adenine (ASH), thymine (TSH), uracil (USH), guanine (GSH), and cytosine (CSH) as spreader-bars adenine, cytosine, thymine, uracil, caffeine, and uric acid were used as analytes. The analyte binding was detected as changes of peak amplitude in cyclic voltammetry or modification of electrochemical impedance; the binding modifies

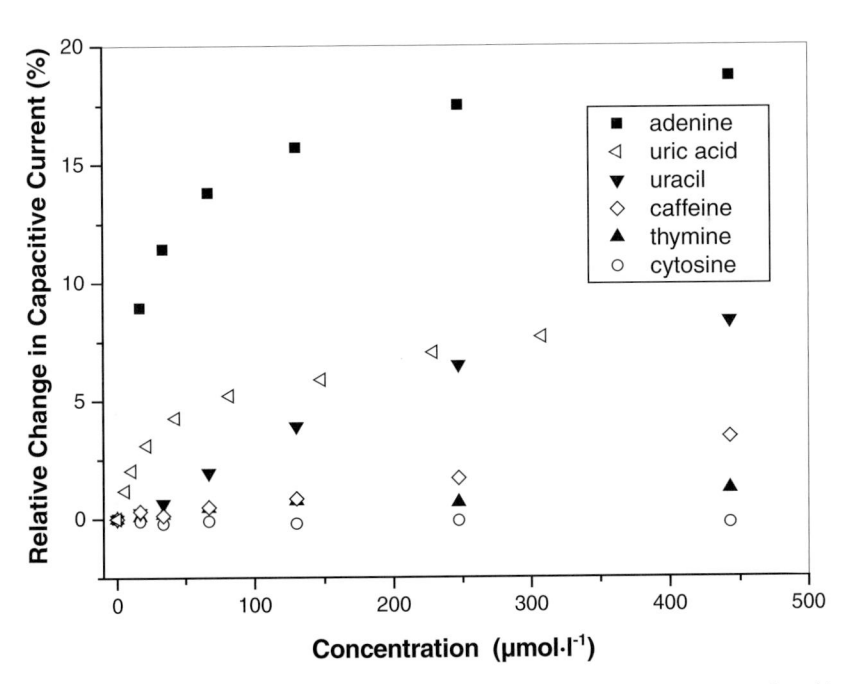

Figure 10.7 Concentration dependence of relative changes in the capacitive current of a gold electrode coated with a mixed monolayer from dodecanethiol and ASH on addition of different analytes.

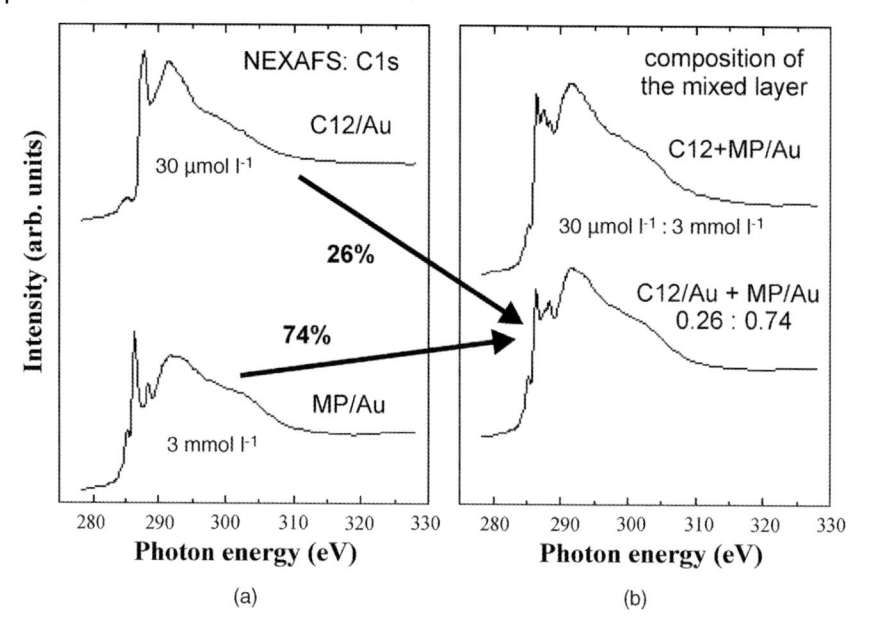

Figure 10.8 (a) C K-edge NEXAFS spectra of SAMs formed from 1-dodecanethiol (C12) and 6-mercatopurine (6-MP) solutions; (b) linear combination of the spectra in (a) (bottom curve) in comparison to the spectrum of the film formed from a mixed C12/6-MP solution. As the spectrum of the mixed film is fully reproduced by the linear combination of the spectra in (a), the content of the 6-MP molecules in the mixed film can be estimated as 74%.

reaction resistance and electrode capacitance while the Warburg impedance does not change. Monitoring of a capacitive current was used as the main detection method. For gold electrodes covered by a monolayer of a single component, either of matrix or spreader-bar molecules, no recognition abilities were found (Figure 10.8).

This application of spreader bar technology in sensor arrays illustrates its high potential in creating a large variety of chemoreceptors with different selectivity, thus fitting the main requirement in the development of modern analytical systems based on pattern recognition.

The observed interaction of adenine with mixed monolayers consisting of ASH was a reason to test this system as an artificial receptor for ATP. The experiment confirmed this possibility: ATP addition resulted in a concentration-dependent decrease of the capacitive current through the mixed monolayer, with saturation at 2.2% and a binding constant of about $2 \times 10^4 \, l \, mol^{-1}$ (Figure 10.9).

The spreader-bar systems were tested as a new method to form chirally sensitive artificial chemoreceptors on a solid support [40]. Gold surfaces coated by various self-assembled monolayers have been investigated by surface plasmon resonance as well as by impediometric techniques in the presence of solutions of racemic mixtures of phenylalanine and model target compound bisnaphthol (BNOH) in different concentrations. Electrodes coated with the conjugates of racemic thioctic acid with different enantiomers of 1,1'-binaphthyl-2,2'-diamine [(R)- and (S)-conjugates

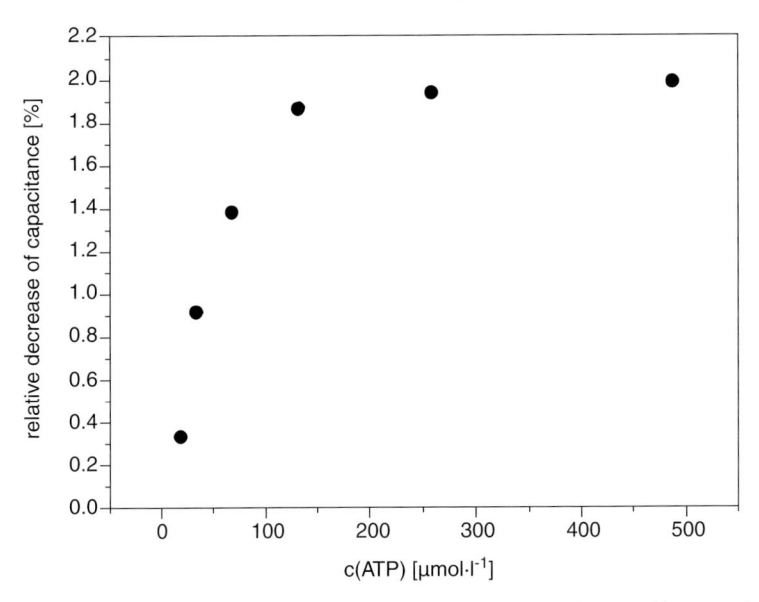

Figure 10.9 Change of electrochemical capacitance of an electrode covered by a mixed monolayer of 6-mercaptopurinethiol and 1-dodecanethiol in the presence of various concentrations of ATP.

correspondingly], being used alone or in a mixture with matrix molecules, displayed no capacitance increase on addition of phenylalanine, even in millimolar concentrations. However, they are very sensitive to the addition of even micromolar concentrations of racemic (R/S)-BNOH solution. Deviations of the recognition properties of the receptors formed by the (R)- and (S)-conjugates from ideal symmetric behavior are most probably caused by preferable binding of definite optical isomers of thioctic acid during the conjugation, thereby leading to the formation of molecules with two chiral centers and loss of the mirror symmetry. The affinity of the model analytes to the conjugates allows their use as spreader-bars for the selective recognition of the individual enantiomers.

The ability of such surfaces to discriminate enantiomers was studied by measuring the change in electrode capacitance on addition of (R)- or (S)-enantiomers of the analyte. The concentration range of BNOH was varied from 6.25 to 50 µM. The ratio of the apparent capacitance changes was assumed to be proportional to the adsorbed amount [23] and therefore to the enantioselectivity, and was taken as a criterion for the chiral recognition properties of each of the tested coatings. The results showed that the concentration ratio between the matrix and the template molecules is the key parameter determining the chiral recognition properties of the modified surfaces.

The sensitivity of the sensors to the target analyte increases on increasing the template concentration in the mixture with the matrix molecule. This is additional evidence for the important role of the template in creating cavities of a specific size that act as a mold for the target analyte. The highest enantioselectivities were obtained for the gold electrodes coated by template/matrix mixtures with molar ratios of 1 : 3.5 and 1 : 2. The effectiveness of artificial receptors was compared for two types of

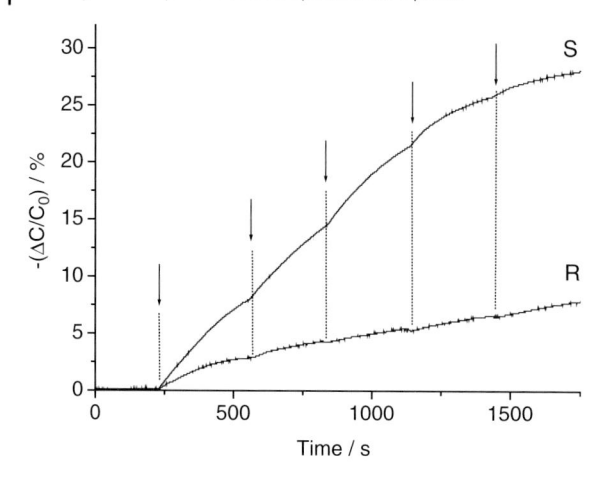

Figure 10.10 Capacitance changes on five successive additions of 10 μM of (R)-BNOH (R) and (S)-BNOH (S). The gold electrodes were modified with a mixture of 1-hexadecanethiol (0.1 mmol l^{-1}) and S-conjugate (0.03 mmol l^{-1}) in ethanol–dioxane (9 : 1 v/v). [41]. Reproduced with permission. Copyright Wiley-VCH Verlag GmbH & Co. KGaA.

matrices, namely, the long-chain alkanethiols (1,16-mercaptohexadecanoic acid, 1-hexadecanethiol) and the shorter one (1-dodecanethiol). A decrease in matrix thickness, realized by the substitution of 1-hexadecanethiol by 1-dodecanethiol, led to the loss of enantioselectivity.

Figure 10.10 shows the kinetics of the decrease in capacitance on additions of analytes.

The chiroselectivity of the spreader-bar structures with S-conjugate as the template was further examined by means of surface plasmon resonance measurements. The ratio of the template and matrix concentrations that had provided the highest enantioselectivity in the capacitive study was used for coating the gold surface. The signal changes upon addition of the (S)- and (R)-BNOH are higher in the case of (S)-BNOH. The enantioselectivity was calculated as a ratio of stationary SPR shifts on addition of corresponding enantiomers; a value of 2.55 was obtained. The unusual kinetics of the SPR signal may reflect some conformational changes in the receptor layer.

The proposed methodology is widely applicable and can be used to form chirally selective receptors for a large variety of species. One can expect a further increase in chiral sensitivity by conjugating a chiral spreader bar to a non-chiral thiolinker. Such sensors may be used for analysis of chiral compounds in complex mixtures, for quality control of chiral drugs and food additives, and in related applications.

The laterally organized surfaces obtained by the spreader-bar technology could be of importance for applications in many fields of biology and medicine, including, for example, development of new bioanalytical methods and new biocompatible surfaces, new approaches for investigation of biological ion pumps, and high-throughput screening of chemical compounds.

References

1. Sagiv, J. (1979) *Isr. J. Chem.*, **18**, 346–353.
2. Moreau, W.M. (1988) *Semiconductor Lithography: Principles and Materials*, Plenum Press, New York.
3. Brambley, D., Martin, B., and Prewett, P.D. (1994) *Adv. Mater. Opt. Electron.*, **4**, 55–74.
4. Rai-Choudhury, P. (1997) *Handbook of Microlithography, Micromachining, and Microfabrication*, SPIE, London.
5. Xia, Y., Zhao, X.-M., and Whitesides, G.M. (1996) *Microelectron. Eng.*, **32**, 255–268.
6. Xia, Y., Kim, E., and Whitesides, G.M. (1996) *J. Electrochem. Soc.*, **143**, 1070–1079.
7. Xia, Y. and Whitesides, G.M. (1998) *Angew. Chem. Int. Ed.*, **37**, 550–575.
8. Zhao, X.M., Xia, Y., and Whitesides, G.M. (1997) *J. Mater. Chem.*, **7**, 1069–1074.
9. Jeon, N.L., Finnie, K., Branshaw, K., and Nuzzo, R.G. (1997) *Langmuir*, **13**, 3382–3391.
10. Xia, Y., Qin, D., and Yin, Y. (2001) *Curr. Opin. Colloid Interface Sci.*, **6**, 54–64.
11. Becker, R.S., Golovchenko, J.A., and Swartzentruber, B.S. (1987) *Nature*, **325**, 419–421.
12. Eigler, D.M. and Schweizer, E.K. (1990) *Nature*, **344**, 524–526.
13. Weiss, P.S. and Eigler, D.M. (1993) *NATO ASI Ser., Ser. E*, **235**, 213–217.
14. Gimzewski, J.K. and Joachim, C. (1999) *Science*, **283**, 1683–1699.
15. Hla, S.-W., Bartels, L., Meyer, G., and Rieder, K.-H. (2000) *Phys. Rev. Lett.*, **85**, 2777–2780.
16. Allara, D.L. (1995) *Biosens. Bioelectron.*, **10**, 771–783.
17. Bain, C.D., Evall, J., and Whitesides, G.M. (1989) *J. Am. Chem. Soc.*, **111**, 7155–7164.
18. Bain, C.D. and Whitesides, G.M. (1989) *Langmuir*, **5**, 1370–1378.
19. Ulman, A. (1991) *An Introduction to Ultrathin Organic Films: from Langmuir-Blodgett to Self-Assembly*, Academic Press, San Diego.
20. Ulman, A. (1996) *Chem. Rev.*, **96**, 1533–1554.
21. Dubois, L.H. and Nuzzo, R.G. (1992) *Annu. Rev. Phys. Chem.*, **43**, 437–463.
22. Poirier, G.E. (1997) *Chem. Rev.*, **97**, 1119–1122.
23. Mirsky, V.M. (2002) *Trends Anal. Chem.*, **21** (6 + 7), 439–450.
24. Hagenström, H., Esplandiú, M.J., and Kolb, D.M. (2001) *Langmuir*, **17**, 839–848.
25. Bain, C.D. and Whitesides, G.M. (1988) *J. Am. Chem. Soc.*, **110**, 6560–6561.
26. Folkers, J.P., Laibinis, P.E., Whitesides, G.M., and Deutch, J. (1994) *J. Phys. Chem.*, **98**, 563–571.
27. Stranick, S.J., Attre, S.V., Parikh, A.N., Wood, M.C., Allara, D.L., Winograd, N., and Weiss, P.S. (1996) *Nanotechnology*, **7**, 438–442.
28. Stranick, S.J., Parikh, A.N., Tao, Y.-T., Allara, D.L., and Weiss, P.S. (1994) *J. Phys. Chem.*, **98**, 7636–7646.
29. Hayes, W.A., Kim, H., Yue, X., Perry, S.S., and Shannon, C. (1997) *Langmuir*, **13**, 2511–2518.
30. Hayes, W.A. and Shannon, C. (1998) *Langmuir*, **14**, 1099–1102.
31. Hirsch, T., Kettenberger, H., Wolfbeis, O.S., and Mirsky, V.M. (2003) *Chem. Commun.*, **3**, 432–433.
32. Bain, C.D. and Whitesides, G.M. (1989) *Angew. Chem. Int. Ed. Engl.*, **28**, 506–512.
33. Folkers, J.P., Laibinis, P.E., and Whitesides, G.M. (1992) *Langmuir*, **8**, 1330–1341.
34. Bain, C.D. and Whitesides, G.M. (1989) *J. Am. Chem. Soc.*, **111**, 7164–7175.
35. Hirsch, T., Zharnikov, M., Shaporenko, A., Stahl, J., Weiss, D., Wolfbeis, O.S., and Mirsky, V.M. (2005) *Angew. Chem. Int. Ed.*, **44** (41), 6775–6778.
36. Schlenoff, J.B., Li, M., and Ly, H. (1995) *J. Am. Chem. Soc.*, **117**, 12528–12536.
37. Hirsch, T., Zharnikov, M., Shaporenko, A., Stahl, J., Weiss, D., Wolfbeis, O.S., and Mirsky, V.M. (2005) *Angew. Chem. Int. Ed.*, **44** (41), 6775–6778.
38. Li, G., Fudickar, W., Skupin, M., Klyszcz, A., Draeger, C., Lauer, M., and Fuhrhop, J.H. (2002) *Angew. Chem. Int. Ed.*, **41**, 1828–1852.
39. Hirsch, T., Shaporenko, A., Mirsky, V.M., and Zharnikov, M. (2007) *Langmuir*, **23** (8), 4373–4377.
40. Heister, K., Zharnikov, M., Grunze, M., and Johansson, L.S.O. (2001) *J. Phys. Chem. B*, **105**, 4058–4061.

41 Prodromidis, M.I., Hirsch, T., Mirsky, V.M., and Wolfbeis, O.S. (2003) *Electroanalysis*, **15** (22), 1795–1798.

42 Mirsky, V.M., Hirsch, T., Piletsky, S.A., and Wolfbeis, O.S. (1999) *Angew. Chem. Int. Ed.*, **38** (8), 1108–1110.

43 Mirsky, V.M., Riepl, M., and Wolfbeis, O.S. (1997) *Biosensors. Bioelectron.*, **12**, 977–989.

44 Lahav, M., Katz, E., and Willner, I. (2001) *Langmuir*, **17**, 7387–7395.

11
Potential of Aptamers as Artificial Receptors in Chemical Sensors
Bettina Appel, Sabine Müller, and Sabine Stingel

11.1
Introduction

Aptamers are short nucleic acid sequences (RNA or DNA, typically 15–60 nts) that bind ligands with high affinity and specificity. The word "aptamer" has its origin in the Latin term *aptus*, meaning "to fit." Despite this general meaning, typically only nucleic acid receptors are termed aptamers. The first aptamers were discovered about 20 years ago. Owing to the powerful technique of SELEX (systematic evolution of ligands by exponential enrichment), it became possible to isolate specific binders out of a combinatorial library of nucleic acid sequences by iterative rounds of selection and amplification (Section 11.2.1) [1, 2]. Aptamers have been developed to bind several different ligands, including non-nucleic acid targets such as amino acids, peptides, proteins, drugs, vitamins, and a large variety of other small organic molecules, metal ions, or even whole cells [3–6]. They form tertiary structures and bind their targets by complementary shape interaction. Aptamer–target interaction is similar to antibody–antigen interaction with the binding affinity of aptamers to their targets being comparable or even superior to that of antibodies. Dissociation constants down to the picomolar range have been obtained [7, 8]. Moreover, aptamers have the advantage of being smaller in size and posses much greater conformational flexibility than their antibody counterparts. Remarkably, aptamers can distinguish between closely related structures. A striking example is the theophylline aptamer, which shows high levels of molecular discrimination against caffeine and theobromine, two analogs of theophylline [7]. Caffeine differs from theophylline by a single methyl group, while theobromine is a structural isomer of theophylline with one methyl group at a different position. Despite these closely related structures, dissociation constants for the aptamer–theophylline complex compared with the aptamer–caffeine or aptamer–theobromine complex differ by 10 000-fold [7]. In the same manner, the L-arginine aptamer shows a 12 000-fold higher affinity for L-arginine than for the corresponding enantiomer D-arginine [9].

Aptamers have great potential for applications in drug discovery, diagnostics, therapy, and analytics. Compared with antibodies, aptamers are advantageous in

Artificial Receptors for Chemical Sensors. Edited by V.M. Mirsky and A.K. Yatsimirsky
Copyright © 2011 WILEY-VCH Verlag GmbH & Co. KGaA, Weinheim
ISBN: 978-3-527-32357-9

several aspects. They can be produced *in vitro*, without the need for animals. This further allows developing aptamers against toxins and non-immunogenic compounds, since the process does not rely on induction of the animal immune system, as required in case of antibody generation. Aptamers are well-defined synthetic molecules, which makes them amenable to reproducible, scalable, and cost-effective production using established synthesis technology. Owing to their rather small size, the structural features of aptamers, and particularly the critical sequence for binding the target, can be easily identified. This in turn allows for functional optimization, such as stabilization by modification, labeling with reporter molecules, immobilization by site-directed chemistry, or integration into nanodevices, without affecting the affinity to the target. Aptamers can be formatted into molecular beacon structures or coupled to ribozymes to generate so-called aptazymes or reporter ribozymes. In general, the change of the aptamer shape upon binding the target can be coupled with various signaling mechanisms for easy monitoring the molecular recognition event. A further advantage is that aptamers can recover their native conformation after one round of sensing and denaturation, making them reusable in analytical devices.

Regarding the application of aptamers as components of chemical sensors, it is of utmost importance that aptamers can be developed by selection *in vitro*. The SELEX process can be carried out at any predefined conditions, allowing selection of aptamers that function, for example, in specific solvents, at a certain pH, temperature, or salt content, thus making it possible to design aptamers for tasks that cannot be met with protein receptors.

Most aptamers are composed of RNA. Since most aptamer–target interactions rely on hydrogen bonding, RNA libraries offer greater diversity than DNA libraries. However, ssDNA has also been shown to be capable of folding into three-dimensional shapes to bind targets with high affinity and specificity. For example, an ssDNA aptamer has been developed that recognizes and binds to cellobiose, but shows no or only little affinity for the related disaccharides lactose, maltose, and gentiobiose [10]. Thus, ssDNA can not only function just as an aptamer, but also can discriminate closely related structures, such as sugar epimers, anomers, and disaccharides differing only by their disaccharide linkages.

Overall, aptamers are promising candidates as components of sensors to be applied in medical diagnostics or bioanalytics. They are easy to produce, stable enough for long-term storage, prone to functional modification without affecting the affinity to the target, and they are reusable after denaturation.

11.2
Generation and Synthesis of Aptamers

11.2.1
Selection of Aptamers from Combinatorial Libraries (SELEX)

Most aptamers have been selected by systematic enrichment of ligands by exponential growth (SELEX), a protocol introduced in 1990 [1, 2]. This procedure involves screening of large libraries of oligonucleotides by iterative rounds of selection and amplification. Libraries are generated by combinatorial synthesis using standard solid-phase chemistry. By theory, the molecular diversity of the library is dependent on the number of randomized nucleotides: for a randomly generated region of length n the number of possible sequences in the library is 4^n. However, due to what is possible to handle in the laboratory, in practice, sequence complexity of oligonucleotide libraries is limited to about 10^{15} molecules. In a typical SELEX experiment, members of the library consist of a variable region flanked by fixed sequences serving as primer binding sites for amplification and reverse transcription. The library is incubated with the target of interest (immobilized, for example, on magnetic beads, affinity column, or microtiter plate [11, 12] or free in solution) in a buffer of choice and at a given temperature. In the initial cycle usually only a very small percentage of individual sequences interact with the target. These sequences need to be isolated from the rest of the library, for example, by affinity chromatography (if the target is immobilized), filter binding, or gel electrophoresis (if the target is free in solution) [13–15]. Subsequently, the isolated population of sequences is amplified to obtain an enriched library, which then is used for further rounds of selection, in which the stringency of binding conditions is increased to identify the tightest-binding sequences (Figure 11.1).

Selectivity can be engineered into aptamers through methods such as counter-SELEX, including rounds of negative selection against another entity during the

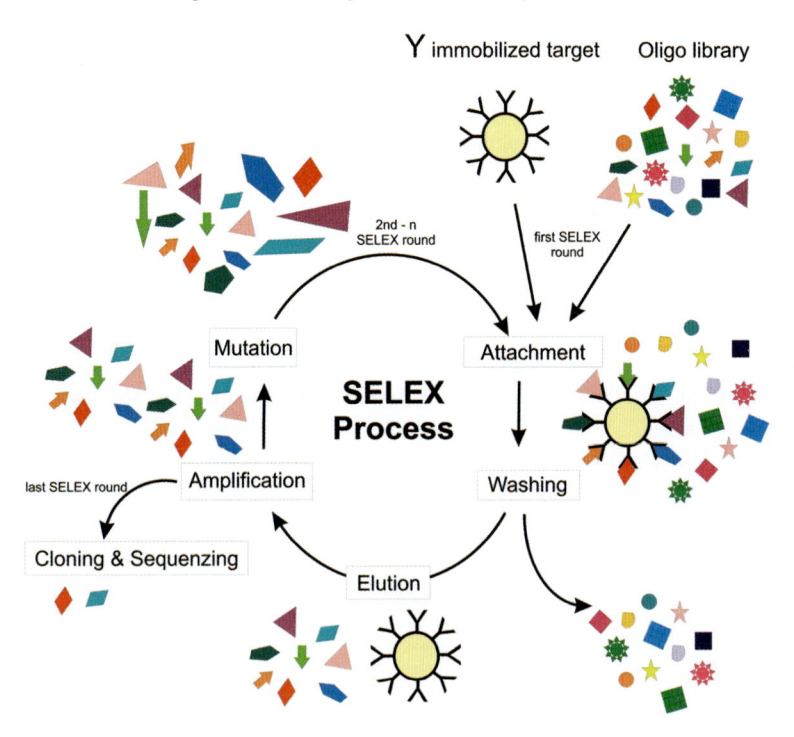

Figure 11.1 SELEX process.

process, such as, for example, the support to which the target is immobilized or structures that are similar to the target of interest.

As mentioned above, it is impossible to produce a fully randomized library of an oligonucleotide with more than 25 variable positions (100 variable positions add up to $4^{100} = 1.6 \times 10^{60}$ possible variants), since production and handling of oligonucleotides are limited to about 10^{15} molecules. However, active molecules that have been selected from this limited pool may be further improved by mutagenic PCR. This requires PCR conditions and polymerases that produce a higher error rate. Typically, error-prone amplification is introduced into the SELEX process only after several rounds.

At the end of a SELEX process (typically after 6 to 15 cycles) a family of sequences with the highest binding affinity to the target remains, which for further analysis are cloned and sequenced. Individual sequences are investigated to define the consensus motif and to eliminate non-functional sequences as, for example, primer binding sites. Once the sequence of an aptamer is identified, it can be synthesized in sizable quantities and where appropriate it can be stabilized by chemical modification (Section 11.2.4).

11.2.2
SELEX Variations

11.2.2.1 SELEX Using Modified Oligonucleotide Libraries

A problem particularly of RNA aptamers is their limited stability in biological media, since RNA is prone to nuclease attack. This requires stabilization of aptamers by chemical modification, which mostly is carried out by chemical synthesis after an aptamer has been selected and characterized. However, this procedure requires extensive experimental work and careful analysis to identify those positions that upon modification do not affect affinity of the aptamer to its target. An alternative approach is the use of modified triphosphates during selection, thus creating a modified oligonucleotide library. The range of modifications is limited by the ability of DNA and RNA polymerases to accept those as substrates for incorporation into the nascent oligomer. So far, only a few reports for enzymatic incorporation of 2'-fluoropyrimidines, 2'-aminopyrimidines, or pyrimidines with modifications at C5 of the heterocyclic base have appeared in the literature [16–19]. Using such modified libraries, aptamers against human IFN-gamma [18], neuropeptide nociceptin/orphanin VG [20], vascular endothelial growth factor (VEGF165) [21], and CD4 Antigen [22] have been developed.

A phosphorothioate DNA library has been generated by the use of thio-substituted deoxynucleoside triphosphates, and this library was successfully screened for aptamers against transcription factor NF-IC5 [21]. In analogy, thio-RNA libraries have also been created [21, 23]. In addition to the advantage of selecting more stable aptamers from modified libraries, SELEX of modified aptamers further widens their sequence space, making in turn the scope of targets and potential specificity even greater.

11.2.2.2 PhotoSELEX

PhotoSELEX is a variation of the SELEX protocol in selecting aptamers by the ability to form a covalent bond with the target by photo-induction [24]. This requires incorporation of a nucleotide derivative in all members of the library that is stable and inactive in the absence of photo-induction, but becomes reactive to the target when exposed to light [25]. A commonly applied photoactive nucleoside derivative is 5-bromo-2′-deoxyuridine, which is used in the form of its triphosphate in place of thymine triphosphate for library generation. For photo-activation, it absorbs UV light of 310 nm, a wavelength at which normal nucleosides are inactive. Thus, after incubation of the immobilized target with the library, it is exposed to light at 310 nm to generate a photo-crosslink between tightly binding members of the library and the target. Non-bound sequences are efficiently washed away in denaturing buffer. Thus, partitioning is improved, since enrichment of sequences that non-specifically bind to any entity of the process is avoided. However, PhotoSELEX is limited to targets with reactive sites, and so far has been used exclusively for selecting aptamers against peptides and proteins, though with dissociation constants down to the picomolar range [11, 26–29]. Bound sequences are released from the target in these cases with proteases, to free aptamer candidates for PCR amplification and further rounds of selection.

Using a related protocol, recently a light-switchable small-molecule–RNA aptamer pair has been selected by photochemical *in vitro* selection. One of these aptamers binds to spiropyran, while the other one binds to the isomer merocyanine (Figure 11.2) [30]. The binding event of each aptamer can be reversibly switched by using light irritation.

11.2.2.3 Automated SELEX

As mentioned above, a typical SELEX experiment with 6–15 cycles carried out manually requires several weeks. This, however, is not consistent with the demand

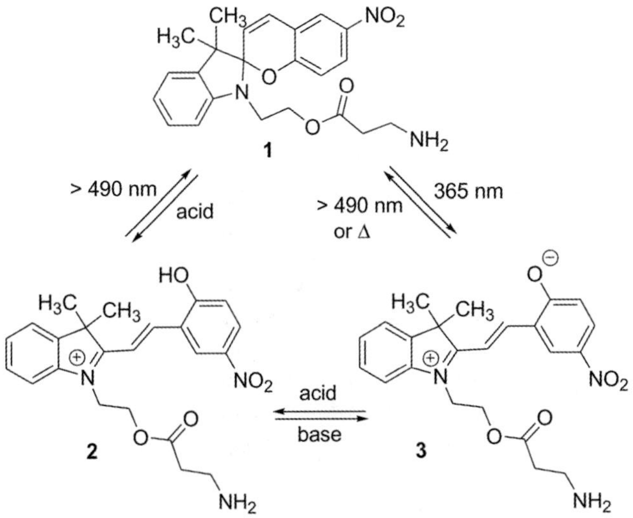

Figure 11.2 Spiropyran **1** (colorless), protonated merocyanine **2** (yellow), and deprotonated meromycine **3** (purple).

for high-throughput methods in aptamer development. Therefore, approaches based on automatic robotic platforms have been developed [31–35]. The first protocol for high-throughput selection was published in 1998, describing a model SELEX experiment for selection of an aptamer against polyT oligonucleotides [31]. The necessary equipment included a pipetting robot coupled to a Thermocycler, a magnetic bead separator, reagent tracks, and a pipetting station. This configuration allowed selecting a polyT aptamer within ten rounds performed on a single day. Methods for high-throughput selection have been further optimized, such that nowadays eight selections can be performed in parallel, allowing the completion of 12 selections in about two days [32]. In this way, aptamers have been developed against, for example, hen egg white lysozyme, tyrosyl transfer RNA synthetase (CYT-18), from the mitochondria of *Neurospora crassa*, human MAP kinase (MEK1), and a transcriptional terminator from *Thermotoga maritima*, all with dissociation constants in the picomolar to nanomolar range [32, 33].

A different approach for automated SELEX is based on aptamer selection on a silicon chip [35]. This procedure has the advantage of requiring extremely small reagent quantities and it is capable of producing aptamers in very short time without the need for extensive and sophisticated equipment. Further optimization of this approach has led to a protocol based on automated microfluidic selection, which was successfully applied for the generation of an anti-lysozyme aptamer [35]. The Microline-based assembly uses LabView-controlled valves and a PCR machine, and has been shown to be capable of the selection and, in addition, the synthesis of aptamers. The microfluidic prototype is a simple apparatus that is relatively inexpensive to assemble. It uses a fluidic pressure gradient for loading and flow of the reagents through microchannels.

11.2.3
Alternative Approaches for Selection of Aptamers from Combinatorial Libraries

11.2.3.1 Capillary Electrophoresis Techniques

With the purpose of developing new methods that allow selection of aptamers in a short time, alternative methods to the traditional SELEX have been established. Capillary electrophoresis has been shown to be a potent approach for the isolation of high-affinity aptamers, requiring only a few cycles of selection/amplification [26]. The procedure is performed free in solution, such that problems like non-specific binding of library members to supports or linker molecules are avoided. Unbound oligonucleotides are separated from target bound oligonucleotides by running the binding reaction through a capillary electrophoresis instrument. A special format of this technique is the so-called non-equilibrium capillary electrophoresis of equilibrium mixtures (NECEEM) [36, 37]. NECEEM allows an increase in the binding constants of more than four orders of magnitude without the need for PCR amplification [37]. In some cases, even a single cycle of NECEEM provides sufficient enrichment to yield a pool of viable aptamers with nanomolar affinities [36]. Furthermore, in addition to equilibrium constants of aptamer target complexes, rate constants can be measured simultaneously. A disadvantage of this approach lies in the fact that basically only aptamers against targets with high molecular masses (i.e., proteins) can be selected. The technique becomes difficult to apply if the molecular mass of the target is less than the molecular size of the library members, because the smaller target does not sufficiently retard the mobility of the aptamer target complex to allow for good separation. Thus, the technique has so far been used for selection of aptamers against proteins (Section 11.4.3.2). For example, a ssDNA aptamer against HIV reverse transcriptase with a K_D of 180 pM was isolated from a combinatorial ssDNA library in only four rounds of selection [14].

11.2.3.2 AFM Techniques

A new approach for aptamer generation relies on a combined atomic force microscope (AFM) and fluorescence microscope to select target-binding aptamers from a small pool of randomized oligonucleotides in a single cycle [38]. A library of small beads, each functionalized with fluorescent oligonucleotides, is created and flowed over immobilized target molecules on a glass cover slip. Aptamers that are tightly bound to the target remain on the substrate surface, while non-bound sequences are washed away. Owing to the fluorescence label, the position of oligonucleotide–target complexes is indicated by a strong fluorescence signal, such that the individual beads plus the attached oligonucleotide can be extracted with the AFM-tip. Isolated sequences are amplified by PCR and immediately sequenced without further rounds of selection. The new technique, registered under the trademark Nanoselection™, has not yet been used to select aptamers from large highly diverse pools such as used in SELEX or NECEEM. However, the potential of this approach has been convincingly demonstrated in a model experiment, selecting aptamers from a binary pool containing a 1:1 mixture of a thrombin aptamer and a nonspecific oligonucleotide [38]. It complements both SELEX and NECEEM, and may develop into a useful alternative to be used for special selection experiments.

11.2.4
Synthesis of Aptamers and Stabilization

11.2.4.1 Chemical Synthesis of Aptamers

After selection of high-affinity aptamers and determination of their sequence, further characterization may include analysis by biochemical and biophysical methods such as surface plasmon resonance, fluorescence correlation spectroscopy, radioactive binding assay, structure mapping, EPR, NMR, or X-ray analysis. For this purpose as well as for further application, aptamers need to be synthesized in sizable amounts. This in most cases is carried out by chemical procedures, since the size of aptamers usually is well within the range required to allow for efficient preparation by automated oligonucleotide synthesis on the solid phase. Nucleic acid chemistry has developed to a degree that allows the synthesis of DNA and RNA oligonucleotides of any desired sequence from the microgram to multi-gram scale. In addition to laboratories being specialized in, or simply capable of, oligonucleotide synthesis, aptamers can be purchased by order from several companies on the open market.

The standard approach for the synthesis of oligodeoxy- and oligoribonucleotides nowadays is based on phosphite triester chemistry mediated on a solid support. The approach consists of the reaction of a 5'-O-dimethoxytritylated nucleoside phosphoramidite with a 3'-O-support coupled nucleoside to furnish a dinucleoside phosphite triester, followed by oxidation to the corresponding phosphotriester. DNA synthesis has been following this standard procedure for many years. However, for RNA synthesis, several procedures based on phosphite triester chemistry have been developed. Methods with modifications of protecting groups and synthesis strategy are used in parallel [39]. In some cases it may be useful to prepare the aptamer by enzymatic synthesis or by combined chemical/enzymatic protocols [39].

Chemical synthesis has the advantage of easy functionalization of the aptamer for further application. Dyes can be attached to the oligonucleotide directly during synthesis or post-synthetically after prior introduction of a functional group (i.e., an aliphatic amino group) that selectively reacts with an activated derivative of the dye. In the same manner, the oligonucleotide can be conjugated to any desired functional elements, such as, for example, further nucleic acid sequences, redox labels, or nanoparticles to be used for sensing the analyte of interest (Section 11.4.3).

11.2.4.2 Stabilization of Aptamers

Stabilization by Chemical Modification As described in Section 11.2.2.1, aptamers, particularly those made of ribonucleotides, are sensitive to nuclease attack and therefore need to be stabilized. This is of paramount interest if the aptamer is supposed to be applied in biological media. The major problem concerning RNA aptamers is the reactivity of the 2'-hydroxy group, not only in the presence of nucleases but also at higher pH, leading to nucleophilic attack of the 2'-OH on the neighboring phosphodiester bond and thus causing a strand break with generation of a 2',3'-cyclic phosphate on one side and a free 5'-hydroxyl group at the other. This reaction is catalyzed by ribonucleases as well as several metal ions and is the reason for the limited life time (<10 min) of RNA aptamers, particularly in biological samples [40]. Therefore, to make aptamers resistant to degradation, modifications of the 2'-hydroxyl group as well as of the phosphate moiety have proven useful [41]. However, great care has to be taken that only such positions are modified that are not involved in aptamer function. Identification of such positions often requires tedious experimental work and analysis. Nevertheless, for example, replacement of natural ribonucleosides at selected positions by 2'-fluoro- and 2'-amino-nucleosides has been shown to give a dramatic increase of the half-lives of aptamers in biological fluids [17, 18].

Spiegelmers An alternative route of producing nuclease-resistant aptamers is based on mirror-image RNA or DNA, a concept introduced in 1996 by Klussmann et al. [42]. Functional mirror-image RNA or DNA oligonucleotides (mirror-image aptamers) are also-called "Spiegelmers," as derived from the German word for mirror *Spiegel*. Spiegelmers are composed of the non-natural L-enantiomer of ribose or 2'-deoxyribose. They bind with high affinity and specificity to a given target molecule and, very importantly, they posses high biological stability [42, 43].

Spiegelmers are produced by an adapted SELEX process (mirror-image SELEX) using the chiral principle. As mentioned above, Spiegelmers are composed of L-nucleotides, which accounts for their high biostability, since nature did not evolve appropriate enzymes to handle mirror-image nucleotides. On the other hand, the lack of Spiegelmer modifying enzymes prevents direct introduction of Spiegelmers into the SELEX process. Therefore, chiral inversion steps need to be introduced and selection is carried out using a combinatorial library of natural oligonucleotides, from which at first a natural aptamer is selected against the synthetic enantiomer of a chosen target. Following the principle of reciprocal chiral specificity, the corresponding mirror-image nucleic acid (L-oligonucleotide) of the selected aptamer, the Spiegelmer, will bind with identical affinity to the other desired enantiomer of the target. Following the selection process and identification of the sequence, the aptamer can be synthesized from L-nucleotides using standard techniques of automated oligonucleotide synthesis.

Apart from their excellent biostability, Spiegelmers induce only minimal immunogenic response and are non-toxic. Spiegelmers have been produced against several small molecules, peptides, and proteins [44]. Their efficacy has been demonstrated in animal models [45], making them ideal candidates for application in biological samples, *in vitro*, and *in vivo* diagnostics.

11.3
Aptamer Arrays

Organization of aptamers in arrays is of utmost interest for high-throughput screening in diagnostics. Basically, the usage of aptamers on a microarray platform instead of in solution encounters similar problems as observed in antibody and DNA arrays, including difficulties associated with multiplexing and interaction with the chip surface as well as cross-reactivity. Furthermore, dependent on the mechanism used for detection, the surface of the array must be either treated or constructed from specific materials, and the aptamer must be positioned by a suitable immobilization method, such that binding of the analyte triggers the desired response or signal. A large variety of immobilization techniques have been described for DNA microarrays (reviewed in Reference [46]), which can be used similarly for attaching RNA aptamers to a surface. Here, it is important to make sure that immobilization of the aptamer does not alter the strength of interaction between target and probe. If linker molecules are used for tethering the aptamer to the surface, care has to be taken regarding the chemical nature and the length of the linker, since those have been shown to affect the signal [47].

One of the first approaches towards using aptamers in a biosensor array was presented by Seetharaman *et al.* [48], who used a combination of aptamers and ribozymes, also called aptazymes (Section 11.4.2). Self-cleaving hammerhead-ribozyme constructs that respond allosterically to six different analytes (Co^{2+}, cGMP, cCMP, cAMP, flavin mononucleotide, and theophylline) were used as detectors. Individual constructs were immobilized in 63 pixels on a polystyrene tissue culture plate that was pre-coated with gold, via a thiotriphosphate moiety attached to the 5'-end of ribozyme strands. It was observed that each hammerhead construct responded individually and cleaved off its 3'-fragment upon addition of the respective effector. Since the catalytically active constructs were internally labeled with ^{32}P, it was possible to observe the resulting loss of radioactivity. A similar switch array was successfully used in measuring qualitatively and quantitatively cAMP concentrations in culture media from *Escherichia coli* strains with sensitivity in the μM range [48]. More recent developments in this area use fluorescence techniques for detection. For example, a technique known as "analyte-dependent oligonucleotide modulation assay" (ADONMA) [49] uses aptamers that are divided into two halves. One half is immobilized on a chip, while the other half is fluorescently labeled and added free in solution. Binding of a specific ligand results in the combination of the two halves, restoring the functional aptamer and generating a fluorescence signal on the surface of the array.

In addition to fluorescence detection, also electrochemical methods [50–52], acoustic methods [53], various optical methods [54–57], and surface plasmon resonance (SPR) [58–61] have been applied to aptamer arrays.

The formation of a stable RNA microarray with the surface ligation chemistry is verified by surface plasmon resonance imaging (SPRI) measurements. A stable RNA microarray can be created using unmodified single-stranded RNA (ssRNA) in a covalent enzymatic ligation reaction with ssDNA immobilized onto a surface [60]. In initial experiments, T4 DNA ligase was used for the surface ligation reaction. This enzyme requires the use of a DNA template that can complicate the fabrication process and limit the surface ligation efficiency. To avoid this problem, T4 RNA ligase was used because RNA ligase does not require a template for DNA-RNA ligation (Figure 11.3). In addition to the use of SPRI measurements, a surface enzyme reaction for the determination of the relative surface densities of this RNA microarray element has been developed. The enzyme RNase H was used to remove the ligated RNA from all of the microarray elements via hydrolysis of RNA-DNA heteroduplexes on the surface. SPRI was used in conjunction with RNase H to measure relative surface densities by the amount of SPRI signal loss at each aptamer array element. To further verify the efficiency of the RNA ligation chemistry, a five-component ssRNA aptamer microarray was created. The five aptamers are loop-stem structures that can potentially bind to the human protein fIXa, involved in the blood coagulation process. A single SPRI measurement identified the best aptamer for fIXa from a set of five potential candidates. In addition to the selection of aptamers, SPRI measurements of RNA microarrays can also be used for the direct detection of multiple proteins from biological samples [60].

While fluorescence techniques require arrays that have no complex surface modifications or complicated technologies, alternative methods for real-time and possibly tag-less detection often require a more complex surface and technology platform. Nevertheless, development of novel approaches for label-free detection is of great interest, since such techniques widen the range of analytes to be sensed and quantified using aptamer arrays. Further critical points to the future of aptamer

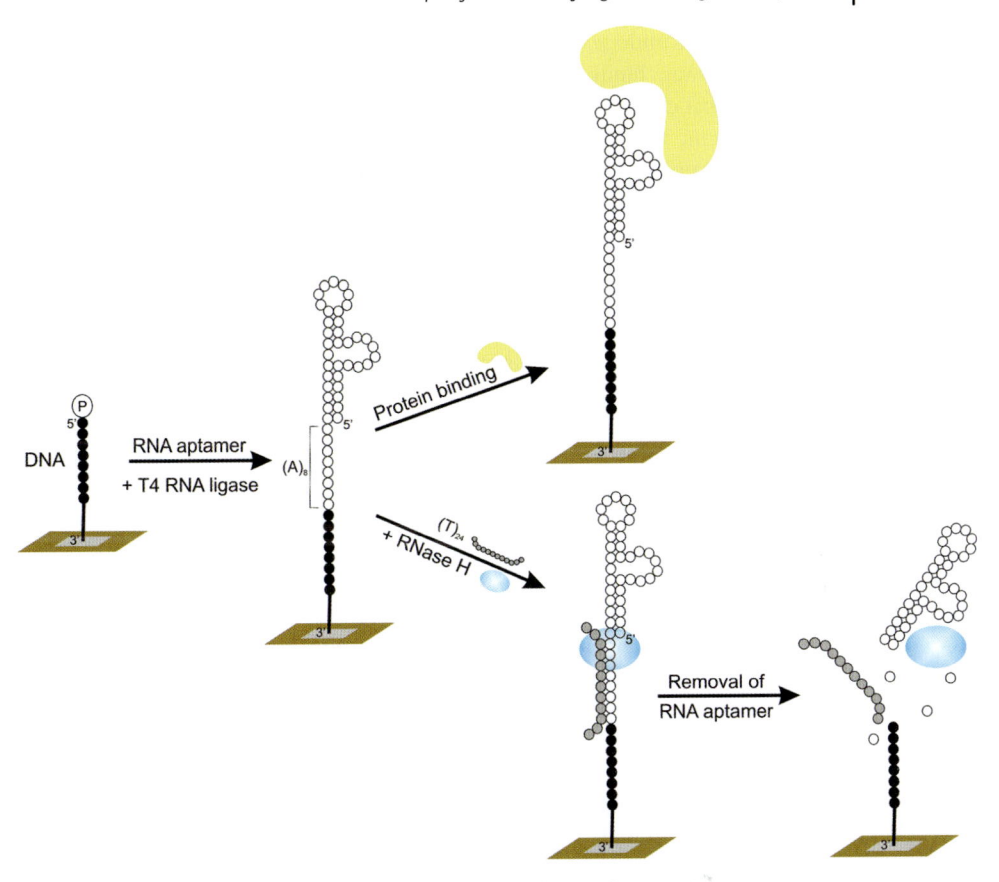

Figure 11.3 T4 RNA ligase and RNase H strategy. Adapted from Reference [60].

arrays are the systematic development of aptamers to a broad variety of targets as well as the on-chip optimization of aptamer sequences and binding conditions. Here, *in silico* studies can also help to explore binding properties of aptamers, to interrogate the sequence landscape, and to develop activity models [62].

11.4
Techniques for Readout of Ligand Binding to the Aptamer

11.4.1
Conformational Effects

Chemical sensors with aptamers as receptors are also called aptasensors. They are primarily classified according to their transduction method (Section 11.4.3). However, based on the structural design, aptasensors also vary in their working mode. Aptamers undergo conformational changes upon target binding and thus, in the simplest way, binding of the target will generate a signal by a significant conformational change. A very popular design for this purpose is organization of the aptamer in a molecular beacon. Molecular beacons contain two structural elements, a loop serving as probe, and two complementary arms that anneal and form a stem flanking the loop. Typically, a fluorophore and a quencher are attached at each, the 5'- and the 3'-end of the stem, thus bringing the two components in close proximity. As a result, fluorescence is quenched and no signal is generated. Binding the target must lead to formation of a probe–target complex that is more stable than the stem in the beacon hairpin, such that the stem is interrupted and the fluorophore and quencher become spatially separated and a fluorescence signal is generated [63–67]. This molecular design has been varied in several similar approaches and can also be coupled with other signal transduction modes, such as, for example, electrochemical signal generation [68]. In all cases, a significant conformational change of the aptamer is responsible for signal generation. A novel and versatile intramolecular signal transduction aptamer probe called an "aptamer switch probe" (ASP) has been developed recently [69]. This probe is composed of three elements: an aptamer, a short DNA sequence partly complementary to the aptamer, and a PEG (polyethylene glycol) linker that connects the two other elements. Furthermore, a fluorophore and a quencher are covalently attached to the termini of the conjugated DNA sequence and the aptamer. In the absence of target molecules, the short DNA is hybridized with a small section of the aptamer, keeping the fluorophore and the quencher in close contact. The fluorescence is "switched off." If a target molecule binds, the conformation of the probe changes and the intramolecular DNA hybridization will be disturbed. The quencher moves away from the fluorophore and the fluorescence is restored. The ASP design has several significant advantages over conventional designs: (i) intramolecular DNA hybridization requires shorter oligonucleotides to achieve the same melting temperature; (ii) a shorter competitor hybridizes to a smaller part of the aptamer and leaves more aptamer sequence free – this increases the binding affinity and sensitivity; (iii) more stable, faster response, and lower background; and (iv) ASP is a robust molecular probe.

11.4.2
Aptazymes

A general problem with molecular beacon design is that target binding is converted into a 1 : 1 signal, and signal amplification has to be part of the signal-detection-unit. A suitable alternative is the construction of reporter ribozymes or aptazymes by combination of aptamers with a catalytic nucleic acid, resulting in ribozymes or DNAzymes, the activity of which is dependent on a specific ligand. Based on the specific structural design, binding of a ligand to the adapter domain can stabilize (positive regulation) or destabilize (negative regulation) the ribozyme domain and the catalytic activity becomes adjustable. The catalytic part of the sensor usually is composed of the hammerhead or the hairpin ribozyme. The sensing unit, the aptamer, is attached to the catalytic part via a bridge element or communication module that is capable of translating the binding event in the aptamer domain to the catalytic part of the sensor and thus of triggering activity [70]. Using aptazymes as part of chemical sensors, the generated signal can be amplified by a readily detectable enzymatic reaction with multiple turnover rate (>1:10). An intriguing feature of such allosterically regulated nucleic acids is their modular design (basically any ribozyme can be combined with any desired aptamer) and the separation of the recognition element from areas where the specific signal for detection is generated. This allows more freedom in sensor design whereby a plurality of target molecules can be detected with the same sensing principle.

11.4.3
Methods of Sensing

11.4.3.1 Optical Sensing

Optical detection by fluorescence spectroscopy is a contemporary and very popular method to detect fluorescently labeled DNA or RNA aptamers. Different strategies to visualize the binding of an aptamer to a target are known. Often the aptamer sequence is labeled with a quencher and a fluorophore molecule. Conformational changes of the aptamer due to the binding of a target effect a change in fluorescence signal. Very common is the use of molecular beacon aptamers wherein a hairpin-like aptamer is end-labeled with a quencher and a fluorophore. Another approach is the use of a dye labeled aptamer and a duplex sequence with a complement dye.

Turro, Tan, and coworkers have designed a wavelength-shifting aptamer by labeling a platelet-derived growth factor (PDGF) binding DNA aptamer with two pyrene molecules to produce a sensor [71]. Upon PDGF binding, the aptamer switches its fluorescence emission from 400 nm (pyrene monomer, with a fluorescence lifetime of $\sim$5 ns) to 480 nm (pyrene excimer, with a lifetime of $\sim$40 ns). This design can handle the problem of the background signal that is inherent with complex biological samples (Figure 11.4).

A self-cleaving DNAzyme that displays a high sensitivity and selectivity for Hg^{2+} has been described by the group of Perrin [72]. The synthetic oligonucleotide contains

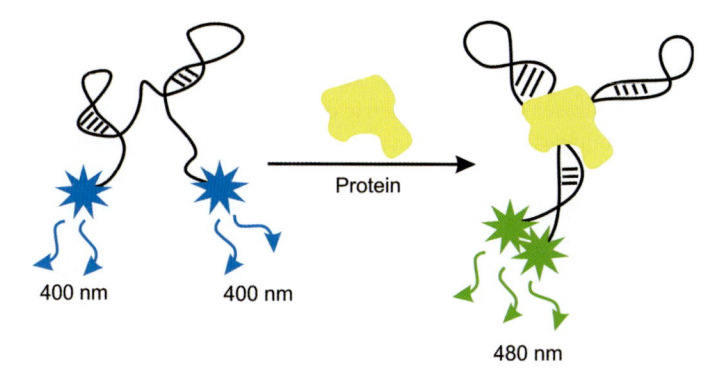

Figure 11.4 Protein binding to a fluorescently labeled aptamer. Adapted from Reference [71].

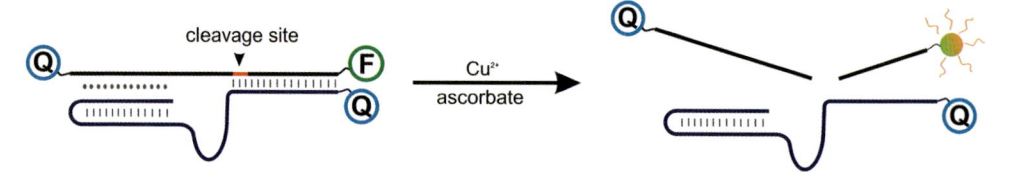

Figure 11.5 DNAzyme cleaves a substrate in the presence of Cu^{2+}. Adapted from Reference [73].

a couple of two different base modifications. These modifications in the side chains of nucleobases are soft ligands that can bind soft metals in an appropriate way. Many other metal cations were tested with the DNAzyme. However, only addition of Hg^{2+} induced enhancement of the cleavage activity up to 65% within 60 min. All other metal ions (e.g., Cu^{2+}, Fe^{2+}, Ba^{2+}, Eu^{3+}) supported only activity <5%. Thus, Hg^{2+} cations could be detected in concentrations between 1 and 10 µM, while higher concentrations of the cation appear to inhibit the system.

Paramagnetic metal ions such as Cu^{2+} have intrinsic fluorescence quenching properties, causing limited success for the design of fluorescent metal sensors in opposition to diamagnetic metals like Hg^{2+}, Pb^{2+}, and Zn^{2+}. Nevertheless, a new DNAzyme catalytic beacon sensor for Cu^{2+} ions in aqueous solution has been presented by Liu and Lu (Figure 11.5) [73]. The group designed a sensor based on a Cu^{2+}-dependent DNA-cleaving DNAzyme previously reported by Breaker [74–76]. The substrate was labeled with a fluorophore (6-carboxyfluorescein) at the 3'-end and a quencher (Iowa Black) at the 5'-end. The DNAzyme contains a 5'-quencher as well. This kind of dual-quencher is used to suppress background signals [77]. The sensor system acts as a catalytic beacon, and contains also 50 µM ascorbate, on the one hand to enhance the reaction rate [76] and, on the other hand, for suppression of quenching. A detection limit of 35 nM (2.3 ppb) was determined. The sensor system is useful for determining the Cu^{2+} concentration in drinking water [73]. The US EPA has defined a maximum contamination level of 20 µM, a value well within the dynamic range of this biosensor. Similarly, a fluorescent aptamer beacon that lights up in the presence of zinc was developed recently by the Ellington laboratory [78]. After 12 total rounds of *in vitro* selection most of the selected aptamer beacons showed Zn^{2+}-dependent elution from the oligonucleotide affinity column and were also specific for Zn^{2+} relative to Mg^{2+}.

A simple, rapid, and ultrasensitive colorimetric detection method for proteins using a dot-blot gold nanoparticle (AuNP) assay has been developed by Wang et al. [55]. α-Thrombin was used as a model protein. The AuNPs were functionalized with α-thrombin-binding aptamer (TBA), which binds with high affinity to the active site of α-thrombin, immobilized on a nitrocellulose membrane. If the aptamer binds to the α-thrombin a red dot appears on the membrane, which can be observed with the naked eye without any instrument. The detection sensitivity of this system could be improved by the use of a silver enhancement solution, lowering the detection limit of the aptamer to 14 fmol. To control the selectivity and specificity of the reaction, BSA, β-thrombin, and γ-thrombin were tested instead of α-thrombin. With the foreign proteins, no color change was detected, even at silver enhancement. Furthermore, α-thrombin was successfully detected in human plasma.

By using α-thrombin as a model target protein Ahn and Yang employed a new fluorescence amplification strategy for an aptamer biosensor system in 2007 [79]. In their system the DNA aptamer domain is part of a DNA duplex. If thrombin is added the aptamer domain bound to the target and the complementary DNA strand build a duplex with an added complementary RNA strand that is labeled with a fluorophore and a quencher. The fluorescence-signal amplification was given by RNase H treatment, which disrupts the RNA.

Very recently, a general strategy to convert the aptamer–target recognition event into an optical signal was introduced by Li and Ho [80]. An unmodified aptamer that binds specifically to the target serves as the molecular recognition element, and a competitor oligo serves as the transduction element.

C-reactive protein (CRP) is an important clinical biomarker. It is the first acute-phase protein (induced by cytokines) to be discovered (1930), and it is a sensitive marker for inflammation and tissue damage. It plays a role in the development of arthrosclerosis plaques. The average concentration of CRP in human serum is 0.8 ppm. After an acute-phase stimulus, the levels increase to more than 500 ppm. Although immunosensors showed very interesting performances, the availability of a CRP RNA aptamer encouraged the development of a RNA-based aptasensor [81]. The affinity constant for this aptamer, expressed as K_D, is 125 nM. In comparison to the aptamer, CRP antibodies have an affinity between 440 and 720 nM. The specific RNA aptamer for CRP was immobilized on a chip surface via streptavidin–biotin binding. CRP is a soluble Ca^{2+}-dependent ligand-binding serum protein, and the Ca^{2+} stabilizes the protein in its pentameric form. Aptamer–CRP binding was tested in the presence and absence of Ca^{2+}. In the absence of calcium, CRP slowly dissociated, whereas in the presence of calcium the aptamer–CRP complex was more stable, and an increase of sensitivity was recorded. The K_D value for this system is 0.5 nM and the detection limit under the best working conditions is 0.005 ppm.

A new paradigm for the development of small molecule-based RNA sensors has been described recently by Sparano and Koide [82]. They prepared a series of photoinduced electron transfer (PET) sensors based on 2′,7′-dichlorofluorescein (DCF) as fluorophore conjugated with aniline derivatives as electron donors (quenchers). Suitable quenchers for the fluorophore were found via NMR and fluorescent spectroscopic analysis. The RNA aptamers were raised against the aniline-based quencher via SELEX (Figure 11.6). One aptamer was found that enhanced the fluorescence intensity of the DCF–aniline conjugate in a concentration-dependent manner. In addition, other aptamers with similar activities were selected to demonstrate the generality of this approach. This report shows that one can develop fluorescence-inducing reporter RNA and morph it into remotely related sequences without prior structural insight into RNA–ligand binding.

Very attractive are colorimetric methods in allowing read out with the naked eye. Lee et al. have presented a method for detecting mercuric ions (Hg^{2+}) in aqueous

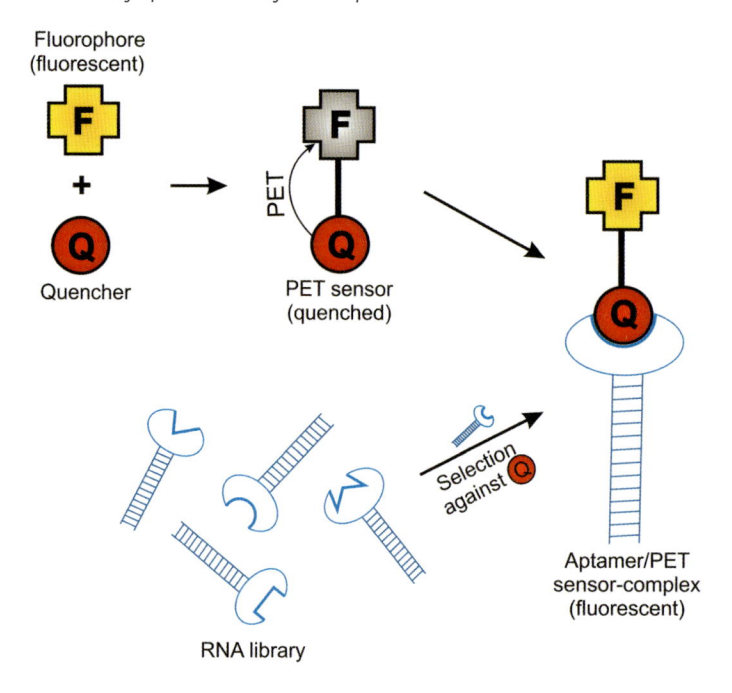

Figure 11.6 Fluorophore, quencher, and aptamer. Adapted from Reference [82].

media by using DNA-functionalized gold nanoparticles (DNA-AuNPs) [83]. The sensing principle is based on thymidine–Hg^{2+}–thymidine coordination and complementary DNA-AuNPs with designed T-T mismatches. Two complementary DNA-AuNPs form DNA-linked aggregates. These aggregates are colorless. If the solution of aggregates is heated up to the melting temperature (47 °C) they dissociate reversibly, associated with a simultaneous purple-to-red color change (Figure 11.7). In the presence of Hg^{2+} specific aggregates with higher melting point were formed and no color occurs by heating up to 47 °C. The detection method does not require specialized equipment other than a temperature control unit. Each increase in concentration steps of 1 μM results in an increase in the melting temperature by approximately 5 °C. The actual range of detection for the system is from 100 nM (20 ppb) up to low micromolar Hg^{2+} concentrations.

The first examples of modular aptameric sensors that transduce recognition events into fluorescence changes through allosteric regulation of noncovalent interactions with a fluorophore have been reported by Stojanovic and Kolpashchikov [84], using the triphenylmethane dye malachite green (MG, Figure 11.8a) as a fluorophore. They tested recognition regions specific for FMN, ATP, and theophylline in combination with the MG binding aptamer as a signaling domain.

MG aptamers were also used by Kubo and coworkers [85] to explore a new technique that can be potentially used in high-throughput *in vitro* selection for

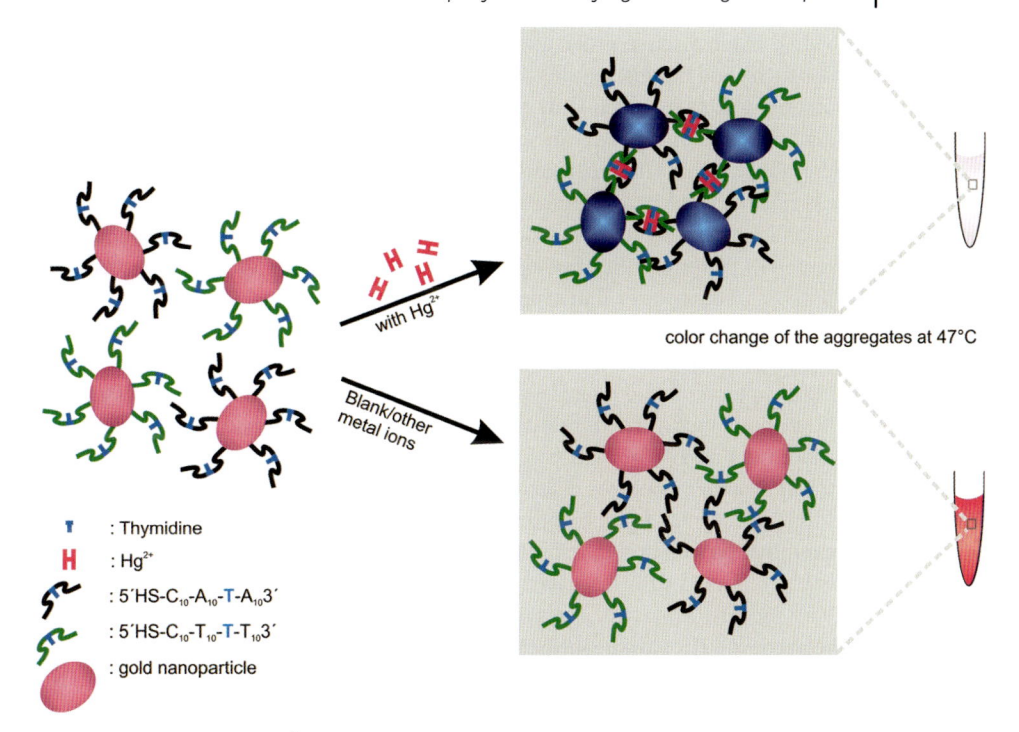

11.4 Techniques for Readout of Ligand Binding to the Aptamer | 349

- T : Thymidine
- H : Hg^{2+}
- ⌇ : 5'HS-C_{10}-A_{10}-T-A_{10}3'
- ⌇ : 5'HS-C_{10}-T_{10}-T-T_{10}3'
- ● : gold nanoparticle

Figure 11.7 Detection of Hg^{2+} with DNA-functionalized gold nanoparticles. Adapted from Reference [83].

(a) malachite green strepdavidin conjugate

(b) thiazole orange (TO)

Figure 11.8 (a) Malachite green–streptavidin conjugate; (b) thiazole orange (TO).

proteins. The malachite green could be connected via a thiourea bond to a streptavidin matrix and the matrix could be immobilized on a kinesin-driven microtubule. By adding a RNA mixture, only MG aptamers bound to the MG-microtubule. By using an excess of free MG in solution the aptamers were washed out from the microtubule. Malachite green-conjugated microtubules were used to demonstrate the performance of kinesin-microtubule systems as mobile bioprobes.

Another label-free RNA based biosensor was developed by Leontis and coworkers [86]. Here too, the triphenylmethane dye malachite green (MG) was used as a fluorescent reporter that stabilizes the recognition complex between two RNA oligomers in the presence of Mg^{2+}, and is thus capable of sensing magnesium ions.

Different analytes, such as tobramycin, theophylline, or cAMP, have been sensed using an aptamer-based thiazole orange (TO, Figure 11.8b) displacement assay [87]. In a previous study, the dimer of thiazole orange (TOTO) had been used in dye displacement assays for detection of the platelet-derived growth factor BB (PDFG-BB) [88]. Another label-free detection method is turbidimetric detection. For example, the molecular recognition of adenosine 5'-triphosphate (ATP) by a DNA aptamer in combination with non-crosslinking aggregation of DNA-linked polymeric micelles could be observed with the naked eye [89].

Very recently a biosensor system for detecting trinitrotoluene (TNT) has been developed [90]. The working principle of this sensor is based on a silanized glass-fiber surface, to which TNT derivatives are immobilized. Fluorescently labeled TNT aptamers bind to the target molecules on the surface, such that fluorescence can be detected through the glass surface. If a sample containing free TNT is added to the flow cell, the aptamer binds to the free TNT and the light signal at the surface is reduced (Figure 11.9).

11.4.3.2 Electrochemical Sensing

Biosensors that detect the presence of a target analyte via electrochemical signals require a sample delivery, a molecular recognition subunit, and a transducer, translating the binding or reaction of target molecules into a measurable physical signal (mass, charge, heat, light; reviewed in Reference [91]). For example, an RNA aptamer-based assay for the detection of the aminoglycoside neomycin B (Figure 11.10) has been built up by Alvarez et al. [92] using faradaic impedance spectroscopy (FIS) as transducer. The RNA was modified with 2'-O-Me-nucleotides. Neomycin B was immobilized on a self-assembled monolayer (SAM) of mercaptopropionic acid (MPA) on Au electrodes through carbodiimide chemistry. Immobilization of the aptamer to the surface-linked neomycin B was carried out by affinity binding. Adding a solution-phase with free neomycin leads to an aptamer displacement on the surface and the free neomycin effects a change of impedance. The faradaic impedance is measured by a Nyquist plot and the data were adjusted to a Randles equivalent circuit. The covalent attachment of the aptamer-free neomycin B at the surface causes a decrease in the electron-transfer resistance because the positively charged amino groups of neomycin B attract the redox probe $[Fe(CN)_6]^{3-/4-}$, facilitating the electron-transfer reaction. High ionic strength incubation conditions were selected to ensure a 1:1 stoichiometry and to avoid

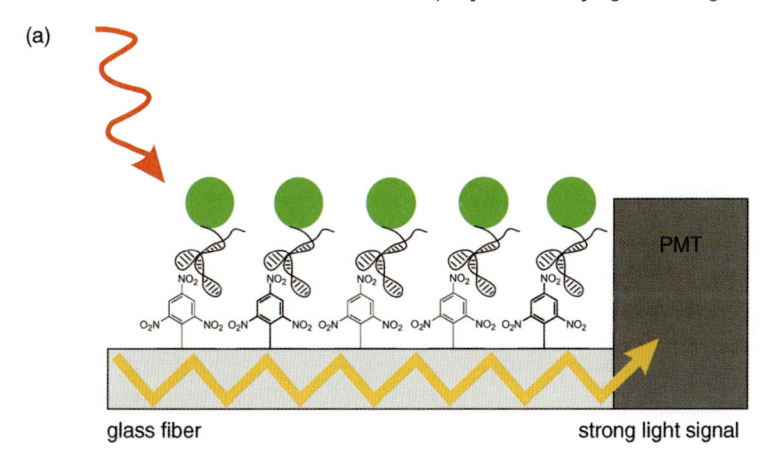

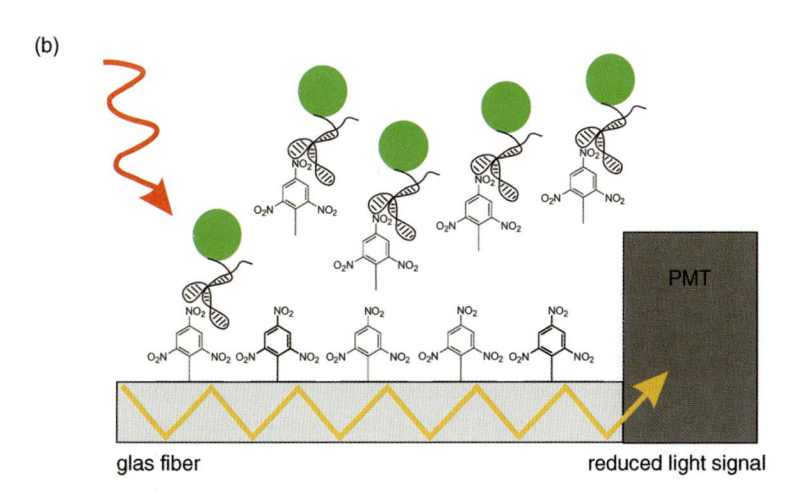

Figure 11.9 TNT glass-fiber system (a) in the absence of free TNT, and (b) after addition of free TNT. Adapted from Reference [90].

electrostatic interactions that can increase cross-reactivity with other antibiotics. The impedimetric aptasensor is completely recovered by treatment with a concentrated solution of neomycin. The small amount of aptamer consumed in each round and the increased stability due to the modified bases minimized the costs of the assay. Using this assay, neomycin B can be detected in milk in a range between 25 and 2500 µM [92].

Wang and colleagues have designed an electrochemical sensor suitable for detection of proteins with large pI values [93]. This sensor consists of an electrode modified with a DNA aptamer that binds lysozyme (pI = 11). The negatively charged aptamer hinders the redox reaction of $[Fe(CN)_6]^{3-/4-}$ on the electrode. Owing to the selective binding of a positively charged protein to the aptamer, the negative charges

Figure 11.10 Neomycin B.

on the electrode are reduced and the interfacial electron transfer is lowered (Figure 11.11).

A different electrochemical sensing method has been described by Kim and coworkers [94]. Their electrode is modified with a thrombin-binding DNA aptamer, which is configured into a hairpin structure capable of binding methylene blue via intercalation. This method can be applied to any aptamer–protein interaction. The presence of the cognate target disrupts the hairpin stem and releases the intercalator, resulting in a decrease of the electrochemical signal.

The design of a deoxyribosensor for detection and quantitation of the plasma protein thrombin has been described by Huang et al. [95]. Here, the sensor is composed of a ligand-binding-receptor (a DNA aptamer) that is functionally linked to a helical DNA charge conduction path. Deoxyribosensors work by showing greater charge flow through them in the presence of bound ligand (analyte) than in its absence. After testing the sensor in a biochemical manner, a chip was constructed, where the whole system was coupled to a gold surface. The density of the DNA

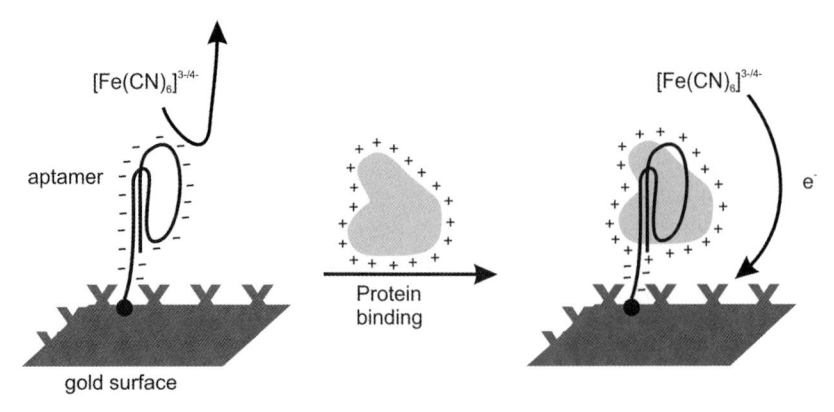

Figure 11.11 Detection of protein binding via electrochemical sensing. Adapted from Reference [93].

constructs on the gold chip was only 2.6 ± 0.3 pmol cm^{-2} and the electrode surface small (0.126 cm^2). Reportedly, this sensor can measure even picomolar concentrations of thrombin in solution or in diluted serum.

Tuberculosis is the most frequent cause of infection-related death worldwide. Therefore, a simple and label-free method to detect interferon-γ (IFN-γ, a selective marker for tuberculosis pleurisy) using aptamer-based electrochemical impedance spectroscopy was developed [96]. Aptamers that first were selected via SELEX were immobilized as a monolayer on a gold electrode. For this, the RNA and DNA aptamers were modified with 5′-thiol groups and immobilized via pentane thiol linkers. Unbound aptamers were eliminated by the use of β-mercapto-ethanol. Then, IFN-γ was added to the ordered self-assembled aptamer monolayer. These immobilization processes were characterized through changes in R_{ct} in a Nyquist plot. Measurements were carried out in 10 mM sodium phosphate (pH 10.3) containing $[Fe(CN)_6]^{3-/4-}$. The formation of the aptamer–IFN-γ complex led to an increase in R_{ct} reflecting electrochemical and physical blocking of electro-active $[Fe(CN)_6]^{3-/4-}$ ions. Since electrochemical detection was processed under high-pH conditions, additionally the conformational changes in the secondary structure of aptamers and IFN-γ in buffers with various pH values were tested using CD spectroscopy. The conformation of aptamers was not significantly changed at high pH. The detection limit for the RNA aptamer was tested down to 100 fM and calculated down to 1.43 fM, whereas the detection limit for the DNA aptamer was tested down to 1 pM and calculated down to 191 fM. The detection limit of each aptamer was obtained by quartz crystal microbalance (QCM) analysis (Section 11.4.3.4) and impedance analysis. The dissociation constants found by electrophoretic mobility shift assay (EMSA) may have caused the different detection limits of the RNA compared with DNA aptamers. The K_D values of RNA and DNA aptamers were 18.7 ± 1.2 and 63.8 ± 7.4 nM. Furthermore, a cross-linking assay revealed that IFN-γ can exist in high multimeric forms with the RNA aptamer; this might also be a reason why the RNA aptamer probe can detect IFN-γ more sensitively than the DNA aptamer can do. In a next approach, the authors tested their systems in fetal bovine serum (FBS) solution, because this has similar components as pleural fluid. Only the DNA aptamer probe worked in FBS solution, while the RNA aptamer presumably is destroyed by RNases. For the DNA aptamer the detection limit in FBS solution was measured at 10 pM and calculated to 1.21 pM. Although the DNA aptamer is less sensitive than the RNA aptamer, the sensitivity (10 pM) is sufficient to detect IFN-γ, considering that IFN-γ is present at a mean concentration of 89 pM in tuberculosis patients.

11.4.3.3 Acoustic Sensing

Acoustic sensing methods for detection of analytes are based on mass changes that are measured through a surface-acoustic wave. For example, an acoustic sensor coated with human thrombin-binding RNA or DNA aptamers can detect the intended protein target and can report protein–protein interactions in real time. In this way, a system for monitoring a blood-coagulation cascade was employed by Famulok and coworkers [97]. Mascini and colleagues have compared aptamer and antibody piezoelectric acoustic biosensors for HIV-1 Tat protein and found that the aptamer sensor is equally specific and sensitive as the antibody sensor, but has the advantage of reusability [98]. Recently, a so-called S-sens K5 surface acoustic wave biosensor was coupled with mass spectrometry (SAW-MS) for the analysis of a protein complex consisting of the human blood clotting cascade factor α-thrombin and human antithrombin III, a specific blood plasma inhibitor of thrombin [99]. Specific binding of antithrombin III to thrombin was recorded. Three of five elements of the sensor chip were coated with RNA *anti*-thrombin aptamers, the other two were used as references. The biosensor measured mass changes on the chip surface, showing that 20% of about 400 fmol cm^{-2} thrombin formed a complex with the 1.7-times larger antithrombin III. Mass spectrometry was applied to identify the bound proteins either directly from the chip surface (thrombin) or after separation by nanocapillary-HPLC (thrombin–antithrombin III complex). In both cases the proteins were first digested with proteases on the sensor element. For following the ligand–analyte (in this case thrombin–antithrombin III) interactions in real-time, the S-sens K5 is a superior and reliable device. It allows coupling of various biomolecules.

11.4.3.4 Quartz Crystal Microbalance Based Sensing

One method of forming a self-assembled monolayer on a gold surface is the use of poly(amidoamine) dendrimers and 1-hexadecanethiol (HDT). These monolayers can be used for preparation of a quartz crystal microbalance (QCM) immunosensor for detection of, for example, human immunoglobulin G (IgG) [100]. Recently, the acoustic transverse shear mode (TSM) method was used to study the surface properties of a DNA aptasensor that specifically binds human immunoglobulin E (IgE) [101]. With this method, the contribution of intermolecular friction between the sensor surface and the buffer to the frequency of crystal oscillation in all stages of sensor preparation can be estimated.

11.4.3.5 Cantilever Based Sensing

The cantilever sensing method has been employed by Gerber *et al.* [102]. This label-free biosensors could detect biomolecular interactions via the bending of microfabricated cantilevers coated with DNA oligonucleotides or antibodies as molecular recognition elements. One example is a DNA aptamer that binds *Taq* DNA polymerase. The addition of *Taq* polymerase to the device in a concentration-dependent manner produces a specific mechanical response [103].

11.5
Outlook/Summary

There are many examples of RNA and DNA aptamers having the potential to serve as receptors in biosensors. Advances in biosensor engineering and the demonstration of functionality of a large variety of nucleic acid sensors convincingly support the potential of nucleic acids for fast and reliable detection of specific molecular structures. Furthermore, the powerful and nowadays well-established method of *in vitro* selection paves the way to the development of a nucleic acid sensor for virtually any desired compound. Several reviews on aptamer development and biosensor construction have appeared in the literature [104–108].

Very importantly, aptamers are destined to play an important role in diagnostic and therapeutic medicine. For example, radiolabeled aptamers for tumor imaging and therapy have been reported [109], and recently an aptamer–antibody sandwich ELISA for the early diagnosis of epithelial tumors has been described [110]. In addition, structural motives in the 5′-untranslated region of certain messenger RNAs that bind metabolites with high specificity and thus exert control over expression of genes they encode have been discovered in nature [111]. These motifes may be easily turned into tools for metabolite sensing: binding of the analyte would be detected by genetic readout. So far, only a few papers have been published reporting a real application of aptamer-based biosensors for the detection of environmental analysis-related molecules [112]. Nevertheless, based on the features of nucleic acids and on the powerful techniques of nucleic acid engineering, further developments in the field may be anticipated.

References

1 Tuerck, C. and Gold, L. (1990) Systematic evolution of ligands by exponential enrichment: RNA ligands to bacteriophage T4 DNA polymerase. *Science*, **249**, 505–510.

2 Ellington, A.D. and Szostak, J.W. (1990) In vitro selection of RNA molecules that bind specific ligands. *Nature*, **346**, 818–822.

3 Famulok, M. (1994) Molecular recognition of amino acids by RNA-aptamers: an L-citrulline binding RNA motif and its evolution into an L-arginine binder. *J. Am. Chem. Soc.*, **116**, 1698–1706.

4 Haller, A.A. and Sarnow, P. (1997) In vitro selection of a 7-methyl-guanosine binding RNA that inhibits translation of capped mRNA molecules. *Proc. Natl. Acad. Sci. USA*, **94**, 8521–8526.

5 Gebhardt, K., Shokraei, A., Babane, G., and Lindquist, B.H. (2000) RNA aptamers to S-adenosylhomocysteine: kinetic properties, divalent cation dependency, and comparison with anti-S-adenosylhomocysteine antibody. *Biochemistry*, **39**, 7255–7265.

6 Wilson, D.S., Keefe, A.D., and Szostak, J.W. (2001) The use of mRNA display to select high-affinity protein-binding peptides. *Proc. Natl. Acad. Sci. USA*, **98**, 3750–3755.

7 Jenison, R.D., Gill, S.C., Pardi, A., and Polisky, B. (1994) High-resolution molecular discrimination by RNA. *Science*, **263**, 1425–1429.

8 Win, M.N., Klein, J.S., and Smolke, C.D. (2006) Codeine-binding RNA aptamers and rapid determination of their binding constants using a direct coupling surface plasmon resonance assay. *Nucleic Acids Res.*, **34**, 5670–5682.

9 Geiger, A., Burgstaller, P., von der Eltz, H., Roeder, A., and Famulok, M. (1996) RNA aptamers that bind L-arginine with sub-micromolar dissociation constants and high enantioselectivity. *Nucleic Acids Res.*, **24**, 1029–1036.

10 Yang, Q., Goldstein, I.J., Mey, H.Y., and Engelke, D.R. (1998) DNA ligands that bind tightly and selectively to cellobiose. *Proc. Natl. Acad. Sci. USA*, **95**, 5462–5467.

11 Zhan, L.S., Shao, N.S., Peng, J.C., Sun, H.Y., and Wang, Q.L. (2003) A procedure for SELEX screening aptamers from ssDNA random library. *Prog. Biochem. Biophys.*, **30**, 151–155.

12 Drolet, D.W., Jenison, R.D., Smith, D.E., Pratt, D., and Hicke, B.J. (1999) A high

throughput platform for systematic evolution of ligands by exponential enrichment (SELEX™). *Comb. Chem. High Throughput Screen*, **2**, 271–278.

13 Burke, D. and Gold, L. (1997) RNA aptamers to the adenosine moiety of S-adenosyl methionine: structural inferences from variations on a theme and the reproducibility of SELEX. *Nucleic Acids Res.*, **25**, 2020–2024.

14 Holeman, L.A., Robinson, S.L., Szostak, J.W., and Wilson, C. (1998) Isolation and characterization of fluorophore-binding RNA aptamers. *Fold. Des.*, **3**, 423–431.

15 Stoltenburg, R., Reimann, C., and Strehlitz, B. (2005) FluMag-SELEX as an advantageous method for DNA aptamer selection. *Anal. Bioanal. Chem*, **383**, 83–91.

16 Padilla, R. and Sousa, R. (1999) Efficient synthesis of nucleic acids heavily modified with non-canonical ribose 2′-groups using a mutantT7 RNA polymerase (RNAP). *Nucleic Acids Res.*, **27**, 1561–1563.

17 Kujau, M.J. and Wölfl, S. (1998) Intramolecular derivatization of 2′-amino-pyrimidine modified RNA with functional groups that is compatible with re-amplification. *Nucleic Acids Res.*, **26**, 1851–1853.

18 Kubik, M.F., Bell, C., Fitzwater, T., Watson, S.R., and Tasset, D.M. (1997) Isolation and characterization of 2′-fluoro-, 2′-amino-, and 2′-fluoro-/amino-modified RNA ligands to human IFN-gamma that inhibit receptor binding. *J. Immunol.*, **159**, 259–267.

19 Thum, O., Jäger, S., and Famulok, M. (2001) Functionalized DNA: a new replicable biopolymer. *Angew. Chem Int. Ed.*, **40**, 3990–3993.

20 Faulhammer, D., Eschgfaller, B., Stark, S. et al. (2004) Biostable aptamers with antagonistic properties to the neuropeptide nociceptin/orphanin FQ. *RNA*, **10**, 516–527.

21 Ruckman, J., Green, L.S., Beeson, J., Waugh, S., Gilette, W.L., Henninger, D.D., Claesson-Welsh, L., and Janjic, N. (1998) 2′-Fluoropyrimidine RNA-based aptamers to the 165-amino acid form of vascular endothelial growth factor (VEGF165). Inhibition of receptor binding and VEGF-induced vascular permeability through interactions requiring the exon 7-encoded domain. *J. Biol. Chem.*, **273**, 20556–20567.

22 Kraus, E., James, W., and Barclay, A.N. (1998) Cutting edge: novel RNA ligands able to bind CD4 antigen and inhibit CD4 + T lymphocyte function. *J. Immunol*, **160**, 5209–5212.

23 Jhaveri, S., Olwin, B., and Ellington, A.D. (1998) In vitro selection of phosphorothiolated aptamers. *Bioorg. Med. Chem. Lett.*, **8**, 2285–2290.

24 Meisenhammer, K.M. and Koch, T.H. (1997) Photocross-linking of nucleic acids to associated proteins. *Crit. Rev. Biochem. Mol.*, **32**, 101–140.

25 Koch, T.H., Smith, D., Tabacman, E., and Zichi, D.A. (2004) Kinetic analysis of site-specific photoaptamer-protein cross-linking. *J. Mol. Biol*, **336**, 1159–1173.

26 Krylov, S.S.N. and Berezovski, M. (2003) Non-equilibrium capillary electrophoresis of equilibrium mixtures–appreciation of kinetics in capillary electrophoresis. *Analyst*, **128**, 571–575.

27 Tang, J.J., Xie, J.W., Shao, N.S., and Yan, Y. (2006) The DNA aptamers that specifically recognize ricin toxin are selected by two *in vitro* selection methods. *Electrophoresis*, **27**, 1303–1311.

28 Mendonsa, S.D. and Bowser, M.T. (2005) In vitro selection of aptamers with affinity for neuropeptide Y using capillary electrophoresis. *J. Am. Chem. Soc*, **127**, 9382–9383.

29 Krylov, S.N. (2006) Nonequilibrium capillary electrophoresis of equilibrium mixtures (NECEEM): a novel method for biomolecular screening. *J. Biomol. Screen*, **11**, 115–122.

30 Young, D.D. and Deiters, A. (2008) Light-regulated RNA-small molecule interactions. *ChemBioChem*, **9**, 1225–1228.

31 Cox, J.C., Rudolph, P., and Ellington, A.D. (1998) Automated RNA selection. *Biotechnol. Progr.*, **14**, 845–850.

32 Cox, J.C. and Ellington, A.D. (2001) Automated selection of anti-protein aptamers. *Bioorg. Med. Chem.*, **9**, 2525–2531.

33 Cox, J.C., Rajendran, M., Riedell, T., Davidson, E.A., Sooter, L.J., Bayer, T.S., Schmitz-Brown, M., and Ellington, A.D. (2002) Automated acquisition of aptamer sequences. *Comb. Chem. High Throughput Screen*, **5**, 289–299.

34 Cox, J.C., Hayhurst, A., Hesselberth, J., Bayer, T.S., Georgiou, G., and Ellington, A.D. (2002) Automated selection of aptamers against protein targets translated *in vitro*: from gene to aptamer. *Nucleic Acids Res.*, **30**, e108.

35 Hybarger, G., Bynum, J., Williams, R.F., Valdes, J.J., and Chambers, J.P. (2006) A microfluidic SELEX prototype. *Anal. Bioanal. Chem*, **384**, 191–198.

36 Berezovski, M., Drabovich, A., Krylova, S.M., Musheev, M., Okhonin, V., Petrov, A., and Krylov, S.N. (2005) Nonequilibrium capillary electrophoresis of equilibrium mixtures: a universal tool for development of aptamers. *J. Am. Chem. Soc.*, **127**, 3165–3171.

37 Berezovski, M., Musheev, M., Drabovich, A., and Krylov, S.N. (2006) Non-SELEX selection of aptamers. *J. Am. Chem. Soc.*, **128**, 1410–1411.

38 Peng, L., Stephens, B.J., Bonin, K., Cubicciotti, R., and Guthold, M. (2007) A combined atomic force/fluorescence microscopy technique to select aptamers in a single cycle from a small pool of random oligonucleotides. *Microsc. Res. Tech.*, **70**, 372–381.

39 Müller, S., Wolf, J., and Ivanov, S.A. (2004) Current strategies for the synthesis of RNA. *Curr. Org. Synth.*, **1**, 111–121.

40 Kusser, W. (2000) Chemically modified nucleic acid aptamers for in vitro selections: evolving evolution. *J. Biotechnol.*, **74**, 27–38.

41 Drude, I., Dombos, V., Vauléon, S., and Müller, S. (2007) Drugs made of RNA: development and application of engineered RNAs for gene therapy. *Minirev. Med. Chem.*, **7**, 912–931.

42 Klussmann, S., Nolte, A., Bald, R., Erdmann, V.A., and Fürste, J.P. (1996) Mirror-image RNA that binds D-adenosine. *Nat. Biotechnol*, **14**, 1112–1115.

43 Nolte, A., Klußmann, S., Bald, R., Erdmann, V.A., and Fürste, J.P. (1996) Mirror-design of L-oligonucleotide ligands binding to L-arginine. *Nat. Biotechnol.*, **14**, 1116–1119.

44 Eulberg, D. and Klussmann, S. (2003) Spiegelmers: biostable aptamers. *ChemBioChem*, **4**, 979–983.

45 Purschke, W.G., Eulberg, D., Buchner, K., Vonhoff, S., and Klussmann, S. (2006) An L-RNA-based aquaretic agent that inhibits vasopressin in vivo. *Proc. Natl. Acad. Sci. USA*, **103**, 5173–5178.

46 Balamurugan, S., Obubuafo, A., Soper, S., and Spivak, D. (2008) Surface immobilization methods for aptamer diagnostic applications. *Anal. Bioanal. Chem.*, **390**, 1009–1021.

47 Balamurugan, S., Obubuafo, A., Soper, S.A., McCarley, R.L., and Spivak, D.A. (2006) Designing highly specific biosensing surfaces using aptamer monolayers on gold. *Langmuir*, **22**, 6446–6453.

48 Seetharaman, S., Zivarts, M., Sudarsan, N., and Breaker, R.R. (2001) Immobilized RNA switches for the analysis of complex chemical and biological mixtures. *Nat. Biotechnol.*, **19**, 336–341.

49 Yamamoto-Fujita, R. and Kumar, P.K.R. (2005) Aptamer-derived nucleic acid oligos: applications to develop nucleic acid chips to analyze proteins and small ligands. *Anal. Chem.*, **77**, 5460–5466; Gopinath, S.C.B., Misono, T.S., and Kumar, P.K.P. (2008) Prospects of ligand-induced aptamers. *Crit. Rev. Anal. Chem.*, **38**, 34–47.

50 Drummond, T.G., Hill, G., and Barton, J.K. (2003) Electrochemical DNA sensors. *Nat. Biotechnol.*, **21**, 1192–1199.

51 Yoon, H., June-Hyung, K., Nahum, L., Byung-Gee, K., and Jang, J. (2008) A novel sensor platform based on aptamer-conjugated polypyrrole nanotubes for label-free electrochemical protein detection. *ChemBioChem*, **9**, 634–641.

52 Kim, S.N., Rusling, J.F., and Papadimitrakopoulus, F. (2007) Carbon nanotubes for electronic and electrochemical detection of biomolecules. *Adv. Mater.*, **19**, 3214–3228.

53 Schlensog, M.D., Gronewold, T.M., Tewes, M., Famulok, M., and Quandt, E. (2004) A Love-wave biosensor using nucleic acids as ligands. *Sens. Actuators, B*, **101**, 308–315.

54 Cho, E.J., Collet, J.R., Szafranska, A.E., and Ellington, A.D. (2006) Optimization of aptamer microarray technology for multiple protein targets. *Anal. Chim. Acta*, **564**, 82–90.

55 Wang, Y., Li, D., Ren, W., Liu, Z., Dong, S., and Wang, E. (2008) Ultrasensitive colorimetric detection of protein by aptamer–Au nanoparticles conjugates based on a dot-blot assay. *Chem. Commun.*, 2520–2522.

56 Lin, Y.C., Liu, Y., and Yan, H. (2007) Self-assembled combinatorial encoding nanoarrays for multiplexed biosensing. *Nano Lett.*, **7**, 507–512.

57 Bock, C., Coleman, M., Collins, B., Davis, J., Foulds, G., Gold, L., Greef, C., Heil, J., Heilig, J.S. et al. (2004) Photoaptamer arrays applied to multiplexed proteomic analysis. *Proteomics*, **4**, 609–618.

58 Wang, Z., Wilkon, T., Xu, D., Dong, Y., Ma, G., and Cheng, Q. (2007) Surface plasmon resonance imaging for affinity analysis of aptamer-protein interactions with PDMS microfluidic chips. *Anal. Bioanal. Chem.*, **389**, 819–825.

59 Corne, C., Fichte, J.B., Gasparutto, D., Cunin, V., Straniti, E., Buhot, A., Fuchs, J., Calemczuk, R., Livache, T., and Favier, A. (2008) SPR imaging for label-free multiplexed analyses of DNA N-glycosylase interactions with damaged DNA duplexes. *Analyst*, **133**, 1036–1045.

60 Li, Y., Lee, H.J., and Corn, R.M. (2006) Fabrication and characterization of RNA aptamer microarrays for the study of protein-aptamer interactions with SPR imaging. *Nucleic Acids Res.*, **34**, 6416–6424.

61 Li, Y., Lee, H.J., and Corn, R.M. (2007) Detection of protein biomarkers using RNA aptamer microarrays and enzymatically amplified surface plasmon resonance imaging. *Anal. Chem.*, **79**, 1082–1088.

62 Warren, C.L., Kratochvil, N.C.S., Hauschild, K.E., Foister, S., Brezinski, M.L., Dervan, P.B., Phillips, G.N., and Ansari, A.Z. (2006) Defining the sequence-recognition profile of DNA-binding molecules. *Proc. Natl. Acad. Sci. USA*, **103**, 867–872.

63 Tyagi, S. and Kramer, F.R. (1996) Molecular beacons: probes that fluorescence upon hybridization. *Nat. Biotechnol.*, **14**, 303–308.

64 Fang, X., Liu, X., Schuster, S., and Tan, W. (1999) Designing a novel molecular beacon for surface-immobilized DNA hybridization studies. *J. Am. Chem. Soc.*, **121**, 2921–2922.

65 Liu, X. and Tan, W. (1999) A fiber-optic evanescent wave DNA biosensor based on novel molecular beacons. *Anal. Chem.*, **71**, 829–834.

66 Fang, X., Sen, A., Vicens, M., and Tan, W. (2003) Synthetic DNA aptamers to detect protein molecular variants in a high-throughput fluorescence quenching assay. *ChemBioChem*, **4**, 829–843.

67 Morse, D.P. (2007) Direct selection of RNA beacon aptamers. *Biochem. Biophys. Res. Commun.*, **359**, 94–101.

68 Wang, X., Zhou, I., Yun, W., Xiao, S., Chang, Z., He, P., and Fang, Y. (2007) Detection of thrombin using electrogenerated chemiluminescence based on Ru(bpy)$_3$(2+)-doped silica nanoparticle aptasensor via target protein-induced strand displacement. *Anal. Chim. Acta*, **598**, 242–248.

69 Zhiwen, T., Prabodhika, M., Ronghua, Y., Youngmi, K., Zhi, Z., Hui, W., and Weihong, T. (2008) Aptamer switch probe based on intramolecular displacement. *J. Am. Chem. Soc.*, **130**, 11268–11269.

70 Müller, S., Strohbach, D., and Wolf, J. (2006) Sensors made of RNA: tailored ribozymes for detection of small organic molecules, metals, nucleic acids and proteins. *IEE Proc. Nanobiotechnol.*, **153**, 31–40.

71 Yang, C.J., Jockusch, S., Vicens, M., Turro, N.J., and Tan, W. (2005) Light-switching excimer probes for rapid protein monitoring in complex biological fluids. *Proc. Natl. Acad. Sci. USA*, **102**, 17278–17283.

72 Hollenstein, M., Hipolito, C., Lam, C., Dietrich, D., and Perrin, D.M. (2008) A highly selective DNAzyme sensor for mercuric ions. *Angew. Chem.*, **120**, 4418–4422.

73 Liu, J. and Lu, Y. (2007) A DNAzyme catalytic beacon sensor for paramagnetic Cu^{2+} ions in aqueous solution with high

sensitivity and selectivity. *J. Am. Chem. Soc.*, **129**, 9838–9839.

74 Carmi, N., Shultz, L.A., and Breaker, R.R. (1996) In vitro selection of self-cleaving DNAs. *Chem. Biol.*, **3**, 1039–1046.

75 Carmi, N., Balkhi, S.R., and Breaker, R.R. (1998) Cleaving DNA with DNA. *Proc. Natl. Acad. Sci. USA*, **95**, 2233–2237.

76 Carmi, N. and Breaker, R.R. (2001) Characterization of a DNA-cleaving deoxyribozyme. *Bioorg. Med. Chem.*, **9**, 2589–2600.

77 Liu, J. and Lu, Y. (2003) Improving Fluorescent DNAzyme Biosensors by combining inter- and intramolecular quenchers. *Anal. Chem.*, **75**, 6666–6672.

78 Rajendran, M. and Ellington, A.D. (2008) Selection of fluorescent aptamer beacons that light up in the presence of zinc. *Anal. Bioanal. Chem.*, **390**, 1067–1075.

79 Ahn, D.-R., and Yang, E.G. (2007) An RNase H-assisted fluorescent biosensor for aptamers. *ChemBioChem*, **8**, 1347–1350.

80 Li, N. and Ho, C.-M. (2008) Aptamer-based optical probes with separated molecular recognition and signal transduction modules. *J. Am. Chem. Soc.*, **130**, 2380–2381.

81 Bini, A., Centi, S., Tombelli, S., Minunni, M., and Mascini, M. (2008) Development of an optical RNA-based aptasensor for C-reactive protein. *Anal. Bioanal. Chem.*, **390**, 1077–1086.

82 Sparano, B.A. and Koide, K. (2007) Fluorescent sensors for specific RNA: a general paradigm using chemistry and combinatorial biology. *J. Am. Chem. Soc.*, **129**, 4785–4794.

83 Lee, J.-S., Han, M.S., and Mirkin, A.A. (2007) Colorimetric detection of mercuric ion (Hg^{2+}) in aqueous media using DNA-functionalized gold nanoparticles. *Angew. Chem.*, **119**, 4171–4174.

84 Stojanovic, M.N. and Kolpashchikov, D.M. (2004) Modular aptameric sensors. *J. Am. Chem. Soc.*, **126**, 9266–9270.

85 Hirabayashi, M., Taira, S., Kobayashi, S., Konishi, K., Katoh, K., Hiratsuka, Y., Kodaka, M., Uyeda, T.Q.P., Yumoto, N., and Kubo, T. (2006) Malachite green-conjugated microtubules as mobile bioprobes selective for malachite green aptamers with capturing/releasing ability. *Biotechnol. Bioeng.*, **94**, 473–480.

86 Afonin, K.A., Danilov, E.O., Novikova, I.V., and Leontis, N.B. (2008) TokenRNA: a new type of sequence-specific, label-free fluorescent biosensor for folded RNA molecules. *ChemBioChem*, **9**, 1902–1905.

87 Pei, R. and Stojanovic, M.N. (2008) Study of thiazole orange in aptamer-based dye-displacement assays. *Anal. Bioanal. Chem.*, **390**, 1093–1099.

88 Zhou, C., Jiang, Y., Hou, S., Ma, B., Fang, X., and Li, M. (2006) Detection of oncoprotein platelet-derived growth factor using a fluorescent signaling complex of an aptamer and TOTO. *Anal. Bioanal. Chem.*, **384**, 1175–1180.

89 Miyamoto, D., Tang, Z., Takarada, T., and Maeda, M. (2007) Turbidimetric detection of ATP using polymeric micelles and DNA aptamers. *Chem. Commun.*, 4743–4745.

90 Ehrenreich-Förster, A., Orgel, D., Krause-Griep, A., Cech, B., Erdmann, V.A., Bier, F., Scheller, F.W., and Rimmele, M. (2008) Biosensor-based on-site explosive detection using aptamers as recognition elements. *Anal. Bioanal. Chem.*, **391**, 1793–1800.

91 Lee, J.-O., So, H.-M., Jeon, E.-K., Chang, H., Won, K., and Kim, Y.H. (2008) Aptamers as molecular recognition elements for electrical nanobiosensors. *Anal. Bioanal. Chem.*, **390**, 1023–1032.

92 de-los-Santos-Alvarez, N., Lobo-Castanon, M.J., Miranda-Ordieres, A.J., and Tunon-Blanco, P. (2007) Modified-RNA aptamer-based sensor for competitive impedimetric assay of neomycin B. *J. Am. Chem. Soc.*, **129**, 3808–3809.

93 Rodriguez, M.C., Kawde, A.N., and Wang, J. (2005) Aptamer biosensor for label free impedance spectroscopy detection of proteins based on recognition-induced switching of the surface charge. *Chem. Commun.*, 4267–4269.

94 Bang, G.S., Cho, S., and Kim, B.G. (2005) A novel electrochemical detection method for aptamer biosensors. *Biosens. Bioelectron*, **21**, 863–870.

95 Huang, Y.C., Ge, B., Sen, D., and Yu, H.-Z. (2008) Immobilized DNA switches as electronic sensors for picomolar detection of plasma proteins. *J. Am. Chem. Soc.*, **130**, 8023–8029.

96 Min, K., Cho, M., Han, S.-Y., Shim, Y.-B., Ku, J., and Ban, C. (2008) A simple and direct electrochemical detection of interferon-γ using its RNA and DNA aptamers. *Biosens. Bioelectron.*, **23**, 1819–1824.

97 Gronewold, T.M., Glass, S., Quandt, E., and Famulok, M. (2005) Monitoring complex formation in the blood-coagulation cascade using aptamer-coated SAW sensors. *Biosens. Bioelectron.*, **20**, 2044–2052.

98 Tombelli, S., Minunni, M., Luzi, E., and Mascini, M. (2005) Aptamer-based biosensors for the detection of HIV-1 tat protein. *Bioelectrochemistry*, **67**, 135–141.

99 Treitz, G., Gronewold, T.M.A., Quandt, E., and Zabe-Kühn, M. (2008) Combination of a SAW-biosensor with MALDI mass spectrometric analysis. *Biosens. Bioelectron.*, **23**, 1496–1502.

100 Svobodová, L., Šejdárková, M., Polohová, V., Grman, I., Rybár, P., and Hianik, T. (2006) QCM immunosensor based on polyamidoamine dendrimers. *Electroanalysis*, **18**, 1943–1949.

101 Šnejdárková, M., Svobodová, L., Polohová, V., and Hianik, T. (2008) The study of surface properties of an IgE-sensitive aptasensor using an acoustic method. *Anal. Bioanal. Biochem.*, **390**, 1087–1091.

102 Fritz, J., Baller, M.K., Lang, H.P., Rothuizen, H., Vettiger, P., Meyer, E., Guntherodt, H., Gerber, C., and Gimzewski, J.K. (2000) Translating biomolecular recognition into nanomechanics. *Science*, **288**, 316–318.

103 Savran, C.A., Knudsen, S.M., Ellington, A.D., and Manalis, S.R. (2004) Micromechanical detection of proteins using aptamer-based receptor molecules. *Anal. Chem.*, **76**, 3194–3198.

104 Navani, N.K. and Li, Y. (2006) Nucleic acid aptamers and enzymes as sensors. *Curr. Opin. Chem. Biol.*, **10**, 272–281.

105 Li, Y.L., Guo, L., Zhang, Z.Y., Tang, J.J., and Xie, J.W. (2008) Recent advances of aptamer sensors. *Sci. Chin., Ser. B: Chem.*, **3**, 193–204.

106 Mairal, T., Özalp, V.C., Sánchez, P.L., Mir, M., Katakis, I., and O'Sullivan, C.K. (2008) Aptamers: molecular tools for analytical application. *Anal. Bioanal. Chem.*, **390**, 989–1007.

107 Willner, I., Shlyahovsky, B., Zayats, M., and Willner, B. (2008) DNAzymes for sensing, nanobiotechnology and logic gate applications. *Chem. Soc. Rev.*, **37**, 1153–1165.

108 James, W. (2007) Aptamers in the virologists' toolkit. *J. Gen. Virol.*, **88**, 351–364.

109 Perkins, A.C. and Missailidis, S. (2007) Radiolabelled aptamers for tumor imaging and therapy. *Q. J. Med. Mol. Imag.*, **51**, 292–296.

110 Ferreira, C.S.M., Papamichael, K., Guilbault, G., Schwarzacher, T., Gariepy, J., and Missailidis, S. (2008) DNA aptamers against the MUC1 tumor marker: design of aptamer-antibody sandwich ELISA for the early diagnosis of epithelial tumours. *Anal. Bioanal. Chem.*, **390**, 1039–1050.

111 Tucker, B.J. and Breaker, R.R. (2005) Riboswitches as versatile gene control elements. *Curr. Opin. Struct. Biol.*, **15**, 342–348.

112 Palachetti, I. and Mascini, M. (2008) Nucleic acid biosensors for environmental pollution monitoring. *Analyst*, **133**, 846–854.

12
Conducting Polymers as Artificial Receptors in Chemical Sensors
Ulrich Lange, Nataliya V. Roznyatovskaya, Qingli Hao, and Vladimir M. Mirsky

12.1
Introduction

Conducting polymers are commonly used materials in chemical and biological sensors. These materials possess receptor as well as transducer functions [1]. Typical examples of conducting polymers (CPs) include polythiophenes, polypyrroles, polyanilines, and polyphenylenes. Such polymers show some intrinsic receptor properties towards analytes that can react/interact with the polymer, like, for example, redox active or acidic/basic gases. Oxidation/reduction and in some cases protonation/deprotonation introduces/removes charge carriers in the conjugated backbone, whereas conformational changes alter the planarity of the conjugated backbone, which leads to changes in the mobility of the charge carriers. Figure 12.1 demonstrates the example of PANI (polyaniline), showing the influence of redox potential and pH on the interconversion between the different forms of PANI.

These changes can be detected by *in situ* resistance, UV-vis, IR, or fluorescence measurements. Direct interactions with the π-system often produce very large signal changes; however, they are quite non-specific. A possible way to introduce some selectivity in these polymers is to modify their conjugated backbone with receptors.

Usually, receptors can be introduced into a polymer backbone either after polymerization or directly to monomer, which is polymerized or copolymerized. The synthetic availability and the ability to polymerize the monomer under certain conditions are the most common limitations in the choice of a derivative with receptor group. Some classes of these receptors are discussed in Section 12.4.

Doping of CPs by counter ions bearing the required receptor [2–8], the inclusion of ionophores into the polymer matrix [9], and the chemical linkage of ligands or receptor units to the polymer backbone [10] are also applied to enhance the selectivity and sensitivity of CPs. Receptors attached to the π-system via some short side chain can communicate with the π-system of the polymer by, for example, electron donating/withdrawing, electrostatic interactions or by inducing conformational changes upon analyte binding. As there is no direct interaction with the polymer backbone, the signal change may be smaller, but also much more specific.

Artificial Receptors for Chemical Sensors. Edited by V.M. Mirsky and A.K. Yatsimirsky
Copyright © 2011 WILEY-VCH Verlag GmbH & Co. KGaA, Weinheim
ISBN: 978-3-527-32357-9

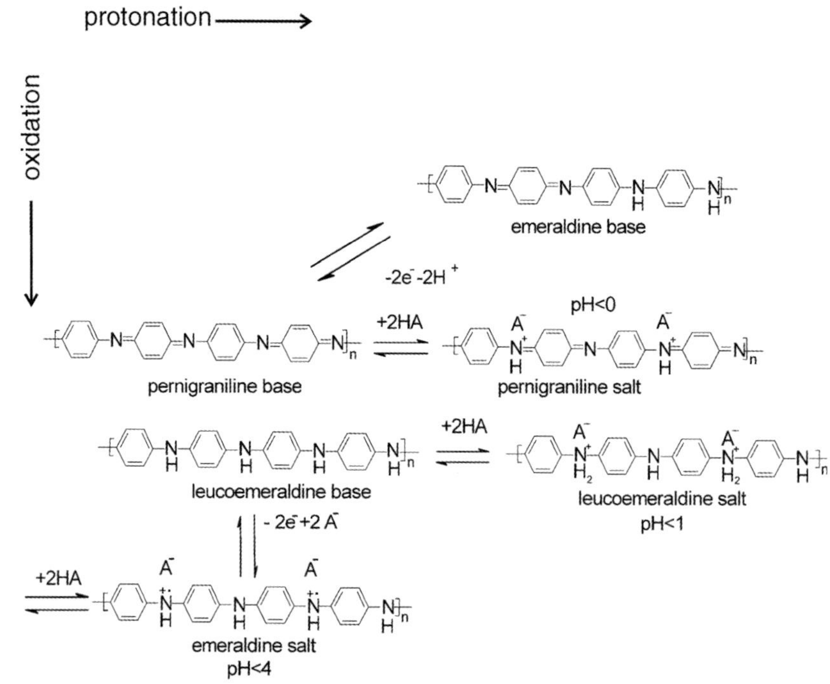

Figure 12.1 Conversion between different states of PANI. Reprinted with permission from Reference [1]. Copyright 2008 Elsevier.

There have been several recent reviews on the application of conducting polymers in sensors. One review [1] includes a broad discussion of applications as well as a design of such sensors and the use of combinatorial techniques for evaluation of sensor materials. A detailed review by Thomas et al. on chemical sensors based on amplifying fluorescent conjugated polymers has been published [11]. Detection of various analytes ranging from ions to proteins is discussed in this review. A short review on linear or macrocyclic polyether polythiophenes functionalized for cation detection and molecular actuation has been published [12]. The review describes an evolution of such receptors from simple linear polyethers towards more sophisticated systems with improved communication between the receptor and the conjugated polymer backbone. Other aspects are reviewed in References [13–16].

12.2
Transducers for Artificial Receptors Based on Conducting Polymers

General approaches to detect analyte binding to artificial receptors are based on the application of mechano-acoustical (quartz crystal microbalance, surface

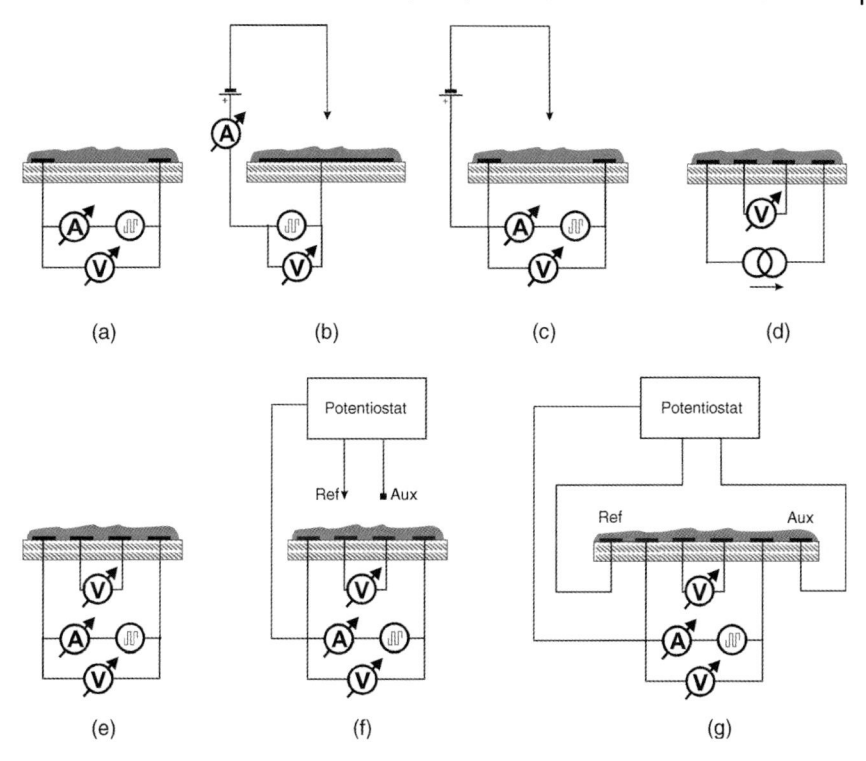

Figure 12.2 Main configurations used for measurements of resistance of conducting polymers: (a) two-point configuration without fixation of polymer potential; (b) typical configuration used in electrochemical experiments; (c) two-point configuration with fixation of polymer potential; (d) "classical" four-point technique with a current source; (e) s24-configuration providing simultaneous two- and four-point measurements without fixation of polymer potential; (f) s24-configuration with fixation of the electrode potential (six-point technique); (g) six-point technique realized at one chip. U. Lange and V. M. Mirsky, paper in preparation.

acoustic waves) or refractometric techniques. These approaches can also be used for CPs, but unique properties of these materials allow one to apply some other approaches which can be not applied for the most of other types of artificial receptors.

The most exciting property of CPs is their electrical conductivity, which can be switched by modification of the polymer redox state or by polymer protonation/deprotonation. Consequently, conductometric/impedometric transducing is the most common transducing technique used with CP. Several configurations have been used for such measurements (Figure 12.2). The simplest approach (Figure 12.2a, c) is based on measurements of lateral resistance. It can be performed without (Figure 12.2a) or with (Figure 12.2c) fixation of the polymer potential [17–30]. The measurement can be performed at DC mode by application of constant current or voltage. However, it may lead to some irreversible polymer damage (Q. Hao, V.

Kulikov, and V. M. Mirsky, unpublished). This can be minimized by limitation of the magnitude of the applied voltage below 100 mV or/and by measurements in AC or quasi-AC (application of symmetrical symmetric voltage pulses) mode. The measurement value is the sum of the polymer resistance between the electrodes and resistances of two electrode–polymer contacts. Typically, interdigitated electrodes are used for these measurements. A deposited polymer layer should overlap the gap between the electrodes. Electrochemists often use the measurement configuration presented in Figure 12.2b. It can be realized with standard electrochemical instrumentation. A polymer layer is deposited on the surface of a solid electrode. Impedance spectroscopy in such a configuration and subsequent analysis of equivalent circuits allows extraction of information on bulk polymer resistance [31–33]. However, this approach is complicated and in many cases does not provide useful information (it is not always possible to find a reasonable equivalent circuit), but it also has several advantages: (i) no electrodes of special design are needed, (ii) it can be realized on commercial electrochemical devices, and (iii) due to the high ratio of polymer area (the coated surface) to the distance between two conducting phases used for the measurement (solid electrode and electrolyte), it can be applied also to poor conductive polymers. The latter point is especially important because it essentially increases possible materials that can be tested. A contribution of contact polymer resistance to the measurement resistance value can be excluded by application of a four-point technique (Figure 12.2d–f); in classical configuration this is based on electrometrical measurement of the potential between two inner electrodes on application of a constant current between two outer electrodes [34, 35] (Figure 12.2d). This approach was improved by using quasi-AC voltage instead of constant current and by combination of two- and four-point configurations (simultaneous two- or four-point measurements or s24-technique) [36, 37]. This configuration can be realized with (Figure 12.2f) or without (Figure 12.2e) fixation of the electrode potential. The latter is typical for measurements in gas phases [36, 38]. The measurement with fixation of the electrode potential is performed in electrolytes [19, 25, 27, 30, 33, 39]. The potential fixation can be performed directly or by using of potentiostats. The configuration Figure 12.2f was recently realized with a solid-phase reference electrode, at one chip. This leads to a six-point measurement configuration (Figure 12.2g).

Another transducing principle often realized with CPs is potentiometric transducing. The intrinsic electrochemical activity of CPs provides a possibility to use them as an intermediate layer for coupling of electrical and ionic exchange at the interface between a metal and a selective membrane. Owing to ion-selectivity of some CPs, which can be either intrinsic or introduced through functional groups, they can also be used as these selective layers. Electrical current caused by oxidation/reduction of electrochemically active analytes can be detected by voltammetric transducing; in this case the selectivity is reached either by selective electrocatalytic effects or by particular electrochemical properties of the analyte. Chemosensitivity of a CP leads to chemosensitivity of diodes and transistors prepared on the basis of these materials. Alternatively, receptor layers based on CP can be deposited on the gate of usual (inorganic) field effect transistors.

A strong modification of electronic band structure of CP due to changes in its redox or protonation states leads to changes of optical spectra in UV and visual range. Changes of optical parameters can also be detected by surface plasmon resonance. Another approach can be based on Raman spectroscopy. In recent years nanocomposites of CP and metallic (Au, Ag, and other) nanoparticles have been studied intensively; in this respect an application of surface-enhanced Raman spectroscopy may be a very promising approach. Fluorescence measurements can be applied either directly, for detection of redox and protonation changes in fluorescent conducting polymers, or as detection schemes based on the energy transfer.

Table 12.1 presents examples of various transducing principles realized on the bases of conducting polymers.

Table 12.1 Examples of transducers for chemosensors with artificial receptors based on conducting polymers.

Transducing principle		Analyte	Polymer (+ receptor)	Reference
Electrochemical and electrical	Conductometric	pH	PPY/PANI	[40, 41]
		HCl	PANI	[36, 42]
		NH_3	PANI/PPY	[43–45]
		NO_2	PANI/P3HTH	[46–48]
		N_2H_4	PANI	[44]
		Cu^{2+}	Polycarbazole	[49]
		Hg^{2+}	PANI + cryptand	[50]
	Potentiometric	pH	PANI/PPY	[51–53]
		Glucose, saccharides	Poly(3-aminophenylboronic acid)	[54, 55]
		Ag^+	P3OTH, PEDOT + sulfonated thiophenes	[56, 57]
		Zn^{2+}	PPY + tetraphenylborate	[6]
		Cu^{2+}	PI, polycarbazole	[58]
		Ca^{2+}, Mg^{2+}	PPY + ATP	[3]
	Amperometric	$Cr_2O_7^{2-}$	PANI	[59]
		NH_4^+	PPY	[60]
		NO_2^-	PPY	[61]
	Voltammetric	Ag^+	PPY + Eriochrome B	[8]
		AA, dopamine	P3MTH, PEDOT, PANI	[62–66]
		Morphine	PEDOT + MIP polymer particles	[67]
		$NADH/NAD^+$	PANI + FAD,	[68]
		Serotonin, dopamine, uric acid	Overox.PPY + Au-nanoparticle	[69, 70]
		Chlorpromazine, dopamine, L-dopa	P3MTH + cyclodextrin	[71]
	Impediometric	K^+, NH_4^+ in presence of Na^+	PPY, PANI, PEDOT	[9, 72]

(Continued)

Table 12.1 (Continued)

Transducing principle		Analyte	Polymer (+ receptor)	Reference
	Chemotransistors and FETs		PANI, PPY	[73–78]
	Chemodiodes	NO_x	P3OTH, PPY	[79–81]
		Methane	PANI	[82]
		Ammonia	P3OTH, PANI	[79, 83]
		Alcohol and water vapors	P3OTH	[79]
Optical	UV-vis	pH	PANI	[84–88]
		Ozone	PANI; m-chloropolyaniline	[89, 90]
		ATP	Poly[3-(3'-N,N,N-trimethylamino-1'-propyloxy)-4-methyl-2-5-thiophene hydrochloride]	[91]
	Near-infrared spectroscopy	Saccharides	Poly(3-aminophenylboronic acid)/PANI copolymer	[92]
		pH	PPY, PANI	[86, 93–95]
	Fluorescence	Nitroaromatics	PPEs/PPVs/PPs	[96, 97]
		Hg^{2+}	PMNT + mercury specific oligonucleotide	[98]
		ATP	Poly[3-(3'-N,N,N-trimethylamino-1'-propyloxy)-4-methyl-2-5-thiophene hydrochloride]	[91]
	SPR	HCl	PANI	[99]
		H_2S, NO_2	PANI	[100]
		H_2O_2	PANI + HRP	[101]
	Raman spectroscopy	pH	PANI	[102]

12.3
Intrinsic Sensitivity of Conducting Polymers

12.3.1
Sensitivity to pH Changes

The sensitivity of CPs to pH changes is based on their backbone structure, that is, on the availability of acidic/basic groups that can be protonated/deprotonated. Thus, a shift of pH leads to a redistribution (uptake or release) of the amount of charge carried along the polymer and hence the change of conductivity and optical properties. As conductivity can be of ionic and/or electronic nature, CPs differ in the influence of pH on their redox properties.

The oxidation or reduction of PPY (polypyrrole) is not coupled with the proton exchange (release/uptake). Simultaneously, the protonation enhances the conductivity of PPY, whereas deprotonation decreases conductivity [103]. The pK_a of the PPY forms is in the range from 2 to 4 and 9 to 11 [93, 103] and its behavior follows the same law as for regular electrodes of second type with the proton as a potential determining ion [51].

In contrast to PPY, the pK_a of PANI and PI is very sensitive to the redox state and a description of pH equilibria is more complicated [104]. The influence of pH on the redox properties of PANI explored in Reference [104] results in a shift of potential of the second redox transition of 0.12 V per pH unit. This is explained by the release of protons in the redox conversion from emeraldine salt into pernigraniline base (Figure 12.1).

Both PPY and PANI show very high proton permeability in aqueous solution [51–53]: films of these polymers on glassy carbon and platinum electrodes demonstrate almost nernstian behavior relative to pH. High proton permeability and the electronic component of conductivity make PANI and PPY perspective materials for solid-state pH sensors development (Table 12.2).

The spectral changes of PANI due to protonation between pH 5 and pH 8 are observed in the visible range [53]. However, the working range of optical pH sensors based on PANI is limited to pH 3–6 by the hysteresis of transition between PANI forms [53]. Optical sensors are usually prepared by the deposition of CP onto an optical transparent support (e.g., ITO electrodes, microtiter plates, cuvettes [86], optical fiber [105]) followed by the registration of adsorption at

Table 12.2 Selected examples of pH sensors based on CPs.

pH range	Polymer/polymerization method	Type of transducing	Reference
5–10	PPY/chemical	Conductometric	[106]
4–10	PPY/electrochemical	Potentiometric	[107]
2–11		Potentiometric	[52]
2–12	PANI/chemical	Spectrophotometric	[84]
3–6	PANI/electrochemical	Spectrophotometric	[102]
1–13	PPY or PANI-CNTs[a]/electrochemical	Potentiometric	[108, 109]
2–9 9–12	PANI, aniline–anthranilic acid copolymer	QCM, spectrophotometric	[110]
2–11	PANI-RVB-PS3[b]/screen-printing/deep coating	Conductometric	[111]
3–10	PANI, poly(o-methoxyaniline)/electrochemical	Potentiometric	[112]

a) CNT = carbon nanotube.
b) RVB-PS3 = poly(vinyl butyral), hypermer surfactant.

800–900 nm [86, 94, 95], or at 500–600 nm [84–88] in the case of PANI, and at 650 nm with PPY [93].

12.3.2
Affinity to Inorganic Ions

Complexation of metal ions by some polymer moieties explains the affinity of these polymers to ions [7, 113]. However, the examples of selective responses are not numerous, because a polymer should have a certain structure to be a strong ligand for ions. In most cases the selectivity and efficiency are enhanced by synthetic receptors introduced into the polymer backbone (Section 12.4)

A selective potentiometric response to Cu^{2+} has been reported for films of PI and polycarbazole [58]. The complexation of Cu^{2+} to polycarbazole enhances the conductivity of the polymer, presumably by changing of the polymer conformation from a compact coil to a higher conducting expanded coil [49]. Selective nernstian response towards Ag^+ ions was found with films of poly-3-octylthiophene [7].

12.3.3
Affinity to Gases and Vapors

Chemical interaction or physical adsorption of a gaseous analyte provides many possibilities for chemosensor construction [114]. Chemical reactions, oxidation/reduction and protonation/deprotonation, change the amount of charge carries in the polymer and, therefore, alter the physical properties of polymers (conductivity and optical adsorption). CPs can be oxidized by gaseous analytes with higher electron affinity like NO_2, I_2, O_3, and O_2 [46–48, 89, 90, 115]. NO_2 interacts with PANI [47] and poly-3-octylthiophene [46] and decreases their resistivity by oxidative doping. A decrease of PPY resistance was observed also with SO_2 [115], having similar mechanism as with NO_2. Optical detection of ozone can be performed with PANI, PNMA [poly(N-methylaniline)], and m-chloropolyaniline, which change their adsorption in the range 500–800 nm after oxidation and protonation with analyte [89, 90]. A partial charge transfer explains the quenching of fluorescence of PPV, PPE, and PP in the presence of nitroaromatics [96, 97]. An interaction of CO with PANI is also considered as an example of partial electron transfer from the polymer [116].

A decrease of CP conductivity is observed by exposure to electron-donating gases like H_2S [117], NH_3 [118, 119], and N_2H_4 [44, 120–122], which reduce and dedope PANI, PPY, and P3HTH (poly(3-octyl-thiophene)).

Several conductometric sensors for HCl and NH_3 [43, 44] vapors based on PANI, PPY [36, 38], or multilayers of PANI with EDOT (ethylenedioxythiophene) [42] have been explored. The operating principle of these sensors is protonation of CP leading to the change of conductivity.

Changes of CP resistance were also observed after exposure of PANI, PPY, PTH (polythiophene), and polythiophene derivatives films [43, 44, 76, 123] to volatile organic vapors (chloroform, acetone, aliphatic alcohols, benzene, toluene, etc.), though the mechanism was not studied in detail. A change of the potential barrier at the boundary between the polymer grains [124] or conformation of polymer chains [44, 125], a change of the dielectric constant of CP [126], the influence on electron jumping along the CP chain [124], or a change in several charge carries [127] are assumed to explain the sensitivity of CP to the organic vapors.

However, the high sensitivity to humidity and low selectivity remain a main drawback of such gas sensors. To enhance the operating of these sensors, composites of CP and metal oxides [46, 128–133] and nanoparticles and carbon nanotubes [134–136] have been proposed and tested.

12.4
Conducting Polymers Modified with Receptor Groups

Receptors based on conducting polymers offer several advantages over monomer based receptors. It was described in Reference [14] that using multiple receptors on a CP-wire leads to a signal amplification in comparison to single receptors, especially if charge transport properties are measured. The reason for this amplification is the collective system response (Figure 12.3): as electrons, holes, or excitations can move along the conjugated π-system, a binding event on just one receptor will influence the properties (resistance, fluorescence) of the whole strain.

Swager et al. have demonstrated that a conjugated polymeric receptor for methyl viologen shows a 65-fold signal amplification in comparison to the monomer based receptor. The amplification depends on the molecular weight. Higher molecular weight gives a higher amplification; however, only up to a certain limit.

Another advantage of conducting polymer based receptors is that the polymer can work not only as receptor but also as transducer. As binding events can lead to changes in resistance, fluorescence, absorption, work function, and so on [1], the most suitable detection method regarding possible applications can be used for sensor design.

12.4.1
Conducting Polymers with Receptor Groups Attached to the Monomer

12.4.1.1 Receptors for Ions
The most frequently used receptor group attached to the conjugated backbone of conducting polymers is various crown ethers. Complexing of cations by these crown ethers leads to different effects, like, for example, changes in redox potential or optical absorption, depending on how the crown ether was attached to the backbone. Crown ether functionalized polythiophenes have been reviewed recently [12] and therefore are

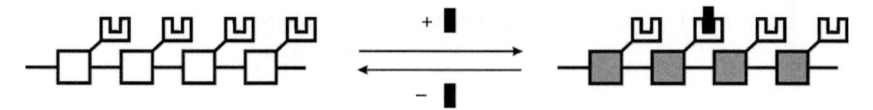

Figure 12.3 Cooperative signal amplification in receptor-modified conducting polymers.

only discussed briefly here. Polythiophenes with crown ethers attached via an alkyl spacer at the 3 position of the ring show a positive shift of the anodic peak potential [137], while crown ethers attached via an alkoxy spacer show a redshift of the absorption maxima upon complexing alkali metals [138]. Better communication between the complexing site and the π-conjugated system was achieved by attaching the polyether loop between the 3 and 4 positions of the thiophene ring [139, 140]. Such systems showed a large positive shift of the anodic peak potential upon complexing Na^+.

16-C5-ether rings have been attached either coplanar or perpendicular to a cyclopentabithiophene precursor [141]. Only the former shows an anodic shift upon complexing Na^+, whereas the later is quite insensitive. If crown ethers are attached to the conjugated backbone via two or more thiophene rings (Scheme 12.1), complexing of metal ions leads to conformational changes of the polymer backbone and therefore to changes in the optical and electrochemical response [142, 143].

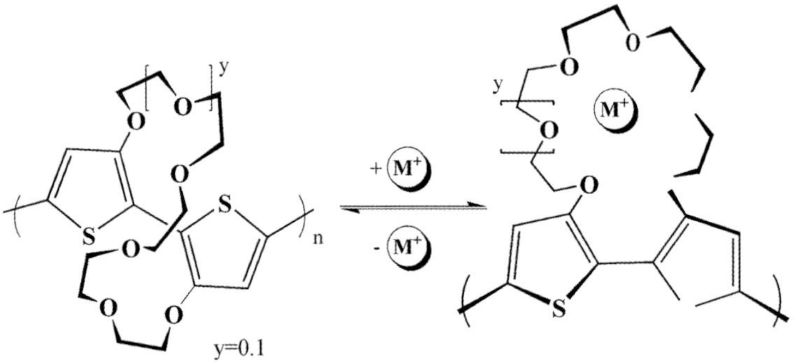

Scheme 12.1

A further improvement was achieved by attaching a calix[4]arene to a bithiophene monomer instead of a crown ether (Scheme 12.2) [144]. The resulting polymer

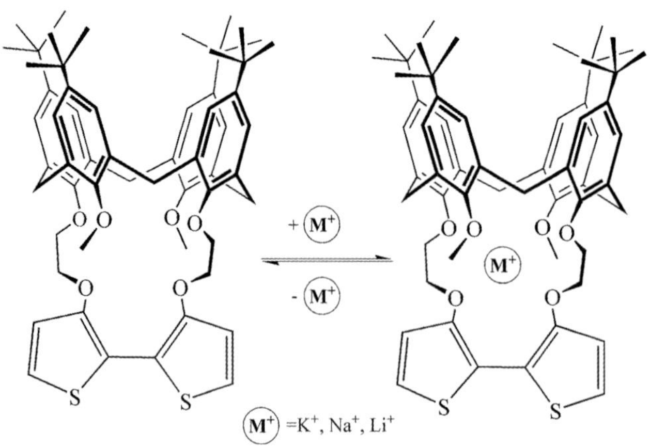

$(M^+) = K^+, Na^+, Li^+$

Scheme 12.2

12.4 Conducting Polymers Modified with Receptor Groups

showed a strong optical and resistive response towards Na^+, with good selectivity compared to Li^+ and K^+. Binding of Na^+ leads to conformational changes, leading to a higher conjugation length. Furthermore, the complexation leads to a higher oxidation potential of the occupied binding sites, due to electrostatic repulsion and decreased electronic donation from the oxygen lone-pairs.

Another way to create complexing sites within a polythiophene film is to polymerize a precursor consisting of two polymerizable groups linked by a flexible polyether chain. In this way complexing cavities are created within the film. Again a positive shift of the anodic peak potential is observed upon coordination of monovalent or divalent cations.

A quite different response upon coordination of K^+ ions is used by the crown ether functionalized polymer (Scheme 12.3) [145]. If K^+ ions are present in solutions they lead to an aggregation of the single polymer strains as shown in Scheme 12.3.

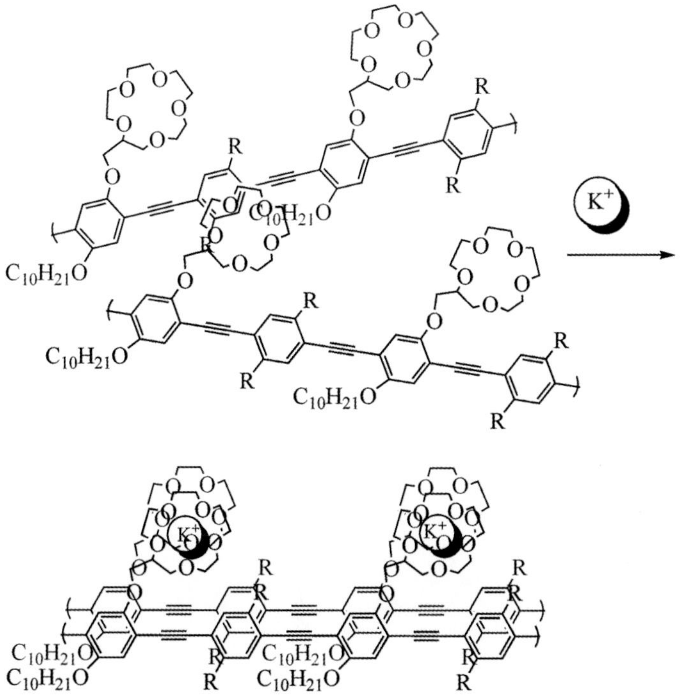

Scheme 12.3

The fluorescence of the crown ether functionalized poly(phenylenevinylene) **1** decreased upon addition of alkali metals [146]. No selectivity was found for the different alkali metals.

[Structure 1: poly(phenylenevinylene) with crown-6 ether groups]

A similar poly(phenylenevinylene) with a crown 5 ether instead of a crown 6 ether showed an increase in photoluminescence upon addition of Na^+, Ca^{2+}, and Eu^{3+}. The strongest increase was observed for Eu^{3+}, the weakest for Ca^{2+} [147].

A crown ether functionalized polypyrrole shows some voltammetric response towards Li^+ and Na^+ cations; however, sensitivity was quite low [148].

Polymerization of a pyrrole substituted crown ether ferrocene led to a polymer that showed a voltammetric response towards Ba^{2+} and Ca^{2+} ions [149, 150].

Polymers **2a** and **2b** showed a decrease of their fluorescence upon addition of alkali metals. The strongest decrease was observed for Li^+, the weakest for K^+; **2a** had a higher selectivity towards Li^+ than **2b**.

[Structures 2a and 2b]

Bipyridines are another functional group that can be incorporated into a conducting polymer and show some receptor properties [151–157]. In most cases the bipyridine groups are directly incorporated in the conjugated backbone of the polymer by coupling the bipyridine unit to fluorene or phenylenevinylene units; however, some oligopyridines have also been linked via a conjugated or a nonconjugated spacer to the polymer backbone [154]. Coordination of transitions metals by the bipyridine unit leads to conformational changes (flattening) of the conjugated π-system and therefore to changes in their optical absorption and emission. The influence of the polymer π-system rigidity on sensitivity towards transition metals has been investigated [153]; it was concluded that lower rigidity leads to a higher sensitivity. A bipyridine modified poly(phenylenevinylene) (**3**), treated with Cu^{2+} ions, has been used as a NO sensor. NO reduced the complexed Cu^{2+} ions to Cu^{+} ions, which led to a turn on in fluorescence [157].

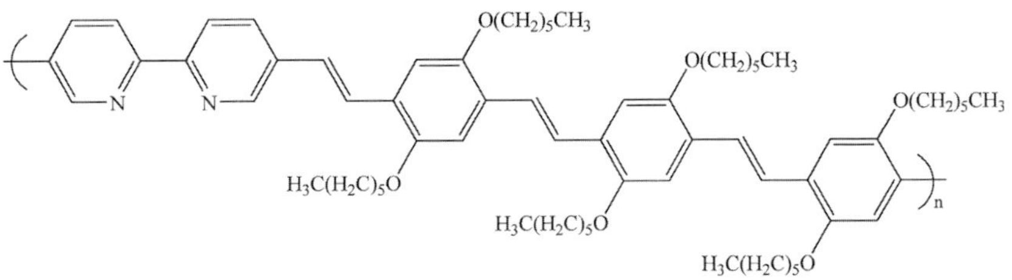

3

A series of conjugated polymers containing bipyridine units has been tested towards their fluorescence response upon addition of Mg^{2+} ions. The fluorescence of polymer **4c** was quenched most effectively by Mg^{2+} binding, whereas no response was observed for polymer **4d** and a weaker response for polymers **4a** and (**4b**).

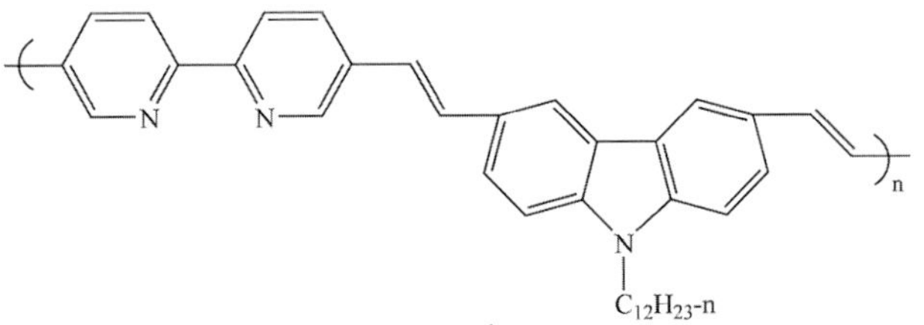

4a

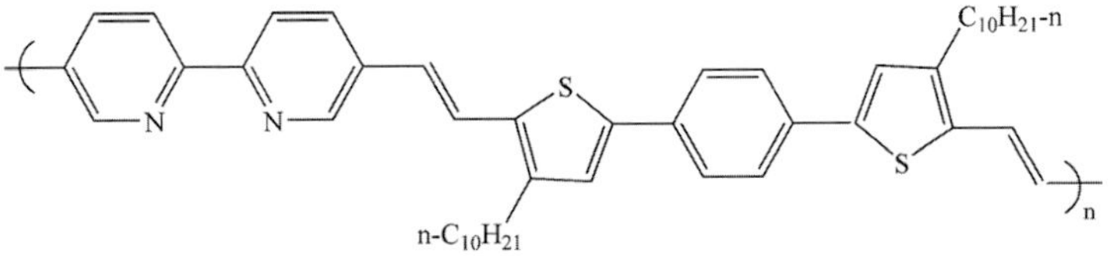

4b

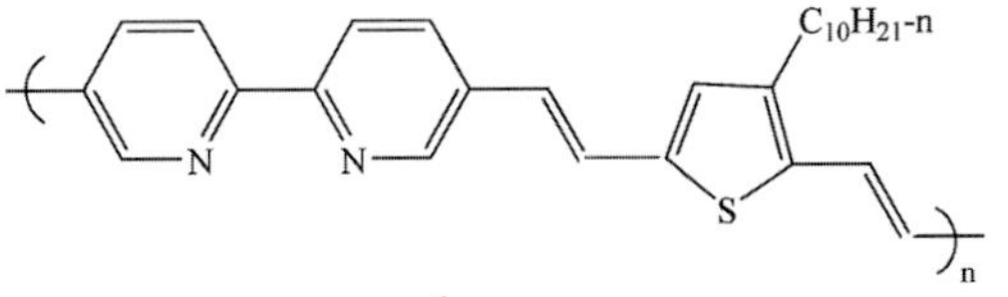

4c

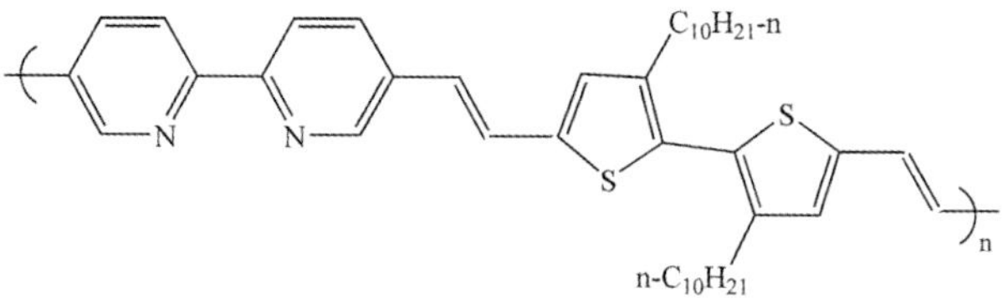

4d

Similar to bipyridine groups, phenanthroline groups can be incorporated in the polymer [158, 159]. Such polymers showed various changes of absorption and emission on adding transition metals and lanthanides. A comparative study has demonstrated that the more rigid phenanthroline unit shows a higher sensitivity towards transition metal ions than the twisted bipyridyl unit [159].

The influence of protons, metal ions, and solvents on the optical absorption and photoluminescence of 1,10-phenanthroline units containing polyphenylene and pyridine units containing polyphenylene has been investigated in [160]. Protonation leads for both polymers to a strong decrease in fluorescence. Both polymers are able to bind transition metal ions, resulting in polymer fluorescence quenching.

Similar behavior was found for the polymers **5a** and **5b**, which consist of alternating pyridine and phenylene units. Again protonation and binding of metal ions lead to fluorescence quenching [161].

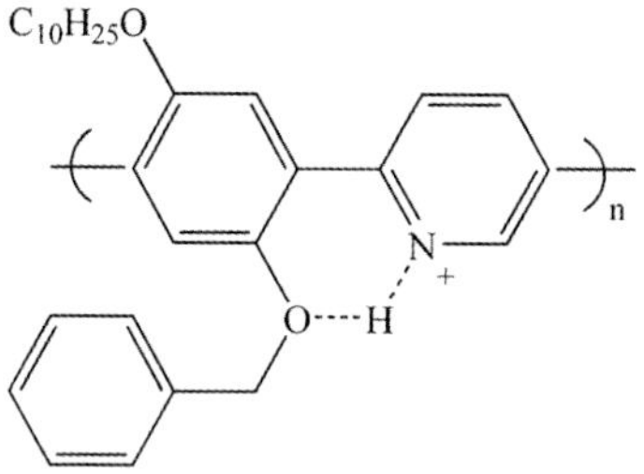

5a 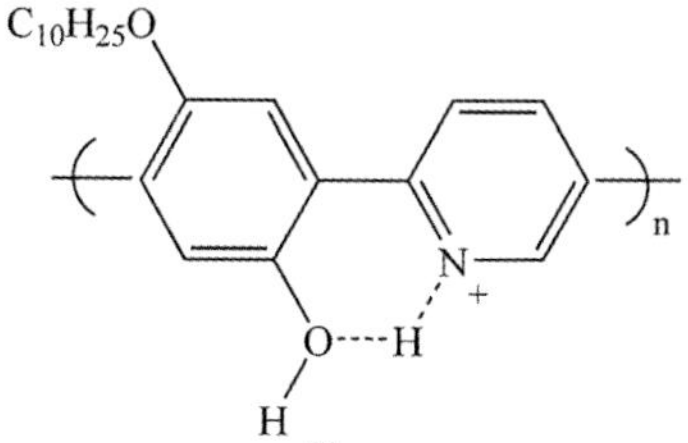   **5b**

A similar receptor domain was used in polymer **6** [162]. Additionally it contained a polyphenylene-ethynylene signaling domain, for efficient signal amplification. Upon addition of Pd^{2+} ions the fluorescence of the polymer was quenched. Polymers consisting of only one domain, either the receptor or the signaling domain, have a much lower Stern–Volmer constant.

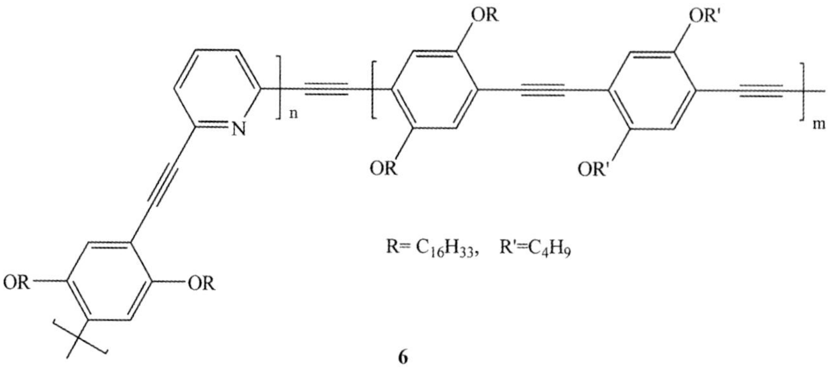

$R = C_{16}H_{33}$, $R' = C_4H_9$

6

12.4 Conducting Polymers Modified with Receptor Groups

Bipyridine and phenanthroline groups have been used to design a polymetallorotaxane, with could be deposited as a film on interdigitated electrodes [163, 164]. As the oxidation potential of the resulting polymer matches the potential of the redox couple Cu^{2+}/Cu^{+}, binding of Cu^{2+} ions leads to oxidation of the polymer, resulting in a strong drop of its resistance.

The influence of transition metals on the fluorescence of polymers **7a** and **7b**, which contain bipyridinophane units, has been investigated in THF solution [165]. The fluorescence of both polymers was quenched by Cu^{2+}, Co^{2+}, and Ni^{2+} ions. The Stern–Volmer constant was larger for **7a**; however, it has a lower selectivity as its fluorescence was also quenched by Ag^{+}, Mn^{2+}, and Zn^{2+}.

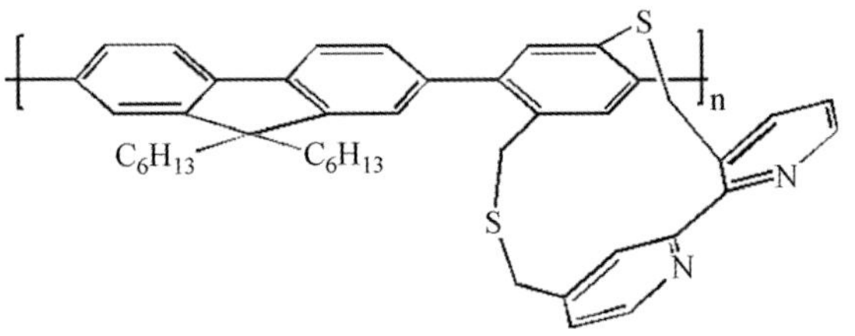

7a

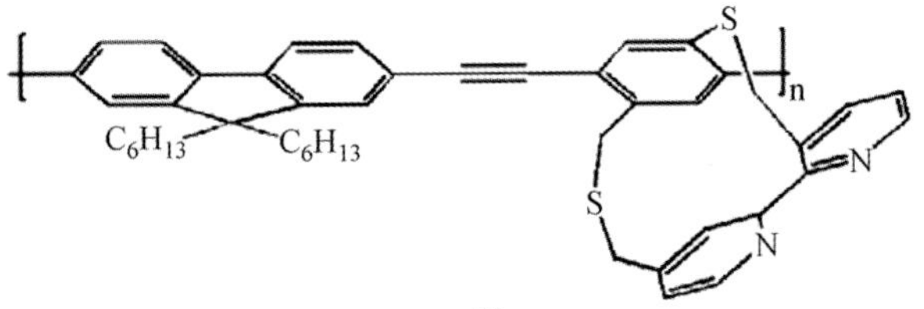

7b

A thymine modified polythiophene has been used as a probe for mercury ions detection [166]. Upon addition of Hg^{2+} to a solution of the polymer in DMSO–EtOH (1: 4) aggregation of the polymer strains occurred, because of the specific thymine-Hg^{2+}-thymine base pair formation. The aggregation was accompanied by a decrease in fluorescence intensity.

Polyfluorenes with phosphonate side chains of variable length (**8a,8b**) have been tested for their sensing properties towards various metal cations. It was found that the polymers' fluorescence is sensitive and selective for Fe^{3+} with an emission quenching up to 210-fold upon addition of Fe^{3+}. The quenching was much more efficient for polymers **8a** and **8b** than for the corresponding monomer model compounds **8c** and **8d**.

The effect of a wide variety of monovalent metal ions (K^+, Na^+, Cu^+, Ag^+) and bivalent metal ions (Zn^{2+}, Cu^{2+}, Cd^{2+}, Pd^{2+}, Hg^{2+}, Ni^{2+}, Co^{2+}, Ca^{2+}) on the fluorescence intensity of **9a** and **9b** at the emission maxima have been studied. Upon addition of most metal ions, they show almost no change in their emission spectra. In contrast, addition of Ag^+ ion to a solution of **9a** and **9b** leads to significant fluorescence quenching. Moreover, **9b** responds much more strongly to the presence of Ag^+ ion than its monomer model compound **9a**. The reason for the quenching is probably due to interpolymer aggregation upon Ag^+ addition, as Ag^+-induced quenching is much less efficient for **9c** ($K_{sv} = 3.9 \times 10^4$ M^{-1}) than **9b** ($K_{sv} = 1.4 \times 10^5$

M^{-1}), which indicates that the bulky side groups stifle interpolymer aggregation and lower the response to silver ion.

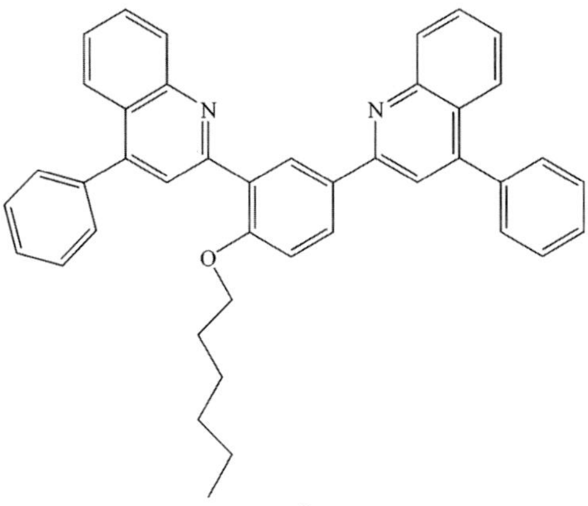

9a

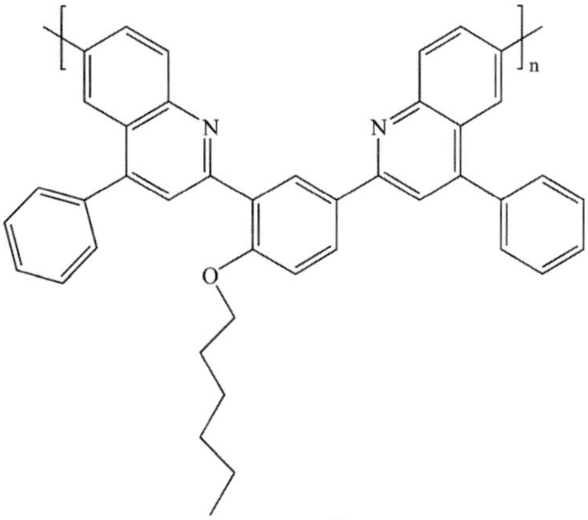

9b

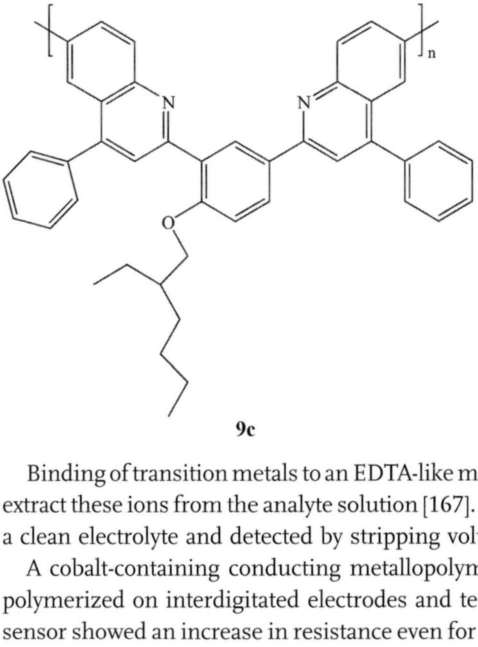

9c

Binding of transition metals to an EDTA-like modified polypyrrole has been used to extract these ions from the analyte solution [167]. Subsequently, they were reduced in a clean electrolyte and detected by stripping voltammetry.

A cobalt-containing conducting metallopolymer (Figure 12.4) has been electropolymerized on interdigitated electrodes and tested as a sensor for NO [168]. The sensor showed an increase in resistance even for concentrations as low as 1 ppm. No reactivity towards O_2, CO, and CO_2 was observed; however, some irreversible resistance decrease was found on exposure to NO_2.

The fluorescence of the carboxylate modified polyphenylene-ethynylene **10a** is quenched in the presence of Pb^{2+} ions [169]. It was shown that the Stern–Volmer constant of the polymer is much higher than that of the monomer model compound **10b**. If papain is added to a solution of this polymer, the resulting complex shows a high sensitivity towards mercury ions. In the presence of mercury

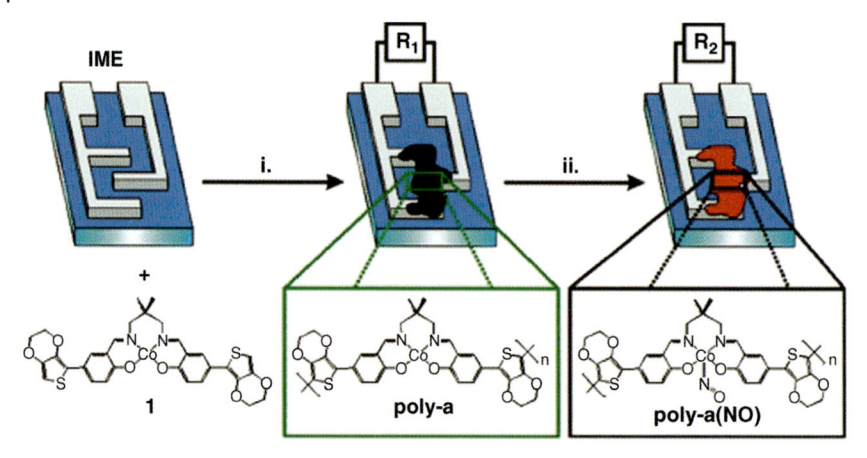

Figure 12.4 Conductometric detection of NO binding to a cobalt-containing metallopolymer. Reprinted with permission from Reference [168]. Copyright 2006 The American Chemical Society.

ions aggregation of the complexes takes place, resulting in a strong decrease of fluorescence [170].

The fluorescence quenching of polymers **11a** and **11b** upon addition of different mercury and lead salts has been compared [171]. It was found that polymer 2, in which a sugar was linked via a triethylene glycol spacer to the polymer, shows the largest Stern–Volmer constant.

The fluorescence of imidazole-functionalized polyfluorenes was strongly quenched by traces of Cu^{2+} ions [172, 173]. It was suggested that the imidazole groups show some affinity towards transition metal ions, leading to an energy transfer from the conjugated polymer backbone over the imidazole groups to the copper ions. This system was also used to detect cyanide ions, which show a stronger affinity towards the copper ions then the imidazole group does [172]. Therefore, the presence of cyanide ions leads to a removal of the copper ions from the polymer and a "turn on" of the fluorescence.

In contrast to conducting polymer based receptors for cations, far fewer studies have dealt with such receptors for anions. A quite selective voltammetric sensor for fluoride has been achieved using a boronic acid functionalized polypyrrole [174]. Binding of fluoride leads to a 250 mV decrease of the potential of the oxidation–reduction couple of the polymer. The other halides showed no effect.

A cationic poly(3-alkoxy-4-methyl-thiophene) has been used as a selective probe for iodine detection [175]. The presence of iodine anions leads to aggregation and planarization of the polymer chains, resulting in changes of the optical absorption and emission.

A sensor for phosphate-type anions has been designed by modifying a quinoxaline with two pyrrole units [176]. As this monomer does not polymerize, EDOT monomers are coupled to it. This so-modified new monomer can be electropolymerized on ITO electrodes, yielding an optical sensor for phosphate anions, which also responded to fluoride but shows no sensitivity to chloride. Upon complexing phosphate ions the absorption at around 620 nm increases, while it decreases between 800 and 1100 nm. This behavior was also dependent on the polymer oxidation state: The binding constant was higher at 0.7 V compared to at 0.0 V. A very similar polymer (12) has been synthesized using phenylene ethynylene monomers instead of EDOT. The quenching constants of fluoride and phosphate anions were evaluated [177]. It was stated that, upon anion complexing, one of the pyrrole units on the quinoxaline is deprotonated, resulting in the generation of non-fluorescent trapping sites, which quench the fluorescence of the excited state.

12

A double strapped porphyrin (Scheme 12.4) contains two small cavities that are able to bind fluoride [178]. Binding of fluoride leads to a decrease of the absorption

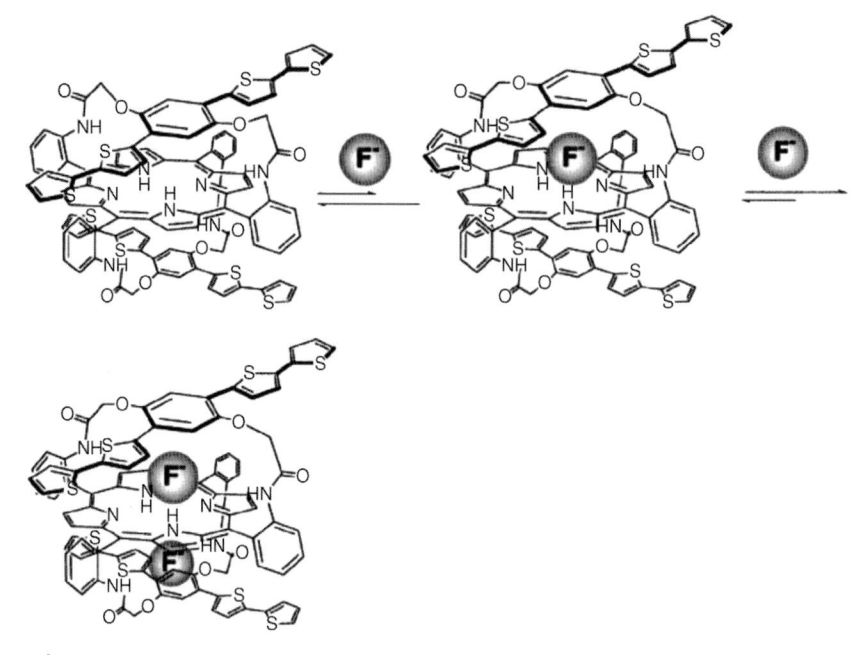

Scheme 12.4

at 424 nm. The absorption change upon fluoride shows a sigmoidal behavior, which indicates a cooperative binding of fluoride. This monomer can be electropolymerized on interdigitated electrodes. Such films show a 50-fold decrease in conductivity and a shift of the porphyrin redox couple to lower voltages upon binding of fluoride.

The sensing behavior of a series of conjugated polymers (**13a,b**) towards various anions that is, F^-, Cl^-, Br^-, I^-, BF_4^-, PF_6^-, and $H_2PO_4^-$ has been tested [179]. The polymers showed a selective colorimetric response towards F^- and $H_2PO_4^-$. Their fluorescence is quenched more effectively by F^- than by $H_2PO_4^-$ for **i–iv** and **vi–viii**; however, polymers **i** and **ii** have a higher selectivity towards F^- than **vi–viii**. The highest selectivity was found for **iii**.

(i) x=0,
(ii) x=0.2
(iii) x=0.4
(iv) x=0.6
(v) x=1

$R_1=C_8H_7$
$R_2=C_4H_9$

13a

(vi) x=0.2
(vii) x=0.4
(viii) x=0.6
(ix) x=1

$R_1 = C_8H_7$
$R_2 = C_4H_9$

13b

12.4.1.2 Receptors for Organic/Bioorganic Molecules

Conducting polymer based receptors for organic and bioorganic molecules make use of various interactions like ionic interactions, π–π interactions, covalent binding, or complexation.

For ionic interactions, cationic polythiophenes [91, 180–187] are often used. Upon binding negatively charged analytes containing phosphate groups (ssDNA, ATP) the conjugated backbone of these polymers is planarized, leading to π–π stacking between the formed complexes and an efficient fluorescence quenching. If an ssDNA bound to this polymer hybridizes with a complementary DNA strain the polymer changes again its conformation and fluorescence is turned on.

Modification of such a cationic polythiophene with a ferrocene group leads to a probe for voltammetric detection of analyte binding [185]. If a receptor modified electrode surface changes its charge due to analyte binding from either negative to neutral or the other way round, binding of the probe to the surface is changed, leading to a difference in the voltammetric signal. Such polymers can also be used to create surfaces for a more specific receptor (DNA, antibody) immobilization. Binding of the receptors and analytes to such surfaces can be detected using a surface plasmon resonance technique [180].

Conducting polymers modified with boronic acid groups can be used as receptor for organic compounds containing diols. Binding of sugars to poly(amino-phenyl-boronic-acid) causes optical and electrical changes in the polymer, which can be detected by NIR spectroscopy, potentiometry, and impedance spectroscopy [54, 55, 92, 188–191].

Polyphenylenes, polyphenylene-vinylenes, and polyphenylene-ethynylenes have been suggested as sensors for nitroaromatic explosives [192–194]. The nitroaromatics bind to the polymer via a π–π-complex, allowing a charge transfer from the excited polymer to the analyte, resulting in fluorescence quenching. The introduction of rigid pentiptycene moieties into the conjugated polymer backbone prevented π-stacking of the polymer chains, resulting in a lower self-quenching of the polymer film, and created cavities for fast diffusion of the analytes in the film.

Carbohydrate functionalized fluorescent conducting polymers can be used to stain bacteria and other pathogens [195–198]. The pathogens can bind to the carbohydrates bound to the polymer and form fluorescent aggregates or quench the fluorescence of the polymer [198]. As polymers, polythiophene, polyphenylenes, and polyphenylene-ethynylene were used.

14

A chiral carbohydrate functionalized polyphenylene-ethynylene (**14**) shows some selectivity between (−)-menthol and (+)-menthol [199]. A water-soluble folate substituted polyphenylene-ethynylene has been used for cancer cell imaging [200]. The folate was used to deliver the polymer to the cells. The polymer targets and images cancer cells *in vitro* with high selectivity.

12.4.2
Conducting Polymers Doped with Receptor

Charged receptors or receptors becoming charged after binding the analyte can be incorporated into CPs by electrostatic doping. To interact with the polymer chain after polymerization, the receptors should compete with counterions, which can be already bound to the backbone. To alleviate this interference, the receptor is used as counterion during polymerization. A PEDOT film prepared by galvanostatic polymerization in the presence of 7,8,9,10,11,12-hexabromocarborane ($Ag^+ CB_{11}H_6Br_6^-$) shows a selective potentiometric for Ag^+ ions compared to several alkali, alkaline-earth, and transition-metal cations [5]. The accumulation of Ag^+ into polymer film was confirmed by cyclic voltammetry. The selective binding of Hg^{2+} ions has been detected with PANI modified by cryptand-222 sensor (Scheme 12.5) [201]. Modification of PANI results in a deprotonation of the polymer backbone by receptor, whereas analyte binding causes deprotonation of receptor and electrostatic interaction of the receptor–analyte complex with the PANI chain. The response of the sensor is pH dependent and maximal at pH 2; the maximal sensitivity

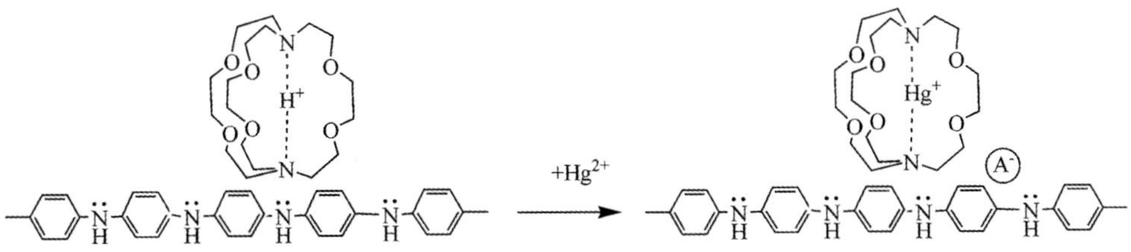

Scheme 12.5

is attained at pH 4. Tetraphenylborate ion-doped PPY shows some sensitivity to Zn^{+2} ions, which is dependent on polymeric microstructure [6].

12.4.3
Molecular Imprinting of Conducting Polymers

Molecularly imprinted synthesis of conducting polymers provides another possibility to modify chemosensitive properties of CP. This approach was developed independently by several groups in 1998–1999 [202–206] and then reproduced in many other laboratories [189, 207–216]. Films of conjugated polymers were formed by electropolymerization of pyrrole [203, 211], aniline [213], and anilineboronic acid [189] and probably also by polymerization of porphyrin [205] and dyes [216]. Analyte detection was performed by voltammetry, measurements of electrical capacitance or by quartz crystal microbalance. Advantages of applications of CPs for fabrication of artificial receptors by molecular imprinting are a simple control of the layer thickness (by electropolymerization) and an electrical access to analyte, allowing the use of electrochemical detection. However, an application of "classical" polymers based on acrylic compounds with different functional groups provides much higher flexibility in the selection of polymerizable monomers and crosslinkers.

12.5
Conclusion

Although this chapter has focused on application of CPs (conducting polymers) as artificial receptors, the multifunctionality of these materials does not allow us to separate receptor properties from transducer properties. A wide multifunctionality of CPs results in very different possibilities in applications of CPs in chemical and biological sensors. CPs can be used for creation of each component of a chemosensor, including a receptor layer, a transducer, a protective coating, or even an electronic circuit for data proceeding.

One can distinguish intrinsic and induced receptor properties of CPs. Intrinsic receptor properties are based on chemosensitivity of the polymer backbone. Induced receptor properties can also be formed by pre- or post-synthetic chemical modifications of polymers, by noncovalent immobilization of receptors, or by molecularly imprinted polymerization. The main drawbacks of CP as receptors for chemical sensors are poor stability of some types of CP in air or in aqueous solutions, detachment of CP from support during storage or repetitive drying, and a strong sensitivity to small variations of conditions during synthesis and treatment. However, intensive development of this field allows us to consider the prospects for development of CPs for applications in chemical sensors as very promising.

List of Abbreviations

CP	conducting polymer(s)
EDOT	ethylenedioxythiophene
P3HTH	poly(3-octyl-thiophene)
PANI	polyaniline
PPY	polypyrrole
PTH	polythiophene

References

1 Lange, U., Roznyatovskaya, N.V., and Mirsky, V.M. (2008) *Anal. Chim. Acta*, **614**, 1–26.
2 Migdalski, J., Blaz, T., and Lewenstam, A. (1996) *Anal. Chim. Acta*, **322**, 141–149.
3 Migdalski, J., Blaz, T., Paczosa, B., and Lewenstam, A. (2003) *Microchim. Acta*, **143**, 177–185.
4 Mousavi, Z., Bobacka, J., and Ivaska, A. (2005) *Electroanalysis*, **17**, 1609–1615.
5 Mousavi, Z., Bobacka, J., Lewenstam, A., and Ivaska, A. (2006) *J. Electroanal. Chem.*, **593**, 219–226.
6 Pandey, P.C., Singh, G., and Srivastava, P.K. (2002) *Electroanalysis*, **14**, 427–432.
7 Vazquez, M., Bobacka, J., Luostarinen, M., Rissanen, K., Lewenstam, A., and Ivaska, A. (2005) *J. Solid State Electrochem.*, **9**, 312–319.
8 Zanganeh, A.R. and Amini, M.K. (2007) *Electrochim. Acta*, **52**, 3822–3830.
9 Cortina-Puig, M., Munoz-Berbel, X., del Valle, M., Munoz, F.J., and Alonso-Lomillo, M.A. (2007) *Anal. Chim. Acta*, **597**, 231–237.
10 Goddard, J.M. and Hotchkiss, J.H. (2007) *Prog. Polym. Sci.*, **32**, 698–725.
11 Thomas, S.W., Joly, G.D., and Swager, T.M. (2007) *Chem. Rev.*, **107**, 1339–1386.
12 Demeter, D., Blanchard, P., Grosu, I., and Roncali, J. (2008) *J. Inclusion Phenom. Macrocyclic Chem.*, **61**, 227–239.
13 Zheng, J. and Swager, T.M. (2005) Poly (arylene ethynylene)s in chemosensing and biosensing, in *Poly(arylene Etynylene)s* (ed. C. Weder), Springer, Berlin, Heidelberg.
14 Swager, T.M. (1998) *Acc. Chem. Res.*, **31**, 201–207.
15 Leclerc, M. (1999) *Adv. Mater.*, **11**, 1491–1498.
16 Fan, L.J., Zhang, Y., Murphy, C.B., Angell, S.E., Parker, M.F.L., Flynn, B.R., and Jones, J. (2009) *Coord. Chem. Rev.*, **253**, 410–422.
17 Yadav, K.L., Narula, A.K., Singh, R., and Chandra, S. (2001) *Appl. Biochem. Biotechnol.*, **96**, 119–124.
18 Narula, A., Singh, R., and Chandra, S. (2000) *Bull. Mater. Sci.*, **23**, 227–232.
19 Holze, R. and Lippe, J. (1990) *Synth. Met.*, **38**, 99–105.
20 Kankare, J. and Kupila, E.L. (1992) *J. Electroanal. Chem.*, **322**, 167–181.

21 Kruszka, J., Nechtschein, M., and Santier, C. (1991) *Rev. Sci. Instrum.*, **62**, 695–699.
22 Genies, E.M., Hany, P., Lapkowski, M., Santier, C., and Olmedo, L. (1988) *Synth. Met.*, **25**, 29–37.
23 Paul, E.W., Ricco, A.J., and Wrighton, M.S. (1985) *J. Phys. Chem.*, **89**, 1441–1447.
24 Schiavon, G., Sitran, S., and Zotti, G. (1989) *Synth. Met.*, **32**, 209–217.
25 Thackeray, J.W., White, H.S., and Wrighton, M.S. (1985) *J. Phys. Chem.*, **89**, 5133–5140.
26 Lippe, J. and Holze, R. (1991) *Mol. Cryst. Liq. Cryst.*, **208**, 99–108.
27 Zotti, G. (1998) *Synth. Met.*, **97**, 267–272.
28 Wei, D., Espindola, P., Lindfors, T., Kvarnstrom, C., Heinze, J., and Ivaska, A. (2007) *J. Electroanal. Chem.*, **602**, 203–209.
29 Pagels, M., Heinze, J., Geschke, B., and Rang, V. (2001) *Electrochim. Acta*, **46**, 3943–3954.
30 Csahok, E., Vieil, E., and Inzelt, G. (2000) *J. Electroanal. Chem.*, **482**, 168–177.
31 Albery, W.J. and Mount, A.R. (1994) *J. Chem. Soc., Faraday Trans.*, **90**, 1115–1119.
32 Deslouis, C., Musiani, M.M., Tribollet, B., and Vorotyntsev, M.A. (1995) *J. Electrochem. Soc.*, **142**, 1902–1908.
33 Lange, U. and Mirsky, V.M. (2008) *J. Electroanal. Chem.*, **622**, 246–251.
34 Genies, E.M., Hany, P., Lapkowski, M., Santier, C., and Olmedo, L. (1988) *Synth. Met.*, **25**, 29–37.
35 Cai, L.T., Yao, S.B., and Zhou, S.M. (1997) *J. Electroanal. Chem.*, **421**, 45–48.
36 Hao, Q., Kulikov, V., and Mirsky, V.M. (2003) *Sens. Actuators, B*, **94**, 352–357.
37 Kulikov, V., Mirsky, V.M., Delaney, T.L., Donoval, D., Koch, A.W., and Wolfbeis, O.S. (2005) *Meas. Sci. Technol.*, **16**, 95–99.
38 Krondak, M., Broncova, G., Anikin, S., Merz, A., and Mirsky, V. (2006) *J. Solid State Electrochem.*, **10**, 185–191.
39 Lankinen, E., Sundholm, G., Talonen, P., Laitinen, T., and Saario, T. (1998) *J. Electroanal. Chem.*, **447**, 135–145.
40 Hailin, G., Yue, F., and Ngin, T.S. (1996) *Sens. Actuators, B*, **32**, 33–39.
41 Talaie, A., Lee, J.Y., Lee, Y.K., Jang, J., Romagnoli, J.A., Taguchi, T., and Maeder, E. (2000) *Thin Solid Films*, **363**, 163–166.
42 Hao, Q., Wang, X., Lu, L., Yang, X., and Mirsky, V.M. (2005) *Macromol. Rapid Commun.*, **26**, 1099–1103.
43 Reemts, J., Parisi, J., and Schlettwein, D. (2004) *Thin Solid Films*, **466**, 320–325.
44 Virji, S., Huang, J., Kaner, R.B., and Weiller, B.H. (2004) *Nano Lett.*, **4**, 491–496.
45 Han, G. and Shi, G. (2007) *Thin Solid Films*, **515**, 6986–6991.
46 Ram, M.K., Yavuz, O., and Aldissi, M. (2005) *Synth. Met.*, **151**, 77–84.
47 Xie, D., Jiang, Y., Pan, W., Li, D., Wu, Z., and Li, Y. (2002) *Sens. Actuators, B*, **81**, 158–164.
48 Yan, X.B., Han, Z.J., Yang, Y., and Tay, B.K. (2007) *Sens. Actuators, B*, **123**, 107–113.
49 Saxena, V., Shirodkar, V., and Prakash, R. (2001) *Appl. Biochem. Biotechnol.*, **96**, 63–69.
50 Muthukumar, C., Kesarkar, S.D., and Srivastava, D.N. (2007) *J. Electroanal. Chem.*, **602**, 172–180.
51 Lakard, B., Herlem, G., Lakard, S., Guyetant, R., and Fahys, B. (2005) *Polymer*, **46**, 12233–12239.
52 Lakard, B., Segut, O., Lakard, S., Herlem, G., and Gharbi, T. (2007) *Sens. Actuators, B*, **122**, 101–108.
53 Lindfors, T. and Ivaska, A. (2002) *J. Electroanal. Chem.*, **531**, 43–52.
54 Shoji, E. and Freund, M.S. (2001) *J. Am. Chem. Soc.*, **123**, 3383–3384.
55 Shoji, E. and Freund, M.S. (2002) *J. Am. Chem. Soc.*, **124**, 12486–12493.
56 Mousavi, Z., Alaviuhkola, T., Bobacka, J., Latonen, R.M., Pursiainen, J., and Ivaska, A. (2008) *Electrochim. Acta*, **53**, 3755–3762.
57 Vázquez, M., Bobacka, J., and Ivaska, A. (2005) *J. Solid State Electrochem.*, **9**, 865–873.
58 Prakash, R., Srivastava, R., and Pandey, P. (2002) *J. Solid State Electrochem.*, **6**, 203–208.
59 Yang, Y.J. and Huang, H.J. (2001) *Anal. Chem.*, **73**, 1377–1381.
60 Lahdesmaki, I., Kubiak, W.W., Lewenstam, A., and Ivaska, A. (2000) *Talanta*, **52**, 269–275.

61 Tian, Y., Wang, J., Wang, Z., and Wang, S. (2004) *Synth. Met.*, **143**, 309–313.
62 Jureviciute, I., Brazdziuviene, K., Bernotaite, L., Salkus, B., and Malinauskas, A. (2005) *Sens. Actuators, B*, **107**, 716–721.
63 Kelley, A., Angolia, B., and Marawi, I. (2006) *J. Solid State Electrochem.*, **10**, 397–404.
64 Kumar, S.S., Mathiyarasu, J., and Phani, K.L. (2005) *J. Electroanal. Chem.*, **578**, 95–103.
65 Mark, H.B., Atta, N., Ma, Y.L., Petticrew, K.L., Zimmer, H., Shi, Y., Lunsford, S.K., Rubinson, J.F., and Galal, A. (1995) *Bioelectrochem. Bioenerg.*, **38**, 229–245.
66 Mathiyarasu, J., Senthilkumar, S., Phani, K.L.N., and Yegnaraman, V. (2008) *Mater. Lett.*, **62**, 571–573.
67 Ho, K.C., Yeh, W.M., Tung, T.S., and Liao, J.Y. (2005) *Anal. Chim. Acta*, **542**, 90–96.
68 Kumar, S.A. and Chen, S.M. (2007) *Sens. Actuators, B*, **123**, 964–977.
69 Li, J. and Lin, X.Q. (2007) *Anal. Chim. Acta*, **596**, 222–230.
70 Li, J. and Lin, X. (2007) *Sens. Actuators, B*, **124**, 486–493.
71 Bouchta, D., Izaoumen, N., Zejli, H., Kaoutit, M.E., and Temsamani, K.R. (2005) *Biosens. Bioelectron.*, **20**, 2228–2235.
72 Parra, V., Arrieta, A.A., Fernandez-Escudero, J.A., Rodriguez-Mendez, M.L., and De Saja, J.A. (2006) *Sens. Actuators, B*, **118**, 448–453.
73 Bufon, C.C.B. and Heinzel, T. (2006) *Appl. Phys. Lett.*, **89**, 012104–012113.
74 Torsi, L., Tanese, M.C., Cioffi, N., Gallazzi, M.C., Sabbatini, L., Zambonin, P.G., Raos, G., Meille, S.V., and Giangregorio, M.M. (2003) *J. Phys. Chem. B*, **107**, 7589–7594.
75 Torsi, L., Tafuri, A., Cioffi, N., Gallazzi, M.C., Sassella, A., Sabbatini, L., and Zambonin, P.G. (2003) *Sens. Actuators, B*, **93**, 257–262.
76 Torsi, L., Tanese, M.C., Cioffi, N., Gallazzi, M., Sabbatini, L., and Zambonin, P.G. (2004) *Sens. Actuators, B*, **98**, 204–207.
77 Hatfield, J.V., Covington, J.A., and Gardner, J.W. (2000) *Sens. Actuators, B*, **65**, 253–256.
78 Meijerink, M.G.H., Koudelka-Hep, M., de Rooij, N.F., Strike, D.J., Hendrikse, J., Olthuis, W., and Bergveld, P. (1999) *Electrochem. Solid-State Lett.*, **2**, 138–139.
79 Assadi, A., Spetz, A., Willander, M., Svensson, C., Lundstrom, I., and Inganas, O. (1994) *Sens. Actuators, B*, **20**, 71–77.
80 Nguyen, V.C. and Potje-Kamloth, K. (1999) *Thin Solid Films*, **338**, 142–148.
81 Nguyen, V.C. and Potje-Kamloth, K. (2000) *J. Phys. D: Appl. Phys.*, **33**, 2230–2238.
82 Campos, M., Bulhoes, L.O.S., and Lindino, C.A. (2000) *Sens. Actuators, A*, **87**, 67–71.
83 Pinto, N.J., Gonzalez, R., Johnson, J., and MacDiarmid, A.G. (2006) *Appl. Phys. Lett.*, **89**, 033505–033513.
84 Jin, Z., Su, Y., and Duan, Y. (2000) *Sens. Actuators, B*, **71**, 118–122.
85 Piletsky, S.A., Panasyuk, T.L., Piletskaya, E.V., Sergeeva, T.A., El'skaya, A.V., Pringsheim, E., and Wolfbeis, O.S. (2000) *Fresenius' J. Anal. Chem.*, **366**, 807–810.
86 Pringsheim, E., Terpetschnig, E., and Wolfbeis, O.S. (1997) *Anal. Chim. Acta*, **357**, 247–252.
87 Sotomayor, M.D.P.T., De Paoli, M.A., and de Oliveira, W.A. (1997) *Anal. Chim. Acta*, **353**, 275–280.
88 Sotomayor, P.T., Raimundo, I.M., Zarbin, A.J.G., Rohwedder, J.J.R., Neto, G.O., and Alves, O.L. (2001) *Sens. Actuators, B*, **74**, 157–162.
89 Ando, M., Swart, C., Pringsheim, E., Mirsky, V.M., and Wolfbeis, O.S. (2002) *Solid State Ionics*, **152–153**, 819–822.
90 Ando, M., Swart, C., Pringsheim, E., Mirsky, V.M., and Wolfbeis, O.S. (2005) *Sens. Actuators, B*, **108**, 528–534.
91 Li, C., Numata, M., Takeuchi, M., and Shinkai, S. (2005) *Angew. Chem.*, **117**, 6529–6532.
92 Pringsheim, E., Terpetschnig, E., Piletsky, S.A., and Wolfbeis, O.S. (1999) *Adv. Mater.*, **11**, 865–868.
93 de Marcos, S. and Wolfbeis, O.S. (1996) *Anal. Chim. Acta*, **334**, 149–153.
94 Ge, C., Armstrong, N.R., and Saavedra, S.S. (2007) *Anal. Chem.*, **79**, 1401–1410.

95 Grummt, U.W., Pron, A., Zagorska, M., and Lefrant, S. (1997) *Anal. Chim. Acta*, **357**, 253–259.

96 Thomas, S.W., III, Amara, J.P., Bjork, R.E., and Swager, T.M. (2005) *Chem. Commun.*, 4572–4574.

97 Toal, S.J. and Trogler, W.C. (2006) *J. Mater. Chem.*, **16**, 2871–2883.

98 Liu, X., Tang, Y., Wang, L., Zhang, J., Song, S., Fan, C., and Wang, S. (2007) *Adv. Mater.*, **19**, 1471–1474.

99 Samoylov, A.V., Mirsky, V.M., Hao, Q., Swart, C., Shirshov, Y.M., and Wolfbeis, O.S. (2005) *Sens. Actuators, B*, **106**, 369–372.

100 Agbor, N.E., Cresswell, J.P., Petty, M.C., and Monkman, A.P. (1997) *Sens. Actuators, B*, **41**, 137–141.

101 Kang, X., Cheng, G., and Dong, S. (2001) *Electrochem. Commun.*, **3**, 489–493.

102 Lindfors, T. and Ivaska, A. (2005) *J. Electroanal. Chem.*, **580**, 320–329.

103 Pei, Q. and Qian, R. (1991) *Synth. Met.*, **45**, 35–48.

104 Huang, W.-S., Humphrey, B.D., and MacDiarmid, A.G. (1986) *J. Chem. Soc., Faraday Trans. 1*, **82**, 2385–2400.

105 Alabanza, H., Bergantin, J., Sevilla, F., and Narayanaswamy, R. (2003) in *Electrochromic Materials and Applications*, The Electrochemical Society.

106 Yue, F., Ngin, T.S., and Hailin, G. (1996) *Sens. Actuators, B*, **32**, 33–39.

107 Carquigny, S., Segut, O., Lakard, B., Lallemand, F., and Fievet, P. (2008) *Synth. Met.*, **158**, 453–461.

108 Kaempgen, M. and Roth, S. (2006) *J. Electroanal. Chem.*, **586**, 72–76.

109 Ferrer-Anglada, N., Kaempgen, M., and Roth, S. (2006) *Phys. Status Solidi B*, **243**, 3519–3523.

110 Ayad, M.M., Abou-Seif, A.K., Salahuddin, N.A., and Alghaysh, M.O. (2008) *Polym. Adv. Technol.*, **19**, 1142–1148.

111 Gill, E., Arshak, A., Arshak, K., and Korostynska, O. (2007) *Sensors*, **7**, 3329–3346.

112 Slim, C., Ktari, N., Cakara, D., Kanoufi, F., and Combellas, C. (2008) *J. Electroanal. Chem.*, **612**, 53–62.

113 Pandey, P.C., Prakash, R., Singh, G., Tiwari, I., and Tripathi, V.S. (2000) *J. Appl. Polym. Sci.*, **75**, 1749–1759.

114 Bai, H. and Shi, G. (2007) *Sensors*, **7**, 267–307.

115 Prissanaroon, W., Ruangchuay, L., Sirivat, A., and Schwank, J. (2000) *Synth. Met.*, **114**, 65–72.

116 Watcharaphalakorn, S., Ruangchuay, L., Chotpattananont, D., Sirivat, A., and Schwank, J. (2005) *Polym. Int.*, **54**, 1126–1133.

117 Hanawa, T., Kuwabata, S., Hashimoto, H., and Yoneyama, H. (1989) *Synth. Met.*, **30**, 173–181.

118 Mohammad, F. (1998) *J. Phys. D: Appl. Phys.*, **31**, 951–959.

119 Sakurai, Y., Jung, H.S., Shimanouchi, T., Inoguchi, T., Morita, S., Kuboi, R., and Natsukawa, K. (2002) *Sens. Actuators, B*, **83**, 270–275.

120 Ellis, D.L., Zakin, M.R., Bernstein, L.S., and Rubner, M.F. (1996) *Anal. Chem.*, **68**, 817–822.

121 Ratcliffe, N.M. (1990) *Anal. Chim. Acta*, **239**, 257–262.

122 Yang, H., Wan, J., Shu, H., Liu, X., Lakshmanan, R.S., Guntupalli, R., Hu, J., Howard, W., and Chin, B.A. (2006) *Proc. SPIE*, **6222**, 62220S–622208.

123 Li, B., Santhanam, S., Schultz, L., Jeffries-EL, M., Iovu, M.C., Sauve, G., Cooper, J., Zhang, R., Revelli, J.C., and Kusne, A.G. (2007) *Sens. Actuators, B*, **123**, 651–660.

124 Ruangchuay, L., Sirivat, A., and Schwank, J. (2004) *Synth. Met.*, **140**, 15–21.

125 Athawale, A.A. and Kulkarni, M.V. (2000) *Sens. Actuators, B*, **67**, 173–177.

126 Vercelli, B., Zecchin, S., Comisso, N., Zotti, G., Berlin, A., Dalcanale, E., and Groenendaal, L.B. (2002) *Chem. Mater.*, **14**, 4768–4774.

127 Tan, C.K. and Blackwood, D.J. (2000) *Sens. Actuators, B*, **71**, 184–191.

128 Conn, C., Sestak, S., Baker, A.T., and Unsworth, J. (1998) *Electroanalysis*, **10**, 1137–1141.

129 Geng, L., Zhao, Y., Huang, X., Wang, S., Zhang, S., and Wu, S. (2007) *Sens. Actuators, B*, **120**, 568–572.

130 Ram, M.K., Yavuz, O., Lahsangah, V., and Aldissi, M. (2005) *Sens. Actuators, B*, **106**, 750–757.

131 Tai, H., Jiang, Y., Xie, G., Yu, J., and Chen, X. (2007) *Sens. Actuators, B*, **125**, 644–650.

132 Tandon, R.P., Tripathy, M.R., Arora, A.K., and Hotchandani, S. (2006) *Sens. Actuators, B*, **114**, 768–773.

133 Vibha, S., Aswal, D.K., Manmeet, K., Koiry, S.P., Gupta, S.K., Yakhmi, J.V., Kshirsagar, R.J., and Deshpande, S.K. (2007) *Appl. Phys. Lett.*, **90**, 043516.

134 Bavastrello, V., Stura, E., Carrara, S., Erokhin, V., and Nicolini, C. (2004) *Sens. Actuators, B*, **98**, 247–253.

135 Chen, Y., Li, Y., Wang, H., and Yang, M. (2007) *Carbon*, **45**, 357–363.

136 Ting, Z., Syed, M., Elena, B., Bong, Y.Y., Robert, C.H., Nosang, V.M., and Marc, A.D. (2007) *Nanotechnology*, **18**, 165504.

137 Bäuerle, P. and Scheib, S. (1993) *Adv. Mater.*, **5**, 848–853.

138 Boldea, A., Lévesque, I., and Leclerc, M. (1999) *J. Mater. Chem.*, **9**, 2133–2138.

139 Bäuerle, P. and Scheib, S. (1995) *Acta Polym.*, **46**, 124–129.

140 Scheib, S. and Bäuerle, P. (1999) *J. Mater. Chem.*, **9**, 2139–2150.

141 Sannicolo, F., Brenna, E., Benincori, T., Zotti, G., Zecchin, S., Schiavon, G., and Pilati, T. (1998) *Chem. Mater.*, **10**, 2167–2176.

142 Marsella, M.J. and Swager, T.M. (1993) *J. Am. Chem. Soc.*, **115**, 12214–12215.

143 Demeter, D., Blanchard, P., Grosu, I., and Roncali, J. (2007) *Electrochem. Commun.*, **9**, 1587–1591.

144 Marsella, M.J., Newland, R.J., Carroll, P.J., and Swager, T.M. (1995) *J. Am. Chem. Soc.*, **117**, 9842–9848.

145 Kim, J., McQuade, D.T., McHuge, S.K., and Swager, T.M. (2000) *Angew. Chem.*, **39**, 3868–3872.

146 Chen, Z., Xue, C., Shi, W., Luo, F.T., Green, S., Chen, J., and Liu, H. (2004) *Anal. Chem.*, **76**, 6513–6518.

147 Ramachandran, G., Simon, G., Cheng, Y., Smith, T.A., and Dai, L. (2003) *J. Fluoresc.*, **13**, 427–436.

148 Pernaut, J.M., Zong, K., and Reynolds, J.R. (2002) *Synth. Met.*, **130**, 1–8.

149 Ion, A., Ion, I., Popescu, A., Ungureanu, E.M., Moutet, J.C., and Saint-Aman, E. (1997) *Adv. Mater.*, **9**, 711–713.

150 Ion, A.C., Moutet, J.C., Pailleret, A., Popescu, A., Saint-Aman, E., Siebert, E., and Ungureanu, E.M. (1999) *J. Electroanal. Chem.*, **464**, 24–30.

151 Chen, L.X., Jager, W.J.H., Gosztola, D.J., Niemczyk, M.P., and Wasielewski, M.R. (2000) *J. Phys. Chem. B*, **104**, 1950–1960.

152 Chen, Y., Fan, Q.L., Wang, P., Zhang, B., Huang, Y.Q., Zhang, G.W., Lu, X.M., Chan, H.S.O., and Huang, W. (2006) *Polymer*, **47**, 5228–5232.

153 Liu, B., Yu, W.L., Pei, J., Liu, S.Y., Lai, Y.H., and Huang, W. (2001) *Macromolecules*, **34**, 7932–7940.

154 Zhang, Y., Murphy, C.B., and Jones, W.E. (2002) *Macromolecules*, **35**, 630–636.

155 Ding, A.L., Pei, J., Yu, W.L., Lai, Y.H., and Huang, W. (2002) *Thin Solid Films*, **417**, 198–201.

156 Wang, B. and Wasielewski, M.R. (1997) *J. Am. Chem. Soc.*, **119**, 12–21.

157 Smith, R.C., Tennyson, A.G., Lim, M.H., and Lippard, S.J. (2005) *Org. Lett.*, **7**, 3573–3575.

158 Bonacchi, S., Dolci, L.S., Sterzo, C.L., Micozzi, A., Montalti, M., Prodi, L., Ricci, A., and Zaccheroni, N. (2008) *J. Photochem. Photobiol. A*, **198**, 237–241.

159 Zhang, M., Lu, P., Ma, Y., and Shen, J. (2003) *J. Phys. Chem. B*, **107**, 6535–6538.

160 Yasuda, T. and Yamamoto, T. (2003) *Macromolecules*, **36**, 7513–7519.

161 Vetrichelvan, M. and Valiyaveettil, S. (2005) *Chem. Eur. J.*, **11**, 5889–5898.

162 Huang, H., Wang, K., Tan, W., An, D., Yang, X., Huang, S., Zhai, Q., Zhou, L., and Jin, Y. (2004) *Angew. Chem. Int. Ed.*, **43**, 5635–5638.

163 Zhu, S.S., Carroll, P.J., and Swager, T.M. (1996) *J. Am. Chem. Soc.*, **118**, 8713–8714.

164 Zhu, S.S. and Swager, T.M. (1997) *J. Am. Chem. Soc.*, **119**, 12568–12577.

165 Wang, W.-L., Xu, J.-W., and Lai, Y.-H. (2006) *J. Polym. Sci. A: Polym. Chem.*, **44**, 4154–4164.

166 Tang, Y., He, F., Yu, M., Feng, F., An, L., Sun, H., Wang, S., Li, Y., and Zhu, D. (2006) *Macromol. Rapid Commun.*, **27**, 389–392.

167 Heitzmann, M., Bucher, C., Moutet, J.C., Pereira, E., Rivas, B.L., Royal, G., and Saint-Aman, E. (2007) *Electrochim. Acta*, **52**, 3082–3087.

168 Holliday, B.J., Stanford, T.B., and Swager, T.M. (2006) *Chem. Mater.*, **18**, 5649–5651.

169 Kim, I.B., Dunkhorst, A., Gilbert, J., and Bunz, U.H.F. (2005) *Macromolecules*, **38**, 4560–4562.
170 Kim, I.B. and Bunz, U.H.F. (2006) *J. Am. Chem. Soc.*, **128**, 2818–2819.
171 Kim, I.-B., Erdogan, B., Wilson, J.N., and Bunz, U.H.F. (2004) *Chem. Eur. J.*, **10**, 6247–6254.
172 Li, Z., Lou, X., Yu, H., Li, Z., and Qin, J. (2008) *Macromolecules*, **41**, 7433–7439.
173 Zhou, X.H., Yan, J.C., and Pei, J. (2004) *Macromolecules*, **37**, 7078–7080.
174 Nicolas, M., Fabre, B., and Simonet, J. (2001) *J. Electroanal. Chem.*, **509**, 73–79.
175 Ho, H.A. and Leclerc, M. (2003) *J. Am. Chem. Soc.*, **125**, 4412–4413.
176 Anzenbacher, J., Jursikova, K., Aldakov, D., Marquez, M., and Pohl, R. (2004) *Tetrahedron*, **60**, 11163–11168.
177 Wu, C.-Y., Chen, M.-S., Lin, C.-A., Lin, S.-C., and Sun, S.-S. (2006) *Chem. Eur. J.*, **12**, 2263–2269.
178 Takeuchi, M., Shioya, T., and Swager, T.M. (2001) *Angew. Chem.*, **113**, 3476–3480.
179 Zhou, G., Cheng, Y., Wang, L., Jing, X., and Wang, F. (2005) *Macromolecules*, **38**, 2148–2153.
180 Björk, P., Persson, N.K., Peter, K., Nilsson, R., Asberg, P., and Inganäs, O. (2005) *Biosens. Bioelectron.*, **20**, 1764–1771.
181 Bera-Aberem, M., Ho, H.A., and Leclerc, M. (2004) *Tetrahedron*, **60**, 11169–11173.
182 Dore, K., Dubus, S., Ho, H.A., Levesque, I., Brunette, M., CORBEIL, G., Boissinot, M., Boivin, G., Bergeron, M.G., Boudreau, D., and Leclerc, M. (2004) *J. Am. Chem. Soc.*, **126**, 4240–4244.
183 Ho, H.A., Boissinot, M., Bergeron, M.G., Corbeil, G., Dore, K., Boudreau, D., and Leclerc, M. (2002) *Angew. Chem. Int. Ed.*, **41**, 1548–1551.
184 Ho, H.A., Dore, K., Boissinot, M., Bergeron, M.G., Tanguay, R.M., Boudreau, D., and Leclerc, M. (2005) *J. Am. Chem. Soc.*, **127**, 12673–12676.
185 LeFloch, F., Ho, H.A., and Leclerc, M. (2006) *Anal. Chem.*, **78**, 4727–4731.
186 Nilsson, K.P. and Inganas, O. (2003) *Nat. Mater.*, **2**, 419–424.
187 Peng, H., Soeller, C., and Travas-Sejdic, J. (2006) *Chem. Commun.*, 3735–3737.
188 Liu, S., Bakovic, L., and Chen, A. (2006) *J. Electroanal. Chem.*, **591**, 210–216.
189 Deore, B. and Freund, M.S. (2003) *Analyst*, **128**, 803–806.
190 Deore, B.A., Braun, M.D., and Freund, M.S. (2006) *Macromol. Chem. Phys.*, **207**, 660–664.
191 Ma, Y. and Yang, X. (2005) *J. Electroanal. Chem.*, **580**, 348–352.
192 Chang, C.P., Chao, C.Y., Huang, J.H., Li, A.K., Hsu, C.S., Lin, M.S., Hsieh, B.R., and Su, A.C. (2004) *Synth. Met.*, **144**, 297–301.
193 Yang, J.S. and Swager, T.M. (1998) *J. Am. Chem. Soc.*, **120**, 11864–11873.
194 Yang, J.S. and Swager, T.M. (1998) *J. Am. Chem. Soc.*, **120**, 5321–5322.
195 Disney, M.D., Zheng, J., Swager, T.M., and Seeberger, P.H. (2004) *J. Am. Chem. Soc.*, **126**, 13343–13346.
196 Baek, M.G., Stevens, R.C., and Charych, D.H. (2000) *Bioconjugate Chem.*, **11**, 777–788.
197 Xue, C., Jog, S.P., Murthy, P., and Liu, H. (2006) *Biomacromolecules*, **7**, 2470–2474.
198 Kim, I.B., Wilson, J.N., and Bunz, U.H.F. (2005) *Chem. Commun.*, 1273–1275.
199 Tanese, M.C., Torsi, L., Cioffi, N., Zotti, L.A., Colangiuli, D., Farinola, G.M., Babudri, F., Naso, F., Giangregorio, M.M., Sabbatini, L., and Zambonin, P.G. (2004) *Sens. Actuators, B*, **100**, 17–21.
200 Kim, I.B., Shin, H., Garcia, A.J., and Bunz, U.H.F. (2007) *Bioconjugate Chem.*, **18**, 815–820.
201 Muthukumar, C., Kesarkar, S.D., and Srivastava, D.N. (2007) *J. Electroanal. Chem.*, **602**, 172–180.
202 Delaney, T.L., Zimin, D., Rahm, M., Weiss, D., Wolfbeis, O.S., and Mirsky, V.M. (2007) *Anal. Chem.*, **79**, 3220–3225.
203 Deore, B., Chen, Z., and Nagaoka, T. (1999) *Anal. Sci.*, **15**, 827–828.
204 Malitesta, C., Losito, I., and Zambonin, P.G. (1999) *Anal. Chem.*, **71**, 1366–1370.
205 Panasyuk, T., Dall'Orto, V.C., Marrazza, G., El'skaya, A., Piletsky, S., Rezzano, I., and Mascini, M. (1998) *Anal. Lett.*, **31**, 1809–1824.
206 Panasyuk, T.L., Mirsky, V.M., Piletsky, S.A., and Wolfbeis, O.S. (1999) *Anal. Chem.*, **71**, 4609–4613.

207 Blanco-Lopez, M.C., Gutierrez-Fernandez, S., Lobo-Castanon, M.J., Miranda-Ordieres, A.J., and Tunon-Blanco, P. (2004) *Anal. Bioanal. Chem.*, **378**, 1922–1928.

208 Cheng, Z., Wang, E., and Yang, X. (2001) *Biosens. Bioelectron.*, **16**, 179–185.

209 Gomez-Caballero, A., Goicolea, M.A., and Barrio, R.J. (2005) *Analyst*, **130**, 1012–1018.

210 Liu, K., Wei, W.Z., Zeng, J.X., Liu, X.Y., and Gao, Y.P. (2006) *Anal. Bioanal. Chem.*, **385**, 724–729.

211 Oezcan, L. and Sahin, Y. (2007) *Sens. Actuators, B*, **127**, 362–369.

212 Peng, H., Zhang, J., Nie, L., Yao, S., Zhang, Y., and Xie, Q. (2001) *Analyst*, **126**, 189–194.

213 Sreenivasan, K. (2007) *J. Mater. Sci.*, **42**, 7575–7578.

214 Ulyanova, Y.V., Blackwell, A.E., and Minteer, S.D. (2006) *Analyst*, **131**, 257–261.

215 Yu, J.C.C., Krushkova, S., Lai, E.P.C., and Dabek-Zlotorzynska, E. (2005) *Anal. Bioanal. Chem.*, **382**, 1534–1540.

216 Blackwell, A. and Minteer, S.D. (2007) Abstracts of Papers. 233rd ACS National Meeting, Chicago, IL, United States.

13
Molecularly Imprinted Polymers as Artificial Receptors
Florian Meier and Boris Mizaikoff

13.1
Introduction

The principles of molecular recognition play a crucial role in virtually any biochemical process vital to living organisms. The elegance of how nature controls complex interactions at a molecular level such as, for example, hormone–receptor, antibody–antigen, or enzyme–substrate interactions has always attracted the attention of researchers aiming at not only a theoretical understanding these basic schemes but also at applying their knowledge to create artificial receptors with similar or preferably superior synergetic properties capable of mimicking these naturally occurring processes.

The technique of molecular imprinting offers a facile and straightforward strategy towards such artificial receptors. While there have been some early reports applying the concept of molecular imprinting predominantly in silica matrices, the technique and terminology itself did not achieve its scientific breakthrough until the early 1980s due to the initial work primarily performed in the research groups of G. Wulff, K. J. Shea, and K. Mosbach, who were among the first to emphasize the role and utility of organic polymer networks serving as matrices for molecular imprints. With their pioneering studies in the field of molecular imprinting, they have paved the way towards a novel and rapidly growing field of research that still enjoys increasing interest, as reflected in the number of annual publications, which has continuously increased since the first mention of the terminology "molecular imprinting" in the literature in 1984 (Figure 13.1).

Along with the increasing interest of the scientific community, the range of applications for artificial receptors prepared by molecular imprinting has expanded greatly. Nowadays, molecularly imprinted polymers (MIPs) have applications in various fields of research, including for example, antibody mimicking pseudo-immunoassays, enzyme mimicking catalysis, imprinted membranes, sensing devices, and intelligent drug delivery. Furthermore, molecularly selective separation

Artificial Receptors for Chemical Sensors. Edited by V.M. Mirsky and A. Yatsimirsky
Copyright © 2011 WILEY-VCH Verlag GmbH & Co. KGaA, Weinheim
ISBN: 978-3-527-32357-9

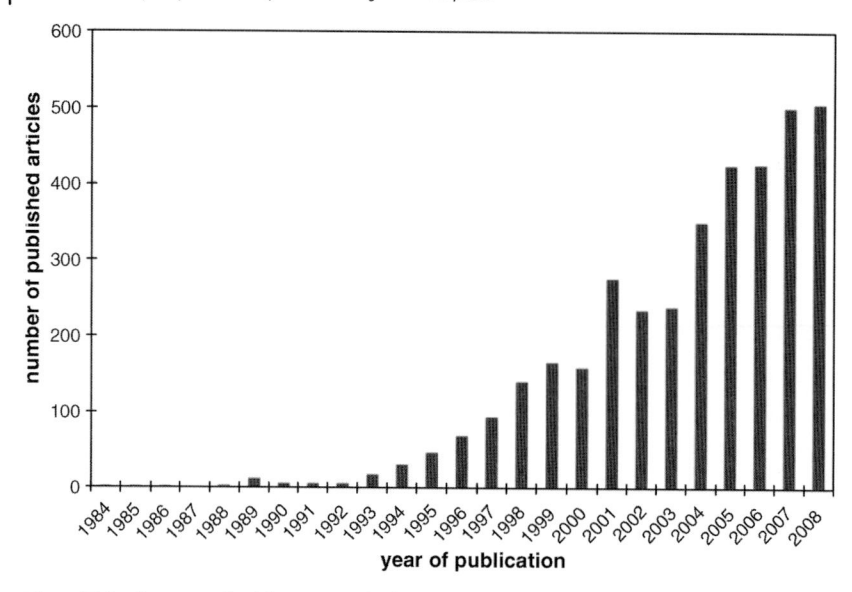

Figure 13.1 Increase of publications in the field of molecular imprinting up to 2008, derived from a search using SciFinder Scholar® based on the terminology "molecular imprinting."

applications applying MIPs as stationary phases include liquid chromatography (LC), capillary electrochromatography (CEC), and solid-phase-extraction (molecularly imprinted solid-phase extraction – MISPE) (Figure 13.2). In the latter case, molecularly imprinted polymers have already evolved into commercially available cartridges offering selective extraction of various analytes such as, for example, clenbuterol, chloramphenicol, or amphetamines from complex matrices. This development clearly indicates the potential of this maturing technology, having already developed into a serious alternative, or at least a supplement, for selected analytical procedures. Consequently, it is fair to assume that MIP technology will continue to gain in importance during the coming decades.

In the following sections we provide a fundamental understanding of the nature of MIPs, how they work, and how close in terms of functional properties they already are in comparison to their natural analogues, such as, for example, antibodies and enzymes. Thus, we initially focus on the fundamentals of molecular imprinting, including the general principles and most prevalent synthesis approaches for MIPs, provide a selection of typical reagents and solvents that are applied in the preparation of MIPs, and give an overview of popular MIP formats. In addition, we discuss the general applicability of MIPs, with particular focus on their utility as artificial receptors, including relevant performance parameters such as the binding capacity, binding affinity, and binding selectivity. In particular, we provide a quantitative overview of these critical performance parameters derived from selected examples in the literature, and emphasize the comparison with the performance of natural receptors. Finally, an outlook towards the rational design of MIPs and future challenges in MIP technology concludes this chapter.

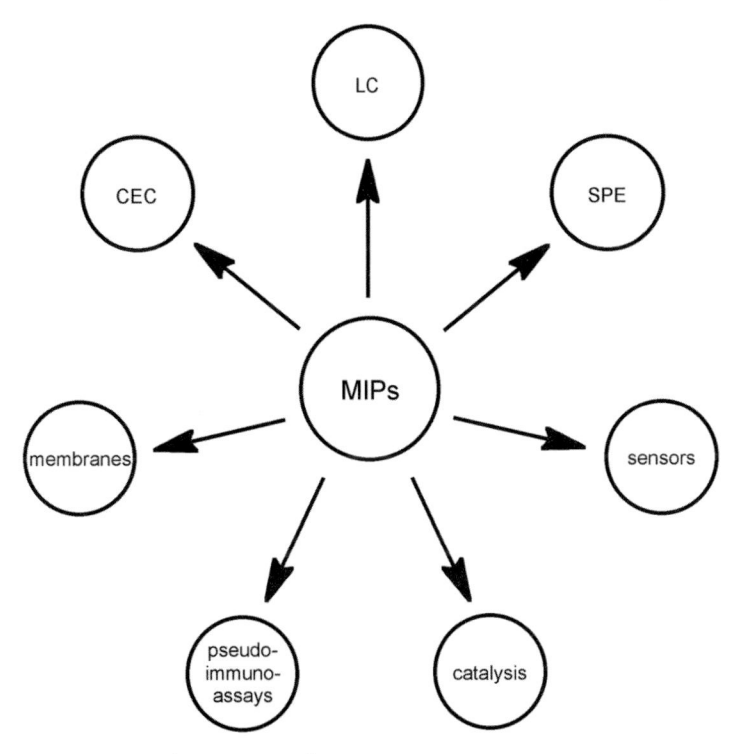

Figure 13.2 Application areas of MIPs.

13.2
Fundamentals of Molecular Imprinting

13.2.1
What are MIPs?

The general concept behind molecular imprinting entails generating selective recognition sites for a certain target molecule, usually referred to as template, within a synthetic polymer network. The corresponding recognition processes in MIPs shows similarities to biochemical recognition processes occurring in nature, as they are taking advantage of similar fundamental interactions such as, for example, hydrogen bonding, electrostatic interactions, and hydrophobic interactions. However, simply taking these fundamental interactions into consideration is not sufficient to explain recognition processes, neither in nature nor in MIPs. Additionally, the steric arrangement of these interactions around a given substrate, that is, template, is a crucial aspect necessary for the creation of binding pockets providing complementary size, shape, and functionality for preferentially facilitating selective recognition along with a high affinity toward the target. Thus, the recognition process in MIPs may be described in analogy with mechanisms established for enzyme–substrate-complexes such as, for example, the "lock-and-key" principle [1].

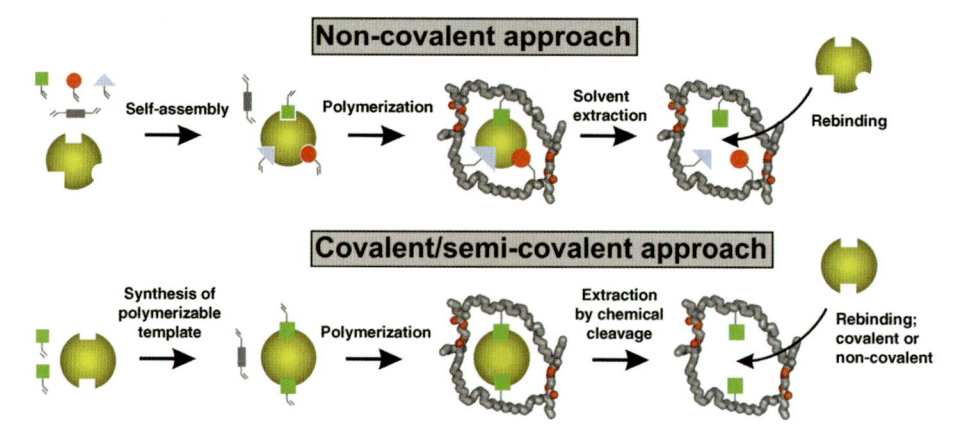

Figure 13.3 General synthesis scheme of molecularly imprinted polymers.

During the generation of MIPs, functional monomers are responsible for the introduction of complementary functionality into the polymer matrix, offering either preferably stable noncovalent, semi-covalent, or covalent interactions with the template, thereby establishing a so-called prepolymerization complex. To maintain this sterical arrangement and, hence, retain complementary size, shape, and functionality resulting in the required selectivity of the final polymer matrix, the prepolymerization complexes are copolymerized in the presence of a crosslinker (CL), which provides the polymeric backbone of the MIP. After the polymerization, the template has to be extracted to expose the binding cavities, thereby activating the MIP for selective rebinding. Figure 13.3 shows schematically the general principles of molecular imprinting using the most prevalent noncovalent or the covalent approach.

13.2.2
Approaches toward Molecular Imprinted Polymers

As already mentioned, there are two general approaches for generating MIPs, differing mainly in the type of interaction between the functional monomer and the template in the prepolymerization solution, namely, noncovalent and covalent imprinting. In addition, some derivatives of these approaches have been described in the literature such as, for example, the semi-covalent approach. In the following, we provide insight into the basics of these general synthesis strategies, highlighting the advantages and disadvantages of each approach, as well as presenting some commonly applied reagents and solvents for the preparation of MIPs.

13.2.2.1 Noncovalent Imprinting (Self-Assembly Approach)
K. Mosbach *et al.* [2] have pioneered the noncovalent imprinting approach, taking advantage of noncovalent interactions primarily including hydrogen-bonding and electrostatic interactions, but also van der Waals forces and hydrophobic interactions

(e.g., π–π stacking), which are considered the dominating driving forces during the formation of the prepolymerization complex between functional monomer and template. Therefore, this approach most closely resembles the mechanism of molecular recognition in nature, where hydrogen-bonding and electrostatic interactions are considered the most prevalent noncovalent interactions. Since noncovalent interactions only provide comparatively weak binding strengths, the functional monomer is usually introduced in molar excess (at least fourfold) [3] versus the template molecule within the prepolymerization solution. Thereby, the equilibrium is shifted toward forming prepolymerization complexes according to the principle of Le Chatelier. However, the probability of creating non-selective binding sites caused by an excess of functional monomers randomly incorporated into the polymer matrix, and which are not associated with the template, is increased [4]. Furthermore, due to the low binding constants, particular attention has to be paid to the selection of an appropriate functional monomer, and also to the properties of the solvent to ensure a preferably stable prepolymerization complex. This is a critical aspect, as the number of stable template-functional monomer complexes in the prepolymerization solution usually amounts to only 0.3–0.6% [4] and this number certainly critically affects the number of selective binding sites generated within the final polymer matrix [5].

13.2.2.2 Covalent Imprinting (Preorganized Approach)

In the covalent imprinting approach, introduced by G. Wulff et al. [6], the functional monomer is covalently bound to the template, thus forming the prepolymerization complex. After the polymerization, the covalent bonds have to be cleaved during the extraction process in order to remove the template, and to expose the binding sites within the polymer matrix. Any rebinding of the template molecule then occurs by either re-establishing this covalent bond or alternatively by noncovalent interactions (see below). The high binding strength of the covalent bonds usually leads to MIPs with superior selectivity in comparison to MIPs prepared by the noncovalent approach, as it is more likely that the prepolymerization complex survives the rather harsh conditions during the usually applied free-radical polymerization process. Furthermore, stoichiometric interactions between functional monomer and template in covalent imprinting provide a narrower binding site affinity distribution, and reduce the probability of non-selective binding [7]. However, the higher binding energy of covalent bonds in comparison to noncovalent binding results in slow binding kinetics during the rebinding process, and – if rebinding is based on covalent interactions – more effort is needed in regenerating the MIP for repeated usage [8].

13.2.2.3 Semi-Covalent Imprinting

Besides the two main approaches for synthesizing MIPs, Sellergren and Andersson [9] have developed an approach called semi-covalent imprinting, which can be regarded as a hybrid between the two approaches mentioned before. In this approach, the benefit of the covalent approach is used for the formation of a covalently bound prepolymerization complex, while the rebinding process is governed by noncovalent interactions assuring accelerated rebinding kinetics.

13.2.2.4 Advantages and Disadvantages of Different Imprinting Approaches

The comparative ease and straightforwardness of the preparation together with the large variety of possible templates renders the noncovalent imprinting approach among the most popular strategies in molecular imprinting technology. However, the covalent imprinting approach also offers some significant benefits while being restricted to a limited number of templates such as, for example, alcohols, amines, aldehydes, ketones, and carboxylic acids. Table 13.1 summarizes the main advantages and disadvantages of both approaches.

The selection of a particular approach strongly depends on the final application of the respective MIP. As MIPs prepared by noncovalent imprinting usually offer fast binding kinetics, they are ideally suited for applications such as, for example, pseudo-immunoassays and solid-phase extraction, where a fast mass transfer is favorable. However, MIPs prepared by covalent imprinting are suitable for applications where a highly selective signal generation is required, such as, for example, in sensing devices and where the MIP may have a scavenger function rather than being repeatedly used.

13.2.3
Reagents and Solvents in Molecular Imprinting Technology

As mentioned earlier, the components required for the preparation of MIPs consist of a template of interest, a functional monomer(s) (either for noncovalent or covalent imprinting), a crosslinker(s), an initiator, and an appropriate solvent. In the following, we elaborate the particular role and influence of each component, and highlight selected examples.

13.2.3.1 Functional Monomers

A crucial component in the preparation of MIPs is the appropriate selection of the functional monomer, as it strongly affects the final overall performance of the

Table 13.1 General *advantages* and disadvantages of noncovalent and covalent imprinting.

	Noncovalent imprinting	Covalent imprinting
Choice of the template and functional monomer	*Wide variety*	Restricted to templates and functional monomers with appropriate functionalities
Preparation of the MIP	*Easy and straightforward*	Preparation of a covalently bound prepolymerization complex necessary
Polymerization conditions	Critical; only noncovalent interactions govern the prepolymerization complex	*Less critical; the prepolymerization complex is more stable*
Binding sites (affinity)	Heterogeneous	Homogeneous
Selectivity	Lower selectivity	*Higher selectivity*
Removal of the template after polymerization	*Easy and at mild conditions*	Cleavage of the monomer-template linkage necessary
Rebinding kinetics	Fast	Slow

synthetic receptor. Particularly in noncovalent imprinting – where only weak interactions between functional monomer and template molecule govern the stability of the prepolymerization complex and subsequently the rebinding process – the choice of functional monomer frequently determines the utility of the finally obtained MIP. A widely used functional monomer in noncovalent imprinting is methacrylic acid, which has been applied extensively to imprint various templates, including, for example, derivatives of L-phenylalanine [10], 17β-estradiol [11], or very recently – for establishing a matrix with controlled release properties – acetyl salicylic acid [12]. Further popular functional monomers in noncovalent imprinting are 4-vinylpyridine, acrylamide, and 2-hydroxyethyl methacrylate (Table 13.2).

Not only in noncovalent but also in covalent imprinting the proper choice of functional monomer plays an important role in the successful preparation of a MIP. Since the functional monomers in covalent imprinting should provide reversible covalent bonds with preferentially high binding kinetics, 1,3-diols, amines, or boronic acids are frequently applied (Table 13.2). Thus, molecular imprinting of various templates could be achieved, including, for example, several ketones (via formation of a ketal bond) [13], dialdehydes (via formation of a Schiff base bond) [14], and fructose and galactose (via formation of a boronic ester bond) [15]. In the latter example, even racemic resolution of the respective saccharides could be achieved, revealing the potential of MIPs in the area of enantiomeric separations.

As well as providing functional groups capable of interacting with the template in a distinct fashion, functional monomers also provide at least one polymerizable group, usually a vinyl group, assuring robust incorporation into the polymer matrix during the polymerization and, therefore, creating stable binding moieties that maintain their steric arrangement.

Most of the functional monomers applied in molecular imprinting are nowadays commercially available, although there have been some examples in the literature applying self-synthesized functional monomers tailored to a particular template molecule. Turkewitsch *et al.* have for instance prepared a MIP for cAMP using a self-synthesized fluorescent functional monomer (*trans*-4-[*p*-(*N,N*-dimethylamino)styryl]-*N*-vinylbenzylpyridinium chloride) (Table 13.2), allowing highly selective and sensitive detection [16] of cAMP in aqueous environments.

13.2.3.2 Crosslinkers

The main role of the crosslinker in noncovalent as well as in covalent imprinting is to provide a rigid polymer matrix enabling the incorporation of binding cavities established via the prepolymerization complex. Thus, the crosslinker holds the functional monomers in place to ensure the stability of the binding pocket, which is essential for selective recognition of the template. To ensure high rigidity, the crosslinker is usually added in very high proportions in comparison to the functional monomer and the template molecule. Consequently, the choice and the amount of the crosslinker have a substantial influence on the morphology, stability, and chemical and physical properties of the MIP; thus, careful selection of the crosslinker is advisable.

Various research groups have investigated the influence of the choice of crosslinker and the degree of crosslinking on the performance of MIPs. Kempe *et al.* [17], for

Table 13.2 Summary of selected functional monomers frequently applied for the preparation of MIPs.

Name	Functional monomers		
	Chemical structure	Class	Predominant interactions
Acrylic acid (AA)		Acid	Hydrogen bonding
Methacrylic acid (MAA)		Acid	Hydrogen bonding
4-Vinylpyridine (4-VP)		Basic	Hydrogen bonding, π–π stacking
Acrylamide (AAm)		Neutral	Hydrogen bonding
2-Hydroxyethyl methacrylate (HEMA)		Neutral	Hydrogen bonding

Monomer	Character	Interaction
4-Vinylphenyl-boronic acid (4-VPBA)	Acid	Formation of boronic ester bonds
2-(4-Vinylphenyl)propane-1,3-diol	Neutral	Formation of acetal bonds
4-Vinylbenzylamine	Basic	Formation of Schiff base bonds
trans-4-[p-(N,N-Dimethylamino)styryl]-N-vinylbenzyl-pyridinium chloride	Neutral	Electrostatic interactions

example, have investigated the effects of the crosslinker on the performance of MIPs for peptide and amino acid derivatives. The use of TRIM, a crosslinker with three polymerizable acrylate groups, leads to MIPs with superior selectivity and capacity in comparison to MIPs prepared with EDMA, which offers only two polymerizable acrylate groups, relating these results to the increased rigidity of the MIP prepared with TRIM. Likewise, Yilmaz et al. [18] have observed that a higher degree of crosslinking in MIPs for theophylline prepared with DVB led to improved separation in chromatographic applications. Here, the rigidity of the obtained polymeric network has been correlated with the respective imprinting effect of the MIP.

Besides the common approaches for MIP synthesis generally prepared in organic solvents, there are several template molecules that are incompatible with this environment, that is, they are predominantly soluble in water. Prominent examples are peptides or proteins. In this case, the choice of the respective crosslinker – and the corresponding functional monomer – has to be adapted utilizing water-soluble crosslinkers (Table 13.3). Hart et al. [19], for example, have prepared MIPs in aqueous media using the crosslinker EBA to obtain selective recognition properties for peptides.

Table 13.3 Selected crosslinkers commonly used in molecular imprinting.

Name	Chemical structure	Characteristics
Ethylene glycol dimethacrylate (EDMA or EGDMA)		Slightly polar
1,4-Divinylbenzene (DVB)		Nonpolar
Trimethylolpropane trimethacrylate (TRIM)		Three polymerizable vinyl groups
N-N'-Ethylene-bisacrylamide (EBA or EBAAm)		Water soluble

13.2.3.3 Radical Initiators

As MIPs are most commonly prepared by either heat- or UV-induced free radical polymerization, radical initiators derived from conventional radical polymerization reactions in macromolecular chemistry are usually applied. Among others, azobis-nitriles are the most commonly used sources of free radicals (Table 13.4). Especially, 2,2'-azobis-(isobutyronitrile) (AIBN) has found frequent applications in the preparation of MIPs. In addition to AIBN, 2,2'-azobis-(2,4-dimethylvaleronitrile) (ADVN) has been applied in several MIP approaches, providing a lower polymerization temperature as its thermal decomposition temperature is below that of AIBN (approx. 40 and 60 °C, respectively). In an application of MIPs as stationary phase in liquid chromatography, a lower polymerization temperature was beneficial to the resulting separation factor during, for example, enantiomer separation of phenylalanine anilide, as shown by O'Shannessy et al. [20]. In this study, MIPs were prepared with AIBN, ADVN, and other radical initiators using either thermally (at 30–60 °C) or photochemically (0 °C) induced polymerization. It was concluded that a lower initial temperature is favorable for the stability of the prepolymerization complex prior to and during the polymerization process. Radical polymerizations performed in water as solvent require appropriate water-soluble radical initiators. Therefore, persulfates are commonly applied, such as for example, ammonium persulfate (APS) [21, 22].

Table 13.4 Common radical initiators used for the preparation of MIPs.

Name	Chemical structure	Characteristic
2,2'-Azobis(isobutyronitrile) (AIBN)		Thermal decomposition at approx. 60 °C, also suitable for photochemical initiation
2,2'-Azobis(2,4-dimethylvaleronitrile) (ADVN or ADBV)		Thermal decomposition at approx. 40 °C, also suitable for photochemical initiation
Ammonium persulfate (APS)		Water-soluble

13.2.3.4 Solvents

Next to dissolving all reaction components prior to the polymerization, the solvent plays an essential role in molecular imprinting. In combination with the selected crosslinker, the solvent has a significant impact on the morphology of the finally obtained MIP, not only governing the porosity (e.g., the availability of pores and the pore-size distribution) but also the format of the resulting MIP (i.e., bulk copolymer, nanospheres, microspheres, rigid or soft polymer, etc.). The solvent in molecular imprinting is frequently referred to as porogenic solvent or simply porogen, which indicates its role in establishing appropriate structures providing access to the binding moieties. Sufficient porosity is desirable to ensure adequate binding kinetics and rapid mass transfer of the template molecules toward and away from the binding cavities.

Furthermore, the choice of the solvent strongly affects the recognition properties of the MIP, as the stability of the prepolymerization complex depends on the chemical environment. For example, highly polar protic solvents such as, for example, methanol or water adversely affect hydrogen bonding interactions between template and functional monomer, thereby potentially destabilizing the prepolymerization complex, which in turn may lead to MIPs with reduced selectivity. These aspects have to be especially considered during the preparation of noncovalent MIPs.

Early studies on the influence of the solvent performed by Sellergren *et al.* [23] showed the superior properties of aprotic solvents such as, for example, dichloromethane or acetonitrile in comparison to polar protic solvents such as, for example, methanol or acetic acid for the preparation of MIPs by observing an improved chromatographic separation of two phenylalanine anilide enantiomers on MAA/EDMA-MIP stationary phases prepared in aprotic solvents. Likewise, in a theoretical and experimental approach Wu *et al.* [24] have studied the imprinting efficiency of MAA/EDMA-MIPs for nicotinamide prepared in various solvents. In these studies, a high correlation between the dielectric constant as well as the hydrogen bond donor/acceptor ability of the respective solvent and the selectivity of the resulting MIP was determined. In summary, it was concluded that with increasing dielectric constant of an apolar solvent the selectivity of the resulting MIP decreases. Moreover, in highly polar protic solvents the hydrogen bond donor/acceptor ability together with the dielectric constant strongly affect the formation of the prepolymerization complex, again leading to MIPs with lower selectivity. The highest selectivity was observed for MIPs prepared in apolar solvents with low dielectric constants.

Table 13.5 gives a summary of selected solvents with respect to their dielectric constants and hydrogen bond donor/acceptor ability.

Despite the drawbacks mentioned above, there is substantial interest in the preparation of MIPs in polar protic solvents, especially in water. Besides providing a cheap and environmentally acceptable solution matrix, there is a growing interest in imprinting water-soluble molecules, especially biomolecules such as, for example, peptides and proteins, which are insoluble or frequently denature in organic media. Therefore, the development of water-based imprinting protocols is considered a major milestone for MIP technology toward antibody-like receptors applicable in biologically native environments.

13.2 Fundamentals of Molecular Imprinting

Table 13.5 Common solvents used in molecular imprinting; dielectric constants are derived from the CRC Handbook of Chemistry and Physics [25].

	Solvents		
Name	Chemical structure	Dielectric constant (20 °C)	H-bonding
Toluene		2.4 (23 °C)	Poor
Chloroform		4.8	Poor
Acetone		21.0	Moderate
Acetonitrile		36.6	Moderate
Dimethyl sulfoxide		47.2	Moderate
Methanol	—OH	33.0	Strong
Water		80.1	Strong

In the literature, there are some examples of successful molecular imprinting in aqueous, or at least highly polar protic, environments. Nevertheless, successful imprinting approaches in neat aqueous solution are still comparatively rare. Haupt et al. [26] have imprinted 2,4-dichlorophenoxyacetic acid in a 4-VP/EDMA matrix using a mixture of methanol and water (4:1) as solvent. The resulting MIP revealed high selectivity toward the imprinted template in comparison to various structural analogs. The obtained results were attributed to ionic and hydrophobic interactions (π–π stacking) as the predominant driving forces for stabilizing the prepolymerization complex, and for rebinding of the template in polar solutions. Especially, the rather non-specific hydrophobic interactions turned out to be notably strong under these conditions. More recently, Rachkov et al. [27] have reported the preparation of a peptide-selective MIP in a 5 mM aqueous phosphate buffer (at pH 7) using [Sar^1,Ala^8] angiotensin II as template molecule in a sodium acrylate/poly(ethylene glycol) diacrylate copolymer matrix. Thereby, it was observed that the ionic interactions between the template and the functional monomer are the predominant interactions providing the selectivity of the respective MIP.

For further information on molecular imprinting in aqueous media the reader is referred to an elaborate survey by Janiak et al. [28] of several approaches toward molecular imprinting of proteins and peptides in aqueous solutions developed in recent years.

In general, one can observe that optimum rebinding of the template is achieved in the same solvent used during the polymerization process, to avoid swelling problems. However, there are various approaches showing MIPs prepared in apolar organic solvents but finally successfully applied in aqueous environments. Andersson [29], for example, has prepared a MIP for *(S)*-propranolol in toluene that also showed remarkable recognition properties in water. However, different recognition profiles were observed, which were attributed to a different balance between hydrophobic and polar interactions in both solvents. Likewise, Janotta *et al.* [30] have successfully prepared a MIP for 4-nitrophenol that offered selective recognition in organic as well as in aqueous media. In contrast, Weiss *et al.* [31], for example, have reported that the imprinting effect of a stationary phase imprinted against mycotoxins is markedly decreased with increasing amounts of water in the mobile phase, whereas in pure acetonitrile as mobile phase, which was also used as porogenic solvent during synthesis of the MIP, a distinct imprinting effect could be observed.

13.2.4
How are MIPs Prepared?

The preparation of a MIP (Figure 13.4) starts by mixing of all reaction components, including template (T), functional monomer (FM), crosslinker (CL), and radical initiator, within the porogenic solvent. Typical ratios of T: FM: CL are in the range 1: 4: 20 or 1: 8: 40. Even ratios of template to functional monomer of 1: 100 [32] or 1: 1000 [18] have been reported without significant decrease in the performance of the resulting MIP. After removal of the dissolved oxygen from the reaction mixture – which otherwise may affect the polymerization process – by, usually, bubbling with nitrogen or argon or by sonication, the polymerization process is either thermally or photochemically initiated. Typical polymerization durations are in the range 8–24 h, whereby thermal polymerization is routinely performed at 45–70 °C, depending on the respective radical initiator. In contrast, photochemically induced polymerization is usually performed at lower temperatures (e.g., 0 °C) using UV-irradiation (e.g., mercury vapor lamp) to decompose the radical initiator. After the polymerization process has finished, the obtained MIP has to be prepared for application. Bulk MIPs have to be subsequently crushed, ground, sieved, and the fines have to be removed, whereas MIPs prepared by other polymerization methods (e.g., precipitation or

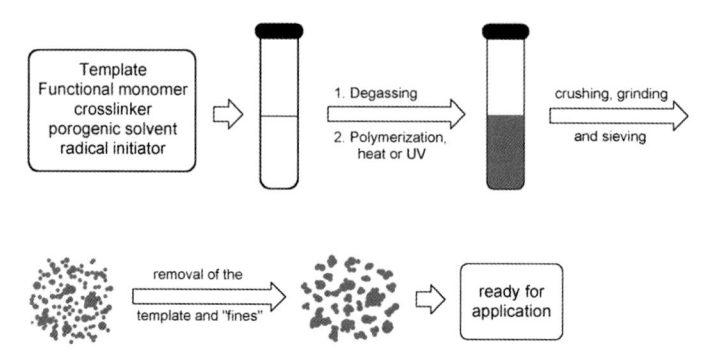

Figure 13.4 General scheme for the preparation of bulk MIPs.

suspension polymerization) may be directly applicable after removal of the template (e.g., via Soxhlet extraction).

13.3
Polymer Formats and Polymerization Methods for MIPs

Several – mainly radically initiated – polymerization methods have been applied in molecular imprinting, leading to various polymer formats. While most MIPs are still prepared as bulk copolymers (Figure 13.5a), a significant focus has recently been on the preparation of more sophisticated polymer formats readily adapted to the final application needs of the obtained MIP such as, for example, microspherical beads (Figure 13.5b) for chromatographic applications or SPE (solid-phase extraction), nanospherical beads (Figure 13.5c) for binding assays, microfluidic separations, or thin films, and membranes for sensing applications or coatings for selective removal purposes (Figure 13.5d). Thereby, several advanced polymerization methods have been introduced, demonstrating the versatility of the free radical polymerization method in preparing such various MIP formats.

The following provides an overview of the most important polymerization methods and MIP formats, complemented by selected examples.

13.3.1
Bulk Polymers

Today, bulk polymerization can be considered the most universal approach for generating MIPs. By far the majority of the MIPs are prepared via this method, given that it is straightforward and experimentally easy to handle. Even though the initial preparation is simple, subsequent processing of bulk MIPs is wasteful and time consuming, as the resulting monolithic structure has to be ground and sieved for separating into particle sizes of interest, usually in the range 10–25 μm, for example, for chromatographic applications. Evidently, grinding bulk polymers leads to a wide-range particle size distributions and irregular particle shapes, which may be unfavorable properties for some applications. In addition, a considerable fraction of

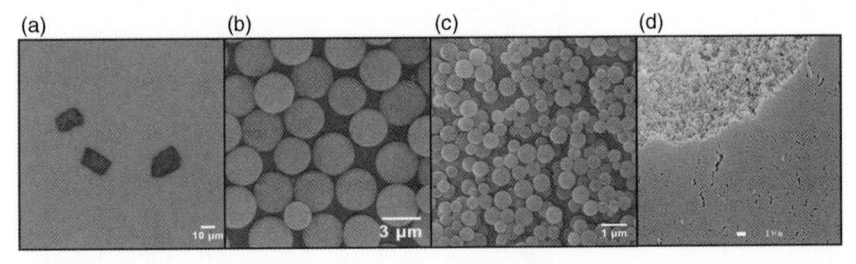

Figure 13.5 Common MIP formats: (a) bulk MIP particles after grinding and sieving (<25 μm); (b) MIP microspheres; (c) MIP nanospheres; (d) MIP film grafted at a substrate surface. All images are SEM micrographs except (a), which is an optical micrograph.

particles end up as so-called fines (particle size < 1 μm), which significantly reduces the yield of the desired μm-sized MIP fraction. In general, less than 50% of the MIP prepared via bulk polymerization will be available for application as stationary phase, for example, in chromatography [33], thus rendering this method not only tedious and wasteful but also costly, especially if expensive templates are imprinted.

Nonetheless, bulk polymerization has proven to be a powerful preparation technique for MIPs for a large number of templates, including, for example, drugs such as penicillin G [34] and ibuprofen [35], herbicides such as phenoxy acetic acids [36] and atrazine [37], flavonoids such as quercetin [38, 39], and even various mycotoxins such as deoxynivalenol and zearalenone [31, 40] moniliformin [41], and ochratoxin A [42, 43] (for chemical structures of templates see Figure 13.6).

A method that has been introduced to overcome some of the drawbacks of bulk MIPs is the *in situ* preparation of monolithic MIPs, which offers the opportunity of establishing MIP-based stationary phases, for example, HPLC or capillary electrochromatography without grinding, sieving, and packing of particulate material into a column or capillary. Matsui *et al.* [44], for example, have prepared MIP rods for cinchona alkaloids directly inside a chromatographic column simply by filling a stainless steel column (50 × 4.6 i.d.) with a previously degassed prepolymerization solution consisting of the template, EDMA, ADVN, MAA, or (trifluoromethyl) acrylic acid (TFMAA) and a cyclohexanol–1-dodecanol mixture as porogenic solvent. After thermal polymerization at 45 °C for 6 h and subsequent extraction of the template, the MIP was instantaneously ready for application, revealing excellent recognition characteristics. Furthermore, Schweitz *et al.* [45] have

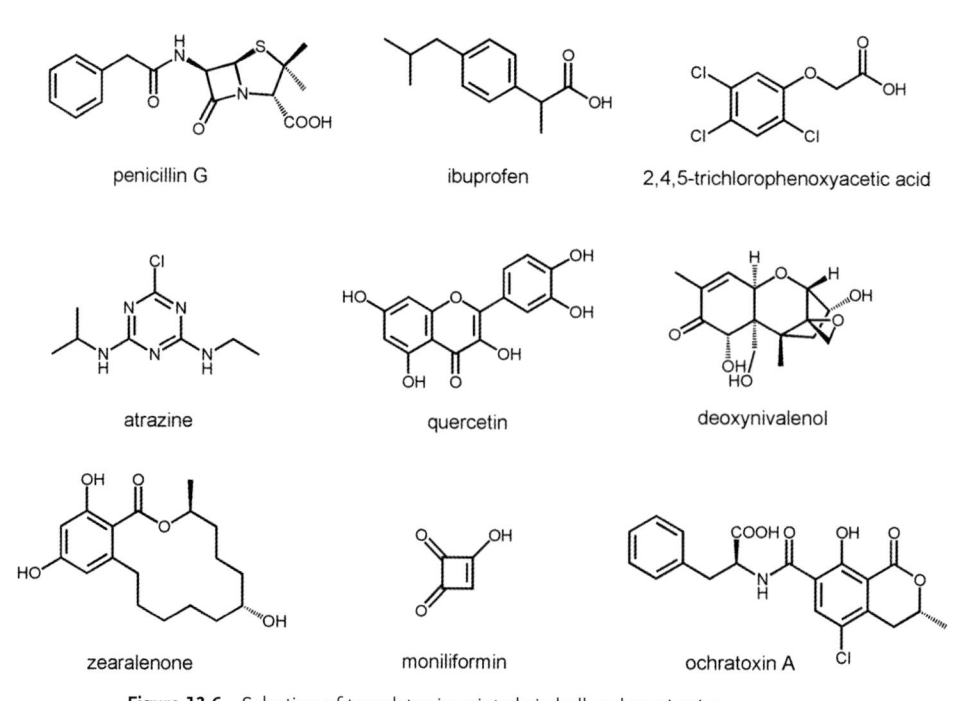

Figure 13.6 Selection of templates imprinted via bulk polymerization.

described the preparation of a monolithic MIP directly inside a fused silica capillary (75 μm i.d.) by *in situ* copolymerization of MAA and TRIM, thereby achieving chiral separation of a racemic mixture of the templates propranolol and metoprolol within less than 120 s.

The ease and straightforwardness of the *in situ* preparation for monolithic MIPs serving as stationary phases certainly has its merits over particulate polymers. However, the preparation of high-affinity binding sites, as well as ensuring sufficient porosity during direct synthesis inside a separation system, remains a challenging task, and is frequently lacking in control of the synthesis and, therefore, of the physical parameters of the obtained monolith [46]. An elegant alternative to overcome the drawbacks of MIPs prepared by bulk polymerization is the preparation of MIPs as micro- or nanobeads, which will be discussed in the next section.

13.3.2
Micro- and Nanobeads

In liquid chromatography, the size and shape of the particles comprising the stationary phase may have a strong impact on the efficiency of the separation system. Superior chromatographic results are usually achieved with spherical particles offering a narrow size distribution in the range of 10 μm or smaller, since the contribution of the eddy diffusion to peak broadening is diminished, and the backpressure of the column is not yet too high, thus still enabling high flow rates and consequently short and efficient analysis times. Hence, novel polymerization techniques have been introduced to form MIPs, including, for example, suspension, precipitation, seed, emulsion, or multistep swelling polymerization (Figure 13.7).

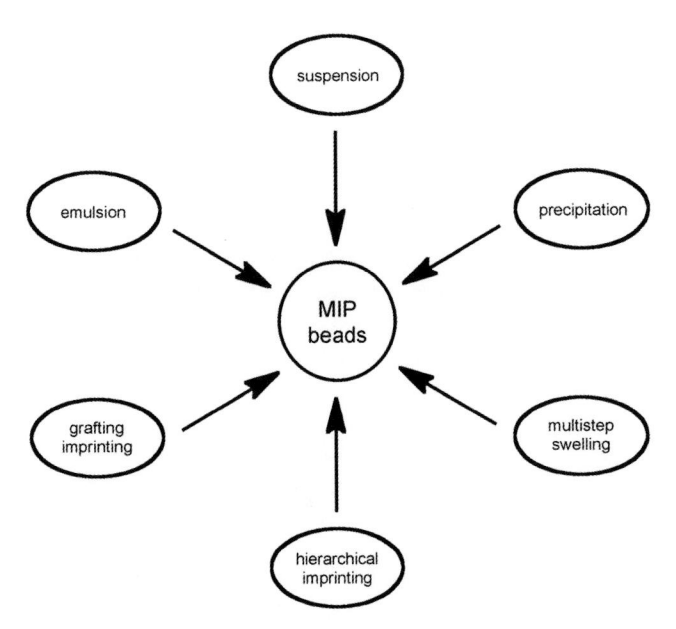

Figure 13.7 Polymerization methods generating micro- or nanobead MIPs.

Moreover, grafting imprinted coatings onto spherical solid supports such as, for example, silica or organic polymer particles like latex, and hierarchical imprinting where the solid silica support is finally dissolved have been reported to overcome the limitations of bulk MIP particles without affecting, for example, enantioselectivity, with particular improvements in mass transfer and yield of useful material (>90%) [47].

Especially for microfluidic systems such as, for example, capillary electrochromatography, spherical nanoparticles (average particle diameter: 160 nm) [48] have been prepared, nowadays providing access to MIP beads with particle diameters ranging from few tens of nanometers up to 100 µm and more. Table 13.6 provides an overview of selected examples of MIP bead preparations.

A further notable method is *in situ* grafting polymerization reported by Zhuang et al. [60]. Here, a MIP is grafted as a layer onto activated hollow glass microspheres (diameter 7–20 µm) directly within a HPLC column (50 × 4.6 mm i.d.). After purging with nitrogen for 5 min, an *in situ* polymerization was performed at 60 °C for 24 h with AAm and TRIM in THF using emodin as template and AIBN as radical initiator. Subsequently, chromatographic comparison of the grafted glass microspheres with an equally prepared bulk MIP column revealed an improved mass transfer rate, lower back pressure, and higher efficiencies for the grafted MIP; however, cross-reactivity towards some structural analogues was observed.

In recent years, the introduction of novel polymerization techniques for the preparation of MIPs has led to various new polymer formats. Especially in affinity-based chromatography, microsphere MIPs are increasingly adopted as the format of choice, as they offer some distinct advantages over conventionally prepared bulk MIPs. However, a phenomenon regularly observed in MIP-based chromatography, especially with MIPs prepared via noncovalent interactions (independent of whether applied as ground bulk polymer particles or monodisperse microspheres), is the asymmetry of the template peak, which usually shows a strong tailing behavior along with a very broad peak width. This observation is mainly associated with the heterogeneous binding site affinity distribution, and slow desorption kinetics [61]. However, a heterogeneous binding site distribution – and thus lower selectivity – is not *per se* disadvantageous, and depends on the application of the synthesized MIP (Figure 13.8). For example, for a one-step simultaneous selection and/or enrichment of various structurally analogous templates such as, for example, metabolites a MIP with lower selectivity may offer advantages. Likewise, MIPs with a relatively low selectivity though offering a high binding capacity may provide desirable properties for the removal of certain constituents. Last but not least, slow desorption kinetics may further favor applications involving scavenging and removing certain constituents from a sample matrix.

13.3.3
MIP Films and Membranes

Next to MIPs with well-defined size and shape, imprinted materials represented as films and membranes are of increasing importance. The possibility of generating

Table 13.6 Selected examples of MIPs prepared in bead format.

	MIP beads			
Polymerization method	Template	Matrix	Particle size (μm)	Reference
Precipitation	Theophylline/caffeine	MAA/DVB-80	3–10 (monodisperse)	Wang et al. [49]
Precipitation	Di-(2-ethylhexyl) phthalate	MAA/EDMA MAA/TRIM	3–3.5 2–2.5	Lai et al. [50]
Precipitation	(R/S)-Propranolol	MAA/DVB-55, MAA/TRIM	0.13–2.4 (monodisperse)	Yoshimatsu et al. [51]
Suspension	Boc-L-Phe	MAA/EDMA, MAA/TRIM	1–100 (polydisperse)	Mayes and Mosbach [33]
Suspension	(R/S)-Propranolol	MAA/TRIM	1–100 (polydisperse)	Kempe and Kempe [52]
Suspension	(S)-Propranolol	Magnetic iron oxide core, MAA/TRIM shell	1	Ansell and Mosbach [53]
Two-step swelling	Bisphenol A	4-VP/EDMA	8	Watabe et al. [54]
Multistep swelling	Green tea catechins	2-VP/EDMA	5–6	Haginaka et al. [55]
Photografting	L-Phenylalanine anilide	MAA/EDMA onto spherical silica support	10 (film thickness 1–2 nm)	Titirici and Sellergren [56]
Photografting	Pyrimethanil	MAA/EDMA onto chloro-methylated polystyrene beads	25–40	Baggiani et al. [57]
Hierarchical imprinting	(−)-Isoproterenol	TFMAA/DVB-80	13	Yilmaz et al. [47]
Hierarchical imprinting	Dipeptide	MAA/EDMA	10	Titirici and Sellergren [58]
Core–shell emulsion	cholesterol	Polystyrene core, poly-DVB shell	0.03–0.209 (monodisperse)	Perez et al. [59]

thin MIP films and membranes with controlled thickness and porosity at planar or corrugated surfaces, or as freestanding membrane renders MIPs ideally suitable as selective recognition layers, for example, in sensing devices, if appropriate binding kinetics for rapid signal generation are ensured. Moreover, the application of MIPs as sensing membrane has some persuasive advantages, including, for example, high chemical and thermal stability accompanied with straightforward preparation procedures, thus providing a viable alternative to biological receptors usually applied in (bio)sensing devices. In the following, we discuss selected preparation and application aspects of MIP-based films and membranes, providing insight into one of the most promising fields of MIP research.

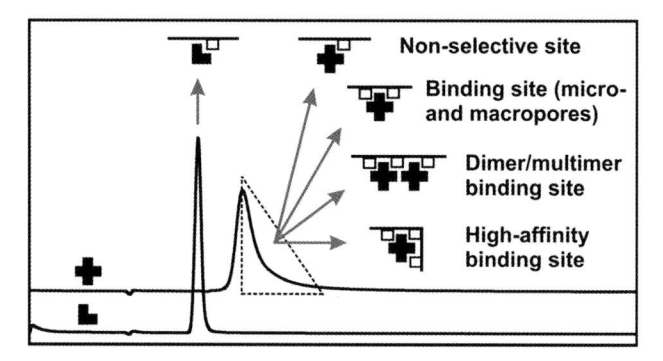

Figure 13.8 Different types of binding sites in noncovalently imprinted polymers contributing to the binding site/affinity heterogeneity.

One crucial aspect for the preparation of MIP-based sensing membranes is the attachment or immobilization of the MIP layer at the transducer surface, as this strongly affects a reproducible signal generation. Panasyuk-Delaney et al. [62], for example, have utilized grafting polymerization for the preparation of an impedometric herbicide chemosensor (Figure 13.9). Here, circular polypropylene membranes as well as self-assembled mercaptohexadecane coated gold electrodes served as solid support material, at which a grafting polymerization process based on 2-acrylamido-2-methyl-1-propane sulfonic acid (AMPS) and N,N'-methylene-diacrylamide (MBA) in the presence of the template desmetryn was initiated. The resulting polymer layer was approx. 10 nm thick. Thus prepared sensors revealed a useful response time of approx. 5 min along with a highly reproducible and reversible response. Moreover, the sensor showed high selectivity toward the template molecule desmetryn in comparison to other triazines, with the exception of metribuzine, which showed comparable response; in addition, a shelf life of at least 6 months was demonstrated when stored at ambient conditions.

Belmont et al. [63] have reported the preparation of a MIP film of controlled porosity onto a glass transducer for spectroscopic sensors utilizing reflectometric interference by either spin-coating followed by *in situ* polymerization or autoassembly of MIP nanoparticles assisted by an associative linear polymer. In the latter case, in the first step the MIP nanoparticles were prepared by mini-emulsion polymerization of MAA with EDMA or TRIM in the presence of the herbicide atrazine, which served as template, thereby leading to particle diameters of 264 and 330 nm for EDMA and TRIM copolymers, respectively, using 1,1,2,2-tetrachloroethane as a solvent. After purification, the nanoparticles were coated onto the transducer surface using poly(ethylene imine) as linker (Figure 13.10). Likewise, spin coating was performed using the same ingredients, resulting in MIP films 1–2 μm thick; however, the addition of poly(vinyl acetate) was necessary to ensure acceptable recognition properties. In conclusion, both MIP-based sensors revealed reproducible and reversible results for the detection of atrazine at concentrations of approx. 8 μM (for the nanoparticle-based MIP film). However, poor sensor-to-sensor reproducibility and stability issues have been reported for these concepts.

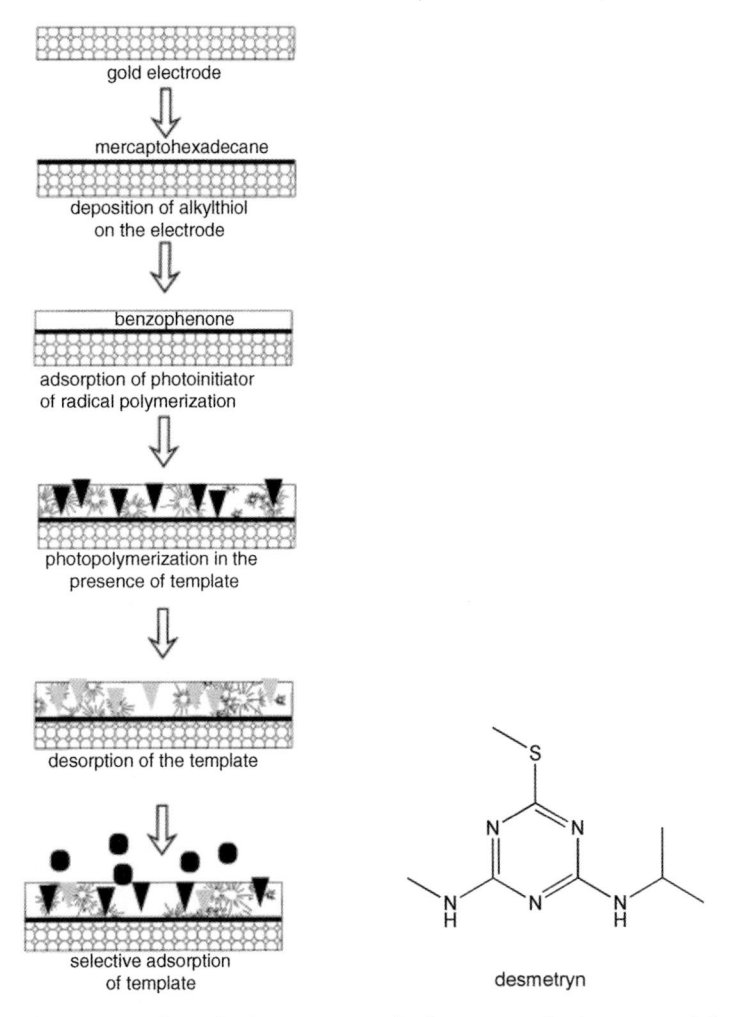

Figure 13.9 Scheme for the preparation of a chemosensor for desmetryn and the chemical structure of the template. Reproduced with permission from Reference [62].

Besides the two examples discussed herein, various alternative concepts for the preparation and application of MIP films and membranes have been reported in the literature recently, and have mainly been coupled with sensing devices based on mass sensitive detection [e.g., quartz crystal microbalance (QCM)] [64–66], optical detection [e.g., FT-IR spectroscopy, surface plasmon resonance (SPR), fluorescence or luminescence] [67–70], or electrochemical transduction (e.g., conductometry, potentiometry, or voltammetry) [71–73], demonstrating the interest in interfacing sensor research with MIP technology. Furthermore, results on MIP membranes for selective analyte transport [74–76] or serving as SPE material [77–82] have been reported.

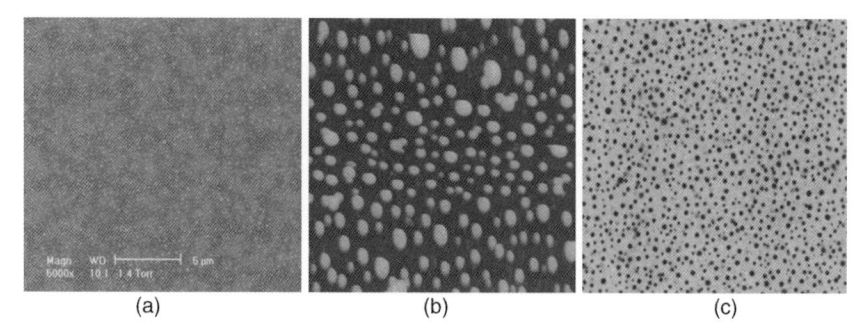

Figure 13.10 (a) SEM image of surface-coated nanoparticles via poly(ethylene imine) linkage; (b) AFM obtained at the same sample (size: 5 μm × 5 μm, z-range: 400 nm); (c) contact mode AFM image of a MIP layer prepared via spin coating (size: 15 μm × 15 μm, z-range: 25 nm). Adapted with permission from Belmont et al. [63].

13.3.4
Comparison of MIP Formats Prepared by Different Polymerization Methods

Following the introduction of some preparation strategies for MIPs, this section focuses on the evaluation of MIP performance, if similarly templated materials are prepared by different polymerization methods, with particular emphasis on their application as a chromatographic separation matrix. Hence, selected examples from the literature are presented, highlighting potential effects of the polymerization method on the performance of the resulting MIP.

A first comparison of MIPs prepared by different polymerization methods was reported 25 years ago by Norrlöw et al. [83], who early on recognized the influence of particle size and shape on the chromatographic performance of MIPs. In this study, several dyes, including, for example, rhodanile blue and Safranine O, were imprinted via copolymerization of the functional monomer methyl methacrylate in the presence of the two crosslinkers, N,N'-1,4-phenylenediacrylamide and N,N'-methylenediacrylamide, in a 1:1 mixture of DMF and water either by conventional bulk polymerization or by grafting onto highly porous silica particles with a diameter of 10 μm. The bulk MIP was then crushed and sieved to obtain particle dimensions of 300–500 μm, whereas for the MIP-coated silica particles no further processing was applied except for the extraction of the template and the removal of agglutinated particles. Subsequent chromatographic comparison of the ground bulk polymer and the MIP-coated silica particles revealed only minute differences in terms of selectivity factors, and the overall efficiency of the obtained separation columns was comparatively low (100–200 theoretical plates per m). However, it was shown that the chromatographic performance of MIP-coated particles enabled at least a 5–10 times faster separation than when using crushed bulk polymer particles.

Fairhurst et al. [84] have also studied the influence of the polymerization method on the performance of MIPs in HPLC applications. MAA/EDMA-MIPs of similar composition imprinted for (S)-propanolol were prepared by bulk or suspension polymerization or by grafting onto spherical silica particles. During subsequent chromatographic measurements, the obtained MIP beads offered significantly better flow properties and chromatographic peak shapes than ground bulk MIPs. Furthermore, the silica-grafted MIP revealed the narrowest peak shape and shortest analysis time among the three approaches, which once again demonstrated the evident superiority of – ideally monodisperse – spherical MIP particles over ground bulk MIPs in chromatographic applications.

Probably the most comprehensive evaluation of different polymerization methods was published by Pérez-Moral and Mayes [85], as they have compared MIPs prepared under similar conditions using a wide variety of polymerization methods, including bulk, suspension, emulsion core–shell, two-step swelling, and precipitation polymerization. The most intriguing aspect of this study is the direct comparison of MIP beads prepared by different polymerization methods but with comparable physical properties such as, for example, bead diameter or morphology. Hence, a direct comparison of the imprinting efficiency and, thus, an evaluation of the selected polymerization method is enabled for (S)-propranolol as the template. This study, especially for beads prepared by precipitation polymerization, reported that the amount of selectively rebound template is apparently greatly influenced by the solvent, in which rebinding occurs. In organic solvents, MIP beads prepared via this method offered the highest fraction of selectively rebound template (50%) among all investigated polymerization methods, whereas in aqueous buffer no selective rebinding was observed at all. For all other polymerization methods, a general trend of decreasing fractions of selective rebinding in aqueous buffer in comparison to the organic solvent could be observed. However, Moral and Mayes pointed out that in order to maintain comparable synthesis conditions among the different methods the optimization of synthesis parameters for individual polymerization methods was sacrificed.

The chromatographic performance of MIPs prepared against 17β-estradiol either by bulk polymerization or by precipitation polymerization have been compared by Wei and Mizaikoff [86], revealing a faster elution of two structurally analogous compounds (estrone and 17α-estradiol) and 17β-estradiol from the bulk-MIP column (Figure 13.11b) in comparison to the microsphere column (Figure 13.11a). Further-

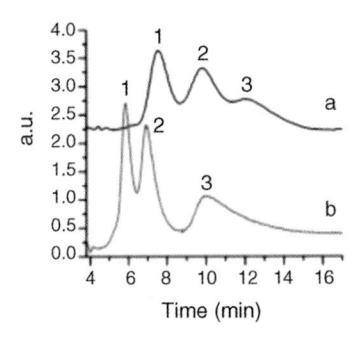

Figure 13.11 Comparison of the chromatographic separation of estrone (peak 1), 17α-estradiol (peak 2) and 17β-estradiol (peak 3) on a microsphere column (a) and on a bulk-MIP column (b). Adapted with permission from Reference [86].

more, a complete separation of 17β-estradiol from its structural analogs could only be achieved on the bulk-MIP column.

In conclusion, it is apparent that individual polymerization methods provide distinct advantages and disadvantages for MIP preparation, and it is evidently difficult to predict which polymerization method may be the most promising strategy for a certain template. Thus, fundamental considerations – including, for example, the solubility of the template in certain solvents or solvent mixtures, which are required for different polymerization methods, the desired morphology of the resulting MIP, the final purpose of the prepared MIP, and related parameters – may point towards a particular polymerization method.

13.4
Evaluation of MIP Performance – Imprinting Efficiency

The performance of MIPs is mainly described by three factors: (i) the binding capacity, (ii) the binding affinity, and (iii) the obtained selectivity, which may be summarized as the overall the imprinting efficiency (Figure 13.12). Binding assays and chromatographic evaluations are among the most widespread methods for the determination of these critical factors, and will be discussed within the following sections.

One crucial point in the evaluation of the performance of MIPs is the contribution of the non-selective to the total binding, since it falsifies the results obtained for the imprinting efficiency of the MIP in every case. This circumstance is regularly taken into account through polymers prepared in the same manner as the MIP but in absence of the template. In the literature, these polymers are often referred to as non-imprinted polymer (NIP or NP), control polymer (CP), or blank polymer (BP).

13.4.1
Binding Capacity and Binding Affinity

In general, the binding capacity describes the maximum amount of template that can be selectively bound by the MIP matrix, for example, during a rebinding experiment. The binding capacity is related to the porosity of the material and, hence, to the specific surface area of a MIP. Usually, one can observe a direct correlation between specific surface area and binding capacity of the MIP; however, the amount of available binding pockets and the pore size distribution may limit the rebinding of the template. Thus, one has to keep in mind that, for example, in noncovalent imprinting usually only approx. 15% of the binding cavities are available for rebinding of the template, for example, due to the shrinking of the binding cavities and so on [3]. Such effects not only decrease the binding capacity of the MIP, but also contribute to so-called template leaching (i.e., continuous leaching of remaining template molecules from the MIP due to incomplete removal of the template during the extraction), which is among the critical aspects when applying MIPs in ultra-trace enrichment and analysis.

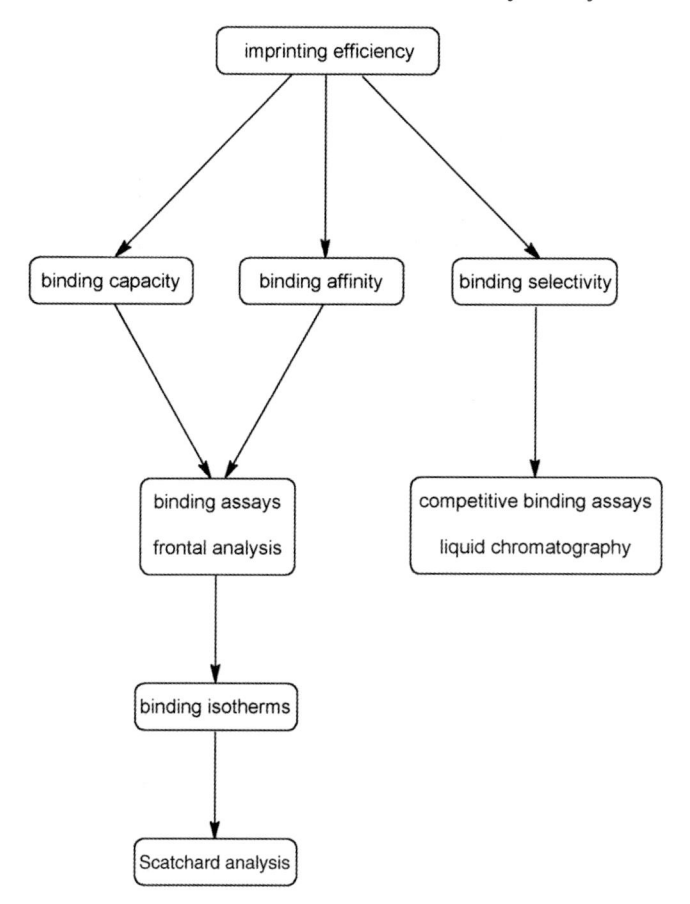

Figure 13.12 Common strategy for evaluation of the imprinting efficiency.

As the binding affinity reflects the binding strength between the template and its respective binding cavity in the MIP, it is directly related to the equilibrium association constant (K_A), and to the equilibrium dissociation constant (K_D), of the template/binding cavity associate, which is defined as:

$$\text{template} + \text{binding cavity} \underset{K_A}{\overset{K_D}{\rightleftarrows}} \text{template/binding cavity associate}$$

(13.1)

Both parameters are most commonly determined by frontal analysis [87–89] or via batch rebinding assays [90–92]. In the latter case (Figure 13.13), a defined amount of MIP is incubated with initial solutions *(I)* of different concentrations of the neat template or – alternatively – a radiolabeled template (i.e., template molecules where, for example, protons ^{1}H are exchanged by radioactive tritium ^{3}H), thereby establishing a radioligand binding assay. After equilibration involving typically 2–24 h of stirring, the supernatant solution is separated from the MIP particles by filtration or

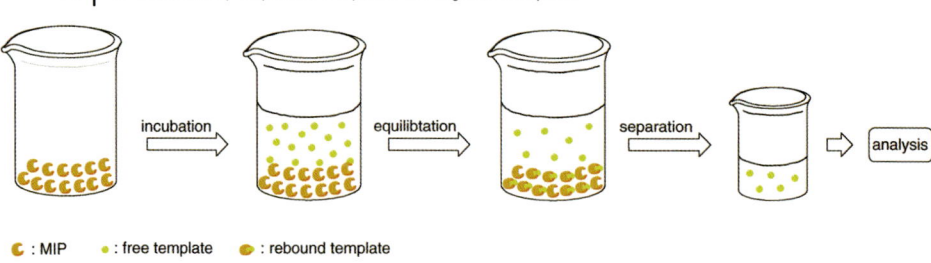

Figure 13.13 Scheme of a batch rebinding assay.

centrifugation, and the concentration of the free template *(F)* left in solution is determined, for example, via HPLC or spectroscopic measurements or by scintillation detection if a radiolabeled template is applied.

The amount of bound template *B* may be calculated using Eq. (13.2):

$$B = I - F \qquad (13.2)$$

From the experimental binding data, an experimental binding isotherm may then be obtained by plotting the amount of template bound to the MIP against the concentration of the free template within the respective individual incubation solutions (*B* versus *F*) (Figure 13.14). By comparing the experimental binding isotherms obtained from a MIP and a NIP in one plot, a qualitative statement on the success of imprinting may be derived simply by comparing the amount of template rebound by the MIP and the NIP.

Moreover, quantitative conclusions on the equilibrium dissociation constant [K_D, which represents the inverted equilibrium association (affinity) constant, K_A] and on the maximum number of binding sites (B_{max}) (i.e., the binding capacity) of the MIP

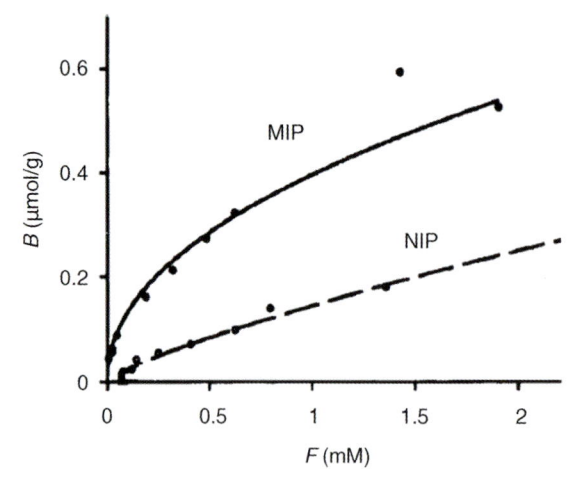

Figure 13.14 Binding isotherms of a MIP and a NIP against adenine-9-acetate. Adapted with permission from Reference [93].

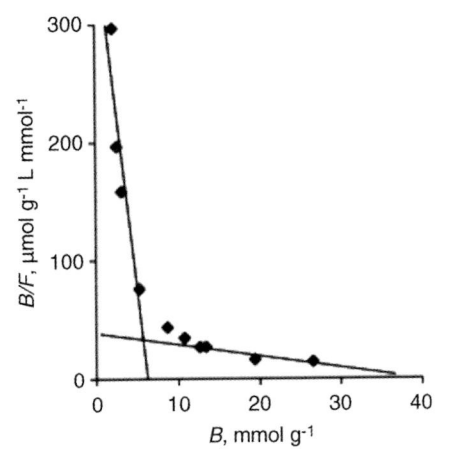

Figure 13.15 Scatchard plot of a MIP against adenine-9-acetate separated into high- and low-affinity binding sites. Adapted with permission from Reference [95].

may be drawn by plotting the ratio of bound to free template (B/F) against the amount of free template (F) according to the Scatchard equation (13.3):

$$B/F = (B_{max} - B)/K_D \qquad (13.3)$$

After linear regression, B_{max} and K_D may be determined by the x-intercept and the slope, respectively. This data representation is usually referred to as Scatchard plot in the literature, and is frequently applied for the determination of the binding parameters of MIPs [91, 94]. However, for analysis of the binding properties of MIPs, usually more than one linear area is observed in the Scatchard plot due to the heterogeneous binding site distribution, which is especially prevalent in MIPs prepared by the noncovalent approach. Thus, a differentiation into high- and low-affinity binding sites is necessary (Figure 13.15).

A Scatchard analysis is based on the Langmuir isotherm, which is a homogeneous binding model assuming that all binding sites are identical, which is the case, for example, for enzymatic systems or monoclonal antibodies. However, low average binding affinities (i.e., high K_D), as well as highly heterogeneous binding site distributions, are most commonly observed in MIPs [88]. Thus, the application of homogeneous binding models, although most commonly applied, might not be the most appropriate model for generically describing the binding site properties of MIPs. Consequently, great emphasis has been laid on the adaptation of theoretical binding models considering the heterogeneity of the binding site distribution encountered for most MIPs during the evaluation of their binding behavior. Therefore, experimental binding isotherms have been adapted to binding models such as, for example, the Freundlich [11, 96, 97] or the Langmuir–Freundlich isotherm [96], which describe the binding adsorption behavior at heterogeneous surfaces. Evaluations of the applicability of different binding site models for describing the binding behavior of MIPs have been reported elsewhere [95, 98]. In particular, the fitting of experimental binding isotherms to the Freundlich isotherm revealed excellent agreement with various MIPs such as, for example, MIPs against L-phenylalanine anilide [99], aminoantipyrine [100], or hemoglobin [101].

13.4.2
Binding Selectivity

Determining the selectivity generally provides information on the ability of a MIP to discriminate between the template and its (structural) analogues or simply any other compound. Therefore, in describing the selectivity of MIPs one has to discriminate between selective and non-selective binding, with the latter also referred to as cross-reactivity in the literature, which is a term again adopted from antibody assay experiments. High selectivity in general indicates that the MIP predominantly rebinds the template molecule used during the preparation of the respective MIP, or the target molecule in cases where a dummy has been used for the imprinting procedure, instead of structural analogues, which are offered, for example, in a competitive rebinding assay. Competitive batch rebinding assays include analogues of the template/target within the respective incubation solution that compete with the template/target for the available binding cavities (Figure 13.16). In addition to a conventional competitive batch rebinding assay, where structural analogs of the template are applied, radioligand binding assays – as described in the previous section – may also be performed to determine the selectivity of a MIP. Here, radiolabeled templates compete with the regular templates and/or structural analogs for the free binding sites of the MIP. After subsequent separation of the MIP particles from the incubation solution, the amount of bound radioligand may be calculated by measuring the radioactive decay in the separated solution using a scintillation counter. Since the radiolabeled template most closely resembles the regular template used during MIP preparation, radioligand binding assays are among the few methods providing information on the absolute selectivity of a MIP. However, radioligands are frequently very expensive and not available for all templates of interest. Furthermore, additional precautionary measures are required for handling radioactive/radiolabeled materials.

As well as the application of competitive batch rebinding assays, the selectivity of MIPs is most commonly determined by liquid chromatography, that is, first and foremost by HPLC measurements. However, using HPLC is limited to MIP particles

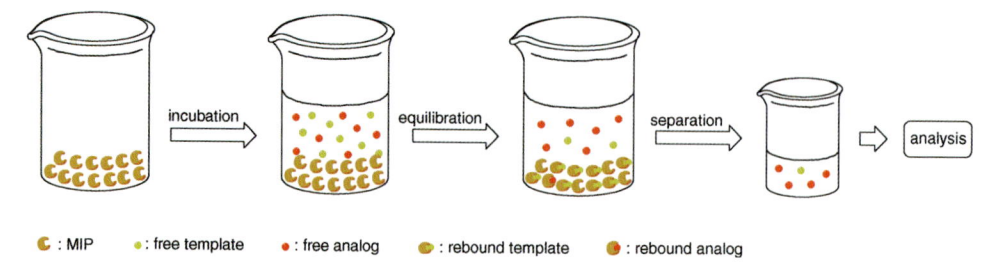

Figure 13.16 Scheme of a competitive batch rebinding assay.

with appropriate particles dimensions (several micrometers and above) to be packed into a HPLC column. To evaluate the selectivity of a MIP, chromatographic data such as the capacity factor $(k)'$, the separation factor (α), the selectivity factor (S), and the imprinting factor (I) are frequently applied. The capacity factor k' indicates how strongly the respective compound is retained by the stationary phase, thus not only providing information on the selectivity but primarily on the binding affinity. The capacity factor $(k)'$ is generally defined by Eq. (13.4), where t_R is the retention time of the respective compound, for example, the template or a structural analog, and t_0 is the retention time of a void marker (e.g., acetone or thiourea):

$$k' = (t_R - t_0)/t_0 \tag{13.4}$$

In the case of successful molecular imprinting, the capacity factor of the template on the MIP column should be remarkably elevated in comparison to the NIP column, whereas the capacity factors of the structural analogs should remain almost identical on both the MIP and the NIP columns.

For direct comparison of the capacity factors of the template and a structural analog, the separation factor (α) has been adopted from conventional chromatographic separations. The separation factor thereby represents the quality of the separation between the template and a structural analog on the same MIP column; in other words, α represents how much better the template binds to the polymer in comparison to its structural analog; α is defined as the ratio of the capacity factor of the template to that of the structural analogue. Consequently, the higher the separation factor the better the discrimination between the template and the structural analog and, therefore, the better the selectivity of the synthesized MIP matrix. However, notably, the peak shape and peak width, especially of the template peak, also need to be considered:

$$\alpha = k'_{\text{template}}/k'_{\text{analog}} \tag{13.5}$$

A direct relation between the separation factors of the template on the MIP and on the NIP is provided by the selectivity factor (S). The retention index (R_I) and the selectivity factor (S) representing the ratio between separation factors ultimately describe the selectivity of the MIP toward its template, thereby numerically removing the amount of non-selective binding. As one can see from the equation below, high selectivity is achieved if the difference between the separation factors is maximized:

$$R_I = \alpha_{\text{NIP}}/\alpha_{\text{MIP}} \tag{13.6}$$

$$S = \alpha_{\text{MIP}}/\alpha_{\text{NIP}} \tag{13.7}$$

An example of an exhaustive chromatographic evaluation of the imprinting efficiency of MIPs prepared against 4-nitrophenol (4-NP) with respect to the capacity factors, separation factors, selectivity factors, and retention indices in both organic and aqueous mobile phase has been provided by Janotta *et al.* (Figure 13.17) [30].

Apart from the typical peak asymmetry of the template peak characterized by distinct tailing due to the heterogeneity of the binding sites, the longer retention time

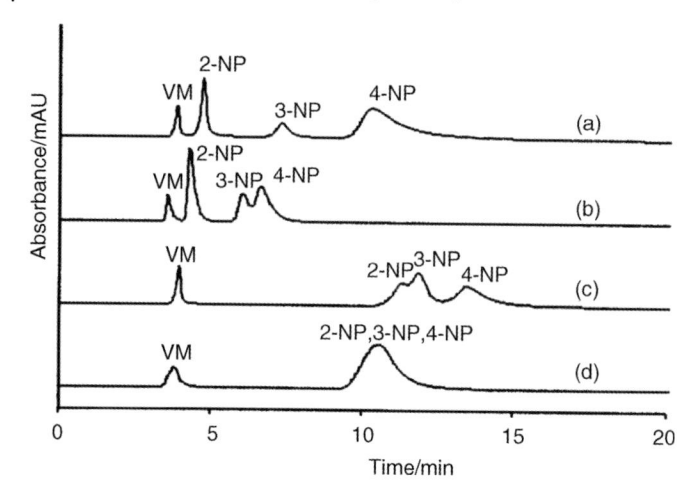

13 Molecularly Imprinted Polymers as Artificial Receptors

Figure 13.17 Chromatograms of a mixture of a void marker (VM), 2-NP, 3-NP, and 4-NP obtained at: (a) MIP-column against 4-NP applying an organic mobile phase, (b) NIP-column applying an organic mobile phase, (c) MIP-column against 4-NP applying an aqueous mobile phase, and (d) NIP-column applying an aqueous mobile phase. Reproduced with permission from Reference [30].

Table 13.7 Capacity factors, separation factors, selectivity factors, and retention indices derived from chromatograms of different analytes obtained at MIP-columns against 4-NP and the respective NIP-columns.

Analytes	Organic mobile phase					Aqueous mobile phase				
	k'_{NIP}	k'_{MIP}	α_{NIP}	α_{MIP}	R_I	k'_{NIP}	k'_{MIP}	α_{NIP}	α_{MIP}	R_I
4-NP	0.84	1.76	1	1	1	1.82	2.42	1	1	1
2-NP	0.19	0.22	4.39	8.07	0.54	1.82	1.83	1.00	1.32	0.76
3-NP	0.66	0.91	1.28	1.94	0.66	1.80	1.98	1.01	1.22	0.82
p-Cresol	0.31	0.41	2.70	4.03	0.62	1.16	1.28	1.56	1.89	0.83
Phenol	0.36	0.44	2.34	4.03	0.58	0.99	1.07	1.82	2.26	0.80
4-Nitrobenzyl alcohol	0.21	0.26	3.93	6.86	0.57	0.70	0.76	2.58	3.17	0.81

for the imprinted template 4-NP on the MIP column is clearly evident in comparison to the NIP column for both mobile phases, while the retention times of the structural analogs of 4-NP (2-NP and 3-NP) are almost unaffected. Moreover, evaluation of the k', α, S and R_I clearly indicates a significant imprinting effect in both organic and aqueous environments (Table 13.7).

13.5
MIPs Mimicking Natural Receptors

In the literature, MIPs are frequently referred to as enzyme mimics, enzyme analogues, artificial antibodies, plastic antibodies, or plastibodies, emphasizing the potential functionality of MIPs as a biomimetic material. Besides selective separation or enrichment, MIPs offer a replacement or at least a complement to natural receptors such as enzymes or antibodies in various applications. This section discusses the advantages and disadvantages of MIPs in comparison to natural receptors, including enzymes and antibodies, and how close MIPs are in mimicking their natural counterparts. Selected examples from the literature are presented, highlighting these properties, and quantitative data on binding parameters such as equilibrium association/dissociation constants (i.e., a measure for the binding affinity), and estimates of the number of binding sites (i.e., a measure for the binding capacity) of MIPs, are provided.

13.5.1
Comparison of MIPs and Antibodies

The relatively low production costs and their inherent stability render MIPs an ideal complement – and in some applications a viable replacement – for natural receptors. However, the economic competitiveness of MIPs equally depends on their performance with respect to binding strength, selectivity, and binding capacity. Hence, a main emphasis in MIP development has been focused on closely mimicking their natural counterparts for these parameters, while retaining the distinct advantages of an artificial polymeric matrix and its inherent robustness. Table 13.8 gives a summarized comparison between MIPs and antibodies.

13.5.2
MIPs as Artificial Antibodies in Pseudo-Immunoassays

The first application of MIPs in a pseudo-immunoassay was reported in 1993, when Vlatakis et al. [103] tested the performance of MAA/EDMA-copolymers prepared against theophylline and diazepam (Figure 13.18).

Here, these MIPs were used as antibody substitutes in a radiolabeled competitive binding assay to determine the amount of theophylline and diazepam in human serum after extraction. The respective experiment was named molecularly imprinted sorbent assay (MIA) to indicate the close relationship to radioimmunoassays (RIAs), which are frequently applied to exploit natural antibodies. A comparison with immunoassays based on the enzyme-multiplied immunoassay technique (EMIT) performed with natural antibodies against both templates revealed excellent correlation between both methods with comparable ranges of linearity (14–224 M and 0.44–28 µM for theophylline and diazepam in the MIA, respectively), sensitivities (detection limits of 3.5 µM for theophylline and 0.2 µM for diazepam), and selectivities (i.e., cross-reactivity toward structural analogs, see Table 13.9) for the MIA. However, the additional extraction step from human serum into an organic phase along with the incubation and the handling of radioactive materials may be considered a major drawback of the MIA method performed this way.

These pioneering results by Vlatakis et al. encouraged many researchers to develop MIAs for a large variety of templates, including, for example, 17β-estradiol [104], *(S)-*

Table 13.8 Comparison of MIPs and antibodies (adapted from Ye and Mosbach [102]).

	MIPs	Antibodies
Structural characteristics	Synthetic polymers where binding sites are stabilized by highly crosslinked 3D network structures	Polypeptides folded into defined 3D structures
Preparation	Synthetic chemistry based on molecular design; MIPs can be produced on a large scale; production costs are low, and are frequently largely determined by the cost of the template molecule	Raised within an animal immune system by triggering an immune response; monoclonal antibodies can be produced on a moderate/large scale using hybridoma techniques
Stability	High stability; can be used in both aqueous and organic solvents	Low stability; usage only at aqueous conditions
Binding site characteristics	Noncovalently imprinted polymers frequently have a heterogeneous binding site distribution with a broad affinity distribution	Monoclonal antibodies have homogeneous binding sites
Structural characterization	Difficult for crosslinked amorphous materials	Possible via crystallization and XRD analysis
Main target molecules	Usually low/medium molecular weight (<1000 Da) molecular compounds; metal ions; imprinting of large molecules is emerging	Biomacromolecules and small immunogenic molecules that can be conjugated to protein carriers

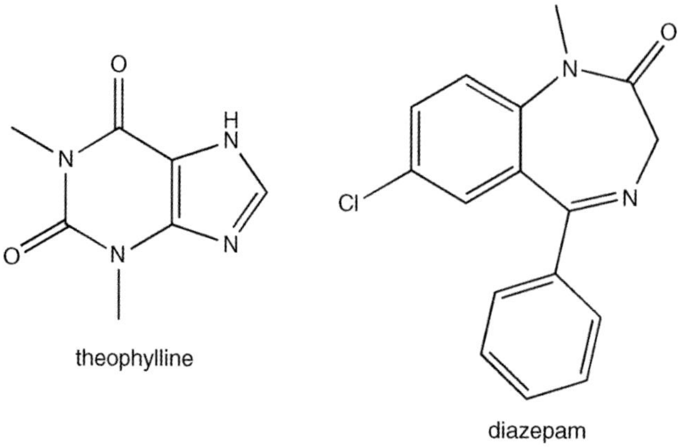

Figure 13.18 Chemical structures of theophylline and diazepam.

Table 13.9 Comparison of the cross-reactivity of the MIA and natural antibodies against theophylline and diazepam (adapted from Vlatakis et al. [103]).

Theophylline antibodies				Diazepam antibodies			
Competitive ligand		Cross-reactivity (%)		Competitive ligand		Cross-reactivity (%)	
Name	Chemical structure	MIP	Antibody	Name	Chemical structure	MIP	Antibody
Theophylline		100	100	Diazepam		100	100
3-Methyl-xanthine		7	2	Alprazolam		40	44

(Continued)

Table 13.9 (Continued)

	Theophylline antibodies				Diazepam antibodies			
	Competitive ligand		Cross-reactivity (%)		Competitive ligand		Cross-reactivity (%)	
Name	Chemical structure		MIP	Antibody	Name	Chemical structure	MIP	Antibody
Caffeine			<1	<1	Desmethyldiazepam		27	32
Theobromine			<1	<1	Clonazepam		9	5
Xanthine			<1	<1	Lorazepam		4	1

propranolol [53], morphine [105], and 2,4-dichlorophenoxyacetic acid [26]. Furthermore, besides radiolabeled MIAs several non-radiolabeled MIAs have emerged to overcome the restrictions when working with radioactive materials using, for example, fluorescent [106], electroactive [107], or enzyme-labeled analytes [108], which further highlights the versatility of MIPs as artificial antibodies.

13.5.3
MIPs as Catalysts with Enzymatic Activity

Besides the application as artificial antibodies, the remarkable recognition properties render MIPs also suitable for mimicking the catalytic properties of enzymes, thereby offering an alternative route for catalyzing reactions over a lower energetic transition state. Hence, a molecular imprint mimicking the transition state of an enzymatic reaction using a so-called transition state analogues (TSAs) was proven to be a viable approach toward MIPs with enzymatic properties.

As shown in Figure 13.19, Wulff et al., for example, have used a monoester of the phosphonic acid **1** as a TSA for the hydrolysis of alkaline esters **2** [109]. The molecularly imprinted catalyst was prepared using an amidine derivate (**3**) as a

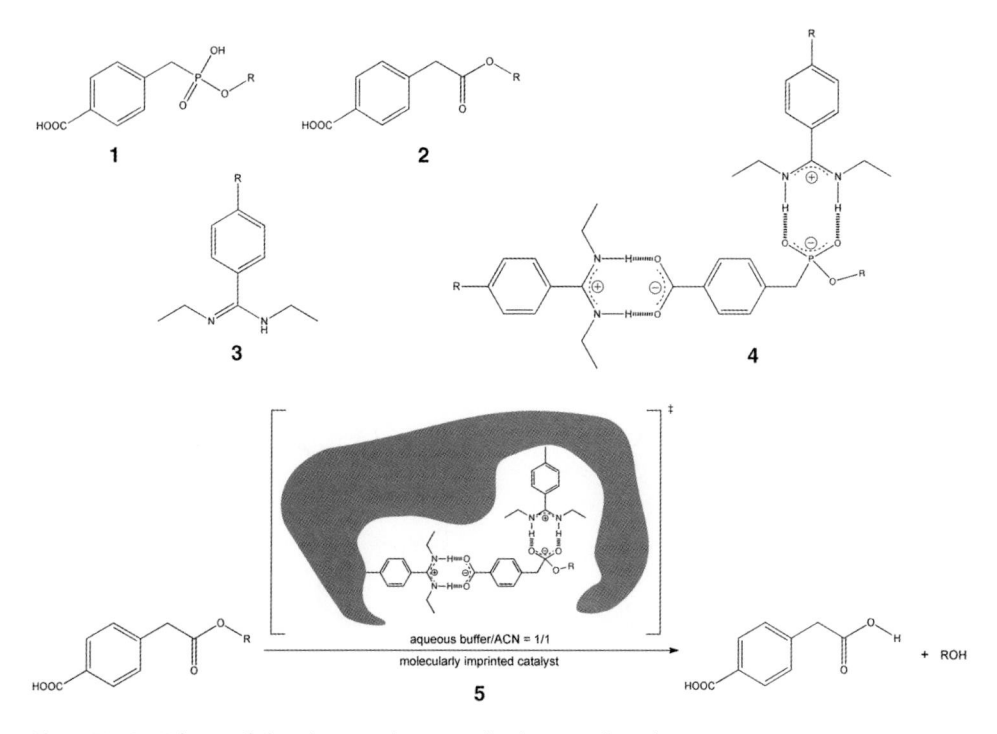

Figure 13.19 Scheme of phosphonic acid monoester **1**, alkaline ester **2**, functional monomer amidine **3**, the complex between the phosphonic acid monoester and two amidine molecules mimicking the transition state (**4**), and, eventually, the hydrolysis reaction, including the respective "real" transition state (**5**).

functional monomer, which forms an almost quantitative stoichiometric 2:1 complex (**4**) with the phosphonic acid monoester in THF, thereby mimicking the transition state of the hydrolysis reaction of the respective alkaline ester. The respective prepolymerization complex was subsequently copolymerized in the presence of EDMA as a crosslinker. After removal of the template, the catalytic MIP was ready for application and – under optimized conditions – was able to accelerate the respective hydrolysis reaction (**5**) by up to a factor of 100 and more, combined with a low cross-reactivity in comparison to analogously performed non-catalyzed reactions.

Moreover, the TSA approach using organic templates toward catalytically active MIPs has also been applied successfully to the hydrolysis of carbonates and carbamates, where an enhancement of the reaction rate by factors of 588 and 1435, respectively, was observed in comparison to reaction rates in a homogeneous buffer solution at the same pH [110]. Likewise, the successful catalysis of a dehydrofluorination [111], Diels–Alder reactions [112, 113], and even enantioselective ester hydrolysis [114, 115] have been reported by imprinting a TSA; however, it has to be noted that the observed activities of the catalysts and the enhancements of the reaction rates are usually far below those of natural enzymes [116, 117].

A possibility to overcome the drawbacks of the TSA approach with organic templates is the incorporation of transition metal catalysts into the polymeric matrix of a MIP, thereby combining the high catalytic activity of a transition metal with the high selectivity of a MIP. Thus, MIPs with properties in analogy to metalloenzymes may be established. For example, a MIP prepared with an organometallic TSA [a diphenylphosphinato ruthenium(II) complex, see Figure 13.20] has been shown to enhance the catalytic reduction of benzophenone by more than a factor of three in comparison to a control catalyst prepared without the diphenylphosphinato ligand, while providing a substantially higher selectivity [116].

Similarly, various alternative transition metals have been introduced successfully for the preparation of catalytically active MIPs, including, for example, palladium for the catalysis of Suzuki or Stille coupling reactions [118], rhodium for the hydro-

(a) transition state analogue   (b) transition state

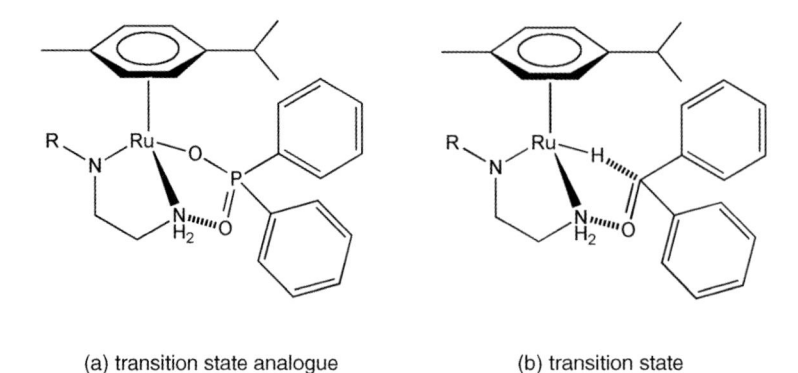

Figure 13.20 Scheme of the transition state analogue (TSA) (a) that has been applied to mimic the transition state (b) of the catalytic reduction of benzophenone in a MIP.

genation of alkenes [119], and manganese for the enantioselective epoxidation of styrene; however, only low enantioselectivity could be observed in the latter case [120]. Recently, the successful preparation of a catalytically active MIP for the determination of *p*-aminophenol has been reported using hemin – a porphyrin ring containing iron (III) – as the catalytically active species to mimic the active center of peroxidases [121].

The examples mentioned herein unambiguously corroborate the significant potential of MIPs for catalytic purposes, offering a cheap and durable alternative to natural enzymes. However, low activity and conversion rates are still a challenge for scaling such applications. The use of highly active transition metal is a viable possibility to improve the activity of catalytic MIPs, and may also pave the way for MIPs in catalytic organic synthesis.

13.5.4
Quantitative Data on the Binding Properties of MIPs

To compare MIPs with natural receptors, besides the achievable selectivity (Section 13.5.2), quantitative data on binding parameters such as the binding affinity and the binding capacity are indispensable. Thus, Table 13.10 provides an overview of these crucial parameters defining the performance of MIPs. Information on selected antibody–antigen binding constants is also included, thereby enabling a side-by-side comparison on the performance of MIPs versus natural receptors such as antibodies.

The data provided in Table 13.10 clearly indicate the binding site heterogeneity of MIPs, as the overall binding affinity is given in a range from low- to high-affinity binding sites. However, there are also examples of MIPs offering a more homogeneous binding site distribution, as indicated by only one value for the K_A in lieu of a bracketed value or several ranges, if different binding models are applied. While the superior binding affinity of natural (monoclonal) antibodies remains evident (Table 13.11), MIPs with K_A values close to their natural counterparts have already been reported.

13.6
Conclusions and Outlook

To date, although already proven a powerful artificial recognition element in many respects, MIPs have yet to reveal their full potential. Still, limited knowledge has been gained on the detailed parameters governing the molecular recognition process in MIPs. While a multitude of analytical tools are available for characterizing each synthesis step, many preparations of MIPs are still based on tedious trial-and-error, combinatorial approaches [52, 130], or adaptations of working MIP protocols derived from the literature. However, the number of attempts toward a more rational design of MIPs has significantly increased during the last decade. Especially, investigation of the prepolymerization complex and its properties has emerged as a very promising starting point toward designing MIPs with superior binding properties, as the stability of the prepolymerization complex strongly affects the recognition properties

Table 13.10 Overview of the binding affinity and binding capacity of MIPs and antibodies for selected examples.

Template	Polymer matrix	Equilibrium association constant K_A (M^{-1})		Number of binding sites B_{max} (μmol per g dry polymer)		Reference
		High affinity	Low affinity	High affinity	Low affinity	
Cortisol	MAA/EDMA	$1.75 \times 10^{6a)}$	$629^{a)}$	0.21 ± 0.05	280 ± 120	Ramström et al. [122]
Corticosterone	MAA/EDMA	$8.13 \times 10^{5a)}$	$1.19 \times 10^{3a)}$	0.37 ± 0.12	130 ± 60	Ramström et al. [122]
17β-Estradiol	MAA/TRIM	$0.469 \pm 0.108 \times 10^6$	$1.03 \pm 0.19 \times 10^4$	2.98 ± 0.75	700 ± 80	Ye et al. [104]
Theophylline	MAA/TRIM	$3.13 \pm 0.72 \times 10^6$	$2.02 \pm 0.47 \times 10^4$	56.8 ± 11.8	2120 ± 340	Ye et al. [104]
Theophylline	MAA/EDMA	$2.86 \times 10^{6a)}$	$1.54 \times 10^{4a)}$	0.016	1.3	Vlatakis et al. [103]
Diazepam	MAA/EDMA	$5.56 \times 10^{7a)}$	$1.67 \times 10^{4a)}$	0.0062 ± 0.0024	1.2 ± 1.0	Vlatakis et al. [103]
4-NP	4-VP/EDMA	2.2×10^4	197	0.44	8.24	Janotta et al. [30]
Morphine	MAA/EDMA	1.09×10^7 (in organic solvent)$^{a)}$ 8.33×10^5 (in aqueous buffer)	1.12×10^5 (in organic solvent)$^{a)}$ 4.17×10^4 (in aqueous buffer)	1.2 ± 0.7 (in organic solvent) 0.78 ± 0.17 (in aqueous buffer)	39 ± 3.4 (in organic solvent) 6.9 ± 0.7 (in aqueous buffer)	Andersson et al. [105]

Leu-enkephalin	MAA/EDMA	7.7×10^6 (organic solvent)[a] 1.0×10^7 (in aqueous buffer)	2.3×10^4 (in organic solvent)[a] 2.3×10^3 (in aqueous buffer)	0.017 ± 0.005 (in organic solvent) $0.0038 \pm 0.00\text{-}18$ (in aqueous buffer)	1.0 ± 2.1 (in organic solvent) 36 ± 6 (in aqueous buffer)	Andersson et al. [105]
Boc-L-Phe-OH	MAA/EDMA	159[a]		28		Kempe and Mosbach [87]
Boc-D-Phe-OH	MAA/EDMA	123[a]		28		Kempe and Mosbach [87]
Vancomycin	Cyclodextrin-vinyl/MBAA	640	44			Asanuma et al. [123]
Metsulfuron-methyl	TFMAA/DVB	$3.10 \times 10^{4\text{[a]}}$	588[a]	9.8	62.9	Zhu et al. [91]
Testosterone	MAA/EDMA	0.94×10^3 (UV) 1.28×10^3 (HPLC)		1.6 (UV) 2.5 (HPLC)		Cheong et al. [124]

a) K_A values calculated from the equilibrium dissociation constant (K_D) using the equation: $K_A = 1/K_D$.

Table 13.11 Examples of selected antibody–antigen binding affinities.

Antibody against	Equilibrium association constant K_A (M^{-1})	Reference
Nitrophenol	1.0×10^7–1.0×10^8	Mariuzza and Strand [125]
β-Estradiol	5.6×10^9	Raam and Cohen [126]
Theophylline	2×10^7	Locascio-Brown et al. [127]
Tetrodotoxin	0.98×10^8	Zhou et al. [128]
Aflatoxin	1×10^9	Groopman et al. [129]

of the finally established MIP matrix. Spectroscopic studies, including NMR [10, 131–135], FT-IR [132], UV-vis [4, 136], X-ray crystallography [133, 137], and very recently Raman investigations [138], are among the most frequently applied methods to investigate the prepolymerization solution so far. Furthermore, isothermal titration calorimetry (ITC) [139, 140] has been shown to be an excellent tool for investigation of the optimum ratio between the template and the functional monomer combining into the prepolymerization complex. Likewise, computational simulations have emerged to complement existing methods on rational MIP design. Several research groups are exploring molecular modeling strategies [141–143] to gain a more detailed understanding of the interactions in the prepolymerization solution, which have very recently culminated in the first simulation of an entire molecularly imprinted prepolymerization system [144].

In future, successful imprinting in aqueous solution will definitely continue to be both one of the most challenging and one of the most promising tasks in molecular imprinting technology. Orchestrating the multitude of available analytical tools accompanied by computational modeling and novel polymerization methods, including, for example, ring-opening metathesis polymerization (ROMP) [145] and imprinting in inorganic silica matrices [146], is certainly a key strategy toward highly efficient MIPs, and might pave the way toward biocompatible MIPs prepared in this environmentally friendly and inexpensive solvent. Molecular imprints prepared and operated in water not only open the door for a multitude of novel applications such as, for example, additives in foods or beverages, selective binding of proteins in biofluids, or serving as selective biocompatible membranes in blood dialysis, but may also generally contribute to a wider acceptance of the concept and principles of molecular imprinting within the scientific community.

References

1 Fischer, E. (1894) Einfluss der Configuration auf die Wirkung der Enzyme. *Ber. Dtsch. Chem. Ges*, **27**, 2985–2993.

2 Arshady, R. and Mosbach, K. (1981) Synthesis of substrate-selective polymers by host-guest polymerization. *Makromol. Chem.*, **182**, 687–692.

3 Wulff, G. (2002) Enzyme-like catalysis by molecularly imprinted polymers. *Chem. Rev.*, **102** (1), 1–27.

4 Andersson, H.S. and Nicholls, I.A. (1997) Spectroscopic evaluation of molecular imprinting polymerization systems. *Bioorg. Chem.*, **25** (3), 203–211.

5 O'Mahony, J. et al. (2005) Molecularly imprinted polymers – potential and challenges in analytical chemistry. *Anal. Chim. Acta*, **534** (1), 31–39.

6 Wulff, G. and Sarhan, A. (1972) The use of polymers with enzyme-analogous structures for the resolution of racemates. *Angew. Chem. Int. Ed. Engl.*, **11** (4), 341.

7 Umpleby, R.J. II, Bode, M., and Shimizu, K.D. (2000) Measurement of the continuous distribution of binding sites in molecularly imprinted polymers. *Analyst*, **125** (7), 1261–1265.

8 Haupt, K. (2001) Molecularly imprinted polymers in analytical chemistry. *Analyst*, **126** (6), 747–756.

9 Sellergren, B. and Andersson, L. (1990) Molecular recognition in macroporous polymers prepared by a substrate analog imprinting strategy. *J. Org. Chem.*, **55** (10), 3381–3383.

10 Sellergren, B., Lepistoe, M., and Mosbach, K. (1988) Highly enantioselective and substrate-selective polymers obtained by interactions. NMR and chromatographic studies on the nature of recognition. *J. Am. Chem. Soc.*, **110**, 5853–5860.

11 Wei, S., Molinelli, A., and Mizaikoff, B. (2006) Molecularly imprinted micro and nanospheres for the selective recognition of 17β-estradiol. *Biosens. Bioelectron.*, **21** (10), 1943–1951.

12 Kan, X. et al. (2009) Magnetic molecularly imprinted polymer for aspirin recognition and controlled release. *Nanotechnology*, **20** (16), 165601.

13 Shea, K.J. and Dougherty, T.K. (1986) Molecular recognition on synthetic amorphous surfaces. The influence of functional group positioning on the effectiveness of molecular recognition. *J. Am. Chem. Soc.*, **108** (5), 1091–1093.

14 Wulff, G., Heide, B., and Helfmeier, G. (1986) Molecular recognition through the exact placement of functional groups on rigid matrices via a template approach. *J. Am. Chem. Soc.*, **108** (5), 1089–1091.

15 Wulff, G. and Haarer, J. (1991) Enzyme-analogue built polymers, 29. The preparation of defined chiral cavities for the racemic resolution of free sugars. *Makromol. Chem.*, **192** (6), 1329–1338.

16 Turkewitsch, P. et al. (1998) Fluorescent functional recognition sites through molecular imprinting. A polymer-based fluorescent chemosensor for aqueous cAMP. *Anal. Chem.*, **70** (10), 2025–2030.

17 Kempe, M. (1996) Antibody-mimicking polymers as chiral stationary phases in HPLC. *Anal. Chem.*, **68**, 1948–1953.

18 Yilmaz, E., Mosbach, K., and Haupt, K. (1999) Influence of functional and cross-linking monomers and the amount of template on the performance of molecularly imprinted polymers in binding assays. *Anal. Commun.*, **36** (5), 167–170.

19 Hart, B.R. and Shea, K.J. (2002) Molecular imprinting for the recognition of N-terminal histidine peptides in aqueous solution. *Macromolecules*, **35** (16), 6192–6201.

20 O'Shannessy, D.J., Ekberg, B., and Mosbach, K. (1989) Molecular imprinting of amino acid derivatives at low temperature (0°C) using photolytic homolysis of azobisnitriles. *Anal. Biochem.*, **177** (1), 144–149.

21 Perez, N., Whitcombe, M.J., and Vulfson, E.N. (2000) Molecularly imprinted nanoparticles prepared by core-shell emulsion polymerization. *J. Appl. Polym. Sci.*, **77** (8), 1851–1859.

22 Lu, Y., Yan, C.L., and Gao, S.Y. (2009) Preparation and recognition of surface molecularly imprinted core/shell microbeads for protein in aqueous solutions. *Appl. Surf. Sci.*, **255** (12), 6061–6066.

23 Sellergren, B. and Shea, K.J. (1993) Influence of polymer morphology on the ability of imprinted network polymers to resolve enantiomers. *J. Chromatogr.*, **635** (1), 31–49.

24 Wu, L. et al. (2005) Theoretical and experimental study of nicotinamide molecularly imprinted polymers with different porogens. *Anal. Chim. Acta*, **549** (1–2), 39–44.

25 Lide, D.R. (ed.) (1998) *CRC Handbook of Chemistry and Physics*, CRC Press, Cleveland, Ohio.

26 Haupt, K., Dzgoev, A., and Mosbach, K. (1998) Assay system for the herbicide 2,4-dichlorophenoxyacetic acid using a molecularly imprinted polymer as an

artificial recognition element. *Anal. Chem.*, **70** (3), 628–631.

27 Rachkov, A. et al. (2004) Molecularly imprinted polymers prepared in aqueous solution selective for [Sar1,Ala8] angiotensin II. *Anal. Chim. Acta*, **504**, 191–197.

28 Janiak, D.S. and Kofinas, P. (2007) Molecular imprinting of peptides and proteins in aqueous media. *Anal. Bioanal. Chem.*, **389** (2), 399–404.

29 Andersson, L.I. (1996) Application of molecular imprinting to the development of aqueous buffer and organic solvent based radioligand binding assays for (S)-propranolol. *Anal. Chem.*, **68** (1), 111–117.

30 Janotta, M. et al. (2001) Molecularly imprinted polymers for nitrophenols – an advanced separation material for environmental analysis. *Int. J. Environ. Anal. Chem.*, **80** (2), 75–86.

31 Weiss, R. et al. (2003) Improving methods of analysis for mycotoxins: molecularly imprinted polymers for deoxynivalenol and zearalenone. *Food Addit. Contam.*, **20** (4), 386–395.

32 Haupt, K. and Mosbach, K. (1998) Plastic antibodies: developments and applications. *Trends Biotechnol.*, **16** (11), 468–475.

33 Mayes, A.G. and Mosbach, K. (1996) Molecularly imprinted polymer beads: suspension polymerization using a liquid perfluorocarbon as the dispersing phase. *Anal. Chem.*, **68**, 3769–3774.

34 Cederfur, J. et al. (2003) Synthesis and screening of a molecularly imprinted polymer library targeted for penicillin G. *J. Comb. Chem.*, **5** (1), 67–72.

35 Hung, C.Y. et al. (2006) Synthesis and molecular recognition of molecularly imprinted polymer with ibuprofen as template. *J. Chin. Chem. Soc.*, **53** (5), 1173–1180.

36 Baggiani, C. et al. (2001) Molecularly imprinted solid-phase extraction sorbent for the clean-up of chlorinated phenoxyacids from aqueous samples. *J. Chromatogr. A*, **938** (1–2), 35–44.

37 Matsui, J., Fujiwara, K., and Takeuchi, T. (2000) Atrazine-selective polymers prepared by molecular imprinting of trialkylmelamines as dummy template species of atrazine. *Anal. Chem.*, **72** (8), 1810–1813.

38 Molinelli, A., Weiss, R., and Mizaikoff, B. (2002) Advanced solid phase extraction using molecularly imprinted polymers for the determination of quercetin in red wine. *J. Agric. Food Chem.*, **50** (7), 1804–1808.

39 Weiss, R. et al. (2002) Molecular imprinting and solid phase extraction of flavonoid compounds. *Bioseparation*, **10** (6), 379–387.

40 Weiss, R., and Mizaikoff, B. (2002) Towards analysis of mykotoxins in beverages with molecularly imprinted polymers for deoxynivalenol and zearalenone. *Mycotoxin Res.*, **18**, 89–93.

41 Appell, M. et al. (2007) Synthesis and evaluation of molecularly imprinted polymers as sorbents of moniliformin. *Food Addit. Contam.*, **24** (1), 43–52.

42 Jodlbauer, J., Maier, N.M., and Lindner, W. (2002) Towards ochratoxin a selective molecularly imprinted polymers for solid-phase extraction. *J. Chromatogr. A*, **945**, 45–63.

43 Turner, N.W. et al. (2004) Effect of the solvent on recognition properties of molecularly imprinted polymer specific for ochratoxin A. *Biosens. Bioelectron.*, **20** (6), 1060–1067.

44 Matsui, J., Nicholls, I.A., and Takeuchi, T. (1998) Molecular recognition in cinchona alkaloid molecular imprinted polymer rods. *Anal. Chim. Acta*, **365** (1–3), 89–93.

45 Schweitz, L., Andersson, L.I., and Nilsson, S. (1997) Capillary electrochromatography with predetermined selectivity obtained through molecular imprinting. *Anal. Chem.*, **69** (6), 1179–1183.

46 Quaglia, M., Sellergren, B., and De Lorenzi, E. (2004) Approaches to imprinted stationary phases for affinity capillary electrochromatography. *J. Chromatogr. A*, **1044** (1–2), 53–66.

47 Yilmaz, E. et al. (2002) A facile method for preparing molecularly imprinted polymer spheres using spherical silica templates. *J. Mater. Chem.*, **12** (5), 1577–1581.

48 Viberg, P. et al. (2002) Nanoparticles as pseudostationary phase in capillary electrochromatography/ESI-MS. *Anal. Chem.*, **74** (18), 4595–4601.

49 Wang, J. et al. (2007) Synthesis and characterization of micrometer-sized molecularly imprinted spherical polymer particulates prepared via precipitation polymerization. *Pure Appl. Chem.*, **79** (9), 1505–1519.

50 Lai, J.P. et al. (2007) Molecularly imprinted microspheres and nanospheres for di(2-ethylhexyl)phthalate prepared by precipitation polymerization. *Anal. Bioanal. Chem.*, **389** (2), 405–412.

51 Yoshimatsu, K. et al. (2007) Uniform molecularly imprinted microspheres and nanoparticles prepared by precipitation polymerization: the control of particle size suitable for different analytical applications. *Anal. Chim. Acta*, **584** (1), 112–121.

52 Kempe, H. and Kempe, M. (2004) Novel method for the synthesis of molecularly imprinted polymer bead libraries. *Macromol. Rapid Commun.*, **25** (1), 315–320.

53 Ansell, R.J. and Mosbach, K. (1998) Magnetic molecularly imprinted polymer beads for drug radioligand binding assay. *Analyst*, **123** (7), 1611–1616.

54 Watabe, Y. et al. (2004) Determination of bisphenol A in environmental water at ultra-low level by high-performance liquid chromatography with an effective on-line pretreatment device. *J. Chromatogr. A*, **1032** (1–2), 45–49.

55 Haginaka, J. et al. (2007) Uniformly-sized, molecularly imprinted polymers for (-)-epigallocatechin gallate, -epicatechin gallate and -gallocatechin gallate by multi-step swelling and polymerization method. *J. Chromatogr. A*, **1156** (1–2), 45–50.

56 Titirici, M.M. and Sellergren, B. (2006) Thin molecularly imprinted polymer films via reversible addition-fragmentation chain transfer polymerization. *Chem. Mater.*, **18** (7), 1773–1779.

57 Baggiani, C. et al. (2007) Molecularly imprinted solid-phase extraction method for the high-performance liquid chromatographic analysis of fungicide pyrimethanil in wine. *J. Chromatogr. A*, **1141** (2), 158–164.

58 Titirici, M.M. and Sellergren, B. (2004) Peptide recognition via hierarchical imprinting. *Anal. Bioanal. Chem.*, **378** (8), 1913–1921.

59 Perez, N., Whitcombe, M.J., and Vulfson, E.N. (2001) Surface imprinting of cholesterol on submicrometer core-shell emulsion particles. *Macromolecules*, **34** (4), 830–836.

60 Zhuang, Y. et al. (2007) In situ synthesis of molecularly imprinted polymers on glass microspheres in a column. *Anal. Bioanal. Chem.*, **389** (4), 1177–1183.

61 Sellergren, B. and Shea, K.J. (1995) Origin of peak asymmetry and the effect of temperature on solute retention in enantiomer separations on imprinted chiral stationary phases. *J. Chromatogr. A*, **690** (1), 29–39.

62 Panasyuk-Delaney, T. et al. (2001) Impedometric herbicide chemosensors based on molecularly imprinted polymers. *Anal. Chim. Acta*, **435** (1), 157–162.

63 Belmont, A.S. et al. (2007) Molecularly imprinted polymer films for reflectometric interference spectroscopic sensors. *Biosens. Bioelectron.*, **22** (12), 3267–3272.

64 Lieberzeit, P.A. et al. (2008) Polymers imprinted with PAH mixtures–comparing fluorescence and QCM sensors. *Anal. Bioanal. Chem.*, **392** (7–8), 1405–1410.

65 Jenik, M. et al. (2009) Pollen-imprinted polyurethanes for QCM allergen sensors. *Anal. Bioanal. Chem.*, **394** (2), 523–528.

66 Guerra, M.R. et al. (2009) Development of piezoelectric sensor for detection of methamphetamine. *Analyst*, **134**, 1565–1570.

67 Jakusch, M. et al. (1999) Molecularly imprinted polymers and infrared evanescent wave spectroscopy. A chemical sensors approach. *Anal. Chem.*, **71**, 4786–4791.

68 Matsui, J. et al. (2009) Molecularly imprinted nanocomposites for highly sensitive SPR detection of a non-aqueous atrazine sample. *Analyst*, **134** (1), 80–86.

69 Navarro-Villoslada, F. et al. (2007) Zearalenone sensing with molecularly imprinted polymers and tailored fluorescent probes. *Sens. Actuators, B*, **121** (1), 67–73.

70 Southard, G.E. et al. (2007) Luminescent sensing of organophosphates using europium(III) containing imprinted polymers prepared by RAFT polymerization. *Anal. Chim. Acta*, **581** (2), 202–207.

71 Sergeyeva, T.A. et al. (1999) Selective recognition of atrazine by molecularly imprinted polymer membranes. Development of conductometric sensor for herbicides detection. *Anal. Chim. Acta*, **392** (2–3), 105–111.

72 Prasad, K. et al. (2007) Molecularly imprinted polymer (biomimetic) based potentiometric sensor for atrazine. *Sens. Actuators, B*, **123** (1), 65–70.

73 Gómez-Caballero, A. et al. (2008) Evaluation of the selective detection of 4,6-dinitro-o-cresol by a molecularly imprinted polymer based microsensor electrosynthesized in a semiorganic media. *Sens. Actuators, B*, **130** (2), 713–722.

74 Mathew-Krotz, J. and Shea, K.J. (1996) Imprinted polymer membranes for the selective transport of targeted neutral molecules. *J. Am. Chem. Soc.*, **118** (34), 8154–8155.

75 Duffy, D.J. et al. (2002) Binding efficiency and transport properties of molecularly imprinted polymer thin films. *J. Am. Chem. Soc.*, **124** (28), 8290–8296.

76 Wang, J.Y. et al. (2009) Binding constant and transport property of S-Naproxen molecularly imprinted composite membrane. *J. Membr. Sci.*, **331** (1–2), 84–90.

77 Sergeyeva, T.A. et al. (2001) Molecularly imprinted polymer membranes for substance-selective solid-phase extraction from water by surface photo-grafting polymerization. *J. Chromatogr. A*, **907** (1–2), 89–99.

78 Zhu, X. et al. (2006) Molecularly imprinted polymer membranes for substance-selective solid-phase extraction from aqueous solutions. *J. Appl. Polym. Sci.*, **101** (6), 4468–4473.

79 Silvestri, D. et al. (2007) Poly(ethylene-co-vinyl alcohol) membranes with specific adsorption properties for potential clinical application. *Sep. Sci. Technol.*, **42**, 2829–2847.

80 Pegoraro, C. et al. (2008) Molecularly imprinted poly(ethylene-co-vinyl alcohol) membranes for the specific recognition of phospholipids. *Biosens. Bioelectron.*, **24** (4), 748–755.

81 Liu, W. and Wang, B. (2009) Preparation and application of Norfloxacin-MIP/polysulfone blending molecular imprinted polymer membrane. *J. Appl. Polym. Sci.*, **113** (2), 1125–1132.

82 Kobayashi, T. et al. (2009) Selective removal of bisphenol A from serum using molecular imprinted polymer membranes. *Ther. Apher. Dial.*, **13** (1), 19–26.

83 Norrlöw, O., Glad, M., and Mosbach, K. (1984) Acrylic polymer preparations containing recognition sites obtained by imprinting with substrates. *J. Chromatogr.*, **299** (1), 29–41.

84 Fairhurst, R.E. et al. (2004) A direct comparison of the performance of ground, beaded and silica-grafted MIPs in HPLC and turbulent flow chromatography applications. *Biosens. Bioelectron.*, **20** (6), 1098–1105.

85 Pérez-Moral, N. and Mayes, A.G. (2004) Comparative study of imprinted polymer particles prepared by different polymerisation methods. *Anal. Chim. Acta*, **504**, 15–21.

86 Wei, S. and Mizaikoff, B. (2007) Binding site characteristics of 17beta-estradiol imprinted polymers. *Biosens. Bioelectron.*, **23** (2), 201–209.

87 Kempe, M. and Mosbach, K. (1991) Binding studies of substrate-and enantio-selective molecularly imprinted polymers. *Anal. Lett.*, **24** (7), 1137–1145.

88 Anderson, H.A.S. et al. (1996) Study of the nature of recognition in molecularly imprinted polymers. *J. Mol. Recognit.*, **9**, 675–682.

89 Li, H., Nie, L., and Yao, S. (2004) Adsorption isotherms and sites distribution of caffeic acid-imprinted polymer monolith from frontal analysis. *Chromatographia*, **60** (7), 425–431.

90 Shea, K.J., Spivak, D.A., and Sellergren, B. (1993) Polymer complements to nucleotide bases. Selective binding of adenine derivatives to imprinted polymers. *J. Am. Chem. Soc.*, **115** (8), 3368–3369.

91 Zhu, Q.Z. et al. (2002) Molecularly imprinted polymer for metsulfuron-methyl and its binding characteristics for sulfonylurea herbicides. *Anal. Chim. Acta*, **468** (2), 217–227.

92 Yao, W. et al. (2008) Adsorption of carbaryl using molecularly imprinted microspheres prepared by precipitation polymerization. *Polym. Adv. Technol.*, **19**, 812–816.

93 Umpleby, R.J. II et al. (2001) Recognition directed site-selective chemical modification of molecularly imprinted polymers. *Macromolecules*, **34** (24), 8446–8452.

94 Matsui, J. et al. (1995) A molecularly imprinted synthetic polymer receptor selective for atrazine. *Anal. Chem.*, **67** (23), 4404–4408.

95 Umpleby, R.J. II et al. (2004) Characterization of the heterogeneous binding site affinity distributions in molecularly imprinted polymers. *J. Chromatogr. B Anal. Technol. Biomed. Life Sci.*, **804** (1), 141–149.

96 Umpleby, R.J. II et al. (2001) Application of the Freundlich adsorption isotherm in the characterization of molecularly imprinted polymers. *Anal. Chim. Acta*, **435** (1), 35–42.

97 Rushton, G.T., Karns, C.L., and Shimizu, K.D. (2005) A critical examination of the use of the Freundlich isotherm in characterizing molecularly imprinted polymers (MIPs). *Anal. Chim. Acta*, **528** (1), 107–113.

98 Li, X. and Husson, S.M. (2006) Adsorption of dansylated amino acids on molecularly imprinted surfaces: a surface plasmon resonance study. *Biosens Bioelectron*, **22** (3), 336–348.

99 Szabelski, P. et al. (2002) Energetic heterogeneity of the surface of a molecularly imprinted polymer studied by high-performance liquid chromatography. *J. Chromatogr. A*, **964** (1–2), 99–111.

100 Yang, G. et al. (2003) Adsorption isotherms on aminoantipyrine imprinted polymer stationary phase. *Chromatographia*, **58** (1), 53–58.

101 Guo, T.Y. et al. (2004) Adsorptive separation of hemoglobin by molecularly imprinted chitosan beads. *Biomaterials*, **25** (27), 5905–5912.

102 Ye, L. and Mosbach, K. (2008) Molecular imprinting: synthetic materials as substitutes for biological antibodies and receptors. *Chem. Mater.*, **20**, 859–868.

103 Vlatakis, G. et al. (1993) Drug assay using antibody mimics made by molecular imprinting. *Nature*, **361** (6413), 645–647.

104 Ye, L., Cormack, P.A.G., and Mosbach, K. (1999) Molecularly imprinted monodisperse microspheres for competitive radioassay. *Anal. Commun.*, **36** (2), 35–38.

105 Andersson, L.I. et al. (1995) Mimics of the binding sites of opioid receptors obtained by molecular imprinting of enkephalin and morphine. *Proc. Natl. Acad. Sci. USA*, **92**, 4788–4792.

106 Piletsky, S.A. et al. (2001) Substitution of antibodies and receptors with molecularly imprinted polymers in enzyme-linked and fluorescent assays. *Biosens. Bioelectron.*, **16** (9–12), 701–707.

107 Haupt, K. (1999) Molecularly imprinted sorbent assays and the use of non-related probes. *React. Funct. Polym.*, **41** (1–3), 125–131.

108 Surugiu, I. et al. (2001) Development of a flow injection capillary chemiluminescent ELISA using an imprinted polymer instead of the antibody. *Anal. Chem.*, **73** (17), 4388–4392.

109 Wulff, G., Gross, T., and Schönfeld, R. (1997) Enzyme models based on molecularly imprinted polymers with strong esterase activity. *Angew. Chem. Int. Ed. Engl.*, **36** (18), 1962–1964.

110 Strikovsky, A.G. et al. (2000) Catalytic molecularly imprinted polymers using conventional bulk polymerization or suspension polymerization: selective hydrolysis of diphenyl carbonate and diphenyl carbamate. *J. Am. Chem. Soc.*, **122** (26), 6295–6296.

111 Beach, J.V. and Shea, K.J. (1994) Designed catalysts. A synthetic network polymer that catalyzes the dehydrofluorination of 4-fluoro-4-(p-nitrophenyl) butan-2-one. *J. Am. Chem. Soc.*, **116** (1), 379–380.

112 Liu, X.C. and Mosbach, K. (1997) Studies towards a tailor-made catalyst for the Diels-Alder reaction using the technique of molecular imprinting.

113 Visnjevski, A. et al. (2005) Catalysis of a Diels-Alder cycloaddition with differently fabricated molecularly imprinted polymers. *Catal. Commun.*, **6** (9), 601–606.

114 Sellergren, B., Karmalkar, R.N., and Shea, K.J. (2000) Enantioselective ester hydrolysis catalyzed by imprinted polymers. *J. Org. Chem.*, **65** (13), 4009–4027.

115 Volkmann, A. and Brüggemann, O. (2006) Catalysis of an ester hydrolysis applying molecularly imprinted polymer shells based on an immobilised chiral template. *React. Funct. Polym.*, **66** (12), 1725–1733.

116 Polborn, K. and Severin, K. (1999) Molecular imprinting with an organometallic transition state analogue. *Chem. Commun.*, **24**, 2481–2482.

117 Severin, K. (2000) Imprinted polymers with transition metal catalysts. *Curr. Opin. Chem. Biol.*, **4** (6), 710–714.

118 Cammidge, A.N., Baines, N.J., and Bellingham, R.K. (2001) Synthesis of heterogeneous palladium catalyst assemblies by molecular imprinting. *Chem. Commun.*, **24**, 2588–2589.

119 Tada, M., Sasaki, T., and Iwasawa, Y. (2002) Novel SiO2-attached molecular-imprinting Rh-monomer catalysts for shape-selective hydrogenation of alkenes; preparation, characterization and performance. *Phys. Chem. Chem. Phys.*, **4** (18), 4561–4574.

120 Disalvo, D., Dellinger, D.B., and Gohdes, J.W. (2002) Catalytic epoxidations of styrene using a manganese functionalized polymer. *React. Funct. Polym.*, **53** (2–3), 103–112.

121 de Jesus Rodrigues Santos, W. et al. (2007) A catalytically active molecularly imprinted polymer that mimics peroxidase based on hemin: application to the determination of *p*-aminophenol. *Anal. Bioanal. Chem.*, **389** (6), 1919–1929.

122 Ramström, O., Ye, L., and Mosbach, K. (1996) Artificial antibodies to corticosteroids prepared by molecular imprinting. *Chem. Biol.*, **3** (6), 471–477.

123 Asanuma, H. et al. (2001) Molecular imprinting of cyclodextrin in water for the recognition of nanometer-scaled guests. *Anal. Chim. Acta*, **435** (1), 25–33.

124 Cheong, S. et al. (1998) Synthesis and binding properties of a noncovalent molecularly imprinted testosterone-specific polymer. *J. Polym. Sci. Part A: Polym. Chem.*, **36**, 1725–1732.

125 Mariuzza, R. and Strand, M. (1981) Chemical basis for diversity in antibody specificity analysed by hapten binding to monoclonal anti-4-hydroxy-3-nitrophenacetyl (NP) immunoglobulins. *Mol. Immunol.*, **18** (9), 847–855.

126 Raam, S. and Cohen, J.L. (1980) Quantitation of oestrogen receptors: use of solid-phase antisteroid antibodies to quantify binding sites and determination of dissociation constant. *J. Clin. Pathol.*, **33** (4), 377–379.

127 Locascio-Brown, L. et al. (1993) Liposome-based flow-injection immunoassay for determining theophylline in serum. *Clin. Chem.*, **39** (3), 386–391.

128 Zhou, Y. et al. (2009) Identification of tetrodotoxin antigens and a monoclonal antibody. *Food Chem.*, **112** (3), 582–586.

129 Groopman, J.D. et al. (1984) High-affinity monoclonal antibodies for aflatoxins and their application to solid-phase immunoassays. *Proc. Natl. Acad. Sci. USA*, **81** (24), 7728–7731.

130 Ramström, O., Ye, L., and Mosbach, K. (1998) Screening of a combinatorial steroid library using molecularly imprinted polymers. *Anal. Commun.*, **35** (1), 9–11.

131 Takeuchi, T., Dobashi, A., and Kimura, K. (2000) Molecular imprinting of biotin derivatives and its application to competitive binding assay using nonisotopic labeled ligands. *Anal. Chem.*, **72** (11), 2418–2422.

132 Molinelli, A. et al. (2005) Analyzing the mechanisms of selectivity in biomimetic self-assemblies via IR and NMR spectroscopy of prepolymerization solutions and molecular dynamics simulations. *Anal. Chem.*, **77**, 5196–5204.

133 O'Mahony, J. et al. (2006) Imprinted polymeric materials. insight into the nature of prepolymerization complexes of quercetin imprinted polymers. *Anal. Chem.*, **78** (17), 6187–6190.

134 O'Mahony, J. *et al.* (2006) Anatomy of a successful imprint: analysing the recognition mechanisms of a molecularly imprinted polymer for quercetin. *Biosens. Bioelectron.*, **21** (7), 1383–1392.

135 O'Mahony, J. *et al.* (2005) Towards the rational development of molecularly imprinted polymers: 1H NMR studies on hydrophobicity and ion-pair interactions as driving forces for selectivity. *Biosens. Bioelectron.*, **20** (9), 1884–1893.

136 Svenson, J. *et al.* (1998) Spectroscopic studies of the molecular imprinting self-assembly process. *J. Mol. Recognit.*, **11**, 83–86.

137 O'Mahony, J. *et al.* (2007) Correlated theoretical, spectroscopic and X-ray crystallographic studies of a non-covalent molecularly imprinted polymerisation system. *Analyst*, **132** (11), 1161–1168.

138 Kantarovich, K. *et al.* (2009) Detection of template binding to molecularly imprinted polymers by Raman microspectroscopy. *Appl. Phys. Lett.*, **94**, 194103.

139 Fish, W.P. *et al.* (2005) Rational design of an imprinted polymer: maximizing selectivity by optimizing the monomer-template ratio for a cinchonidine MIP, prior to polymerization, using microcalorimetry. *J. Liq. Chromatogr. Relat. Technol.*, **28** (1), 1–15.

140 Alvarez-Lorenzo, C. *et al.* (2006) Imprinted soft contact lenses as norfloxacin delivery systems. *J. Control Release*, **113** (3), 236–244.

141 Piletsky, S.A. *et al.* (2001) Recognition of ephedrine enantiomers by molecularly imprinted polymers designed using a computational approach. *Analyst*, **126** (10), 1826–1830.

142 Pavel, D. and Lagowski, J. (2005) Computationally designed monomers and polymers for molecular imprinting of theophylline and its derivatives. Part I. *Polymer*, **46** (18), 7528–7542.

143 Wei, S., Jakusch, M., and Mizaikoff, B. (2007) Investigating the mechanisms of 17beta-estradiol imprinting by computational prediction and spectroscopic analysis. *Anal. Bioanal. Chem.*, **389** (2), 423–431.

144 Karlsson, B.C. *et al.* (2009) Structure and dynamics of monomer-template complexation: an explanation for molecularly imprinted polymer recognition site heterogeneity. *J. Am. Chem. Soc.*, **131** (37), 13297–13304.

145 Enholm, E.J. *et al.* (2006) A comparison of a radical polymerization vs ROMP matrix for molecular imprinting. *Macromolecules*, **39** (23), 7859–7862.

146 Walcarius, A. and Collinson, M.M. (2009) Analytical chemistry with silica sol-gels: traditional routes to new materials for chemical analysis. *Annu. Rev. Anal. Chem.*, **2**, 121–143.

14
Quantitative Affinity Data on Selected Artificial Receptors
Anatoly K. Yatsimirsky and Vladimir M. Mirsky

This chapter collects data on receptor–analyte binding constants taken from the current literature for representative types of analytes, such as nucleotides, sugars, amino acids, inorganic anions, metal ions, and so on (Tables 14.1 and 14.2). The compilation covers most of the receptors discussed in different chapters of this book. For more extensive compilations of thermodynamic data on receptor–ligand interactions and related subjects (protonation, selectivity coefficients), the references given at the end of the chapter may be consulted. In particular, complexation with crown ethers, their analogs, and calixarenes with chelating groups has been covered extensively in a series of reviews by R. M. Izatt *et al.* [72–79] for which reason these receptors are not included in the present compilation.

Only receptors operating in water or aqueous–organic mixtures were considered. The binding constants are the observed or "conditional" constants, which correspond

Table 14.1 Binding constants of molecular receptors.

Receptor	Analyte	Log K (M^{-1})	Method, conditions	Reference
1	ATP	6.0	UV-vis, Fl, at 20 °C	[1]
	ADP	5.7		
	AMP	3.3		
	GTP	6.1		
	GDP	3.6		
	GMP	3.4		
	CTP	5.4		
	CDP	3.9		
	CMP	3.6		
	UTP	5.9		
	UDP	4.3		
	UMP	3.5		

(*Continued*)

Artificial Receptors for Chemical Sensors. Edited by V.M. Mirsky and A.K. Yatsimirsky
Copyright © 2011 WILEY-VCH Verlag GmbH & Co. KGaA, Weinheim
ISBN: 978-3-527-32357-9

Table 14.1 (Continued)

Receptor	Analyte	Log K (M^{-1})	Method, conditions	Reference
2a	ATP	5.4	Fl, UV-vis, at 20 °C	[2]
	ADP	4.4		
	AMP	4.2 (3.2)[a]		
	GTP	6.8		
	GDP	5.2		
	GMP	4.1 (4.5)[a]		
	UTP	4.4		
	UMP	<3		
	CMP	3.8 (3.3)[a]		
	3′,5′-cGMP	3.1 (3.6)[a]		
	2′,3′-cGMP	2.9 (3.7)[a]		
	Guanosine, adenosine, cytidine, uridine	<2		
2b	3′,5′-cGMP	3.0 (3.3)[a]		
	AMP	4.4 (3.0)[a]		
	GMP	3.4 (5.0)[a]		
	CMP	3.9 (2.6)[a]		
	UMP	<3		
2c	3′,5′-cGMP	<3		
	AMP	3.2 (2.6)[a]		
	GMP	3.7 (4.3)[a]		
	CMP	<3		
	UMP	<3		
3	ATP	4.2	Fl, pH 7.4	[3]
	ADP	2.8		
	AMP	2.1		
	GTP	4.9		
4	ATP	6.1	Fl, pH 7.4; no interference with 1 mM AMP, c-GMP, UDP-Glu, ADP-Glu, AcO$^-$, SO$_4^{2-}$, NO$_3^-$, HCO$_3^-$	[4]
	ADP	6.2		
	GTP	6.2		
	CTP	5.8		
	UDP	5.7		
	P$_2$O$_7^{4-}$	7.6		
	Inositol-1,3,4-trisphosphate	6.5		
	c-AMP	3.1		
5	ATP	4.1	Fl, DMSO/20 mM aqueous HEPES pH 7.4, 6/4, v/v	[5]
	GTP	4.9		
	CTP	3.9		
6	ATP	5.3	Fl, indicator Displacement, pH 7.4, 37 °C	[6]
	ADP	4.9		
	GTP	5.3		
	P$_2$O$_7^{4-}$	6.0		
	Citrate	4.8		

Table 14.1 (Continued)

Receptor	Analyte	Log K (M^{-1})	Method, conditions	Reference
7	Sorbitol	2.57	UV-vis, indicator displacement, pH 7.4	[7]
	Fructose	2.20		
	Galactose	1.18		
	Glucose	0.66		
	Mannose	1.11		
8	Glucose	1.8	Fl, 33% MeOH/H$_2$O, pH 7.77	[8]
	Fructose	3.0		
	Allose	2.5		
	Galactose	2.2		
9	Glucose	3.6		
	Fructose	2.5		
	Allose	2.8		
	Galactose	2.2		
10	Sorbitol	3.00	Fl, pH 7.4	[9]
	Fructose	2.82		
	Galactose	1.28		
	Glucose	1.48		
	Mannose	1.43		
11	Sorbitol	3.27		
	Fructose	3.01		
	Galactose	1.98		
	Glucose	1.34		
	Mannose	1.40		
12	Sorbitol	3.66		
	Fructose	3.13		
	Galactose	2.18		
	Glucose	1.58		
	Mannose	1.71		
13	Glucose	1.31	Fl, pH 7.5	[10]
	Fructose	3.19		
14	Glucose	0.00		
	Fructose	2.74		
15	Glucose	0.36		
	Fructose	2.04		
16	Glucose	1.23	^{1}H NMR, D$_2$O, pH 7.4	[11]
	Fructose	2.78		
	Glucose	1.49	UV-vis, indicator displacement, pH 7.4	
	Methyl α-D-glucopyranoside	1.34		
	Methyl β-D-glucopyranoside	0.95		
	Methyl α-D-galactopyranoside	1.46		
	Methyl β-D-galactopyranoside	1.36		
	Methyl α-D-mannopyranoside	1.38		
	Methyl α-D-fucopyranoside	1.40		
	Methyl 6-deoxy-α-D-glucopyranoside	< 0.7		

(Continued)

14 Quantitative Affinity Data on Selected Artificial Receptors

Table 14.1 (Continued)

Receptor	Analyte	Log K (M^{-1})	Method, conditions	Reference
17	Glucose	2.43	UV-vis, indicator displacement, pH 7.4	[12]
18	Fructose	2.89	Fl, pH 8.21, 52.1 wt% MeOH	[13]
	Galactose	2.82		
	Glucose	2.98		
	Mannose	1.87		
19	Ribose	3.38	UV-vis, 90% DMSO, pH 7.4	[14]
	Allose	3.18		
	Fructose	3.04		
	Galactose	2.49		
	Glucose	2.30		
20	Glucuronic acid	3.76	UV-vis, indicator displacement, MeOH/H$_2$O (3:1 v/v)	[15]
	Lactobionic acid	5.44		
	Gluconic acid	6.75		
	N-Acetylneuraminic acid	4.77		
21	Sorbitol	2.36	Fl, pH 7.4	[16]
	Fructose	2.08		
	Tagatose	2.01		
	Galactose	0.90		
	Glucose	0.38		
22	Fructose	1.4	Fl, 1% MeOH, pH 7.4	[17]
	Glucose	0.15		
	Sorbitol	2.04		
23	Fructose	1.66	Fl, indicator displacement, pH 7.4	[18]
	Catechol	2.54		
	Adrenaline	1.98		
	Dopamine	2.88		
24	D-Glucose	4.39	Fl, 50% MeOH, pH 11.7	[19]
	L-Glucose	4.07		
L-25	D-Fructose	1.66	Fl, pH 7.4	[20]
	D-Glucose	0.18		
	D-Sorbitol	2.00		
D-25	D-Fructose	1.74	Fl, pH 7.4	[21]
	D-Glucose	0.20		
	D-Sorbitol	2.00		
26	D-Glucarate	3.71		
	D-Gluconate	3.16		
	D-Sorbitol	3.11		
	D-Glucuronic acid	1.66		
	D-Glucose	1.79		

Table 14.1 (Continued)

Receptor	Analyte	Log K (M^{-1})	Method, conditions	Reference
27	Glucose	2.20	UV-vis, indicator displacement, 75% MeOH–H$_2$O, pH 7.4	[22]
	Fructose	2.48		
	Lactate	2.70		
	Malate	4.68		
	Tartrate	4.74		
	Citrate	5.30		
	Succinate	2.54		
28	Glucose	2.15		
	Fructose	2.60		
	Lactate	2.70		
	Malate	4.17		
	Tartrate	5.15		
	Citrate	5.25		
	Succinate	<2.15		
29	Glucose	2.95		
	Fructose	2.78		
	Lactate	3.04		
	Malate	3.93		
	Tartrate	4.60		
	Citrate	4.43		
	Succinate	No binding		
30	Glucose	No binding		
	Fructose	No binding		
	Lactate	—b)		
	Malate	4.11		
	Tartrate	4.23		
	Citrate	4.79		
	Succinate	3.56		
31	Citrate	5.3	Fl, UV, pH 6.3	[23]
	Glucuronic acid	3.51	UV, pH 6, 80% DMSO	[24]
	Galacturonic acid	3.79	UV, pH 6, 80% DMSO	
	Glucose-1-phosphate	4.41	UV, pH 4, 10% DMSO	
	Galactose-1-phosphate	4.32	UV, pH 4, 10% DMSO	
	Mannose-1-phosphate	4.41	UV, pH 4, 10% DMSO	
	Methyl phosphate	4.10	UV, pH 4, 10% DMSO	
32	Phe-Gly-Gly	1.7	Fl, indicator displacement, pH 7.0	[25]
	Gly-Phe-Gly	1.3		
	Gly-Gly-Phe	1.9		
	Phe-Gly	2.3		
	Gly-Phe	1.9		
	Ac-Gly-Phe	2.8		

(Continued)

Table 14.1 (Continued)

Receptor	Analyte	Log K (M^{-1})	Method, conditions	Reference
33	Gly-Gly	2.93	UV-vis, pH 6.9	[26]
	Gly-Phe	4.71		
	Phe-Phe	4.52		
	Gly-Gly-Gly	3.41		
	Phe-Gly-Gly	4.48		
	Gly-Gly-Gly-Gly	5.02		
34	Me$_4$N$^+$	3.58	Calorimetry, pH 7.8	[27]
	Choline	4.41		
	Acetylcholine	4.16		
	Carnitine	2.18		
35	Malonate	4.97	Fl, 50% MeOH, pH 7.4	[28]
	Succinate	5.24		
	Glutarate	5.47		
	Citrate	5.17		
	N-Acetyl-L-glutamate	6.02		
	L-Glutamate	6.75		
36	Ac-Lys-OMe	3.64	NMR, D$_2$O, pH 7.0	[29]
	Ts-Arg-OEt	3.25		
	KKLVFF	4.58		
	Ac-His-OMe	2.85		
	With derivatives of all amino acids besides Lys, Arg, His	<0.7		
37	Trp-Gly-Gly	5.11	Calorimetry, 27 °C, pH 7.0	[30]
	Gly-Trp-Gly	4.32		
	Gly-Gly-Trp	3.49		
	Trp	4.63		
	Phe	3.72		
	Tyr	3.34		
	His	No binding		
38	L-Ala	3.22	UV-vis, 50% MeOH, pH 7.4	[31]
	D-Ala	3.51		
	L-Val	3.40		
	D-Val	3.70		
	L-Phe	3.71		
	D-Phe	3.96		
39	Ac-L-AlaOH	No binding	UV-vis, 20% DMSO, pH 6.0	[32]
	Ac-L-PheOH	2.90		
	Ac-L-Ala-L-AlaOH	3.68		
	Ac-D-Phe-L-AlaOH	3.70		
	Ac-L-Phe-L-AlaOH	3.00		
	Ac-D-Phe-L-PheOH	3.27		
	Ac-L-Phe-L-PheOH	3.36		
	Ac-D-Ala-D-AlaOH	3.30		
	Ac-D-Ala-D-ValOH	3.00		
	Ac-D-Val-D-AlaOH	3.63		
	Ac-D-val-D-ValOH	3.45		

Table 14.1 (Continued)

Receptor	Analyte	Log K (M^{-1})	Method, conditions	Reference
40	Ac-D-Ala-D-AlaOH	4.52	UV-vis, pH 6.1	[33]
	Ac-L-Ala-L-AlaOH	4.49		
	Ac-D-Ala-D-LacOH	4.27		
	Ac-D-Ala-GlyOH	3.69		
	Ac-Gly-D-AlaOH	3.63		
	Ac-D-Ala-D-ValOH	3.70		
	Ac-D-Val-D-AlaOH	3.80		
	Ac-Gly-GlyOH	3.47		
	Ac-L-AlaOH	<3.0		
41	Ac-RGD-NH$_2$	3.43	UV-vis, Fl, pH 6.1	[34]
42	N-Ac-L-AlaOH	2.66		
	N-Ac-L-LacOH	2.34		
43	N-Ac-L-AlaOH	2.51	NMR, 60% DMSO	[35]
	N-Ac-L-LacOH	2.43		
44	1-Hexanol	1.01	Fl, pH 7.0	[36]
	1-Heptanol	1.40		
	Cyclohexanol	1.59		
	Cycloheptanol	2.19		
45	d-Borneol	4.38	Fl., 30 °C, pH 7.0	[37]
	l-Borneol	4.20		
46	Acetate	2.43	NMR, D$_2$O, pH 6.6	[38]
	Phosphate	2.98		
47	Fluoride	4.5	UV-vis, pH 4.9	[39]
48	Zn(II)	7.36	Fl, pH 7.0; interfere Cu(II), Hg(II)	[40]
49	L-Mandelate	2.76	Fl, 50% MeOH, pH 7.4	[41]
	D-Mandelate	1.58		
50	Cd(II)	2.79	Fl, pH 5.0	[42]
	Ag(I)	4.31		
	Hg(II)	5.42		
51	Oxalate	5.11	Fl, indicator displacement, pH 7.0	[43]
	Malonate	4.49		
	Succinate	3.32		
	Glutarate	2.84		
52	Inositol-1,3,4-trisphosphate	8.66	Fl, indicator displacement, pH 7.0	[44]
53	P$_2$O$_7^{4-}$	5.61	Fl, pH 7.4	[45]
	ATP	5.53		
	ADP	3.96		
54	F$^-$	3.20	UV-vis, pH 5.0	[46]
	Cl$^-$	3.98		
	Br$^-$	3.01		
	I$^-$	2.39		
55	Cu(II)	6.86	UV-vis, 20% EtOH, pH 6.98	[47]

(Continued)

Table 14.1 (Continued)

Receptor	Analyte	Log K (M^{-1})	Method, conditions	Reference
56 ($n=6$)	Cholic acid	4.83	Fl, pH 7.0	[48]
	Deoxycholic acid	4.15		
	1-Adamantol	5.62		
	2-Adamantol	5.76		
56 ($n=7$)	Cholic acid	4.61		
	Deoxycholic acid	5.34		
	1-Adamantol	3.18		
	2-Adamantol	3.20		
57a	SO$_4^{2-}$	5.28	ITC, 50% MeOH	[49, 50]
	I$^-$	3.79		
	Br$^-$	3.45		
	Cl$^-$	2.51		
57b	SO$_4^{2-}$	5.10		
	I$^-$	4.00		
	Br$^-$	3.30		
	Cl$^-$	1.86		
57c	SO$_4^{2-}$	5.32		
	I$^-$	3.61		
	Br$^-$	3.03		
	Cl$^-$	No binding		
57d	SO$_4^{2-}$	5.97		
	I$^-$	4.43		
	Br$^-$	4.01		
	Cl$^-$	3.39		
57e	SO$_4^{2-}$	6.73	2:1 (v/v) MeCN–water	
	I$^-$	4.46		
57f	SO$_4^{2-}$	6.83		
	I$^-$	4.75		
58	HSO$_4^-$	2.78	UV-vis, 50% DMSO, pH 4.6	[51]

a) Log K_2.
b) Complex not 1:1 stoichiometry.

to the given solvent, temperature, and pH and are expressed in terms of total reactant and product concentrations rather than individual species participating in interactions. If the binding constant was not indicated in the publication, it was estimated from concentration dependencies. Uncertainties in K values are typically $\pm 10\%$. If solvent composition is not indicated, the reaction medium is water; if pH is not indicated, the reaction medium is pure unbuffered solvent. The temperature is 22–25 °C if not stated otherwise.

For further information on thermodynamic data on receptor–ligand interactions the reader is referred to the Further Reading section and References [72–79].

Table 14.2 Binding constants of spreader-bar systems and molecularly-imprinted polymers.

Technique	Receptor	Analyte	Log K (M^{-1})	Method, conditions	Reference
Spreader-bar	2-Thiobarbituric acid and 1-dodecanethiol	Barbiturate	4	Capacitive, 100 mM KCl, 5 mM phosphate, pH 5.5	[52]
	6-Mercaptopurine and 1-dodecanethiol	ATP Adenine	4.3 4.4		[53]
	4-Hydroxy-5-methyl-2-mercaptopyrimidine and 1-dodecanethiol	Adenine	5		
Molecularly imprinted polymer[a)]	2-Acrylamido-2-methyl-1-propane sulfonic acid and N,N-methylene-diacrylamide	Creatinine	3.2, 5.0	Capacitive, 25 mM phosphate, 100 mM NaCl, pH 7.5	[54]
		Triazines	3.6		[55]
	2-Acrylamido-2-methyl-1-propane sulfonic acid and methacrylic acid	Microcystin-LR	9.0	Competitive assay, 50 mM phosphate, pH 7.0	[56]
	Methacrylic acid	Biotin	8.9	25 mM phosphate, pH 7.5, ELISA	[57]
	2-(Trifluoromethyl)acrylic acid		8.3		
	2-Acrylamido-2-methyl-1-propane sulfonic acid		7.8		
	Bismethacryloyl-β-cyclodextrin and 4-vinylpyridine	Aciflurofen	3.0, 3.3	Water–MeOH, spectroscopy	[58]
	Methacrylic acid and dimethacrylate	Cotinine	2.4, 3.3	pH 9.0, HPLC	[59]
	β-Cyclodextrin and toluene 2,4-diisocyanate	N-Phenyl-1-naphthyl-amine	~3.5 (3–5)	pH 4–11, HPLC, fluorescence	[60]
	Methacrylic acid and ethylene glycol dimethacrylate	Morphine	3.0, 4.3	Colorimetry	[61]
	N-3,5-Bis(trifluoromethyl)-phenyl-N'-4-vinylphenylurea	Phosphotyrosine	5.94	Adsorption isotherm, 20% MeCN	[62]
	4-Vinylpyridine and ethylene glycol dimethacrylate	Sulfamethoxazole	4.7	MeCN	[63]
	2-(Trifluoromethyl)acrylic acid and ethylene glycol dimethacrylate	Cefalexin	2.6, 3.9	Aqueous solution	[64]

(Continued)

Table 14.2 (Continued)

Technique	Receptor	Analyte	Log K (M^{-1})	Method, conditions	Reference
	Vinyl monomer of α- or β-cyclodextrins with N,N'-methylenebisacrylamide	Vancomycin	3.8	5 mM Tris, pH 8.0, 5 °C	[65]
		Cefazolin	3.5		
		d-Phe-d-Phe	4.8		
	Bismethacryloyl-cyclodextrin, 4-vinylpyridine and ethylene dimethacrylate	Acifluorfen	2.9, 3.1	MeOH, 50%	[66]
	Bismethacryloyl-β-cyclodextrin, 2-(diethylamino)-ethylmethacrylate and ethylene dimethacrylate	Norfloxacin	2.9, 3.5	MeOH, 50%	[67]
	Methacrylic acid and ethylene glycol dimethacrylate	Mepivacaine	2.5	50 mM citrate, 5% ethanol, pH 5.0, chromatography	[68]
		Ethycaine	3.2		
		Ropivacaine	3.8		
		Bupivacaine	5.0		
		Pentycaine	4.4		
	3-[N,N-Bis(9-anthrylmethyl)-amino]-propyltriethoxysilane, tetraethyoxysilane and 3-aminopropyltriethoxysilane	2,4-Dichlorophenoxyacetic acid	5.7	Phosphate, pH 7.0, fluorescence	[69]
	$trans$-4-[p-(N,N-Dimethylamino)styryl]-N-vinylbenzylpyridinium, trimethylolpropane trimethacrylate and 2-hydroxyethyl methacrylate	cAMP	5.2	Fluorescence	[70]
	Methacrylic acid, vinylimidazole and ethylene glycol dimethacrylate	Adenine	3.6	UV spectroscopy	[71]

a) Additional data on molecularly imprinted polymers are presented in Table 13.10.

14.1 Structures of Receptors

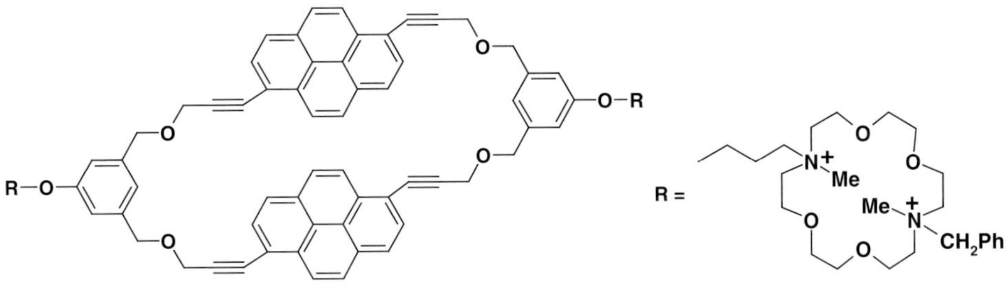

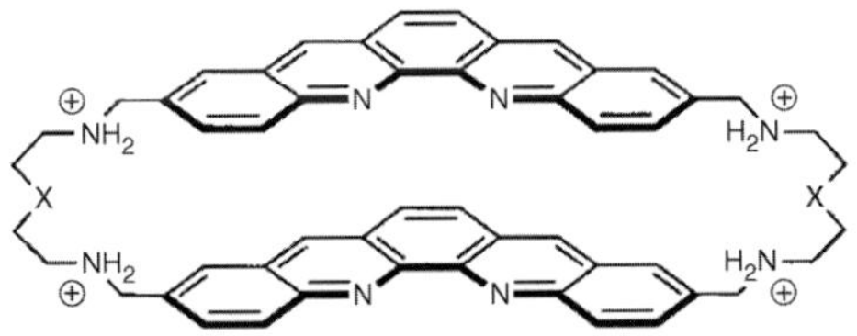

2a, X=NH
2b, X=O
2c, X=CH$_2$NH(Me)$^+$CH$_2$

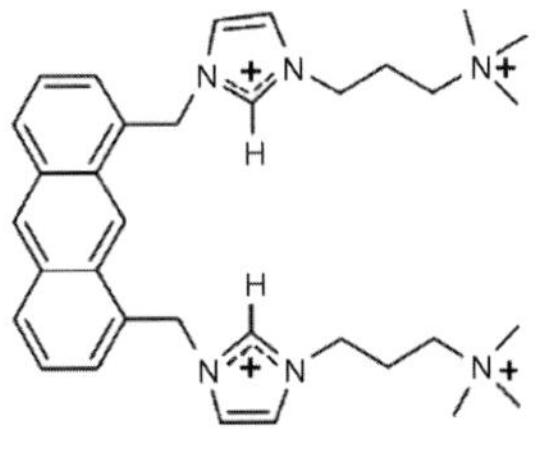

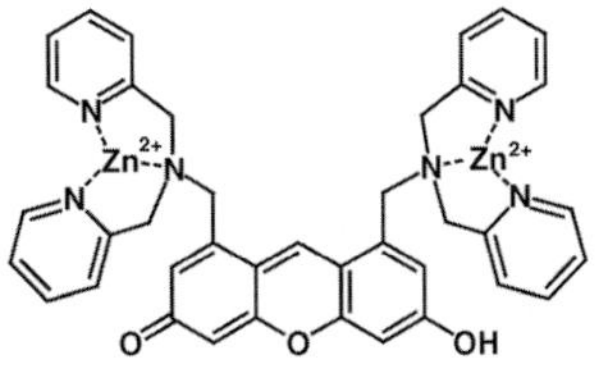

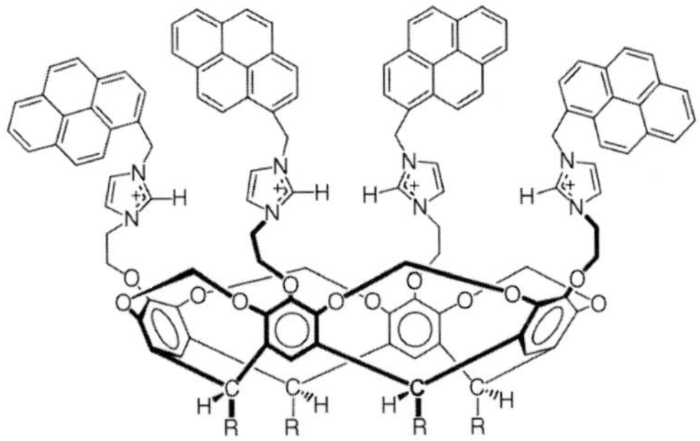

R = CH$_2$CH$_2$Ph

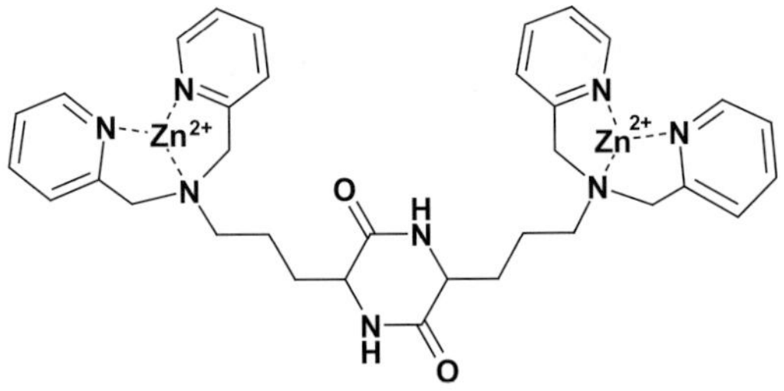

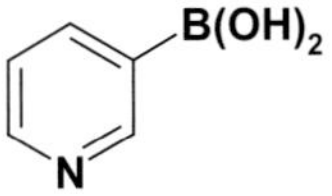

14.1 Structures of Receptors | 451

17

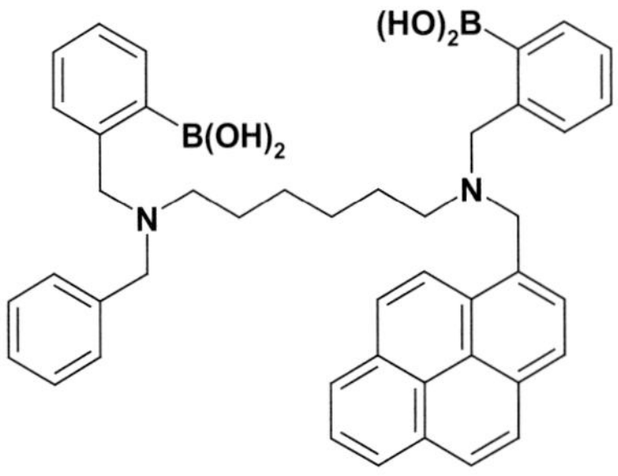

18

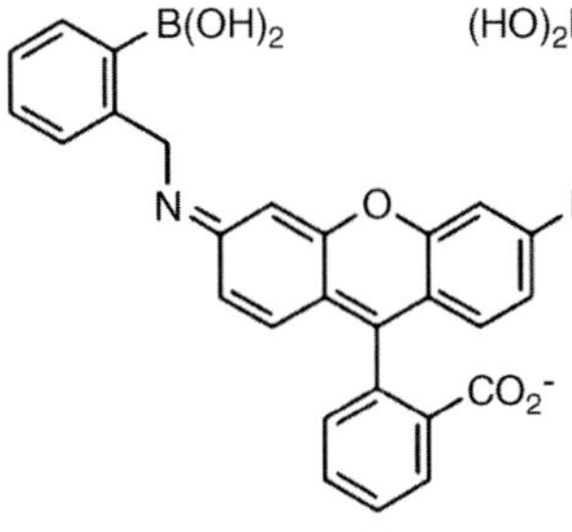

19

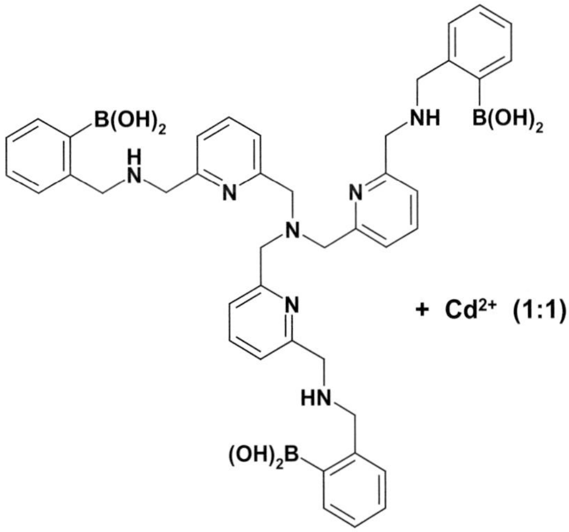

20 + Cd²⁺ (1:1)

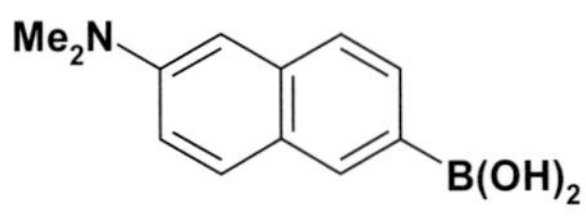

21

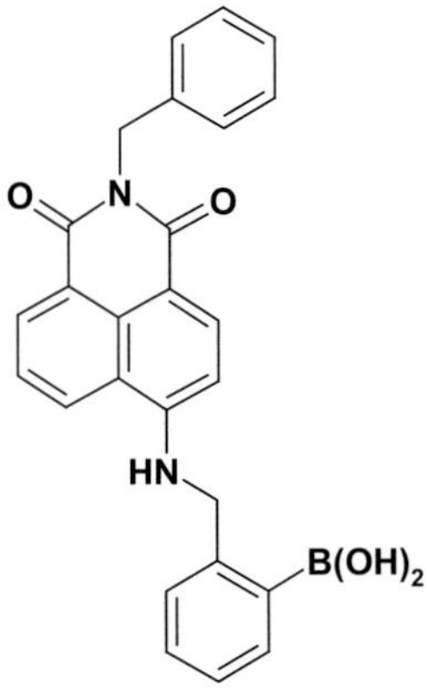

22

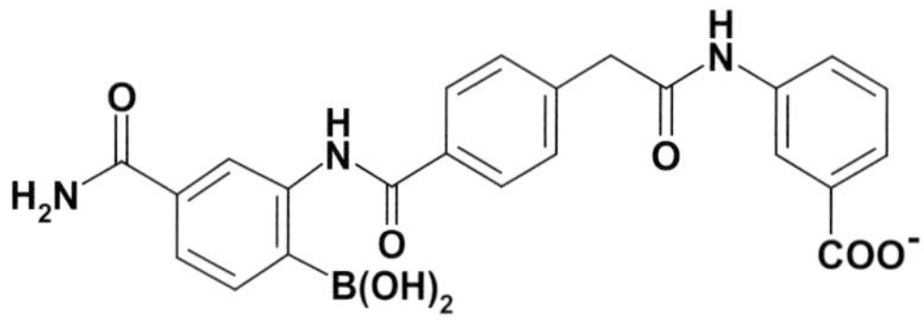

23

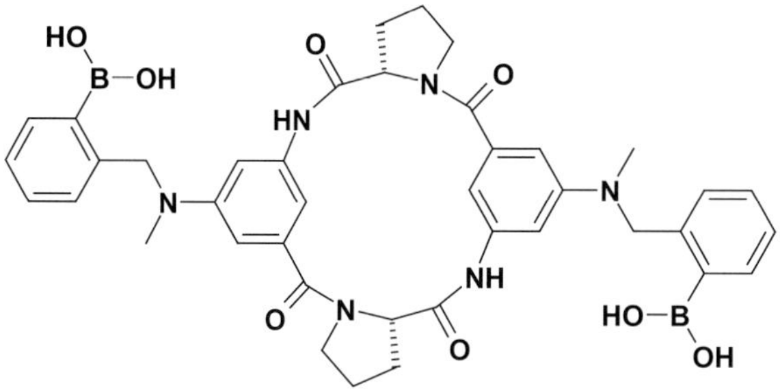

24

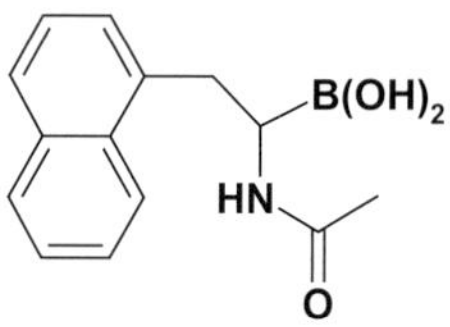

25

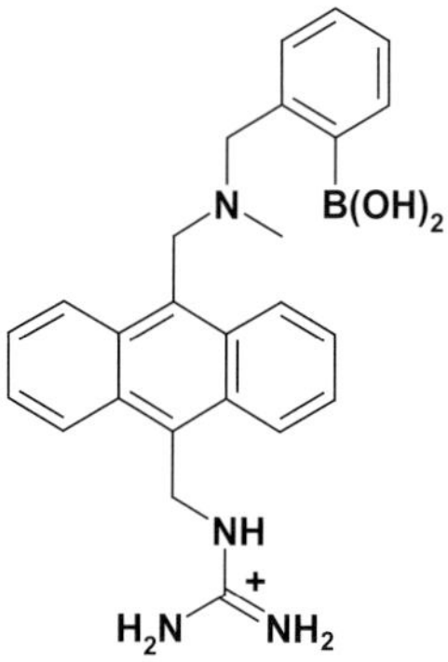

26

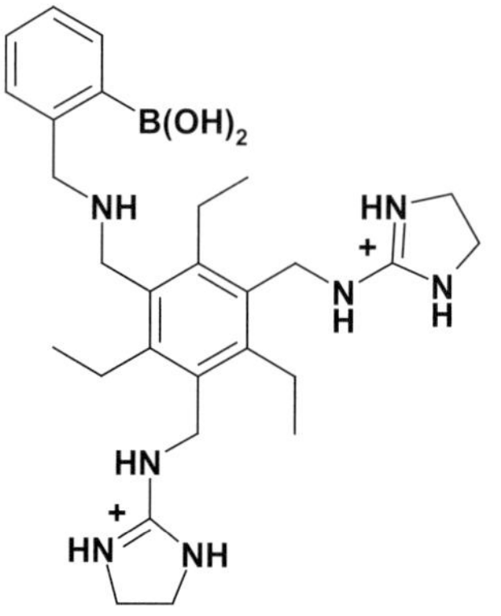

27

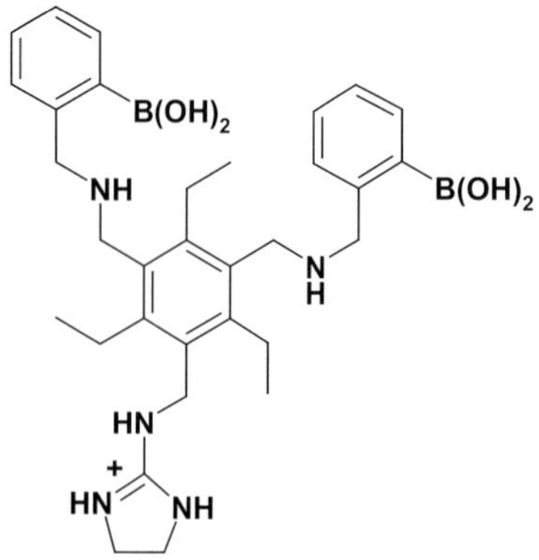

28

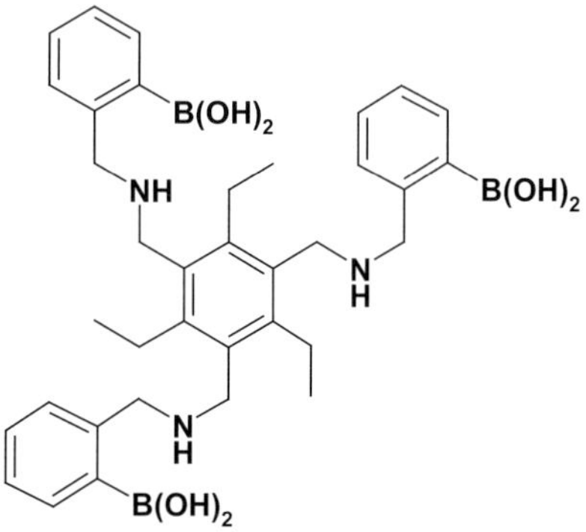

29

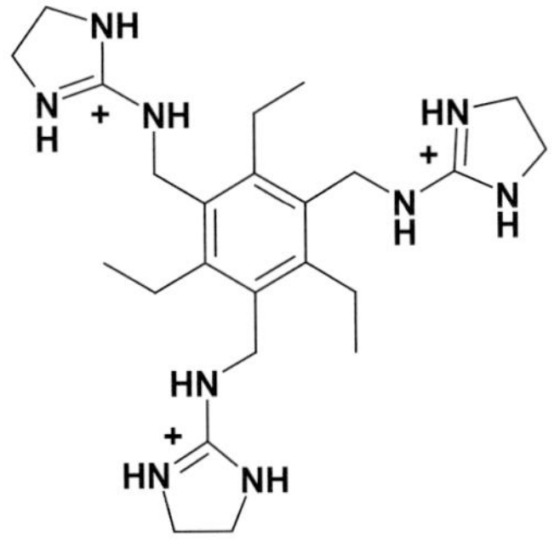

14.1 Structures of Receptors | 453

30

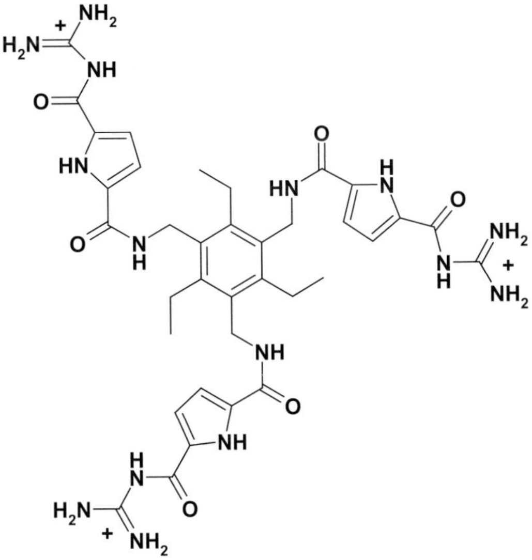

31

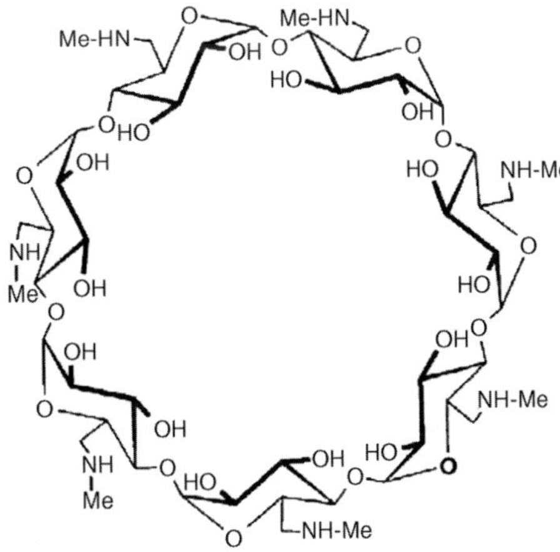

32

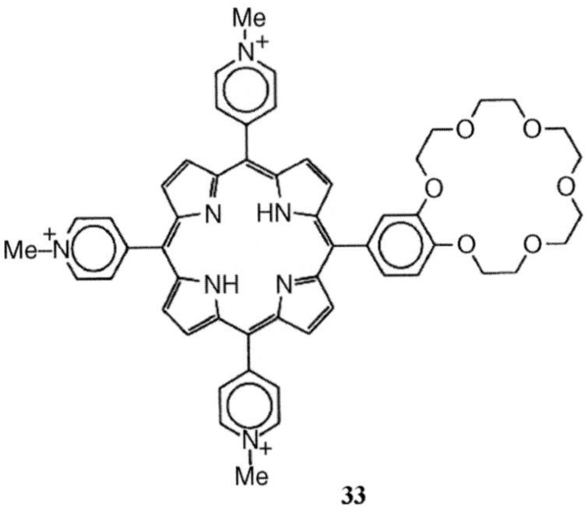

33

34

35

36

37

38

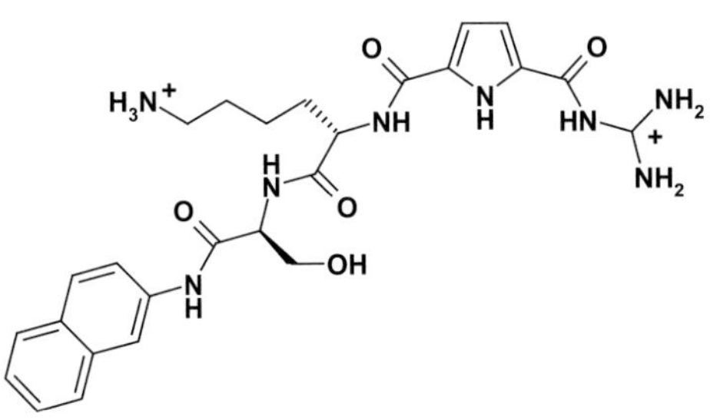

14.1 *Structures of Receptors* | 455

39

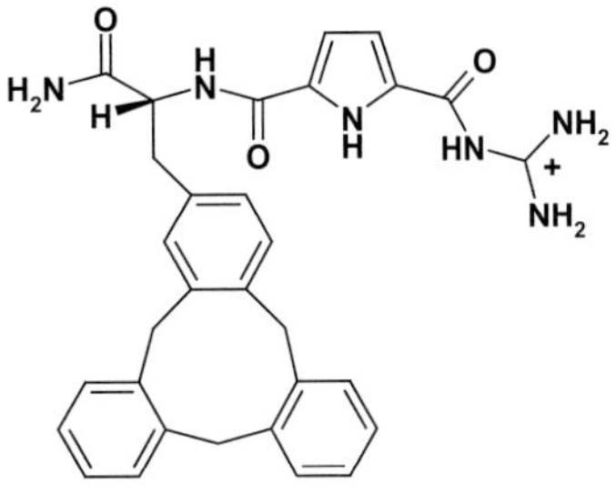

40

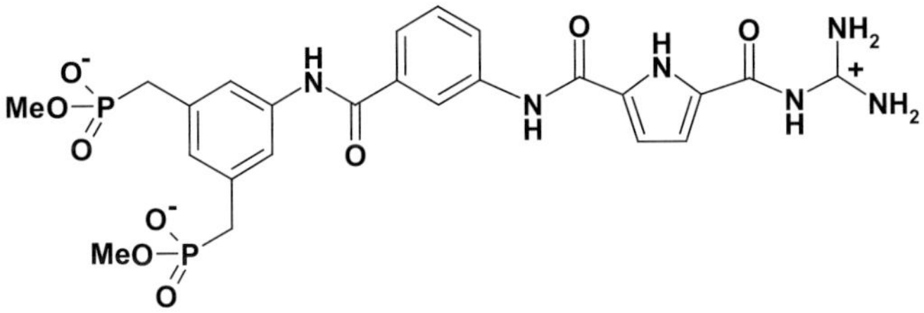

41

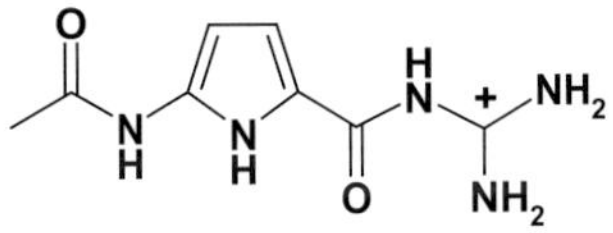

42

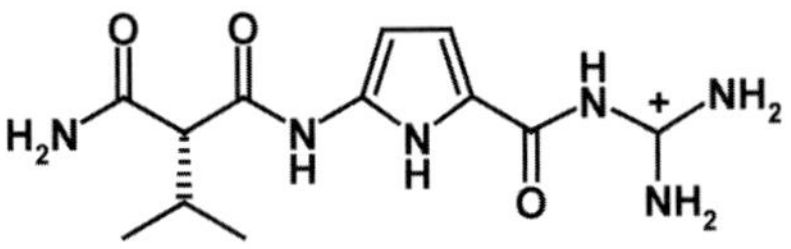

43

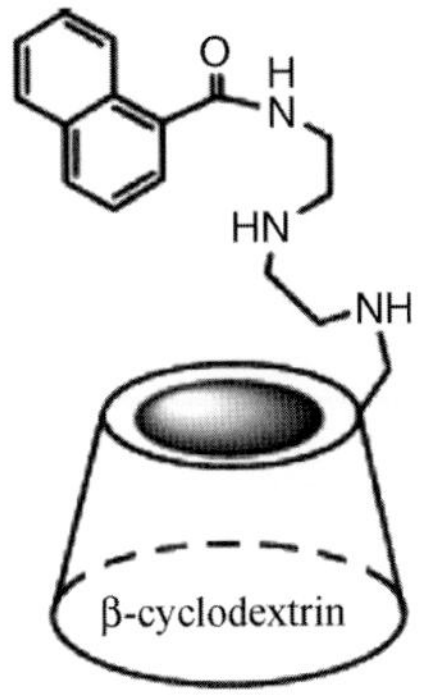

44

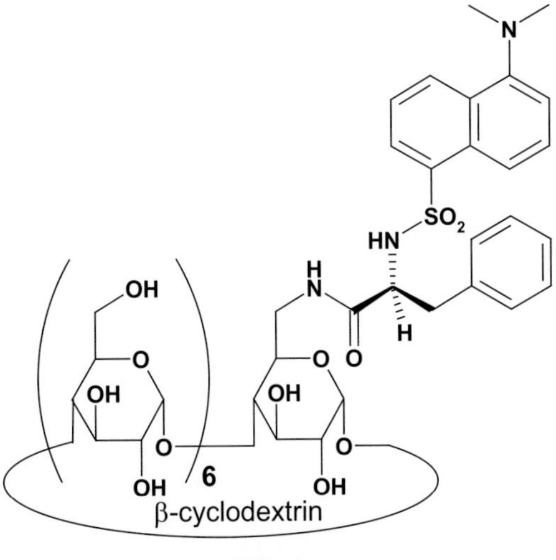

45

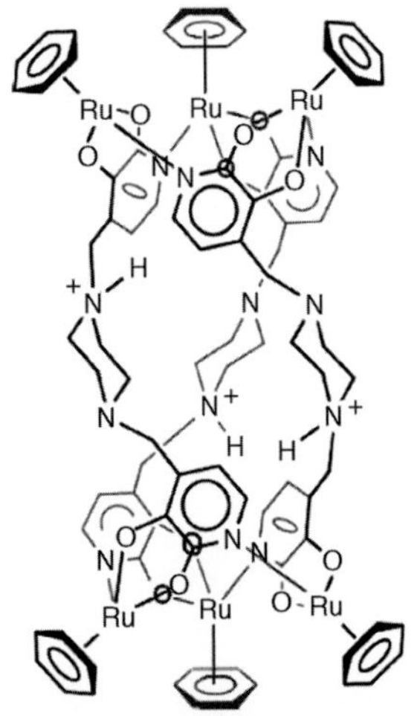

46

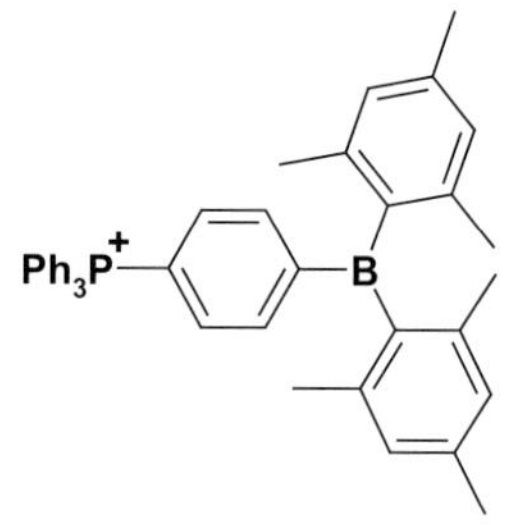

456 | *14 Quantitative Affinity Data on Selected Artificial Receptors*

47

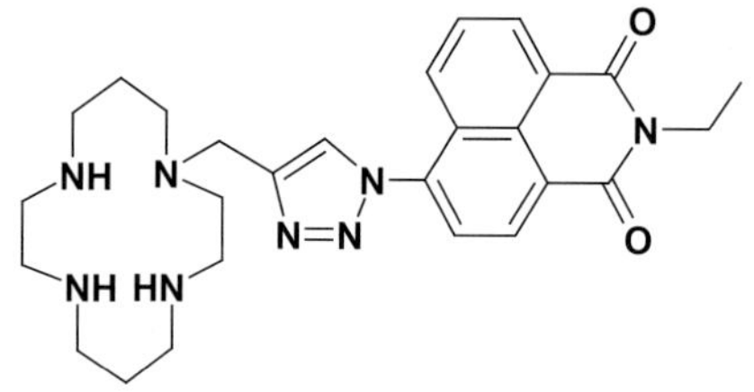

48

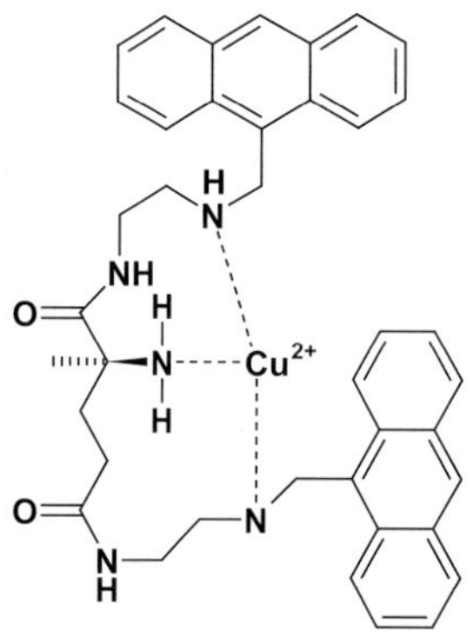

49

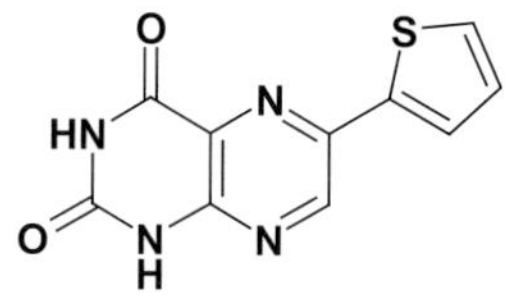

50

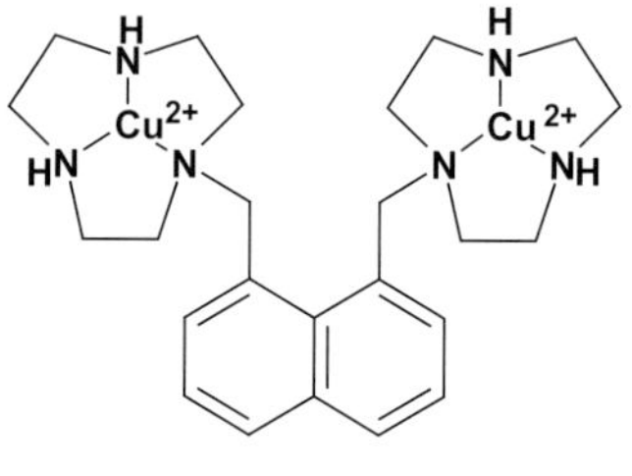

51

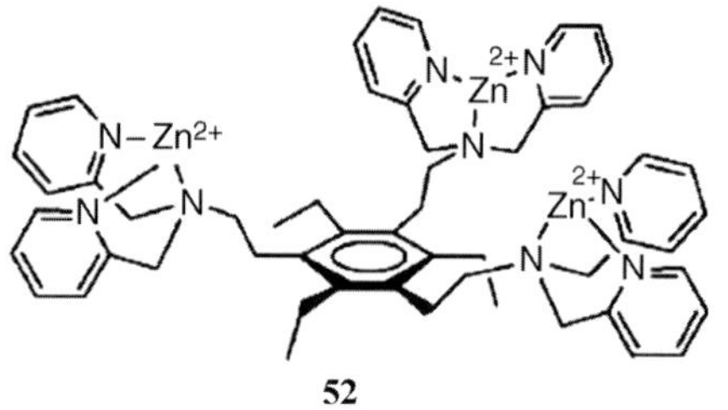

52

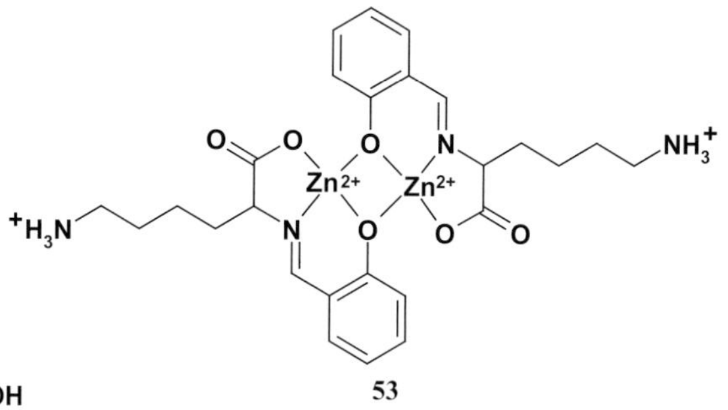

53

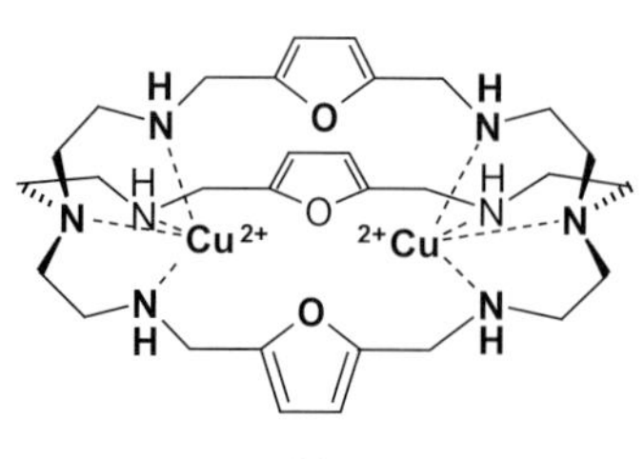

54

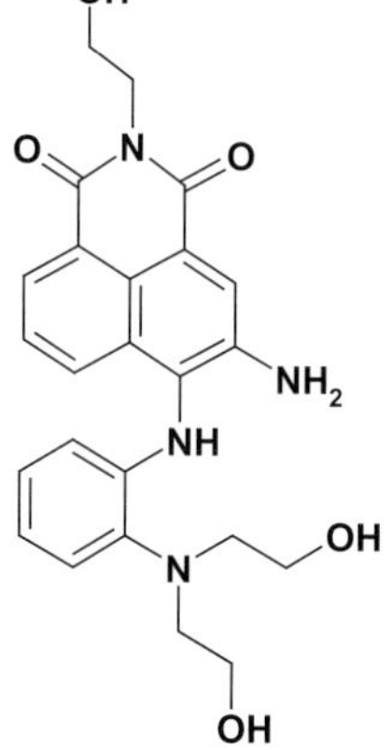

55

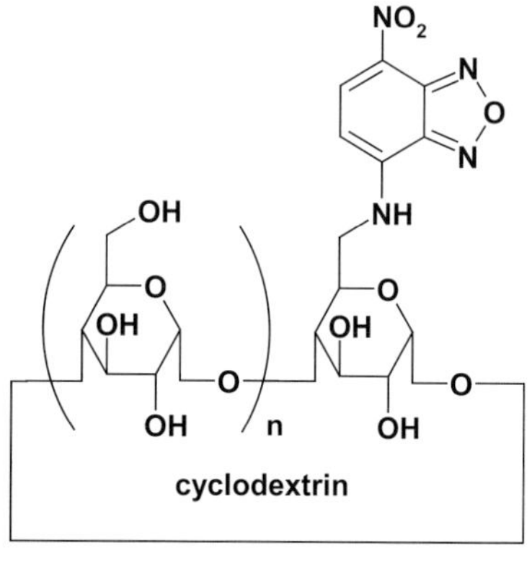

56 cyclodextrin

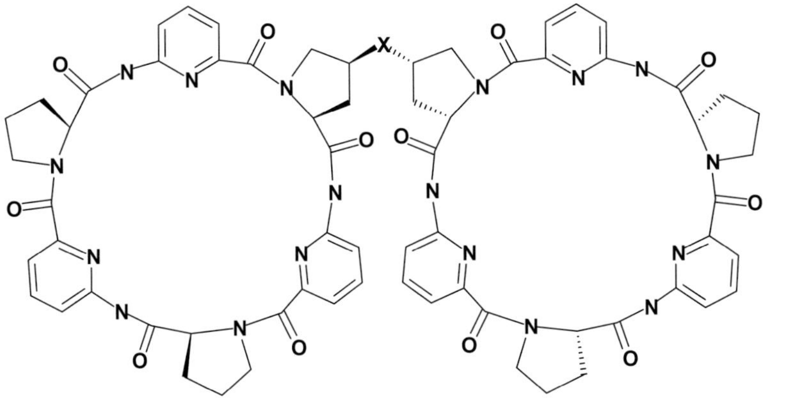

14.1 Structures of Receptors

57

X =

a, b, c, d, e, f

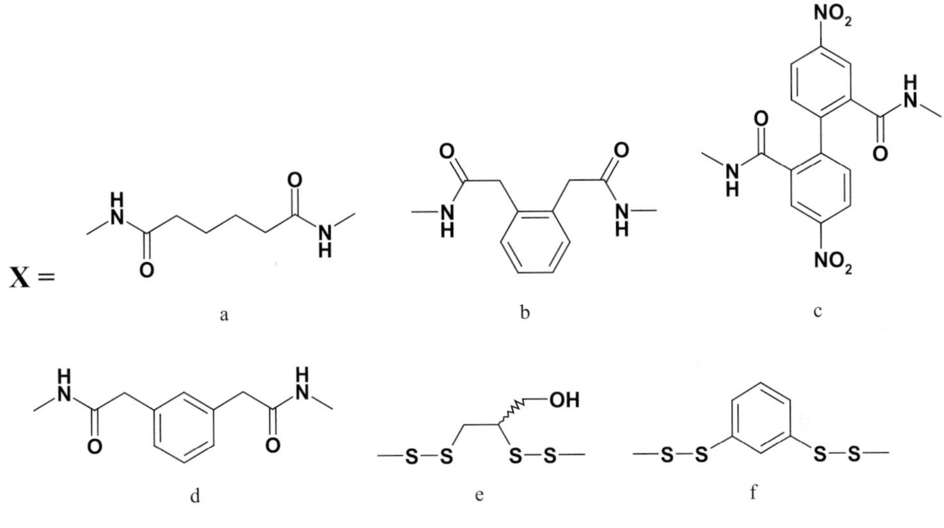

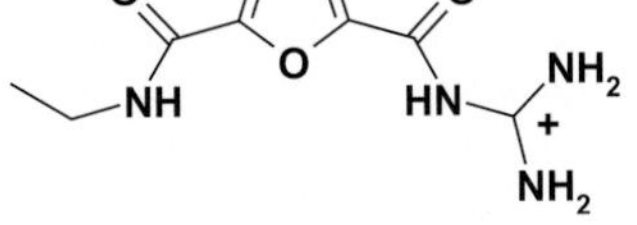

58

References

1. Abe, H., Mawatari, Y., Teraoka, H., Fujimoto, K., and Inouye, M. (2004) *J. Org. Chem.*, **69**, 495.
2. Baudoin, O., Gonnet, F., Teulade-Fichou, M.-P., Vigneron, J.-P., Tabet, J.-C., and Lehn, J.-M. (1999) *Chem. Eur. J.*, **5**, 2762.
3. Kwon, J.Y., Singh, N.J., Kim, H., Kim, S., Kim, K.S., and Yoon, J. (2004) *J. Am. Chem. Soc.*, **126**, 8892.
4. Ojida, A., Takashima, I., Kohira, T., Nonaka, H., and Hamachi, I. (2008) *J. Am. Chem. Soc.*, **130**, 12095.
5. Kim, S.K., Moon, B.-S., Park, J.H., Seo, Y.I., Koh, H.S., Yoon, Y.J., Lee, K.D., and Yoon, J. (2005) *Tetrahedron Lett.*, **46**, 6617.
6. Carolan, J.V., Butler, S.J., and Jolliffe, K.A. (2009) *J. Org. Chem.*, **74** (8), 2992–2996.
7. Springsteen, G. and Wang, B. (2002) *Tetrahedron*, **58**, 5291–5300.
8. James, T.D., Sandanayake, K.R.A.S., Iguchi, R., and Shinkai, R.S. (1995) *J. Am. Chem. Soc.*, **117**, 8982–8987.
9. Akay, S., Yang, W., Wang, J., Lin, L., and Wang, B. (2007) *Chem. Biol. Drug Des.*, **70**, 279–289.
10. Badugu, R., Lakowicz, J.R., and Geddes, C.D. (2005) *Bioorg. Med. Chem.*, **13**, 113–119.
11. Bérubé, M., Dowlut, M., and Hall, D.G. (2008) *J. Org. Chem.*, **73**, 6471–6479.
12. Boduroglu, S., El Khoury, J.M., Reddy, D.V., Rinaldi, P.L., and Hu, J. (2005) *Bioorg. Med. Chem. Lett.*, **15**, 3974–3977.
13. Arimori, S., Bell, M.L., Oh, C.S., Frimata, K.A., and James, T.D. (2001) *Chem. Commun.*, 1836–1837.
14. Jiang, S., Escobedo, J.O., Kim, K.K., Alpturk, O., Samoei, G.K., Fakayode, S.O., Warner, I.M., Rusin, O., and Strongin, R.M. (2006) *J. Am. Chem. Soc.*, **128**, 12221–12228.
15. Zhang, T. and Anslyn, E.V. (2006) *Org. Lett.*, **8**, 1649–1652.
16. Gao, X., Zhang, Y., and Wang, B. (2005) *Tetrahedron*, **61**, 9111–9117.
17. Jin, S., Wang, J., Li, M., and Wang, B. (2008) *Chem. Eur. J.*, **14**, 2795–2804.
18. Jin, S., Li, M., Zhu, C., Tran, V., and Wang, B. (2008) *ChemBioChem*, **9**, 1431–1438.
19. Heinrichs, G., Schellenträger, M., and Kubik, S. (2006) *Eur. J. Org. Chem.*, 4177–4186.
20. Jin, S., Zhu, C., Li, M., and Wang, B. (2009) *Bioorg. Med. Chem. Lett.*, **19**, 1596–1599.
21. Yang, W., Yan, J., Fang, H., and Wang, B. (2003) *Chem. Commun.*, 792–793.
22. Wiskur, S.L., Lavigne, J.J., Metzger, A., Tobey, S.L., Lynch, V., and Anslyn, E.V. (2004) *Chem. Eur. J.*, **10**, 3792–3804.
23. Schmuck, C. and Schwegmann, M. (2005) *J. Am. Chem. Soc.*, **127**, 3373–3379.
24. Schmuck, C. and Schwegmann, M. (2005) *Org. Lett.*, **7**, 3517–3520.
25. Hacket, F., Simova, S., and Schneider, H.-J. (2001) *J. Phys. Org. Chem.*, **14**, 159–170.
26. Sirish, M., Chertkov, V.A., and Schneider, H.-J. (2002) *Chem. Eur. J.*, **8**, 1181–1188.
27. Hof, F., Trembleau, L., Ullrich, E.C., and Rebek, J. Jr. (2003) *Angew. Chem. Int. Ed.*, **42**, 3150–3153.
28. Liu, S., Fang, L., He, Y., Chan, W., Yeung, K., Cheng, Y., and Yang, R. (2005) *Org. Lett.*, **7**, 5825–5828.
29. Fokkens, M., Schrader, T., and Klärner, F.-G. (2005) *J. Am. Chem. Soc.*, **127**, 14415–14421.
30. Bush, M.E., Bouley, N.D., and Urbach, A.R. (2005) *J. Am. Chem. Soc.*, **127**, 14511–14517.
31. Kim, H., Asif, R., Chung, D., and Hong, J. (2003) *Tetrahedron Lett.*, **44**, 4335–4338.
32. Schmuck, C. and Hernandez-Folgado, L. (2007) *Org. Biomol. Chem.*, **5**, 2390–2394.
33. Schmuck, C., Rupprecht, D., and Wienand, W. (2006) *Chem. Eur. J.*, **12**, 9186–9195.
34. Schmuck, C., Rupprecht, D., Junkers, M., and Schrader, T. (2007) *Chem. Eur. J.*, **13**, 6864–6873.
35. Schmuck, C. and Dudaczek, J. (2005) *Tetrahedron Lett.*, **46**, 7101–7105.
36. Nakashima, H. and Yoshida, N. (2006) *Org. Lett.*, **8**, 4997–5000.
37. Ikeda, H., Li, Q., and Ueno, A. (2006) *Bioorg. Med. Chem. Lett.*, **16**, 5420–5423.
38. Olivier, C., Grote, Z., Solari, E., Scopelliti, R., and Severin, K. (2007) *Chem. Commun.*, 4000–4002.
39. Kim, Y. and Gabbai, F.P. (2009) *J. Am. Chem. Soc.*, **131**, 3363–3369.

40 Tamanini, E., Katewa, A., Sedger, L.M., Todd, M.H., and Watkinson, M. (2009) *Inorg. Chem.*, **48**, 319–324.

41 Chen, Z., He, Y., Hu, C., and Huang, X. (2008) *Tetrahedron: Asymmetry*, **19**, 2051–2057.

42 Saleh, N. (2009) *Luminescence*, **24**, 30–34.

43 Tang, L., Park, J., Kim, H., Kim, Y., Kim, S.J., Chin, J., and Kim, K.M. (2008) *J. Am. Chem. Soc.*, **130**, 12606–12607.

44 Oh, D.J. and Ahn, K.H. (2008) *Org. Lett.*, **10**, 3539–3542.

45 Khatua, S., Choi, S.H., Lee, J., Kim, K., Do, Y., and Churchill, D.G. (2009) *Inorg. Chem.*, **48**, 2993–2999.

46 Amendola, V., Bonizzoni, M., Esteban-Gomez, D., Fabbrizzi, L., Licchelli, M., Sancenon, F., and Taglietti, A. (2006) *Coord. Chem. Rev.*, **250**, 1451–1470.

47 Huang, J., Xu, Y., and Qian, X. (2009) *Dalton Trans.*, 1761–1766.

48 Ikeda, H., Murayama, T., and Ueno, A. (2005) *Org. Biomol. Chem.*, **3**, 4262–4267.

49 Reyheller, C., Hayb, B.P., and Kubik, S. (2007) *New J. Chem.*, **31**, 2095–2102.

50 Otto, S. and Kubik, S. (2003) *J. Am. Chem. Soc.*, **125**, 7804–7805.

51 Schmuck, C. and Machon, U. (2006) *Eur. J. Org. Chem.*, 4385–4392.

52 Mirsky, V.M., Hirsch, Th., Piletsky, S.A., and Wolfbeis, O.S. (1999) *Angew. Chem., Int. Ed.*, **38**, 1108–1110.

53 Hirsch, Th, Kettenberger, H., Wolfbeis, O.S., and Mirsky, V.M. (2003) *Chem. Commun.*, 432–433.

54 Panasyuk-Delaney, T., Mirsky, V.M., and Wolfbeis, O.S. (2002) *Electroanalysis*, **14**, 221–224.

55 Panasyuk-Delaney, T., Mirsky, V.M., Ulbricht, M., and Wolfbeis, O.S. (2001) *Anal. Chim. Acta*, **435**, 157–162.

56 Chianella, I., Lotierzo, M., Piletsky, S.A., Tothill, I.E., Chen, B., Karim, K., and Turner, A.P.F. (2002) *Anal. Chem.*, **74**, 1288–1293.

57 Piletska, E., Piletsky, S.A., Karim, K., Terpetschnig, E., and Turner, A.P.F. (2004) *Anal. Chim. Acta*, **504**, 179–183.

58 Xu, Z., Kuang, D., Feng, Y., and Zhang, F. (2010) *Carbohydr. Polym.*, **79**, 642–647.

59 Yang, J., Zhu, X.L., Cai, J.B., Su, Q.D., Gao, Y., and Zhang, L. (2005) *Chin. Chem. Lett.*, **16**, 1503–1506.

60 Ng, S.M. and Narayanaswamy, R. (2009) *Sens. Actuators, B*, **139**, 156–165.

61 Hsu, H.-C., Chen, L.-C., and Ho, K.-C. (2004) *Anal. Chim. Acta*, **504**, 141–147.

62 Emgenbroich, M., Borrelli, C., Shinde, S., Lazraq, I., Vilela, F., Hall, A.J., Oxelbark, J., De Lorenzi, E., Courtois, J., Simanova, A., Verhage, J., Irgum, K., Karim, K., and Sellergren, B. (2008) *Chem. Eur. J.*, **14**, 9516–9529.

63 Liu, X., Ouyang, C., Zhao, R., Shangguan, D., Chen, Y, and Liu, G. (2006) *Anal. Chim. Acta*, **571**, 235–241.

64 Guo, H. and He, X. (2000) *Fresenius' J. Anal. Chem.*, **368**, 461–465.

65 Asanuma, H., Akiyama, T., Kajiya, K., Hishiya, T., and Komiyama, M. (2001) *Anal. Chim. Acta*, **435**, 25–33.

66 Xu, Z., Kuang, D., Feng, Y., and Zhang, F. (2010) *Carbohydr. Polym.*, **79**, 642–647.

67 Xua, Z., Kuang, D., Liu, L., and Deng, Q. (2007) *J. Pharm. Biomed. Anal.*, **45**, 54–61.

68 Karlsson, J.G., Karlssona, B., Andersson, L.I., and Nicholls, I.A. (2004) *Analyst*, **129**, 456–462.

69 Leung, M.K.P., Chow, C.-F., and Lam, M.H.W. (2001) *J. Mater. Chem.*, **11**, 2985–2991.

70 Turkewitsch, P., Wandelt, B., Darling, G.D., and Powell, W.S. (1998) *Anal. Chem.*, **70**, 2025–2030.

71 Mathew, J. and Buchardt, O. (1995) *Bioconjugate Chem.*, **6**, 524–528.

72 Izatt, R.M. (1974) The synthesis and ion bindings of synthetic multidentate macrocyclic compounds. *Chem. Rev.*, **74**, 351–384.

73 Izatt, R.M., Bradshaw, J.S., Nielsen, S.A., Lamb, J.D., Christensen, J.J., and Sen, D. (1985) Thermodynamic and kinetic data for cation-macrocycle interaction. *Chem. Rev.*, **85**, 271–339.

74 Izatt, R.M., Pawlak, K., Bradshaw, J.S., and Bruening, R.L. (1981) Thermodynamic and kinetic data for macrocycle interactions with cations and anions. *Chem. Rev.*, **91**, 1721–2085.

75 Izatt, R.M., Bradshaw, J.S., Pawlak, K., Bruening, R.L., and Tarbet, B.J. (1992) Thermodynamic and kinetic data for macrocycle interaction with neutral molecules. *Chem. Rev.*, **92**, 1261–1354.

76 An, H., Bradshaw, J.S., and Izatt, R.M. (1992) Macropolycyclic polyethers (cages) and related compounds. *Chem. Rev.*, **92**, 543–572.

77 An, H., Bradshaw, J.S., Izatt, R.M., and Yan, Z. (1994) Bis- and oligo(benzocrown ether)s. *Chem. Rev.*, **94**, 939–991.

78 Izatt, R.M., Pawlak, K., Bradshaw, J.S., and Bruening, R.L. (1995) Thermodynamic and kinetic data for macrocycle interaction with cations, anions, and neutral molecules. *Chem. Rev.*, **95**, 2529–2586.

79 Zhang, X.X., Bradshaw, J.S., and Izatt, R.M. (1997) Enantiomeric recognition of amine compounds by chiral macrocyclic receptors. *Chem. Rev.*, **97**, 3313–3362.

Further Reading

Bencini, A., Bianchi, A., Garcia-España, E., Micheloni, M., and Ramirez, J.A. (1999) Proton coordination by polyamine compounds in aqueous solution. *Coord. Chem. Rev.*, **188**, 97–156.

García-España, E., Díaz, P., and Llinares, J.M., and Bianchi, A. (2006) Anion coordination chemistry in aqueous solution of polyammonium receptors. *Coord. Chem. Rev.*, **250**, 2952–2986.

Hazai, E., Hazai, I., Demko, L., Kovacs, S., Malik, D., Akli, P., Hari, P., Szeman, J., Fenyvesi, E., Benes, E., and Bikadi, Z. (2010) Cyclodextrin knowledgebase, a web-based service managing CD-ligand complexation data. *J. Computer-Aided Mol. Design*, **24**, 713–717.

Izatt, R.M., Christensen, J.J., and Rytting, J.H. (1971) Sites and thermodynamic quantities associated with proton and metal ion interaction with ribonucleic acid, deoxyribonucleic acid, and their constituent bases, nucleosides, and nucleotides. *Chem. Rev.*, **71**, 439–481.

Mader, H.S. and Wolfbeis, O.S. (2008) Boronic acid based probes for microdetermination of saccharides and glycosylated biomolecules. *Microchim Acta*, **162**, 1–34.

Nolan, E.M. and Lippard, S.J. (2008) Tools and tactics for the optical detection of mercuric ion. *Chem. Rev.*, **108** (9), 3443–3480.

Rekharsky, M.V. and Inoue, Y. (1998) Complexation thermodynamics of cyclodextrins. *Chem. Rev.*, **98**, 1875–1918.

Talanova, G.G. (2000) Phosphorus-containing macrocyclic ionophores in metal ion separations. *Ind. Eng. Chem. Res.*, **39**, 3550–3565.

Umezawa, Y. (ed.) (1990) *Handbook of Ion-Selective Electrodes: Selectivity Coefficients*, CRC Press, Boca Raton, FL.

Umezawa, Y., Bühlmann, P., Umezawa, K., and Hamada, N. (2002) Potentiometric selectivity coefficients of ion-selective electrodes. Part III. Organic ions. *Pure Appl. Chem.*, **74**, 995–1099.

Umezawa, Y., Bühlmann, P., Umezawa, K., Tohda, K., and Amemiya, S. (2000) Potentiometric selectivity coefficients of ion-selective electrodes. Part I. Inorganic cations. *Pure Appl. Chem.*, **72**, 1851–2082.

Umezawa, Y., Umezawa, K., Bühlmann, P., Hamada, N., Aoki, H., Nakanishi, J., Sato, M., Xiao, K.P., and Nishimura, Y. (2002) Potentiometric selectivity coefficients of ion-selective electrodes. Part II. Inorganic anions. *Pure Appl. Chem.*, **74**, 923–994.

Cyclodextrin Database: www.cyclodextrine.net

Index

a

α-amino alcohols 169
actinides (An) 30
acylated/alkylated cyclodextrins 194
ADONMA. *see* analyte-dependent oligonucleotide modulation assay (ADONMA)
affinity properties
– binding constants
– – of immobilized receptors 7
– – from kinetic measurements 7–10
– experimental techniques 12, 13
– measurements 3
– – under equilibrium conditions 3–7
– – kinetic 7–10
– quantitative characterization of 1
– receptor, chemical immobilization of 2
– selectivity, correlation 21–24
– temperature dependencies, analysis of 10, 11
– – activation energies for binding and dissociation 11
– – entropy production 10
– – free energy 10
– – transition state theory 11
affinity sensors 2, 4, 20
– transducers of 2
AFM techniques 341
α-hydroxyacids 169
alkali metal ions 26
alkylated cyclodextrins 192–194
alkylation 193
aluminum phthalocyanine chloride (AlPC) 324
analyte-dependent oligonucleotide modulation assay (ADONMA) 344
amidinium-carboxylate complex 277
anionic cyclodextrins 196, 197

anionic tetrahedral boronate 170
anion recognition 273
Anslyn's receptor 287, 296
– tripodal receptor 280
aptamers 335
– application of 336
– arrays 343–345
– dissociation constants 179
– generation and synthesis of 336
– – chemical synthesis of 341, 342
– – selection of 340, 341
– – selection of aptamers from combinatorial libraries (SELEX) 336–338
– – SELEX variations 338–340
– – stabilization of 342, 343
– ligand binding, readout techniques
– – aptazymes 346, 347
– – conformational effects 345, 346
– sensing methods
– – acoustic sensing methods 355, 356
– – cantilever 356
– – electrochemical sensing 352–355
– – optical sensing 347–352
– synthetic molecules 336
artificial anion receptors, design 273
artificial receptors
– quantitative affinity data 441–450
– – structures of 451–459
automated SELEX 339, 340
avidin 128
azole-based cyclopeptide 144

b

bibracchial lariat ether 26
bicyclic guanidinium receptors 279
binding constants 27, 28
– of immobilized receptors 7
– from kinetic measurements 7–10

Artificial Receptors for Chemical Sensors. Edited by V.M. Mirsky and A. Yatsimirsky
Copyright © 2011 WILEY-VCH Verlag GmbH & Co. KGaA, Weinheim
ISBN: 978-3-527-32357-9

– of molecular receptors   441–448
– of spreader-bar systems   449, 450
binuclear Zn(II) complex   59
biomimetic receptor   290
biotin-appended cyclodextrins   127
bis cationic receptors   300
– binding constants   300
bisnaphthol (BNOH)   330
– concentration range   331
bis(phenanthridinium) receptor   58
bis-pyrene-appended γ-cyclodextrins   116–118
blood glucose biosensors   71
boron atom   169
boronic acid   169, 172
– receptors and sensors for carbohydrates   169
– sensors for catechol-based neurotransmitters   169
boronic acid-based carbohydrate sensing field   170
boronic acid-based sensor   172
– carbohydrate sensors   171
– de novo design   172–177
– – apparent binding constants   175
– – bis(pyridinium-boronic) acid sensor   175, 176
– – calculated distances among   174
– – CAVEAT program   173
– – chemosensor for dopamine and norepinephrine   176, 177
– – iminium ion, and internal hydrogen bond formation   177
– – selectivity of sensor for   173
– – sensor for dopamine   173, 174
– – structurally analogous sensors   174
– – uronic acid sensor   176
– – vector-based approach   172, 173
boronic acid reporters   170
boronic acid sensors/binders
– designing and selecting   170
boronic ester   171
boronolectins   170
bottom-up approaches   321
Brønsted acid   170

c

calix[n]arenes   50, 137, 209–218, 251
– conformations   252
– features   251
– isolable conformers   251
– as molecular scaffolds   263–266
calix[4]pyrroles   61
calix[4]resorcinarenes   218–224

cantilever based sensing   356
capillary electrochromatography (CEC)   394
carbamoylated cyclodextrins   194
carbohydrate-boronic acid complexes   170
carboxyfluorescein   296
carboxylate binding site (CBS)   281
carboxymethylation   197
cationic cyclodextrins   197, 198
cationic receptors   50
cation–π interactions   48
CBS. see carboxylate binding site (CBS)
C–C bond   171
CD-based chemosensor   127
CD–peptide conjugates   113
– as chemosensors   128–130
CD–protein conjugates   113
CEC. see capillary electrochromatography (CEC)
cellulose membrane   130–132
chelating agents   171
chemical immobilization of receptor   2
chemical receptors, selectivity of   18–21
chemical warfare agents (CWAs)   102
chemoreceptors   2
chemosensor design   172
chemosensors   128
– based on CDs   115
– CD–peptide conjugates as   128–130
chiral bicyclic tetrakis-guanidinium receptor   304
chiral receptor compounds   191
chiral recognition, artificial receptor compounds for   191
combinatorial approaches   177–182
– aptamer-based boronic acid libraries   180
– aptamer dissociation constants   179
– aptamers/NABLs for fibrinogen and PSA   179
– bis-boronic acid, immobilized form on glyoxal agarose beads   180
– bromopyrogallol red (BPR) as indicator for   181
– collection of binding sequences   179
– immobilization and derivatization, of boronic acids using   180
– libraries of solid-supported pentapeptide-based bisboronic acids   182
– linear discriminant analysis (LDA)   181
– PBL libraries   180
– resin-to-resin transfer reactions (RRTRs)   180
– for selection of DNA aptamer   178
– SELEX   178, 179
– solid-supported PBLs derived from   182

– structures of PBL library 181
combinatorial library, receptors in 291
combinatorial materials screening 68, 69
– goals 72
– materials developed using 69
– sensing film compositions using passive RFID sensors 101
– typical process for 70
– wireless technologies 97–102
combinatorial receptor library 310
combinatorial technologies 67
complexation, in solution 257–263
– absorption changes of 257
– association constants, measured by 258, 260
– to complex dumb-bell-shaped dimer of C_{60}-C_{120} fullerene 259
– drawbacks 257
– formation of 2 : 1 complexes in solid state 258
– poor solubility of C_{120} fullerene 262
– provide a sizable cavity 259
complexation-induced strain 30
complex target molecules recognition
– multifunctional receptors
– – amino acids and peptides 49–55
– – nucleotides and nucleosides 55–61
conducting polymers (CPs) 363
– artificial receptors based on 364–368
– intrinsic sensitivity of
– – affinity of 370, 371
– – to pH changes 368–370
– molecular imprinting of 385
– PANI, conversion 364
– with receptor groups
– – electrostatic doping 384, 385
– – to monomer 371–384
conducting polymers modified, with receptor groups 371
– conducting polymers, doped with receptor 384, 385
– molecular imprinting, of conducting polymers 385
– receptors for
– – ions 371–383
– – organic/bioorganic molecules 383, 384
conformational equilibrium, of cyclodextrin–peptide conjugates 130
coumarin 130
CPs. see conducting polymers (CPs)
C-reactive protein (CRP) 349
crown ethers 24–28, 200, 201
– 1,1′-binaphthalene-based crown ethers 201

– carbohydrate-based crown ethers 202, 203
– derivative for selective alkali ion detection 28
– employed as components of 28
– hole size fitting 24–28
– interaction 200
– preparation 201
– with pyridine moieties 208, 209
– for selective extraction and 27
– tartaric acid-based crown ethers 204–206
cryptand complexes 24–28
– hole size fitting 24–28
CWAs. see chemical warfare agents (CWAs)
cyanide ions 169
cyclic oligosaccharides 191
cyclic peptides, as artificial receptors 137
cyclodextrin-based liquid chromatographic phases
– application of 199
cyclodextrin derivatives, with substitution patterns 194
cyclodextrin-DTPA amide derivative 198
cyclodextrin–peptide conjugates 129
cyclodextrins (CDs) 113, 137, 191
– in chiral recognition applications 192
– as optical sensor 149
– with pendant aromatic side chains 52
– properties of 114
– structures of 114
– superimposed with one D-glucose subunit 192
cyclopeptides 135
– derived synthetic receptors 137
– immobilization of 158
– structurally diverse family 143
cyclophane 48, 49, 56, 225, 234
– receptor 57
cyclotriveratrylenes 251
cytosine nucleobase 61

d

dansyl-phenylalanine-appended CDs 122, 123
dansyl-valine-appended 122, 123
– molecular recognition by regioisomers
– – 6-O, 2-O, and 3-O-dansyl-β-CDs 125
– – 6-O, 2-O, and 3-O-dansyl-γ-CDs 123–125
DFT calculation 282, 283
diazepam 424–426
2,6-di-O-pentyl-α, β- or γ-cyclodextrins 193
dipole–dipole interactions 113
ditopic receptors 54
DMSO 27, 39, 154, 280, 281, 285, 300
DNA-RNA ligation 344

donor–donor–acceptor–acceptor (DD-AA)
   complex   277
double gradient sensor microarray   97

e

electrical energy transduction sensors
   84–92
electrochemical capacitance   331
energy transfer mechanism   60
enthalpy   2
1-ethyl-3-(3-dimethylaminopropyl)
   carbodiimide   2

f

fabrication methods   73
FEP. *see* free energy perturbation (FEP)
fluorescent CD, immobilization
– with aldehyde groups on   132
– on cellulose membrane   130–132
fluorescent chemosensors   127
– for molecule recognition   127
– turn-off system   125
– turn-on system   125–127
fluorescent labeled tetrapeptide substrate   309
fluoride   169
fluorophore–amino acid–CD triad systems
– dansyl-leucine-appended CDs   118–122
– – conformational changes of   118
– – conformational equilibrium   120, 121
– – fluorescence decay   120
– – fluorescence lifetimes   119
– – guest dependency   121, 122
– – sensitivity parameters   122
– dansyl-phenylalanine-appended CDs   122, 123
– dansyl-valine-appended CDs   122, 123
– – chiral discrimination abilities   123
free energies, of associations   48, 49
free energy perturbation (FEP)   18
FRET system for chemosensor   130
Frumkin isotherm   4
FTIR spectra   327
fullerenes   249
– applications and practical uses   250
– three-dimensional structures   250

g

Gable-type bisporphyrins   53
galactose-1-phosphate   290
Gibbs free energy   6
Gibbs transfer energies   24
glycobiology   169
glycopeptide antibiotic   137
glycoproteins   169, 177, 178

glycosylation   177, 178
gold electrodes   263
– covered by monolayer of   330
– mercaptohexadecane coated   412
– modified with mixture of
   1-hexadecanethiol   332
– scanning electron microscopy   325
– spreader-bar systems with   328
GTP   55, 59, 291
guanidine-based receptors   284
– sulfate receptor   285
guanidiniocarbonyl pyrrole cations   280
guanidinium based anion receptors
– amino acid recognition   297–301
– dipeptides as substrate   301–303
– inorganic anion recognition, recent
   advances   283–287
– instructive historical examples   275–283
– introduction   273–275
– organic and biological phosphates   287–292
– polycarboxylate binding   292–297
– polypeptide recognition   303–312
guanidinium cation   276, 286
guanidinium chloride   281
– ion, dominant form   276
guanidinium moiety   273
– anion recognition   275, 297
– Anslyn's receptor   296
– Marsura's receptors   292
– positively charged   274
– sulfate recognition   285
– tweezer receptors, combinatorial library
   of   308
– urea moieties   283

h

Hamilton's tweezer receptor   277, 279
– combinatorial library   321
– two-armed library   308
HDCA. *see* hyodeoxycholic acid (HDCA)
heavy metal ions   28–31
– complexation, selectivity of   28–31
heptakis(2,3-di-O-acetyl-6-sulfato)-
   β-cyclodextrin   196
heptakis(2,6-di-Omethyl-3-O-pentyl)-
   β-cyclodextrin   193
heptakis(2,3-di-O-pentyl-6-O-methyl)-
   β-cyclodextrin   193
heptakis(2-O-methyl-3,6-di-O-pentyl)-
   β-cyclodextrin   193
heptakis(2,3,6-tri-O-methyl)-β-
   cyclodextrin   192
heptakis(2,3,6-tri-O-pentyl)-β-
   cyclodextrin   193

hexakis(2,3,6-tri-O-methyl)-α-cyclodextrin 192
hexakis(2,3,6-tri-O-pentyl)-α-cyclodextrin 193
HIV-1 Tat protein
– tertiary structure 274
homooxacalixarenes 251
HostDesigner software 154
host–guest complex 289
human STAT protein 54
hydrogen bond donor 41, 227, 274, 298, 299, 404
hydrogen bonded complexes 37–41
– binding constants for 1:1 complexes for 39
– diphenylurea, with different halides in 38
– hydrogen bonding receptors, selectivity of 37
– of nucleic bases by artificial ligands 40
– with open-chain
– – bis-indole ligand and halide anions 38
– – tetraindole and halide anions 38
– provided by receptor to anion binding. 40
– rotaxane 40
– selective recognition of adenine 41
hydrogen bonding 17, 33, 50, 60, 156, 256, 396, 404
– and stacking interactions, with nucleobase 58, 59
2-hydroxypropylether cyclodextrins 195
hyodeoxycholic acid (HDCA) 130

i
immobilized fluorescent CD, on cellulose membrane 130–132
– membrane fluorescence measurements 132
immobilized receptors
– binding constants of 7
– for sensitive optical methods 13
immunoglobulin G (IgG) 356
intrinsic sensitivity, of conducting polymers 368
– affinity, to gases and vapors 370, 371
– affinity, to inorganic ions 370
– sensitivity, to pH changes 368–370
ionicity 29
ion pairing 50
ITC measurements 297

l
Langmuir adsorption isotherm 3, 4, 5
Langmuir–Blodgett technique 2
Langmuir isotherm
– binding constant, extraction of 6

lanthanides (Ln) 30
lectins 169
Lewis acidity 170
Lewis acids 28, 41, 42, 46, 47
– receptors, with transition metal complexes 42–46
Lewis bases 28, 46, 169, 171, 172, 185
ligand–receptor interactions 1
ligand–receptor systems 2
liquid chromatography (LC) 394
Lissoclinum patella 136, 143
lithographic techniques
– speed and versatility 322

m
macrocycles 61, 284
macrocyclic polyether polythiophenes 364
macrocyclic receptors 275, 293
Marsura's receptors 292
mathematical QSAR model 311
matrix molecules 323
– chain length 328
mechanical energy transduction sensors 79–84
metal-containing receptor compounds 235–237
metal ions 28, 29
metallic nanoparticles, generation of
– spreader-bar systems 324
metallo-receptor 286
microcontact printing (μ-CP) 321
micro-structured ultrathin layers 321
MIPs. *see* molecularly imprinted polymers (MIPs)
mixed cyclodextrin-crown ether systems 199
modern theoretical methods, advantages 303
molecularly imprinted polymers (MIPs) 393, 394
– comparison of 414–416
– liquid chromatography (LC) 394
– membranes 410–414
– mimicking natural receptors 422, 423
– – antibodies, comparison of 423, 424
– – as artificial antibodies in 423
– – binding affinity and binding capacity of 430, 431
– – catalysts, with enzymatic activity 427–429
– – pseudo-immunoassay 423–427
– – quantitative data on the binding properties 429
– – selected antibody–antigen binding affinities 432

– performance evaluation 416
–– binding capacity 416–420
–– binding selectivity 420–422
– polymerization methods for 407
–– bulk polymers 407–409
–– micro- and nanobeads 409, 410
– preparation of 406, 407
– reagents
–– crosslinker 399–402
–– functional monomers 398, 399
–– radical initiators 403
– recognition processes 395
– solvents 404–406
molecularly imprinted polymers (MIPs), approaches for 323, 394, 396
– advantages and disadvantages of 398
– covalent imprinting 397
– general synthesis of 396
– noncovalent imprinting 396, 397
– polymerization methods 407
–– bulk polymers 407–409
–– micro- and nanobeads 409, 410
–– MIP films and membranes 410–413
–– performance evaluation – imprinting efficiency 416
–– vs. different polymerization methods 414–416
– preparation of MIP 404–406
– reagents and solvents in 398
–– crosslinkers 399, 402
–– functional monomers 398–401
–– radical initiators 403
–– solvents 404–406
– semi-covalent imprinting 397
molecularly imprinted solid-phase extraction (MISPE) 394
molecular receptors
– binding constants of 441–448
molecular recognition 17
– chemical sensing methods
–– in water 113
– principles of 393
molecule sensing
– protein environment, effect of 127, 128
monensin 127
monensin-appended cyclodextrins 127
mono-diethylenetriaminepentaacetic acid (DTPA) 198
multifunctional receptors 49
– for recognition of complex target molecules 49
–– of amino acids and peptides 49–55
–– of nucleotides and nucleosides 55–61

n

N-alkylated imidazole 146
NBDamine-appended cyclodextrins 126
– guest-induced conformational changes 127
4-N-dimethylaminopyridine (DMAP) 194
near edge X-ray absorption fine structure (NEXAFS) 324, 325
– data 327
– detection limit 326
NECEEM. see non-equilibrium capillary electrophoresis of equilibrium mixtures (NECEEM)
neomycin B 354
NEXAFS. see near edge X-ray absorption fine structure (NEXAFS)
N-methylquinuclidinium iodide complex 150
non-equilibrium capillary electrophoresis of equilibrium mixtures (NECEEM) 340
nucleobase recognition 60
nucleotide–polyammonium receptor 58

o

n-octadecyltrichlorosilane (OTS) monolayer 321
octakis(3-O-butyryl-2,6-di-O-pentyl)-γ-cyclodextrin 194

p

patellamide D 145
peptide-based receptors 57
phenylboronic acid 170, 171
– binding equilibrium 171
phosphate recognition element 61
PhotoSELEX 338
pKa values 171
p-nitro benzoates 48
polar bisphosphonate precursor 307
polyaniline (PANI) 363
– conversion between different states 364
porphyrin-based receptors 53
π-π interactions 250, 251
propylene carbonate (PC) 26
prostate cancer biomarker 177
prostate specific antigen (PSA) 177
protein, microenvironment 273
protonated macrocyclic polyamines 55
proximity-sensing approaches 98
proximity wireless sensors 98
PSA. see prostate specific antigen (PSA)
p-tert-butylphenol 251
pyrene-appended cyclodextrins 114
– bis-pyrene-appended γ-cyclodextrins 116–118

– – guest-induced conformational
    change   118
– γ-CD, as bicarbonate sensor   115, 116
– γ-CD, as molecule sensors   114, 115
– pyrene-appended γ-cyclodextrins
– – as bicarbonate sensor   115–117
– – as molecule sensors   114, 115
pyrene–pyrene interaction   117

q

quantitative structure–activity relationship
    (QSAR)   311
quartz crystal microbalance (QCM)   13, 356

r

radiant energy transduction sensors   73–79
γ-radiation   71
radio frequency identification (RFID)
    sensors   98
receptor, for length- and sequence-selective
    detection of tripeptides   53
receptor–analyte binding constants, data   441
receptor compounds   225–234
– acyclic receptor   231
– with acyclic thiourea receptor   227
– axial chiral π-electron deficient tetracationic
    receptor   225
– aza macrocycle   233
– bisbinaphthyl macrocyclic species   225
– bischromenylurea macrocyclic compound
    with   228
– cleft-shaped molecule   229
– clip-shaped molecule   233
– compounds with two oxazoline rings
    attached to   232
– compound with two proline moieties   231
– contains naphthyl substituent groups   225
– crown ether appendages   225
– crown moiety of   225
– C_3-symmetric receptor   230
– exhibiting enantiomeric discrimination,
    toward tetrabutylammonium
    mandelate   231
– exhibiting enantioselective binding of
    catechol amines and   228
– form hydrogen bonds to the nitrogen atoms
    of   232
– macrocyclic compounds showing selective
    binding toward   227
– macrocyclic receptor   225
– – containing N,N′-bis(6-acylamino-2-
    pyridinyl)isophthalamide unit, 227
– organic-soluble analog   229
– phosphorylated macrocycle   232
– showing strong binding of monocarboxylates
    in chloroform   227
– tripodal oxazoline   232
– water-soluble cyclophane   234
receptor properties
– for anions   150–157
– – ability of cyclopeptides containing   150
– – aminocoumarin fluorophore
    integration   155
– – anion affinity of rigid macrocyclic
    triamide   154
– – binding affinity   156
– – binding of carboxylates   156
– – bis(cyclopeptide)   154
– – carboxylate binding pocket and complex
    stability   156
– – conformation with all NH groups   151
– – convergent arrangement of all NH groups
    in   151
– – dipicolylamino groups complexed to
    Zn(II)   157
– – evaluation, influence of linker structure
    on   154
– – improving binding properties of   154
– – iodide complex   152, 153
– – NMR spectroscopic investigations   157
– – quantitative evaluation of halide and sulfate
    affinity   152
– – replacement of alanine in   150
– – replacing 3-aminobenzoic acids in   151
– – stabilities of halide and nitrate complexes
    of   155
– – stoichiometries of anion complexes   152
– – structural modification of   154
– – synthetic analogs of vancomycin   156
– for cations   148
– – affinity for transition metals   143
– – affinity of cyclopeptides for quaternary
    ammonium ions   141
– – azole-based cyclopeptides   144
– – binding affinity   139
– – chirality of peptides   142
– – complexation of the enantiomers of   141
– – conformational analysis   140
– – containing alternating D-leucine and
    L-tryptophan   148
– – cyclic octapeptides   143
– – DFT calculations   147
– – interaction of aromatic substituents
    with   142
– – interaction of valinomycin, analysis   138
– – interaction with metal ions   141
– – intramolecular hydrogen bond
    formation   139

– – ion pair affect complex stability 143
– – molecular hinge 147
– – monolayer of dodecanethiol/octadecyl sulfide on 148
– – reduction of conformational flexibility of 141
– – reduction of nucleophilicity 139
– – sodium affinity 139
– – synthetic analogs of patellamide 145
– – synthetic peptide 146
– – in valinomycin analogs 138
– for ion pairs 149, 150
– – ability, peptides to bind 149
– – coordination of anion 150
– – polar environment 149
– – stability 149
– for neutral substrates 157
– – affinity for arginine derivatives in 160
– – application of cyclohexapeptide libraries 158
– – binding of monosaccharides by cyclopeptides 160
– – complexation of neutral guests 158
– – enantiomeric differentiation 158
– – highest affinity for 159, 160
– – orientation of carboxylate groups in 160
– – quartz crystal microbalance induced by 159
– – sensitivity for arginine in 158
– non-homogeneity of 4
– structures of 451–459
recognition, of molecules
– of complex target molecules 49
– of transition metal ions 28–31
– via ion pairing 31–37
– – binding constants of dianions 33, 34
– – binding of symmetrical trianion 33
– – exhibiting in water high binding constants 36
– – isomeric tricarboxylates, selective recognition of 32, 33
– – observed or conditional stability constants 34, 35
– – recognition of isomeric dicarboxylates 33
– – Zwitterionic macrocyclic hosts 37
regioisomers, of dansyl-appended CDs 123
– molecular recognition by 123
– – 6-o, 2-o, and 3-o-dansyl-β-CDs 125
– – 6-o, 2-o, and 3-o-dansyl-γ-CDs 123–125
resorcinarenes 251
RFID sensors 100, 102. see radio frequency identification (RFID) sensors

s

sapphyrins 61
Schmuck's furan based receptor 286
Schmuck's guanidiniocarbonyl-pyrrole, substitution 298
selection of aptamers from combinatorial libraries (SELEX) 336–338
selective recognition of TTP 60
selectivity 18
– of affinity sensors 20
– aspects of binding 18
– of complexation by crown ethers and 24
– of complexation of metal ions 30, 31
– correlation, affinity and 21–23
– – host–guest complexes 21, 24
– for equimolar mixture of analytes 19
– factors, depends on 27, 28
– inverse 26
– with monobenzo derivative 27
– for most synthetic receptors 21
– simulated species distribution plot 20
– single receptor 18
– of tweezer host for N/C-protected amino acids 49
SELEX. see selection of aptamers from combinatorial libraries (SELEX); systematic evolution of ligands by exponential enrichment (SELEX)
self-assembled monolayer (SAM) 321
– C K-edge NEXAFS spectra 330
– multi-component 322
sensing materials
– combinatorial technologies 67
– designs of combinatorial libraries of 71–73
– opportunities for
– – across technology readiness levels 70
– – concept of TRLs 70, 71
– – optimization, using discrete arrays 73
– – electrical energy transduction sensors 84–92
– – mechanical energy transduction sensors 79–84
– – radiant energy transduction sensors 73–79
– optimization, using gradient arrays 92
– – diffusion layer thickness 95–97
– – variable concentration, of reagents 92, 93
– – variable 2D composition 95–97
– – variable thickness, of sensing films 93, 94
– types of arrays 73
silane, adsorption 322
small molecule boronolectins (SMBLs) 170
SMBLs. see small molecule boronolectins (SMBLs)

solid state complexation, by calixarenes 252–257
– 1: 1 stoichiometry 253
– 2: 1 encapsulation 256
– choice of toluene 253
– columnar structural motifs, in crystal packing of 257
– IR measurements 254
– macrocyclic skeleton 255
– proposed structure of 255
– pufirication of C_{60} from crude arc soot 252, 253
– structure of toluene complex 255
– substitution of the upper rim of 256
– X-ray crystallographic studies 255
– X-ray structure of 254
specific electrical capacitance 12
spiegelmers 342, 343
spreader-bar approach 327
spreader-bar systems 330
– affinity study 328
– artificial receptors 321
– binding constants of 449, 450
– metallic nanoparticles, generation of 324
– principle 323
– of 2-thiobarbituric acid 328
stacking/edge-to-face binding 47
steric distortions 48
steric effects 28
Streptomyces fulvissimus 136
sulfated β-cyclodextrin 196
sulfobutyl ether β-cyclodextrin (SBE-CD) 196
surface-acoustic wave (SAW) 13, 98
surface plasmon resonance (SPR) 344
systematic evolution of ligands by exponential enrichment (SELEX) 178

t

TBDMS. *see* tert-butyldimethylsilyl chloride (TBDMS)
technology readiness levels (TRLs) 69
template directed synthesis 182–185
– boronic acid-based fluorescent polymers for 183
– boronic acid monomer 183
– coelectropolymerization of phenol with 184
– to construct carbohydrate sensors/binders 182
– in developing boronic acid-based electrochemical sensors 184
– flow calorimetry 185
– ion-sensitive field effect transistor (ISFET) device, use of 184
– molecular imprinting process 182, 183

– small molecule boronic acid-based sensors (SMBLs) for 185
– steps 182
– template directed boronic acid-based sensor 185
– uses, for preparation of selective recognition sites for 182
tert-butyldimethylsilyl chloride (TBDMS) 193
tert-butyldimethylsilyl chloride-substituted cyclodextrins 195, 196
tetrabutylammonium fluoride (TBAF) 193
tetrahedral boronate 171
5,10,15,20-tetrakis(4-sulfonatophenyl) porphyrin (TMPP)
– spreader-bar molecules 324
theophylline 424–426
thickness shear mode (TSM) 98
α-thrombin-binding aptamer (TBA) 348
thymidine nucleotides 60
transducers
– of affinity sensors 2
– for artificial receptors 364–368
transition
– complexation, selectivity of 28–31
– metal ions, associations 42
– – with other lewis acids 46, 47
– – with transition metal complexes 42–46
trinitrotoluene (TNT) 352
– glass-fiber system 353
tris-cationic receptor 298, 299
TRLs. *see* technology readiness levels (TRLs)
T4 RNA ligase 345
tumor suppressor protein, P53 304
– ribbon representation 304
turn-on fluorescent chemosensors 125–127
– NBDamine, as fluorophore 125, 126

v

vancomycin 137
van der Waals interactions 47–49

w

Watson–Crick recognition 60
wireless highthroughput screening, of materials properties 99, 101

x

X-ray photoelectron spectroscopy (XPS) 324, 325
– data 327

z

Zn(II) polyamine complex 60